Subjective and Objective
Bayesian Statistics

Second Edition

WILEY SERIES IN PROBABILITY AND STATISTICS

Established by WALTER A. SHEWHART and SAMUEL S. WILKS

Editors: *David J. Balding, Peter J. Bloomfield, Noel A. C. Cressie, Nicholas I. Fisher, Iain M. Johnstone, J. B. Kadane, Louise M. Ryan, David W. Scott, Adrian F. M. Smith, Jozef L. Teugels*;
Editors Emeriti: *Vic Barnett, J. Stuart Hunter, David G. Kendall*

A complete list of the titles in this series appears at the end of this volume.

Subjective and Objective Bayesian Statistics

Principles, Models, and Applications

Second Edition

S. JAMES PRESS

with contributions by

SIDDHARTHA CHIB
MERLISE CLYDE
GEORGE WOODWORTH
ALAN ZASLAVSKY

WILEY-INTERSCIENCE
A John Wiley & Sons, Inc., Publication

Published by John Wiley & Sons, Inc., Hoboken, New Jersey.
Published simultaneously in Canada.

Library of Congress Cataloging-in-Publication Data is available.

ISBN 0-471-34843-0

10 9 8 7 6 5 4 3 2 1

To my Family
G, D, S, and all the J's

Reason, Observation, and Experience—The Holy Trinity of Science

—Robert G. Ingersoll (1833–1899)

The Reverend Thomas Bayes

This sketch of the person we believe to be Thomas Bayes was created by Rachel Tanur and is reproduced here by permission of her estate.

CONTENTS

PREFACE

This second edition is intended to be an introduction to Bayesian statistics for students and research workers who have already been exposed to a good preliminary statistics and probability course, probably from a frequentist viewpoint, but who have had a minimal exposure to Bayesian theory and methods. We assume a mathematical level of sophistication that includes a good calculus course and some matrix algebra, but nothing beyond that. We also assume that our audience includes those who are interested in using Bayesian methods to model real problems, in areas that range across the disciplines.

This second edition is really a new book. It is not merely the first edition with a few changes inserted; it is a completely restructured book with major new chapters and material.

The first edition to this book was completed in 1988. Since then the field of Bayesian statistical science has grown so substantially that it has become necessary to rewrite the story in broader terms to account for the changes that have taken place, both in new methodologies that have been developed since that time, and in new techniques that have emerged for implementing the Bayesian paradigm. Moreover, as the fields of computer science, numerical analysis, artificial intelligence, pattern recognition, and machine learning have also made enormous advances in the intervening years, and because their interfaces with Bayesian statistics have steadily increased, it became important to expand our story to include, at least briefly, some of those important interface topics, such as *data mining tree models* and *Bayesian neural networks*. In addition, as the field of Bayesian statistics has expanded, the applications that have been made using the Bayesian approach to learning from experience and analysis of data now span most of the disciplines in the biological, physical, and social sciences. This second edition attempts to tell the broader story that has developed.

One direction of growth in Bayesian statistics that has occurred in recent years resulted from the contributions made by Geman and Geman (1984), Tanner and Wong (1987), and Gelfand and Smith (1990). These papers proposed a new method, now called *Markov chain Monte Carlo* (or just MCMC), for applying and implementing Bayesian procedures numerically. The new method is computer intensive and involves sampling by computer (so-called Monte Carlo sampling) from the posterior distribution to obtain its properties. Usually, Bayesian modeling procedures result in ratios of multiple integrals to be evaluated numerically. Sometimes these multiple integrals are high dimensional. The results of such Bayesian analysis are wonderful theoretically because they arise from a logical, self-consistent, set of axioms for making judgments and decisions. In the past, however, to evaluate such ratios of high-dimensional multiple integrals numerically it was necessary to carry out tedious numerical computations that were difficult to implement for all but the very computer-knowledgeable researcher. With a computer environment steadily advancing from the early 1980s, and with the arrival of computer software to implement the MCMC methodology, Bayesian procedures could finally be implemented rapidly, and accurately, and without the researcher having to possess a sophisticated understanding of numerical methods.

In another important direction of growth of the field, Bayesian methodology has begun to recognize some of the implications of the important distinction between *subjective* and *objective* prior information. This distinction is both philosophical and mathematical. When information based upon underlying theory or historical data is available (subjective prior information), the Bayesian approach suggests that such information be incorporated into the prior distribution for use in Bayesian analysis. If families of prior distributions are used to capture the prior knowledge, such prior distributions will contain their own parameters (called *hyperparameters*) that will need to be assessed on the basis of the available information. For example, many surveys are carried out on the same topic year after year, so that results obtained in earlier years can be used as a best guess for what is likely to be obtained in a new survey in the current year. Such "best available" information can be incorporated into a *prior distribution*. Such prior distributions are always proper (integrate or sum to one), and so behave well mathematically. A Bayesian analysis using such a prior distribution is called *subjective* Bayesian analysis.

In some situations, however, it is difficult to specify appropriate subjective prior information. For example, at the present time, there is usually very little, if any, prior information about the function of particular sequences of nucleotide base pairs in the DNA structure of the human genome. In such situations it is desirable to have meaningful ways to begin the Bayesian learning updating process. A prior distribution adopted for such a situation is called *objective*, and an analysis based upon such an objective prior distribution is called an *objective* Bayesian analysis. Such analyses serve to provide benchmark statistical inferences based upon having inserted as little prior information as possible, prior to taking data. Objective prior distributions correspond to "knowing little" prior to taking data. When such prior distributions are continuous, it is usually the case that these (improper) prior distributions do not integrate to one (although acceptable *posterior* distributions that correspond to these improper prior distributions must integrate to one). Sometimes, in simple cases, posterior inferences based upon objective prior distributions will result in inferences that correspond to those arrived at by frequentist means. The field has begun to focus on the broader implications of the similarities and differences between subjective and objective types of information. We treat this important topic in this edition, and recognize its importance in the title of the book. In many applications of interest, there is not enough information in a problem for classical inference to be carried out. So some researchers resort to subjective Bayesian inference out of necessity. The subjective Bayesian approach is adopted because it is the most promising way to introduce sufficient additional information into the problem so that a real solution can be found.

In earlier years, it was difficult to take into account uncertainty about which model to choose in a Bayesian analysis of data. Now we are learning how to incorporate such uncertainty into the analysis by using *Bayesian model averaging*. Moreover, we have been learning how to use Bayesian modeling in a *hierarchical* way to represent nested degrees of uncertainty about a problem. A whole new framework for *exploratory factor analysis* has been developed based upon the Bayesian paradigm. These topics are new and are discussed in this edition.

In this edition, for the first time, we will present an extensive listing, by field, of some of the broad-ranging applications that have been made of the Bayesian approach.

As Bayesian statistical science has developed and matured, its principal founders and contributors have become apparent. To record and honor them, in this edition we have included a *Bayesian Hall of Fame*, which we developed by means of a special opinion poll taken among senior Bayesian researchers. Following the table of contents is a collection of the portraits and brief biographies of these most important contributors to the development of the field, and there is an appendix devoted to an explanation of how the members of the Hall of Fame were selected.

The first edition of this book contained eight chapters and four appendices; this edition contains 16 chapters, generally quite different from those in the first edition, and seven appendices. The current coverage reflects not only the addition of new topics and the deletion of some old ones, but also the expansion of some previously covered topics into greater depth, and more domains. In addition, there are solutions to some of the exercises.

This second edition has been designed to be used in a year-long course in Bayesian statistics at the senior undergraduate or graduate level. If the academic year is divided into semesters, Chapters 1–8 can be covered in the first semester and Chapters 9–16 in the second semester. If the academic year is divided into quarters, Chapters 1–5 (Part I) can be covered in the fall quarter, Chapters 6–11 (Parts II and III) in the winter quarter, and Chapters 12–16 (Part IV) in the spring quarter.

Three of the sixteen chapters of this second edition have been written with the assistance of four people: Chapter 6 by Professor Siddhartha Chib of Washington University; Complement A to Chapter 6 by Professor George Woodworth of the University of Iowa; Chapter 13 by Professor Merlise Clyde of Duke University; and Chapter 14 by Professor Alan Zaslavsky of Harvard University. I am very grateful for their help. Much of Appendix 7 was written with the help of my former students, Dr. Thomas Ferryman, Dr. Mahmood Ghamsary, and Ms. Dawn Kummer. I am also grateful to Stephen Quigley of John Wiley and Sons, Inc., who encouraged me to prepare this second edition, and to Heather Haselkorn of Wiley, who helped and prodded me until it was done. Dr. Judith Tanur helped me to improve the exposition and to minimize the errors in the manuscript. The remaining errors are totally my responsibility. I am grateful to Rachel Tanur for her sketch of Thomas Bayes at the beginning of the book. Her untimely death prevented her from her intention of also sketching the scientists who appear in the Bayesian Hall of Fame. Dr. Linda Penas solved some of our more complex LaTex editorial problems, while Ms. Peggy Franklin typed some of the chapters in LaTex with indefatigable patience and endurance.

S. JAMES PRESS

Oceanside, CA
September, 2002

PREFACE TO THE FIRST EDITION

This book is intended to be an introduction to Bayesian statistics for students and research workers who have already been exposed to a good preliminary statistics and probability course from a classical (frequentist) point of view but who have had minimal exposure to Bayesian theory and methods. We assume a mathematical level of sophistication that includes a good calculus course and some matrix algebra but nothing beyond that. We also assume that our audience includes those who are interested in using Bayesian methods to model real problems in the various scientific disciplines. Such people usually want to understand enough of the foundational principles so that they will (1) feel comfortable using the procedures, (2) have no compunction about recommending solutions based upon these procedures to decision makers, and (3) be intrigued enough to go to referenced sources to seek additional background and understanding. For this reason we have tried to maximize interpretation of theory and have minimized our dependence upon proof of theorems.

The book is organized in two parts of four chapters each; in addition, the back of the book contains appendixes, a bibliography, and separate author and subject indexes. The first part of the book is devoted to theory; the second part is devoted to models and applications. The appendixes provide some biographical material about Thomas Bayes, along with a reproduction of Bayes's original essay.

Chapter I shows that statistical inference and decision making from a Bayesian point of view is based upon a logical, self-consistent system of axioms; it also shows that violation of the guiding principles will lead to "incoherent" behavior, that is, behavior that would lead to economically unsound decisions in a risky situation.

Chapter II covers the basic principles of the subject. Bayes's theorem is presented for both discrete and absolutely continuous random variables.

We discuss Bayesian estimation, hypothesis testing, and decision theory. It is here that we introduce prior distributions, Bayes' factors, the important theorem of de Finetti, the likelihood principle, and predictive distributions.

Chapter III includes various methods for approximating the sometimes complicated posterior distributions that result from applications of the Bayesian paradigm. We present large-sample theory results as well as Laplacian types of approximations of integrals (representing posterior densities). We will show how *importance sampling* as well as *simulation* of distributions can be used for approximation of posterior densities when the dimensions are large. We will also provide a convenient up-to-date summary of the latest Bayesian computer software available for implementation.

Chapter IV shows how prior distributions can be assessed subjectively using a group of experts. The methodology is applied to the problem of using a group of experts on strategic policy to assess a multivariate prior distribution for the probability of nuclear war during the decade of the 1980s.

Chapter V is concerned with Bayesian inference in both the univariate and multivariate regression models. Here we use vague prior distributions, and we apply the notion of predictive distributions to predicting future observations in regression models.

Chapter VI continues discussion of the general linear model begun in Chapter V, only here we show how to carry out Bayesian analysis of variance and covariance in the multivariate case. We will invoke the de Finetti notion of exchangeability (of the population mean vector distributions).

Chapter VII is devoted to the theory and application of Bayesian classification and discrimination procedures. The methodology is illustrated by appylying it to the sample survey problem of second guessing "undecided" respondents.

Chapter VIII presents a case study of how disputed authorship of some of the Federalist papers was resolved by means of a Bayesian analysis.

The book is easily adapted to a one- or two-quarter sequence or to a one-semester, senior level, or graduate course in Bayesian statistics. The first two chapters and the appendixes could easily fill the first quarter, with Chapters III–VIII devoted to the second quarter. In a one-quarter or one-semester course, certain sections or chapters would need to be deleted; which chapters or sections to delete would depend upon the interests of the students and teacher in terms of the balance desired between (1) theory and (2) models and applications.

The book represents an expansion of a series of lectures presented in South Australia in July 1984 at the University of Adelaide. These lectures were jointly sponsored by the Commonwealth Scientific and Industrial Research Organization (CSIRO), Division of Mathematics and Statistics and by the University of Adelaide's Departments of Economics and Statistics. I am grateful to Drs. Graham Constantine, William Davis, and Terry Speed, all of CSIRO, for their stimulating comments on the original lecture material, for their encouragement and support, and for planting the seeds from which this monograph grew. I am grateful to Dr. John Darroch, Dr. Alastair Fischer, Dr. Alan James, Dr. W. N. Venables, and to other participants of the lecture series for their stimulating questions that helped to put the book into perspective. Dr. John Pratt and Dr. S. L. Zabell helped to clarify the issues about de Finetti's theorem in Section 2.9.3, and Dr. S. K. Sinha suggested an example used in Section 2.7.1. Dr. Persi Diaconis and Dr. Richard Jeffrey presented stimulating discussions about randomness, exchangeability, and some of the foundational issues of the subject in a seminar at Stanford University during winter quarter of 1984–1985, a sabbatical year the author spent visiting Stanford University. I am deeply grateful to Drs. Harry Roberts and Arnold Zellner for exposing me to Bayesian ideas. Dr. Stephen Fienberg provided encouragement and advice regarding publishing the manuscript. I am also grateful to Dr. Stephen Fienberg, Dr. Ingram Olkin, and an anonymous publisher's referee for many helpful suggestions for improving the presentation. I am very grateful for suggestions made by Dr. Judith Tanur who read the entire manuscript; to Dr. Ruben Klein who read Chapters I and II; and to Drs. Frederick Mosteller and David Wallace who read Chapter VIII. I also wish to thank graduate students, James Bentley, David Guy, William Kemple, Thomas Lucas, and Hamid Namini whose questions about the material during class prompted me to revise and clarify various issues. Mrs. Peggy Franklin is to be congratulated for her outstanding typing ability and for her forbearance in seeing me through the many iterations that the manuscript underwent. We think we have eliminated most, if not all, errors in the book, but readers could help the author by calling any additional ones they find to his attention.

S. JAMES PRESS

Riverside, California
January, 1989

A BAYESIAN HALL OF FAME

Bayes, Thomas
1701–1761

DeFinetti, Bruno
1906–1985

DeGroot, Morris
1931–1989

Jeffreys, Harold
1891–1989

Lindley, Dennis V.
1923–

Savage, Leonard J.
1917–1971

THOMAS BAYES (1710–1761)
Bayes' Theorem of inverse probability is attributed to Thomas Bayes. Bayes applied the theorem to binomial data in his essay. The essay was published posthumously in 1763. Bayesian statistical science has developed from applications of his theorem.

BRUNO DE FINETTI (1906–1985)
One of the founders of the subjectivist Bayesian approach to probability theory (*Introduction to Probability Theory*, 1970, 2 volumes). He introduced the notion of *exchangeability* in probability as a weaker concept than independence. De Finetti's theorem on exchangeability yields the strong law of large numbers as a special case.

MORRIS DE GROOT (1931–1989)
Morris DeGroot was a student of Jimmie Savage and a leader in the Bayesian statitics movement. He contributed to the field in a wide-ranging spectrum of fields from economic utility and decision theory (*Optimal Statistical Decisions*, 1970) to genetics and the history of statistics. He founded the Carnegie Mellon University Department of Statistics.

HAROLD JEFFFREYS (1891–1989)
Geophysicist, astronomer (*The Earth: Its Origin, History, and Physical Constitution*), philosopher of science, and statistician (*Theory of Probability*, especially the Third Edition of 1961). The 1961 edition showed, using the principle of invariance, how to generate prior probabilities that reflected "knowing little"). Jeffreys believed that probabilities should be common to all (objectivist Bayesian approach). He advocated placing probabilities on hypotheses and proposed a very general theory of hypothesis testing that provided for probabilities that could be placed on hypotheses.

DENNIS V. LINDLEY (1923–)
Dennis Lindley (*Introduction to Probability and Statistics From a Bayesian Viewpoint*, 1965, 2 volumes), has been a stroong leader of the Bayesian movement. Moreover, several of his students have also become leaders of the movement as well as developers of important results in the field. He started as an objectivist Bayesian and became a subjectivist in his later years. Along with A.F.M. Smith he became a strong advocate of hierarchical Bayesian modeling

LEONARD J. (Jimmie) SAVAGE (1917–1971)
Jimmie Savage began his career as a mathematician, but contributed widely to mathematical statistics, probability theory, and economics. He developed an axiomatic basis for Bayesian probability theory and decision-making (*The Foundations of Statistics*, 1954). He held a subjectivist Bayesian view of probability. Along with W. Allen Wallis he founded the Department of Statistics at the University of Chicago. Along with Edwin Hewitt he demonstrated the widely applicable De Finetti theorem involving exchangeability under a broad set of conditions. They demonstrated its uniqueness property.

The portrait of Leonard J. Savage is reprinted with permission from "the Writings of Leonard Jimmie Savage: A Memorial," Copyright 1981 by the American Statistical Association and the International Mathematical Society. All rights reserved.

The portrait of Morris DeGroot is reprinted with the permission of the Institute of Mathematical Statistics from Statistical Science Journal.

The portrait of Bruno De Finetti is reprinted with the permission of Springer-Verlag from "The Making of Statisticians."

The portrait of Harold Jeffreys is reprinted with the permission of Springer-Verlag from "Statisticians of the Centuries."

The portrait of Dennis V. Lindley is reprinted with his permission and the permission of John Wiley & Sons, Inc., from "A Tribute to Lindley."

Subjective and Objective Bayesian Statistics

Second Edition

PART I

Foundations and Principles

CHAPTER 1

Background

1.1 RATIONALE FOR BAYESIAN INFERENCE AND PRELIMINARY VIEWS OF BAYES' THEOREM

In 1763 an important scientific paper was published in England, authored by a Reformist Presbyterian minister by the name of Thomas Bayes (Bayes, 1763). The implications of the paper indicated how to make statistical inferences that build upon earlier understanding of a phenomenon, and how *formally* to combine that earlier understanding with currently measured data in a way that updates the degree of belief (subjective probability) of the experimenter. The earlier understanding and experience is called the "prior belief" (belief or understanding held prior to observing the current set of data, available either from an experiment or from other sources), and the new belief that results from updating the prior belief is called the "posterior belief" (the belief held after having observed the current data, and having examined those data in light of how well they conform with preconceived notions). This inferential updating process is eponymously called Bayesian inference. The inferential process suggested by Bayes shows us that to find our subjective probability for some event, proposition, or unknown quantity, we need to multiply our prior beliefs about the event by an appropriate summary of the observational data. Thus, Bayesian inference suggests that all formal scientific inference inherently involves two parts, a part that depends upon subjective belief and scientific understanding that the scientist has prior to carrying out an experiment, and a part that depends upon observational data the scientist obtains from the experiment. We present Bayes' theorem compactly here in order to provide an early insight into the development of the book in later chapters.

Briefly, in its most simple form, the form for events (categorical or discrete data), Bayes' theorem or formula asserts that if $P\{A\}$ denotes the probability of an event A, and $P\{B|A\}$ denotes the probability of an event B conditional on knowing A, then:

$$P\{B|A\} = \frac{P\{A|B\}P\{B\}}{P\{A|B\}P\{B\} + P\{A|\bar{B}\}P\{\bar{B}\}},$$

where $\bar{B}$ denotes the complementary event to event B. This simple statement of conditional probability is the basis for all Bayesian analysis. $P(B)$ denotes the prior belief about B, $P\{B|A\}$ denotes the posterior belief about B (once we know A), and $P\{A|B\}$ denotes the model, that is, the process that generates the event A based upon knowing B.

As an example, suppose you take a laboratory test for diabetes. Let A denote the outcome of the test; it is a *positive* outcome if the test finds that you have the tell-tale indicators of diabetes, and it is a *negative* outcome if you do not. But do you really have the disease? Sometimes, although you do not actually have diabetes, the test result is positive because of imperfect characteristics of the laboratory test. Similarly, sometimes when you take the test there is a negative outcome when in fact you do have the disease. Such results are called false positives and false negatives, respectively. Let B denote the event that you actually do have diabetes. You would like to know the chances that you have diabetes in light of your positive test outcome, $P\{B|A\}$. You can check with the laboratory to determine the sensitivity of the test. Suppose you find that when the test is negative, the error rate is 1 percent (false negative error rate), and when the test is positive, its accuracy is 3 percent (the false positive error rate). In terms of the Bayes' formula,

$$\begin{aligned} P\{A = +test|B = diabetes\} &= 1 - P\{\bar{A} = -test|B = diabetes\} \\ &= 1 - 0.01 = 0.099, \end{aligned}$$

and

$$P\{+test|\bar{B} = no\,diabetes\} = \text{probability of a false positive} = 0.03.$$

Bayes' formula then gives:

$$P\{B\,|+\} = \frac{P\{+|B\}P\{B\}}{P\{+|B\}P\{B\} + P\{+\,|\bar{B}\}P\{\bar{B}\}}$$

$$P\{diabetes\,|+\} = \frac{(0.99)P\{B\}}{(0.99)P\{B\} + (0.03)P\{\bar{B}\}}.$$

It only remains to determine $P\{B\}$, the chances of someone having diabetes. Suppose there is no indication that diabetes runs in your family, so the chance of you having diabetes is that of a randomly selected person in the population about your age, say about one chance in one million, that is, $P\{B\} = 10^{-6}$. Substituting in the above formula gives:

$$P\{you\ have\ diabetes\,|positive\ test\ result\} = 0.003 = 0.3\%.$$

If we are concerned with making inferences about an unknown quantity θ, which is continuous, Bayes' theorem takes the form appropriate for continuous θ:

$$h(\theta | x_1, \ldots, x_n) = \frac{f(x_1, \ldots, x_n \,|\, \theta) g(\theta)}{\int f(x_1, \ldots, x_n \,|\, \theta) g(\theta) d\theta},$$

where $h(\cdot)$ denotes the probability density of the unknown θ subsequent to observing data $(x_1, \ldots, x_n)$ that bear on θ, f denotes the likelihood function of the data, and g denotes the probability density of θ prior to observing any data. The integration is taken over the support of θ. This form of the theorem is still just a statement of conditional probability, as we will see in Chapter 4.

A large international school of scientists (some of whom even preceded Bayes) supported, expanded, and developed Bayesian thinking about science. These include such famous scientists as James Bernoulli, writing in 1713, Pierre Simon de Laplace, writing in 1774, and many nineteenth- and twentieth-century scientists. Today, scientists schooled in the Bayesian approach to scientific inference have been changing the way statistical methodology itself has been developing. Many believe that a paradigm shift has been taking place in the way scientific inference is carried out, away from what is sometimes referred to as classical, or frequentist, statistical inference. Many scientists now recognize the advantages of bringing prior beliefs into the inferential process in a formal way from the start, instead of striving, and almost inevitably failing, to achieve total objectivity, and bringing the prior information into the problem anyway, in surreptitious, or even unconscious ways. Subjectivity may enter the scientific process surreptitiously in the form of seemingly arbitrarily imposed constraints in the introduction of initial and boundary conditions in the arbitrary levels of what should be called a significant result (selecting the "level of significance"), and in the de-emphasizing of certain outlying data points that represent suspicious observations.

Scientists will see that Bayes' theorem gives the degree of a person's belief (that person's subjective probability) about some unknown entity once something about it has been observed (i.e., posterior to collecting data about that entity), and shows that this subjective probability is proportional to the product of two types of information. The first type of information characterizes the data that are observed; this is usually thought of as the objective portion of posterior belief, since it involves the collection of data, and data are generally thought to be objectively determined. (We recognize that we do not really mean that data are objective unless we assume that there were no subjective influences surrounding the data collected.) This so-called *objective information* is summarized in the likelihood function. But the likelihood function is of course almost invariably based upon data that has been influenced by the subjectivity of the observer. Moreover, in small or often in even moderate size samples its structural form is not very well determined. So the likelihood function will almost invariably contain substantial subjective influences and uncertainty.

The second type of information used in Bayesian analysis is the person's degree of belief, the subjective probability about the unknown entity, held prior to observing anything related to it. This belief may be based, at least in part, on things that were observed or learned about this unknown quantity prior to this most recent measurement. Using Bayes' theorem, scientific belief about some phenomenon is formally updated by new measurements, the idea being that we learn about something by modifying what we already believe about it (our prior belief) to obtain a posterior belief after new observations are taken.

While it is well known that for a wide variety of reasons there are always some subjective influences in the research of scientists, and always have been, it is less well known that strong major subjective influences have actually been present in some of the work of the most famous scientists in history (see, for example, Press and Tanur, 2001). The personal beliefs and opinions of these scientists have often very strongly influenced the data they collected and the conclusions they drew from those data. While the phenomena these scientists were investigating were generally truly objective phenomena, external to the human mind, nevertheless, the data collected about these phenomena, and the decisions made relative to these phenomena were often driven by substantial subjectivity. Bayesian analysis, had it been available to these scientists, and had it been used, might have permitted these scientists to distinguish between models whose coexistence has caused controversy about their results even hundreds of years later.

Further, several scientists examining the same set of data from an experiment often develop different interpretations. This phenomenon is not unusual in science. When several scientists interpret the same set of data they rarely have *exactly* the same interpretations. Almost invariably, their own prior beliefs about the underlying phenomenon enter their thinking, as do their individual understanding of how meaningful each data point is. Their conclusions regarding the extent to which the data support the hypothesis will generally reflect a mixture of their prior degree of belief about the hypothesis they are studying, and the observed data.

Thus, we see that whether formal Bayesian inference is actually used in dealing with the data in an experiment, or whether other, nonBayesian methods are used, subjective prior belief is used in one way or another by all good scientists in a natural, and sometimes quite informal, way. Science cannot, and should not, be totally objective, but should and does involve a mixture of both subjective and objective procedures, with the one type of procedure feeding back on the other. As the data show the need for modification of the hypothesis, a new hypothesis is entertained, a new experiment is designed, new data are taken, and what was posterior belief in the earlier experiment becomes the prior belief in the new experiment, because the result of the last experiment is now the best understanding the scientist has of what result to expect in a new experiment. To study the future, scientists must learn from the past, and it is important—indeed inevitable—that the learning process be partly subjective.

During the twentieth century, since the development of methods of Bayesian statistical inference, there have been many exciting new scientific discoveries and developments. Some have been simply of the qualitative type where certain phenomena have been discovered that were not previously known (such as the discovery of the existence of the radiation belts that surround the Earth, the discovery of superconductivity, or the discovery of the double helical structure of DNA), and others have been quantitative, establishing relationships not previously established (such as the discoveries of the dose/effect relationships of certain pharmaceutical drugs, vaccines, and antibiotics that would minimize the chances of contracting various infectious diseases, or maximize the chance of a cure).

Considerable scientific advance is based upon finding important phenomena that are sometimes so shrouded in noise that it is extremely difficult to distinguish the phenomenon of interest from other factors and variables. In such cases, prior information about the process, often based upon previous theory, but sometimes on intuition or even wild guesses, can often be profitably brought to bear to improve the chances of detecting the phenomenon in question. A considerable amount of Bayesian statistical inference procedures that formally admit such prior information in the scientific process of data analysis have had to await the advent of modern computer methods of analysis, an advent that did not really occur until the last couple of decades of the twentieth century. However, since the arrival of real-time interactive computers, computational Bayesian methods such as Markov Chain Monte Carlo (MCMC, see Chapter 6) have been very usefully applied to problems in imaging and other problems in physics and engineering (see the series of books edited by different authors, every year since 1980, Maximum Entropy and Bayesian Methods published by Kluwer), problems of meta-analysis to synthesize results in a field—in biology, medicine, economics, physics, sociology, education, and others—and in a variety of scientific fields (see, for example, Appendix 5).

Subjectivity in science implies that we generally arrive at universal scientific truths by a combination of subjective and objective means. In other words, the methodology we use to discover scientific truths benefits greatly from bringing informed scientific judgment to bear on the hypotheses we formulate, and on the inferences we make from data we collect from experiments designed to test these hypotheses. Informed scientific judgment should not be shunned as a nonobjective, and therefore a poor methodological approach; collateral information about the underlying process should be actively sought so that it can be used to improve understanding of the process being studied. Combining informed knowledge with experimental data will generally improve the accuracy of predictions made about future observations.

Subjectivity is an inherent and required part of statistical inference and the scientific method. It is a *sine qua non* in the process of creating new understanding of nature. It must play a fundamental role in how science is carried out.

However, excessive, informal, untested subjectivity in science is also responsible for some basic errors, misrepresentations, overrepresentations, or scientific beliefs that were later shown to be false, that have occurred in science (see, for example, Press and Tanur, 2001). This author's views of subjectivity in science coincide closely with those of Wolpert (1992, p. 18) who wrote:

> ... the idea of scientific objectivity has only limited value, for the way in which scientific ideas are generated can be highly subjective, and scientists will defend their views vigorously... It is, however, an illusion to think that scientists are unemotional in their attachment to their scientific views—they may fail to give them up even in the face

> of evidence against them . . . scientific theories involve a continual interplay with other scientists and previously acquired knowledge . . . and an explanation which the other scientists would accept.

To illustrate the notion that subjectivity underlies experimental science, in Section 1.2 we use a very simple example involving whether or not a desired effect is observed in an experiment to show that merely observing scientific data and forming a likelihood function can involve considerable subjectivity.

1.2 EXAMPLE: OBSERVING A DESIRED EXPERIMENTAL EFFECT

Let us suppose that 100 observations are collected from an experiment replicated 100 times; there is one observation from each replication. These data are sent to five scientists located in five different parts of the world. All five scientists examine the same data set, that is, the same 100 data points. (Note that for the purposes of this example, the subjectivity involved in deciding what data to collect and in making the observations themselves is eliminated by sending the same "objective" data to all five scientists.) Should we expect all five of the scientists to draw the same conclusions from these data?

The answer to this question is a very definite "no". But how can it be that different observers will probably draw different conclusions from precisely the same data? As has been said above, inferences from the data will be a mixture of both subjective judgment (theorizing) and objective observation (empirical verification). Thus, even though the scientists are all looking at the same observational data, they will come to those same data with differing beliefs about what to expect. Consequently, some scientists will tend to weight certain data points more heavily than others, while different scientists are likely to weight experimental errors of measurement differently from one another. Moreover, if scientists decide to carry out formal checks and statistical tests about whether the phenomenon of interest in the experiment was actually demonstrated (to ask how strongly the claimed experimental result was supported by the data), such tests are likely to have different results for different scientists, because different scientists will bring different assumptions to the choice of statistical test. More broadly, scientists often differ on the mathematical and statistical models they choose to analyse a particular data set, and different models usually generate different conclusions. Different assumptions about these models will very often yield different implications for the same data.

These ideas that scientists can differ about the facts are perhaps startling. Let us return to our 100 observations and five scientists to give a very simple and elementary example, with the assurance that analogous arguments will hold generally for more realistic and more complicated situations.

Let us assume that the purpose of the experiment is to determine the probability that a certain genetic effect will take place in the next generation of a given type of simple organism. The question at issue is whether the effect occurs randomly or is subject to certain genetic laws. If the experiment is carried out many times, independently, under approximately the same set of conditions, how frequently will the genetic effect be observed? If the effect is observed say, 50 percent of the time, we will say it occurs merely at random, but it might occur, say 60 percent of the time, or more frequently. More generally, what percentage of the time will it occur under those experimental conditions? Call the (unknown) proportion of the time the effect should be observed, when the experiment is carried out repeatedly and independently, p; $p = 0.5$ if the effect is merely observed at random.

To determine p you repeat the experiment many times, each time record the result, and on the basis of the outcome, determine whether the effect is observed (called a "success"), or not (called a "failure"). In our case the experiment is repeated 100 times, and the outcome of success or failure is recorded for each trial of the experiment. Next, suppose that 90 successes were obtained. The question now is, "What is the value of p"?

These are the data you send to the five scientists in five different locations around the world (three women and two men) to see how they interpret the results. You tell each of them that there were 90 successes out of 100 trials of an experiment (or 90 percent of the experiments had successful outcomes). You also tell them that you plan to publish their estimates and their reasoning behind their estimates in a professional scientific journal, and that their reputations are likely to be enhanced or to suffer, in proportion to their degrees of error in estimating p. As we shall see, it will turn out that they will all have different views about the value of p after having been given the results of the experiment.

Scientist #1 is largely a *theorist* by reputation. She thinks successes are occurring at random since her theoretical understanding of the biochemical mechanism involved suggests to her that there should not be an effect; so for her, $p = 0.5$ no matter what. Her line of reasoning regarding the experimental outcomes is that although it just happened that 90 percent of the first 100 replications of the experiment were successful, it does not mean that if the experiment were to be repeated for another 100 trials, the next 100 trials would not produce, say, 95 failures, or any other large proportion of failures. Scientist #1 has a very strong preconceived belief based upon theory that the effect should not take place ($p = 0.5$), in the face of real data that militates against that belief. For her, unless told otherwise, all such experiments under these preset conditions should not demonstrate any real effect, even if many runs of successes or many runs of failures just happen to occur.

Scientist #2 has the reputation for being an *experimentalist*. He thinks $p = 0.9$, because that is the proportion of successes found in 100 replications. (This estimate of p is actually the maximum likelihood estimate.) Scientist #2's definition of the best estimate available from the data is the fraction of successes actually obtained. While Scientist #1 believed strongly in theory, Scientist #2 is ready to abandon theory in favor of strong belief in data, regardless of theory.

Scientist #3 is also a well-known *skeptic*. She decides that there is something strange about the reported results since they violate her strongly held expectation that the effect really should not be observed, other than at random, so there should be only about 50 successes in 100 replications. Scientist #3 then writes to you and asks you for details about the experimental equipment used. When Scientist #3 receives these details, she decides to replicate the experiment herself another 100 times. Most of the time she finds successes. These efforts convince Scientist #3 that the effect is really being produced, just as Scientist #2 concluded. The effect actually turns up 82 times. But what should Scientist #3 do now? Since her recent collection of replications found 82 successes in her new 100 trials, and the previous collection found 90 successes, Scientist #3 reasons that the experiment has been replicated 200 times, so perhaps it would not be unreasonable to estimate p as $(82 + 90)/200 = 0.86$.

When he actually carries out experiments in his research, which is only occasionally, Scientist #4 is an extremely *thorough and careful experimentalist*. He feels that he should know more about the experiment outcomes than he has been told so far. He writes to you for a copy of the actual sequence of experiment outcomes, studies it, and decides to ignore a run of 50 straight successes that occurred, feeling that such a run must be a mistake in reporting since it is so unlikely. This reduces the sample size (the number of available data points) from 100 down to 50, and out of those 50, 40 were successes. So his conclusion is that $p = 40/50 = 0.8$. Scientist #4 has taken the practical posture that many scientists take of weighting the observations so that some observations that are believed to be errors are discarded or downweighted in favor of those thought to be better measurements or more valid in some experimental sense.

Scientist #5 may be said to be *other-directed*. She badly wants the recognition from her peers that you said would depend on the accuracy of her estimate. She learns from you the names of the other four scientists. She then writes to them to find out the estimates they came up with. Having obtained their results, Scientist #5 decides that the best thing to do would be to use their average value. So for Scientist #5, the estimate of $p = (0.5 + 0.9 + 0.86 + 0.8)/4 = 0.765$. Or perhaps, learning that Scientist #1 made an estimate of p that was not data dependent, Scientist #5 might eliminate Scientist #1's estimate from her average, making the subjective judgment that it was not a valid estimate. Then Scientist #5 would have an estimate of $p = (0.9 + 0.86 + 0.8)/3 = 0.853$. Scientist #5's strategy is used by many scientists all the time so that the values they propose will not be too discrepant with the orthodox views of their peers.

So the five scientists came up with five different estimates of p for the same observed data. All the estimates are valid. Each scientist came to grips with the data with a different perspective and a different belief about p.

Suppose there had been a sixth scientist, a *decision theorist*, driven by a need to estimate unknown quantities on the basis of using them to make good decisions. Such a scientist would be interested in minimizing the costs of making mistakes, and might perhaps decide that overestimating p is as bad as underestimating it, so the costs should be the same for making these two types of errors. Moreover, he wants to select his estimator of p to enhance his reputation, and you have told him that a correct estimate will do just that. So he decides to select his estimator in such a way that the cost of being wrong will be as small as possible—regardless of the true value of p (under such circumstances he would often adopt a "quadratic loss function"). If Scientist #6 were to adopt the subjective belief that all values of p are equally likely, and then he used Bayes' theorem, his resulting estimate of p would be $p = 91/102 = 0.892$.

In summary, the values of p found by the six scientists are thus:

Scientist	#1	#2	#3	#4	#5	#6
Estimated value of p	0.500	0.900	0.860	0.800	0.765 or 0.853	0.892

But what is the true value of p? Note that with the exception of Scientist #1, who refuses to be influenced by the data, all the scientists agree that p must be somewhere between 0.765 and 0.900, agreeing that the effect is not occurring at random, but disagreeing about how often it should occur theoretically.

Suppose the experiment had been replicated 1000 times instead of the 100 times we have just considered, but with analogous results of 90 percent successes. Would this have made any difference? Well perhaps it might have to Scientist #1, because people differ as to the point at which they will decide to switch positions from assuming the effect is occurring only at random, in spite of the experimental outcome results, to a position in which they are willing to assume the effect is actually being generated from the experimental conditions. One scientist may switch after 9 successes out of 10 trials, another after 90 out of 100, whereas another may not switch until there are perhaps 900 successes out of 1000 trials, or might insist on an even more extensive experiment. In any case the scientists may still differ in their views about the true value of p for a very wide variety of reasons, only a few of which are mentioned above.

The biochemical experiment example just discussed was very elementary as far as experiments go, but it was science nevertheless. Similar interpretive and methodological issues arise in all branches of the sciences. In this book we examine the probabilistic and inferential statistical foundations, principles, methods, and applications of Bayesian statistical science, and throughout we adopt the notion of subjective probability, or degree of belief.

1.3 THOMAS BAYES

Thomas Bayes was a Presbyterian minister and mathematician who lived in England in the 1700s (born about 1702 and died April 17, 1761). Richard Price, a friend of Bayes, and interested in Bayes' research, submitted Bayes' manuscript on inverse probability in the binomial distribution to the professional journal, *Philosophical Transactions of the Royal Society*, which published the paper (posthumously) in 1763, an article reproduced in Appendix 4 of this book, along with biographical information about Bayes reproduced in Appendices 1 to 3.

There has been some mystery associated with Thomas Bayes. We are not quite certain about the year of his birth or about the authenticity of his portrait. Moreover, questions have been raised about what his theorem actually says (Stigler, 1982), and who actually wrote the paper generally attributed to him (Stigler, 1983). The common interpretation today is that the paper proposed a method for making probability inferences about the parameter of a binomial distribution conditional on some observations from that distribution. (The theorem attributed to Thomas Bayes is given in Proposition 9, Appendix 4; the Scholium that follows it has been controversial, but has been widely interpreted to mean that "knowing nothing" about a parameter in the unit interval implies we should take a uniform distribution on it.) Common belief is that Bayes assumed that the parameter had a uniform distribution on the unit interval. His proposed method for making inferences about the binomial parameter is now called Bayes' theorem (see Section 2.2) and has been generalized to be applicable beyond the binomial distribution, to any sampling distribution. Bayes appears to have recognized the generality of his result but elected to present it in that restricted binomial form. It was Laplace (1774) who stated the theorem on inverse probability in general form, and who, according to Stigler (1986), probably never saw Bayes' essay, and probably discovered the theorem independently. (Bayes carried out his work in England, where his theorem was largely ignored for over 20 years; Laplace carried out his work in France.) Jeffreys (1939) rediscovered Laplace's work.

Your distribution for the unknown; unobservable parameter is called *your* prior distribution, because it represents the distribution of *your* degree of belief about the parameter prior to *your* observing new data, that is, prior to your carrying out a new experiment that might bear on the value of the parameter. Bayes' theorem gives a mathematical procedure for updating your prior belief about the value of the parameter to produce a posterior distribution for the parameter, one determined subsequent to your having observed the outcome of a new experiment bearing on the value of the unknown parameter. Thus, Bayes' theorem provides a vehicle for changing, or updating, the degree of belief about an unknown quantity (a parameter, or a proposition) in light of more recent information. It is a formal procedure for merging knowledge obtained from experience, or theoretical understanding of a random process, with observational data. Thus, it is a "normative theory" for learning from experience, that is, it is a theory about how people *should* behave, not a theory about how they *actually do* behave, which would be an "empirical theory."

The ideas in the theorem attributed to Bayes were really conceived earlier by James Bernoulli in 1713, in Book 4 of his famous treatise on probability, *Ars Conjectandi* ("The Art of Conjecturing"), published posthumously. In that book, James Bernoulli, or Jakob Bernoulli as he was known in German, not only developed the binomial theorem and laid out the rules for permutations and combinations but also posed the problem of inverse probability of Bayes (who wrote his essay 50 years later). However, Bernoulli did not give it mathematical structure. In *Ars Conjectandi*, James Bernoulli (1713) wrote:

> To illustrate this by an example, I suppose that without your knowledge there are concealed in an urn 3000 white pebbles and 2000 black pebbles, and in trying to determine the numbers of these pebbles you take out one pebble after another (each time replacing the pebble you have drawn before choosing the next, in order not to decrease the number of pebbles in the urn), and that you observe how often a white and how often a black pebble is withdrawn. The question is, can you do this so often that it

becomes ten times, one hundred times, one thousand times, etc., more probable (that is, it be morally certain) that the numbers of whites and blacks chosen are in the same 3 : 2 ratio as the pebbles in the urn, rather than in any other different ratio?

This is the problem of inverse probability, which concerned eighteenth-century mathematicians (Stigler, 1986, Chapter 2). According to Egon Pearson (1978, p. 223), James Bernoulli "...was destined by his father to be a theologian, and [he] devoted himself after taking his M.A. at Basel (Switzerland) to theology." This endeavor involved Bernoulli in philosophical and metaphysical questions (Bayes, of course, received similar training, being a minister). In fact, Maistrov (1974, p. 67), evaluating Bernoulli, believes he was an advocate of "metaphysical determinism," a philosophy very similar to that of Laplace, some of whose work did not appear until 100 years after Bernoulli. Writing in *Ars Conjectandi*, in the first chapter of the fourth part, Bernoulli said:

> For a given composition of the air and given masses, positions, directions, and speed of the winds, vapor, and clouds and also the laws of mechanics which govern all these interactions, tomorrow's weather will be no different from the way it should actually be. So these phenomena follow with no less regularity than the eclipses of heavenly bodies. It is, however, the usual practice to consider an eclipse as a regular event, while (considering) the fall of a die, or tomorrow's weather, as chance events. The reason for this is exclusively that succeeding actions in nature are not sufficiently well known. And even if they were known, our mathematical and physical knowledge is not sufficiently developed, and so, starting from initial causes, we cannot calculate these phenomena, while from the absolute principles of astronomy, eclipses can be pre-calculated and predicted... The chance depends mainly upon our knowledge.

In the preceding excerpt, Bernoulli examined the state of tomorrow's weather, given today's observational data that relate to weather and a belief about tomorrow's weather. He noted that, of necessity, because of our inability to understand precisely the behavior of the forces governing weather, we must treat tomorrow's weather as uncertain and random to some extent, but predictable in terms of chance (probability), in accordance with our knowledge. This is precisely the kind of question addressable by Bayes' theorem of 1763, in terms of degree of belief about tomorrow's weather, given today's observations and a prior belief about tomorrow's weather. Moreover, the development of quantum mechanics and the Heisenberg uncertainty principle have elaborated Bernoulli's view of chance, showing that chance is a fundamental property of nature that goes beyond mere lack of knowledge of the physical laws governing some physical phenomenon.

Additional background on the life of Thomas Bayes may be found in Appendices 1 to 3.

1.4 BRIEF DESCRIPTIONS OF THE CHAPTERS

This book is subdivided into four parts. Part 1 includes Chapters 1 to 5 on foundations and principles, Part 2 includes Chapters 6 and 7 on numerical implementation of the Bayesian paradigm, Part 3 includes Chapters 8 to 11 on Bayesian inference and decision making, and Part 4 includes Chapters 12 to 16 on models and applications.

In this introductory chapter we have shown that the interpretation of scientific data has always involved a mixture of subjective and objective methods. We have pointed out that it is no longer necessary to adopt the informal subjective methods used by scientists in the past, but that now, Bayesian inference is available to formalize the introduction of subjectively based information into the analysis of scientific data, and computational tools have been developed to make such inference practical. Chapter 2 discusses probability as a degree of belief, calibration, and the axiomatic basis of Bayesian inference and decision making. Chapter 3 presents the meaning of the likelihood principle and conditional inference. Chapter 4, the central chapter in the book, presents and interprets Bayes' theorem for both discrete and continuous data and discrete and continuous parameters. Chapter 5 presents a wide variety of subjective and objective prior distributions, both discrete and continuous, for individuals and for groups, in both univariate and multivariate parameter cases.

Chapters 6 and 7 focus on computer routines that sample from the posterior distribution, approximations, and relevant software. We treat Gibbs sampling, the Metropolis–Hastings algorithm, data augmentation, and some computer software that has been developed for use in Bayesian analysis. In Complement A to Chapter 6 we have provided a detailed discussion of the WinBUGS Program, probably the most widely used software program that affords access to the Gibbs sampling paradigm without doing your own programming. Complement B to Chapter 6 provides a listing of other popular Bayesian software. Chapter 7 discusses the role of large sample posterior distributions (normal), the Tierney–Kadane and Naylor–Smith approximations, and importance sampling.

Chapter 8 discusses Bayesian estimation, both point and interval, and Chapter 9 discusses hypothesis testing. Both treat univariate and multivariate cases. Chapter 10 discusses predictivism, including the de Finetti theorem, de Finetti transforms, and maximum entropy. Chapter 11 presents the Bayesian approach to decision making. In the last part of the volume, in Chapter 12, we discuss the Bayesian approach to the analysis of the general linear model, both univariate and multivariate, including regression, analysis of variance and covariance, and multivariate mixed models. In Chapter 13 we discuss Bayesian model averaging to account for model uncertainty. Chapter 14 explicates Bayesian hierarchical modeling, Chapter 15 discusses Bayesian factor analysis, and Chapter 16 concludes with a presentation of Bayesian classification and discrimination methods.

There are seven appendices, the first three of which relate to the life of Thomas Bayes, and the posthumous submission of his famous theorem to a journal. Appendix 4 presents Bayes' original paper in its entirety. Appendix 5 is a list of references, by subject, of applications of the Bayesian paradigm that have been made across diverse disciplines. It is not intended to be an exhaustive list, merely an indication of a few of the applications that have been made. Appendix 6 presents an explanation of how the most important contributors to the development of Bayesian statistics, a Bayesian Hall of Fame, were selected. There are exercises at the end of each chapter. Appendix 7 provides solutions to selected exercises.

SUMMARY

In this chapter we have examined the role of subjectivity in science and have found that subjectivity, judgment, and degree of belief are fundamental parts of the scientific process. We have also seen from the example involving observing a desired experimental effect (Section 1.2) that observing some phenomenon involves an interpretation of the observational data. Different observers will often have different interpretations of the same data, depending upon their own backgrounds, beliefs they brought to the experiment *a priori*, and theoretical expectations of the experimental outcomes.

EXERCISES

1.1 Give examples of subjectivity in medical diagnosis by a physician; in the introduction of "evidence" by an attorney in civil or criminal trials; in studies of the human genome by molecular biologists; in portfolio analysis of financial securities by certified financial analysts; in the design of a building by a civil engineer; in the "reading" of X rays by a radiologist; and in the evaluation of objects found at an ancient archeological site.

1.2* By going to the original reference sources, explain who Stigler believes actually wrote Bayes' theorem, and also, explain what Stigler believes Bayes' theorem was really saying.

1.3 In the observing of a desired experimental effect example of Section 1.2 of this chapter, can you think of other interpretations of the same data?

1.4* Why is the book *Ars Conjectandi* by James Bernoulli of particular interest to people interested in Bayesian statistical inference?

1.5 In the observing of a desired experimental effect example of Section 1.2 of this chapter, which interpretation of the data seems most appropriate to you? Why?

1.6 In an experiment you were doing, or one that you were analysing, how would you decide whether certain observations were too large or too small to be considered as part of the basic data set?

1.7* Distinguish between prior and posterior distributions.

1.8* The methodological approaches to science of some of the most famous scientists in history (Aristotle, Marie Curie, Charles Darwin, Albert Einstein, Michael Faraday, Sigmund Freud, Galileo Galilei, William Harvey, Antoine Lavoisier, Isaac Newton, Louis Pasteur, and Alexander von Humboldt) were sometimes quite subjective. Give examples in the cases of three of these people. (*Hint*: see Press and Tanur, 2001.)

1.9* Explain the meaning of the statement, "Most scientific inference is already partly subjective, and it always has been."

* Solutions for asterisked exercises may be found in Appendix 7.

FURTHER READING

Bayes, T. (1763). "An Essay Towards Solving a Problem in the Doctrine of Chances," *Philos. Trans. Royal Soc. London*, **V61.53**, 370–418. Reprinted in Appendix 4.

Bernoulli, J. (1713). *Ars Conjectandi*, Book 4, Baseae, Impensis Thurnisiorum.

Gatsonis, C., Hodges, J. S., Kass, R. E., and Singpurwalla, N. D., Eds. (1993). *Case Studies in Bayesian Statistics*, New York, Springer-Verlag.

Gatsonis, C., Hodges, J. S., Kass, R. E., and Singpurwalla, N. D., Eds. (1995). *Case Studies in Bayesian Statistics*, Vol. II, New York, Springer-Verlag.

Gatsonis, C., Hodges, J. S., Kass, R. E., McCulloch, R., Rossi, P., and Singpurwalla, N. D. (1997). *Case Studies in Bayesian Statistics*, Vol. III, New York, Springer-Verlag.

Gatsonis, C., Kass, R. E., Carlin, B., Carriquiry, A., Gelman, A., Verdinelli, I., and West, M. (1999). *Case Studies in Bayesian Statistics*, Vol. IV, New York, Springer-Verlag.

Howson, C. and Urbach, P. (1989). *Scientific Reasoning*, La Salle, Illinois, Open Court Publishing Co.

Jeffreys, H. (1939, 1st edition; 1948, 2nd edition; 1961, 3rd edition). *Theory of Probability*, Oxford, Oxford University Press and Clarendon Press.

Laplace, P. S. (1774, 1814). *Essai Philosophique sur les Probabilites*, Paris. This book went through five editions (the fifth was in 1825) revised by Laplace. The sixth edition appeared in English translation by Dover Publications, New York, in 1951. While this philosophical essay appeared separately in 1814, it also appeared as a preface to his earlier work, *Theorie Analytique des Probabilites*.

Maistrov, L. E. (1974). *Probability Theory: A Historical Sketch*, S. Kotz, Trans. and Ed. New York, Academic Press.

Matthews, R. A. J. (1998a). "Fact versus Factions: the Use and Abuse of Subjectivity in Scientific Research", ESEF Working Paper 2/98, *The European Science and Environment Forum*, Cambridge, England, September, 1998.

Matthews, R. A. J. (1998b) "Flukes and Flaws," *Prospect.*, 35, 20–24.

Pearson, E. (1978). *The History of Statistics in the 17th and 18th Centuries*, New York, Macmillan.

Press, S. J. and Tanur, J. M. (2001). *The Subjectivity of Scientists and the Bayesian Approach*, New York, John Wiley and Sons, Inc.

Stigler, S. M. (1982). "Thomas Bayes and Bayesian Inference," *J. Roy. Statist. Soc.* (*A*), **145** (2) 250–258.

Stigler, S. M. (1983). "Who Discovered Bayes' Theorem," *Amer. Stat.*, **37**(4), 290–296.

Stigler, S. M. (1986). *The History of Statistics*, Cambridge, MA, The Belknap Press of Harvard University Press.

Wolpert, L. (1992). *The Unnatural Nature of Science*, Boston, Harvard University Press.

CHAPTER 2

A Bayesian Perspective on Probability

2.1 INTRODUCTION

In this chapter we discuss some fundamental notions associated with probability from a Bayesian perspective, so that our development of statistics will be built upon a solid, fundamentally Bayesian probabilistic foundation. We begin with the notion that probability reflects a degree of randomness.

There are various interpretations of probability. In Bayesian statistics we will place greatest emphasis on *subjective probability* (degree of belief). We will then compare several distinct axiom systems of probability, finally focusing on the Renyi system of conditional probability, and we will adopt that system of axioms as the fundamental system of algebra for subjective probability. We will discuss the concept of coherence, important when assessing prior distributions. We will then explain how to operationalize an individual's subjective probability beliefs. For the convenience of the reader, in the appendices to this volume we have reproduced the original famous probability essay of Bayes (Appendix 4), along with additional biographical and related material (Appendices 1 to 3).

In science in general, and in Bayesian statistical science in particular, we must obviously be able to deal not only with randomness, but also with degrees of randomness. The phenomenon of having "measurable rain" on a particular day in August in Los Angeles, California, is generally treated as random, but anyone who has examined historical records about rainfall in August in Los Angeles knows that rain occurs only rarely. Therefore, it would be foolhardy not to take knowledge about rainfall in August in Los Angeles into account before specifying one's degree of belief about the likelihood of rain next August in Los Angeles. Bayesian analysis permits us to do that in a very simple way.

Most coins we have ever seen tend to fall "heads or tails" with approximately equal frequency in a coin-tossing experiment. They do not have exactly equal frequency because the features carved on each side of the coin to establish identification of the faces cause a weight distribution imbalance. In turn, the imbalance generates a tendency for the coin to fall on its two faces with unequal frequency (but the frequencies are still close enough to equal for most practical purposes). However, some coins are designed with a bias, that is, to fall heads and tails with unequal frequencies. Bayesian analysis permits us to specify what we believe to be the bias of the coin, be it a "fair" or a "biased" coin.

How should the different degrees of randomness that usually occur with random phenomena be specified or quantified? Conventional wisdom on this question suggests we use the notion of probability. We extend this idea to quantities or entities that are unknown to an individual. For example, if a quantity X, such as the opening price tomorrow on the NASDAQ stock exchange of Microsoft (MSFT) common stock, is unknown to you, you can be expected to have a probability distribution for X in your head that we will attempt later to elicit from you. This probability distribution reflects your degree of uncertainty about X. For example, perhaps you feel that it is equally likely that $X > 100$, and that $X < 100$, so that 100 is the median of your probability distribution for X; you might also feel that there is just a 5 percent chance that $X > 110$, and so on. We will assume that any quantity unknown to you can be assigned a personal probability distribution. Note that it is also possible that you feel there is a 60 percent chance that $X > 120$. If so, you are being *incoherent* since the probabilities you are specifying are incompatible (see Section 2.3, below).

As another example, but of a different type, suppose you are a biologist, and you are asked about whether a particular gene expression is part of the genetic regulation of left-handedness. You may not know the answer, but you do have a certain degree of belief as to whether this particular gene could possibly be involved, at least in part, in determining left-handedness. You may believe, for example, based upon what you already know about functionality of genes, that there is a 30 percent chance that this particular gene is so involved (it either is, or it is not, but your degree of belief about it is 30 percent). It will be explained below with a quantitative interpretation of what we mean by your degree of belief.

2.2 TYPES OF PROBABILITY

2.2.1 Axiom Systems

Suppose in a coin-tossing experiment the same coin falls "heads" on each of 10 consecutive tosses. If we do not have knowledge of how the coin was fabricated (knowledge obtained prior to the actual tossing of the coin, called "prior knowledge") and we do not know how the coin has fallen in the past, what are we to believe about the bias of the coin? We consider two possibilities, stated as propositions:

Proposition A. The coin is strongly biased toward falling "heads."

Proposition B. The events of falling "heads" or "tails" are equiprobable, but we happen to have observed a possible, but unusual, run of ten "heads."

Should we believe Proposition A or B? Many people would be inclined to believe A, but some others would believe B. But if we change the level of occurrence from "all heads on 10 consecutive tosses" to "all heads out of 20 consecutive tosses," an even greater number of observers would accept Proposition A instead of B. In fact, there is no way to know with certainty how this coin will fall on the next toss; moreover, even if we were to toss it thousands of times and it always fell "heads," we would only gather an increasing amount of empirical evidence for more "heads" than "tails," and we could gather even more empirical evidence by continuing to toss the coin more and more times. This accumulating empirical evidence affects our degree of belief about how the coin is likely to fall on the next, as yet unobserved, trial. If you wish to calculate the chance of getting say, 3 heads, or 10 heads out of 10 trials, without knowing the bias of the coin, you may use de Finetti's theorem (see Section 10.5).

The outcome of an experiment (even a simple one such as tossing a coin to see whether the coin will fall heads or tails), or the truthfulness of any proposition, such as, "the Los Angeles Dodgers baseball team will win tomorrow's game," is called an "event." Your degree of belief about an event is called your *subjective probability* for that event. It is sometimes called your *personal probability* for that event. This subjective probability of events may be assigned numerical values, as explained below.

The probability of an event A, expressed $P\{A\}$, or $P(A)$, is a nonnegative, real-valued function of events satisfying certain axioms that permit algebraic manipulation of probabilities. Various axiom systems have been proposed for the algebra of the events. Moreover, there have been several interpretations of $P\{A\}$, alternative to subjective probability, such as the frequentist notion of probability, and others. We first mention several other definitions of probability.

2.2.2 Frequency and Long-Run Probability

The kind of probability that most people are familiar with before they study Bayesian statistical science is called *frequency probability*. We assume that the reader is familiar with the conventional notion of probability as a frequency, but we mention it here for comparison purposes, because the comparison is quite important.

If an event can occur in just a *finite* number of ways, the frequency probability of the event is defined as the ratio of the favorable number of ways the particular event can occur compared with the total number of possible events. For example, there are four ways of drawing an ace randomly from a very well shuffled deck of 52 cards (the notion of a very well shuffled deck translates to there being the same chance of drawing any particular, preassigned card in the deck, such as the queen of hearts). If the deck is well shuffled, the chance of drawing an ace at random is $4/52$, since the deck is assumed to contain four randomly located aces. If we were to draw cards repeatedly from this deck, each time replacing the card drawn in a random location in the deck, we would be likely to draw an ace about 4 times in every 52 draws of a card. To define frequency probability requires that we consider events that can be repeated many times so that we can examine the proportion of times favorable events have occurred. If the number of ways the event can occur is *infinite*, the definition of frequency probability is more complicated, but it may still be defined in terms of limits of frequencies (see standard references to classical probability, such as Rohatgi, 1976, or Rice, 1994).

Subjective probability can be defined not only if the number of ways the event can occur is *infinite*, but even for events that will occur only once; it can also be defined for someone's degree of belief about the truthfulness of a proposition. It will be seen that frequency probability is a special case of subjective probability. Frequency probability is also known by the names *objective*, *mathematical*, or *numerical probability*. This frequency interpretation of probability, although limited to events that are repeatable, is also sometimes referred to as the classical interpretation because it has been the most commonly used, but it is not the one we adopt in this book, except as a special case of subjective probability.

The *long run*, or *empirical probability* of an event is the limit of the proportion of times the event has occurred, as the number of trials of an experiment increases ad infinitum (von Mises, 1928). This notion relates the theoretical probability construct directly to events in the real world.

2.2.3 Logical Probability

There is also *logical probability* (Keynes, 1921). This idea relates inversely to the degree to which a proposition can be falsified (Popper, 1959), but this interpretation of probability has been generally abandoned because no prescription was ever given for how logical probabilities were to be assigned and compared (as pointed out by Ramsey, 1926). Keynes believed that not all degrees of belief could be measured numerically. Moreover, he believed that (logical) probabilities are only *partially ordered* (not *well ordered*). So they can not all be compared, even if they can be measured. These notions curtailed further development of his ideas in probability, since what was needed was an ability to compare probabilities. Jeffreys (1939) extended Keynes' ideas of logical probability by introducing a hitherto nonexistent axiom system to buttress the theory. He suggested rules for generating (objective) prior probabilities, but did not generate logically consistent probabilities, a fact that he readily admits himself in his book (p. 37, 1939). Regardless, many of the objective prior probabilities commonly used today are attributable to Jeffreys.

2.2.4 Kolmogorov Axiom System of Frequency Probability

For convenience and reference, we repeat here the probability axiom system introduced by the famous Russian mathematician A. N. Kolmogorov (1933). For any event A:

1. $0 \leq P\{A\}$.

2. $P\{\text{all possible events}\} = 1$.

3. Countable additivity: If $A_1, A_2, \ldots$ are mutually exclusive and exhaustive events, then

$$P\left\{\bigcup_1^{\infty} A_j\right\} = \sum_1^{\infty} P\{A_j\}.$$

Probability functions satisfying this axiom system are frequency or mathematical probabilities. As mentioned above, mathematical probabilities are a special case of subjective probabilities.

De Finetti (1974) suggested that the Kolmogorov system be modified so that Property 3 above need hold only for a finite collection of events (this is called the finite additivity axiom), and he required that the probability function $P\{\cdot\}$ be interpreted subjectively, as a *personal* probability for a given individual.

The first formal mention of probability as a subjective, or personal, probability was made by Ramsey (1926). Kolmogorov seems to have had a more objective interpretation in mind in which the interpretation of $P\{\cdot\}$ did not depend upon the individual.

The Kolmogorov axiom system adopts the *countable* additivity property. (This is discussed further below.) Ramsey (1926), Savage (1954), and de Finetti (1974) all favored the finite additivity axiom system and the subjective probability interpretation. Countable additivity implies that the probability function $P\{\cdot\}$ is continuous, whereas finite additivity does not imply continuity. However, in the countable additivity system, some events (propositions) cannot be assigned a probability (these are called nonmeasureable events), whereas in a finite additivity axiom system, every event can be assigned a probability. In a countable additivity system, at every point of discontinuity x_0, of some cumulative distribution function, $F(x)$, we have

$$\lim_{x \to x_0^+} F(x) = F(x_0)$$

In a finite additivity system, this continuity result will no longer hold true. Recall that the cumulative distribution function, $F(x)$, is defined as $F(x) \equiv P\{X \leq x\}$. That is, it is the probability of the event $\{X \leq x\}$, where X denotes some variable that is random, which is, itself, a function that maps events into the real line. Such a technical difficulty causes problems in developing asymptotic theory, for example. Since we never really pass all the way to the limit in applications, however, there is no difficulty in practice.

2.2.5 Savage System of Axioms of Subjective Probability

Savage (1954) expanded upon the finite additivity axiom system of subjective probability to develop a more broadly based, seven-axiom system that could serve as the basis for a theory of rational decision making using Bayesian principles (and added in the "sure thing principle" as well; see complement to Chapter 2). The Savage theory not only includes a finite additivity system of probability, but also includes axioms appropriate for the mathematical scaling of preferences, and axioms governing the use of lotteries to assess subjective probabilities of a decision maker. Savage's pioneering efforts in decision making were influenced by the earlier work of von Neumann and Morgenstern (1947), who developed a theory of computing economic gain and loss when engaging in a gamble (see Chapter 11). This "theory of utility" they developed became an integral component of Savage's theory. (Utility functions are defined and discussed in the complement to this chapter; their applications are discussed in Chapter 11.) In fact, a major implication of Savage's work, and perhaps the one most often quoted, is that decisions should be made so as to maximize expected utility, conditional on all data already available (see Complement to Chapter 2). That is, if the axioms of Savage's axiom system are satisfied, optimal decision-making behavior results from maximizing the average utility of the decision maker (i.e., averaged over the distribution obtained from Bayes' theorem, called the *posterior distribution*). Sometimes we minimize expected *loss functions* instead of maximizing expected *utility functions*, but loss and utility functions are just linearly related. The seven Savage axioms, as well as his theory of maximization of expected or average utility, are not repeated here (see Savage, 1954, and the complement to Chapter 2 where the axioms are enumerated) because to do so would distract us from the thrust of our effort. The Savage theory is *normative*, in that it is a theory for how people *should* behave, to be logical and rational, instead of a theory for how they *actually* do behave (an *empirical* theory); see also Chapter 1 for this distinction regarding Bayesian analysis in general.

Pratt et al. (1965) developed another formal axiomization of subjective probability as it relates to utility and the scaling of preferences of a decision maker for various consequences of a lottery. The axiom system is logically equivalent to that of Savage (1954) but has, perhaps, more intuitive appeal and is more readily operationalizable. Related work, relying upon utility theory as it relates to subjective probability, may be found in Anscombe and Aumann (1963).

2.2.6 Rényi Axiom System of Probability

Rényi, (1970, p. 70) generalized the Kolmogorov axiom system by developing a countable additivity axiom system based upon conditioning. That is, Rényi assumed that all probabilities must be conditioned on certain prior historical information or evidence and that the entire axiom system should be erected on such a conditional foundation. A decision-making (or judgment-formulating) behavior conditioned on information available prior to taking any data seems closer to real-world behavior than an unconditional system.

Rényi's axiom system requires the same mathematical foundation as Kolmogorov's, but the axioms are conditional, as seen below. In the Kolmogorov system, conditional probability is a concept that is derivable from the basic axioms, whereas in the Rényi system the basic axioms are built upon a conditional probability foundation. The axiom system is given by:

1. For any events A, B, we have $P\{A \mid B\} \geq 0$, *and* $P\{B \mid B\} = 1$.
2. For disjoint events $A_1, \ldots$ (events that cannot occur in the same experiment), and some event B, we have

$$P\left\{\bigcup_1^{\infty} A_i \mid B\right\} = \sum_1^{\infty} P\{A_i \mid B\}.$$

3. For every collection of events (A, B, C), $B \subseteq C$, $0 < P\{B \mid C\}$, we have:

$$P\{A \mid B\} = \frac{P\{(A \bigcap B) \mid C\}}{P\{B \mid C\}}.$$

As with frequency probability, if $P\{A \cap B\}$ does not depend upon B, A and B are said to be independent.

Rényi shows that his generalized axiom system includes the Kolmogorov axiom system as a special case. Thus, the Bayesian is free to interpret a probability as a frequency probability if he feels that it is appropriate in a given situation, but in other situations, the probability can be interpreted as a purely personal one.

In the system we are adopting, continuity of the probability function $P\{\cdot\}$ is subsumed (see Section 2.2.4). In this book, we adopt the Rényi conditional probability axiom system, and we interpret the probability function $P\{A\}$ subjectively, that is, as a degree of belief about the event A. Our rationale for adopting the Rényi system is pragmatic: real-world judgment formulation and decision making is almost invariably conditional upon prior information and experience. Moreover, at the minor cost of introducing an axiom system with the theoretical drawback of having nonmeasurable events, the countable additivity axiom of the Rényi system permits us to have continuity and asymptotic theory as part of our armamentarium.

It will be seen later that for some purposes it will be useful to use probability distributions that spread their masses uniformly over the entire real line. In all of the axiom systems mentioned, such continuous distributions are improper, because their densities do not integrate to unity. We will find that a good use for such improper distributions is to represent an individual's distribution of belief about an unknown quantity that can lie anywhere on the real line, prior to observing the outcome of an experiment. Suppose the individual to be you. If all possible values of the unknown quantity seem to you to be equally likely over a certain interval, you might adopt a uniform distribution over that interval. But how large should the interval be if you are really not knowledgeable about the quantity? As the length of the interval increases to infinity, to reflect a greater degree of ignorance or uncertainty the distribution goes from proper to improper. See Chapter 5 for a detailed discussion of uniform (vague) prior distributions and for an explanation of why they are of such fundamental importance in Bayesian statistics. For our purposes it is useful to appreciate that the Rényi axiom system of probability accommodates (by means of limits of ratios) probabilities on the entire real line and that it is thus closest to the system of probability required for Bayesian inference. It will be seen that we will have frequent occasion to use posterior distributions that are proper but that have resulted, by application of Bayes' theorem, from use of improper prior distributions. The mathematical formalism required to handle probability calculations involving improper distributions in a rigorous way can be accommodated by the Rényi probability structure, but cannot be accommodated very well by any of the other systems. It would be useful to have a formal axiomatic theory of subjective probability that is not only conditional in the sense of Rényi [so that it accommodates probabilities on the entire real line (unbounded measures)] but which also captures the advantages of a finitely additive probability system. Fishburn (1981), reviewed some 30 distinct axiom systems that are attempts to weaken the axioms. The systems are mainly small variations on one another. Such systems include probability axioms as well as preference-ordering axioms. For an up-to-date tour through the various axiom systems, see Fishburn (1986).

2.3 COHERENCE

An individual whose probability statements (beliefs) about a collection of events do not satisfy our (Rényi) system of axioms (for example, if they are mutually inconsistent) is said to be incoherent (see Ramsey, 1926; de Finetti, 1937, p. 111). According to cognitive psychologists, observed human behavior tends to be incoherent (see, for example, Edwards, 1982), whereas our axiom systems and normative interpretation specify ideal, appropriate, Bayesian probability behavior. If an incoherent individual is willing to make a bet based upon these (incoherent) probabilities, and he considers each bet fair or favorable, he will suffer a net loss no matter what happens (such a bet is called a *Dutch book*).

2.3.1 Example of Incoherence

Suppose you claim that your degree of belief about Event A is $P\{A\} = 0.7$. You claim moreover that your degree of belief about Event B is $P\{B\} = 0.5$. Furthermore, you know that A and B are independent, and you believe $P\{\text{both}\, A \,\text{and}\, B\} = 0.1$. In such a case you are being incoherent because by independence,

$$P\{\text{both}\, A \,\text{and}\, B\} = P\{A\}P\{B\} = \{0.7\}\{0.5\} = 0.35,$$

but your belief is that $P\{\text{both}\, A \,\text{and}\, B\} = 0.1$; a contradiction. You cannot have it both ways, so in trying to do so, you are being incoherent.

Whenever multiple subjective probabilities are assessed on the same individual, coherency checks should be run to preclude the possibility of Dutch book, and to ensure internal consistency. This notion becomes extremely important when you are assessing the (prior) probability distribution of a person who has expert knowledge about an unknown quantity. If the expert's assessments are mutually inconsistent, his probability distribution will be incoherent.

In another example, suppose you feel $P\{A\} = 0.75$, and also $P\{\bar{A}\} = 0.75$, where $\bar{A}$ denotes the complementary event to A (that is, you are being incoherent). We define the two lotteries:

Lottery (1): N_1 bets \$3 on event A, and N_2 bets \$1 on event $\bar{A}$.

Lottery (2): N_3 bets \$3 on event $\bar{A}$, and N_4 bets \$1 on event A.

In Lottery (1), you could bet with individual N_1 or individual N_2, individuals who take the two sides of the lottery. Similarly, in Lottery (2), you could bet with N_3 or N_4. However, because you would feel that both lotteries are fair (as a consequence of your belief that $P\{A\} = P\{\bar{A}\} = 0.75$), you would be willing to take either side of both lotteries. Suppose you choose N_1, and N_3, and I choose N_2 and N_4. If A occurs you receive \$1 from Lottery (1), but you must pay \$3 from Lottery (2), with a net loss of \$2. If $\bar{A}$ occurs, you must pay \$3 in Lottery (1), but you can collect \$1 in Lottery (2), for a net loss again of \$2. So you lose \$2 no matter what happens. I have made a Dutch book against you because you were incoherent.

2.4 OPERATIONALIZING SUBJECTIVE PROBABILITY BELIEFS

How should an individual think about an event (Event A) to quantify his/her judgment about how strongly he/she believes the event is likely to occur? Equivalently, how does an individual generate $P\{A\}$?

Operationalizing subjective probability statements can be carried out in terms of lotteries. This idea is attributed to Ramsey (1926), who required the notion of an "ethically neutral proposition" called E, for which degree of belief is 0.5.

Your degree of belief in a (ethically neutral) proposition E is 0.5 if you are indifferent to receiving either:

1. α, if E is true, and β, if E is false, or
2. α, if E is false, and β, if E is true,

for any α and any β. Translating Ramsey's notion to an event occurring, suppose you are interested in some event, say Event A. Denote by $p = P\{A\}$, your degree of belief that Event A will occur. Your degree of belief is illustrated as follows. Suppose you are offered a choice between the following two options:

1. Receiving some small reward R (such as \$1) if Event A occurs, but receiving no reward if A does not occur;
2. Engaging in a lottery in which you win the small reward R with probability p, but you lose and receive no reward with probability $(1 - p)$.

If you are indifferent between choices 1 and 2, your degree of belief in Event A occurring is p. The size of the reward R plays no role as long as it is small. (If the reward were large, there would be a confusion between the individual's preference for a large increase in wealth, and his/her degree of belief about the truthfulness of a proposition, or the likelihood of occurrence of an event.) In the remainder of this book we assume that when you say your degree of belief in Event A is p, you mean you are indifferent to the pair of choices given above.

2.4.1 Example of Subjective Probability Definition and Operationalization

Suppose the local weather forecaster claims the probability of measurable precipitation in your geographical area for tomorrow is 0.7, or 70 percent. One way to interpret the statement of the forecaster is the following. If there were an urn containing seven white balls and three black balls, all randomly mixed in the urn, the forecaster believes that the chance of measurable precipitation in your geographical area for tomorrow is the same as the chance that if someone selects a ball at random from the urn, it will be a white ball. The forecaster should then be indifferent between:

1. Receiving \$1 if measurable precipitation actually occurs, or receiving nothing if measurable precipitation does not occur; compared with
2. Receiving \$1 if a white ball is selected randomly from an urn containing exactly seven white balls and three black balls, very well mixed up, or receiving nothing otherwise.

We could equally well have used an urn with 70 white balls and 30 black balls, or an urn with 700 white balls and 300 black balls, and so on. We could also have used a random number table or a random number generator, or any other random mechanism. Note that this definition of probability affords us the opportunity to score a probability assessor on how well he assesses. Therefore, if measurable precipitation actually occurs, we can give the weather forecaster positive feedback and a reward. If precipitation does not occur, he/she gets negative feedback, and no reward. If the forecasting is carried out repeatedly the forecaster should improve his probability assessing ability and be considered a *good probability assessor* (Winkler, 1967; Winkler and Murphy, 1968). This idea will be elaborated on in Section 2.5.

2.5 CALIBRATION OF PROBABILITY ASSESSORS

The idea of improving subjective probability assessment ability using feedback is referred to as *calibration* of the probability assessor. For example, the weather forecasting assessor is evaluated each day by comparing his actual forecast with the true weather that resulted. If he is given the feedback as to what weather actually occurred, over time he should improve his assessment ability. Adapting from Stael von Holstein (1970), this notion is modeled below.

Let an individual's judgment about an uncertain quantity X be represented by a probability distribution on the outcomes. Let $E_1, \ldots, E_n$ denote a set of mutually exclusive and exhaustive possible outcome events. Define $p_j \equiv P\{X \epsilon E_j\}$, and $\mathbf{p} \equiv (p_1, \ldots, p_n)$. The vector $\mathbf{p}$ represents the assessor's true beliefs. The assessor gives the response vector $\mathbf{r} \equiv (r_1, \ldots, r_n)$. This is his judgment, and $\mathbf{r}$ might not be equal to $\mathbf{p}$. An *honest* assessment is one for which $\mathbf{r} = \mathbf{p}$, and feedback is intended to induce the assessor to bring the two into agreement. A *scoring rule* is a function of $\mathbf{r}$ and the event being assessed that actually occurs. An assessor receives a score $S_k(\mathbf{r})$ if the kth event occurs. Various scoring rules have been proposed. For example:

1. Quadratic scoring rules (Brier, 1950; de Finetti, 1962, 1965), $S_k(\mathbf{r}) = 2r_k - \sum_{i=1}^{n} r_i^2$
2. Spherical scoring rule (Toda, 1963; Roby, 1965), $S_k(\mathbf{r}) = r_k \sqrt{\sum_{i=1}^{n} r_i^2}$
3. Logarithmic scoring rule (Good, 1952), $S_k(\mathbf{r}) = \log r_k$.

As a specific example in the context of weather forecasting (assessing), suppose X denotes the amount of precipitation in inches in Riverside, California, on January 13, 2001. X can be a number in the interval $(0, \infty)$. Suppose the forecaster works for a television channel that requires the forecaster to provide forecasts for the four intervals: $[0, 0.5]$, $(0.5, 1.0]$, $(1.0, 3.0]$, $(3.0, \infty)$. Suppose we adopt the Brier quadratic scoring rule $S_k(\mathbf{r}) = 2r_k - \sum_{i=1}^{n} r_i^2$. The probability forecast $(0.5, 0.3, 0.1, 0.1)$ will receive a score, $S_1(\mathbf{r}) = 0.64$, if the actual amount of precipitation received on that day was in the interval $[0, 0.5]$. If the actual amount of precipitation received were in the second interval $(0.5, 1.0]$, the forecaster would receive a score of $S_2(\mathbf{r}) = 0.24$; analogously, $S_3(\mathbf{r}) = -0.16$; and $S_4(\mathbf{r}) = -0.16$.

Scoring becomes particularly important to society when the subjective probability assessment involves the chance of rare events occurring (one time in one million, one time in one billion, etc., such as with nuclear power plant meltdown disasters, natural gas tank explosions, etc.). In this context it has been found that simultaneously competing risks can often confuse the assessor, and result in much lower perceived risks than those that actually occur (Freudenburg, 1988).

These notions operationalize our way of thinking about subjective probabilities.

2.6 COMPARING PROBABILITY DEFINITIONS

How is our degree-of-belief notion of probability to be reconciled with other interpretations of probability? The notion of probability as a "long-run frequency" is a theoretical construct with no obvious way to be operationalized in a finite number of trials. To clarify this point, our coin-tossing example will serve well again. Every time an experimenter tosses the same coin the initial conditions change. So his belief about "heads" coming up on this toss should depend upon the initial conditions of the toss. I have never personally tossed a coin as many as one million times (nor do I know of anyone else who has), so I do not know what would happen after even so few tosses, but I doubt that the proportion of "heads" found would be precisely 0.5 for almost any normal coin. In spite of such empirically related uncertainty, long-run frequency remains a useful notion for some applications, as long as we remain prepared to recognize that the notion must be abandoned if it contradicts our degree of belief (in small samples in any given situation).

Using the classical Kolmogorov concept of mathematical probability, if we toss a coin 20 times and obtain 20 "heads," we should continue to believe the coin is "fair" if that is what we believed prior to the first toss. We should believe this in spite of the fact that a fair coin could only fall that way with a mathematical probability of about one in a million (actually, one chance in 1,048,576). Here, "fair" means its two faces are equiprobable in the long-run frequency sense. In any real-world situation, however, such as someone in a gaming situation in a casino, or in the case of the National Aeronautics and Space Agency (NASA) noting that the last 20 trials of a rocket launch of a particular type of rocket all ended up in *success*, the individual or agency is unlikely to believe the 20 "heads" (or rocket launch successes) occurred with probability 0.5 on each trial; the individual or agency is much more likely to modify his/its belief about the fairness of the coin, or the equiprobable nature of the successes and failures of rocket launches. (Individuals will, of course, vary according to how many straight successes it will take before they will alter their belief about the bias of the trial results.)

In the case of the coin, if you are willing to accept as prior knowledge (prior to the tossing of the coin) the knowledge that the coin has a particular bias (such as equiprobable outcomes), then our predictions of outcomes coincide with those of Kolmogorov (mathematical probability). For philosophical elaborations of the various definitions and interpretations of probability, such as that of Laplace (1814), as compared with probability as a long-run frequency, logical probability, and subjectivist probability, and for further discussion of the construction of logical axiomatic systems of scientific inference, and arguments for and against the finite, versus countable, additivity axiom, and so on, see Carnap and Jeffrey (1971), Good (1983), a collection of some of his more philosophical papers, Jeffrey (1980), Volume 2 of Carnap and Jeffrey, (1971), Jaynes et al. (2003), Keynes (1921), Lindley (1976), Popper (1968), and Skyrms (1984). Some interesting and important articles on the foundations of Bayesian inference may be found in Corefield and Williamson (2001).

SUMMARY

In this chapter we have tried to summarize the foundational issues upon which Bayesian statistics is based. We started with the elemental notion of randomness, which is intrinsic to nature. We then developed our ideas about probability as a quantitative characterization of randomness. We compared the axiomatic developments of Kolmogorov (mathematical probability) with those of de Finetti, Ramsey, Rényi, and Savage (subjective probability). We discussed coherence as well as how to operationalize our degree of belief about an unknown quantity. We considered the long-run frequency interpretation of probability of von Mises, and finally we discussed the Reverend Thomas Bayes and the intuitive interpretation of his famous theorem. The mathematical treatment of the theorem is presented in Chapter 4. We also discussed how to calibrate subjective probability assessors by means of scoring rules and feedback.

COMPLEMENT TO CHAPTER 2: THE AXIOMATIC FOUNDATION OF DECISION MAKING OF L. J. SAVAGE

Definition 1. **f** will be used to denote an act, that is, a function, attaching the consequence $f(s)$ to the event s.

Axiom 1. The relation "$\prec$" is a simple ordering among acts. Thus, $\mathbf{f} \prec \mathbf{g}$ means that act **f** is not preferred to act **g**, or there is indifference between acts **f** and **g**. So if a person had to decide between acts **f** and **g**, and no other acts were available, he/she would decide upon act **g**.

Axiom 2. For every act **f** and every act **g**, and every event $B, \mathbf{f} \prec \mathbf{g}$, given B, or $\mathbf{g} \prec \mathbf{f}$, given B. Also, if $\mathbf{f} \prec \mathbf{g}$, and $\mathbf{g} \prec \mathbf{h}$, then $\mathbf{f} \prec \mathbf{h}$.

"Sure Thing Principle". If $(\mathbf{f} \prec \mathbf{g}) \mid B$, and $(\mathbf{f} \prec \mathbf{g}) \mid \bar{B}$, then $\mathbf{f} \prec \mathbf{g}$. Here, $\bar{B}$ denotes the event: not B. That is, if your choice is the same regardless of event B, then B is irrelevant to your choice. The principle is often violated empirically. The most famous empirical violation that leads to refutation of this principle is a result of Allais (1953).

Definition 2. $\mathbf{f} \equiv g$ means a constant act. That is, for every event, $s, f(s) = g =$ the consequence of the act, and act **f** results in the same consequence regardless of s.

Axiom 3. If $f(s) = g, f'(s) = g'$, for every $s \in B$, and B is not null, then $\mathbf{f} \prec \mathbf{f}'$, given B, if and only if $g \leq g'$.

Definition 3. $A \lhd B$ means that event A is not more probable than event B (note that "probable" means in a subjective probability sense).

Axiom 4. For every $A, B, A \lhd B$ or $B \lhd A$.

Axiom 5. It is not true that for every f and $f', f \prec f'$. That is, there is at least one pair of consequences f, f' such that f is preferred to f'.

Axiom 6. If $\mathbf{g} \prec \mathbf{h}$, and f is any consequence, then there exists a partition of the space of all events, S, such that if $\mathbf{g}$ or $\mathbf{h}$ is so modified on any one element of the partition as to take the value f at every s there, other values being undisturbed, then the modified $\mathbf{g}$ remains less than $\mathbf{h}$, or $\mathbf{g}$ remains less than the modified $\mathbf{h}$, as the case may require.

Definition 4. $\mathbf{f} \prec g$, given B, if and only if $\mathbf{f} \prec \mathbf{h}$, given B, when $h(s) = g$ for every s.

Axiom 7. If $\mathbf{f} \prec g(s)$, given B, for every $s \in B$, then $\mathbf{f} \prec \mathbf{g}$, given B.

Conclusions

Savage (1954) shows that Axioms 1 to 7 imply the existence of a "utility function," $U(f)$, such that:

1. If $\mathbf{f}$ and $\mathbf{g}$ are bounded, then $\mathbf{f} \prec \mathbf{g}$ if and only if $U(\mathbf{f}) \leq U(\mathbf{g})$;
2. If $\mathbf{f}$ and $\mathbf{g}$ are bounded, and $P(B) > 0$, then $\mathbf{f} \prec \mathbf{g}$, given B, if and only if $0 \leq E\{[U(\mathbf{g}) - U(\mathbf{f})] \mid B\}$.

Utility Functions

A utility function $U(\mathbf{f})$ is a real valued function defined on the set of consequences, such that $\mathbf{f} \prec \mathbf{g}$ if and only if $U(\mathbf{f}) \leq U(\mathbf{g})$. Let $f(s) = g$. Then, to each consequence g there corresponds a numerical valued function $U(g)$ that represents the value you place on the consequence g. Moreover, it is unique only up to a linear transformation, that is, any linear function:

$$U^*(g) = aU(g) + b,$$

where $a > 0$, is also a utility function. In scaling utilities, for convenience and uniqueness, we often define $U(g)$ on the unit interval, and we take the utility of the worst and best possible consequences in a gamble (decision problem) to be zero and unity, respectively.

In summary, in this book we adopt the Rényi axiom system for probability, with a subjective interpretation of the probability function. In addition, for decision making we will adopt the Savage principle of "maximization of expected utility."

EXERCISES

2.1 What is meant by "inverse probability"?

*Solutions for asterisked exercises may be found in Appendix 7.

2.2* Distinguish between finite and countable additivity, and explain the importance of the difference.

2.3* Explain the importance of the Rényi axiom system of probability for Bayesian inference.

2.4* What is meant by "coherence"? Give an example (not from the text) of incoherence.

2.5* Explain what you mean by the statement that YOUR subjective probability, or YOUR degree of belief, that there will be measurable rain tomorrow in Riverside, CA, is 70 percent.

2.6* What is the meaning of the (von Mises) long-run frequency approach to probability?

2.7 By going to the original reference sources, explain who Stigler believes actually wrote Bayes' theorem, and also explain what Bayes' theorem was really saying about how to choose the prior distribution.

2.8* Explain why a frequency interpretation of probability cannot possibly interpret the statement in Exercise 2.5.

2.9 Why is the book *Ars Conjectandi* by James Bernoulli of particular interest to people interested in Bayesian statistical inference?

2.10* What is meant by "Savage's axioms," and what is the context to which this axiom system applies?

2.11* What is meant by utility theory?

2.12* Explain the assertion "Mathematical probability is merely a special case of subjective probability."

2.13* What is meant by a "normative theory of inference and decision-making behavior" as compared with an empirical theory?

FURTHER READING

Allais, M. (1953). "Le Comportement de L'Homme Rationnel Devant Le Risque: Critique Des Postulats et Axiomes de L'Ecole Americaine," *Econometrica*, **21**, 503–546. Translated in *Expected Utility Hypotheses and the Allais Paradox*, M. Allais and Hagen, D., Eds., Dordrecht; Reidel, 1979.

Anscombe, F. J. and Aumann, R. J. (1963). "A Definition of Subjective Probability," *Ann. Math. Stat.*, **34**, 199–205.

Brier, G. W. (1950). "Verification of Forecasts Expressed in Terms of Probabilities," *Monthly Weather Review*, **78**, 1–3.

Carnap, R. and Jeffrey, R. C., Eds. (1971). *Studies in Inductive Logic and Probability*, Vol. 1, Berkeley, University of California Press.

De Finetti, B. (1937). "Le Prevision: Ses Lois Logiques, Ses Sources Subjectives," *Ann. de l'Institut Henri Poincare*, **7**, 1–68. Reprinted in *Studies in Subjective Probability*, Melbourne, FL, Krieger, 1980 (English translation).

De Finetti, B. (1962). "Does it Make Sense to Speak of 'Good Probability Appraisers'?," in I. J. Good, Ed., *The Scientist Speculates—An Anthology of Partly-Baked Ideas*, London, Heinemann, 357–364.

De Finetti, B. (1965). "Methods for Discriminating Levels of Partial Knowledge Concerning a Test Item," *Br. J. Math. and Statist. Psychol.* **18**, 87–123.

De Finetti, B. (1974). *Theory of Probability*, Vols 1 and 2, New York, John Wiley and Sons, Inc.

Edwards, W. (1982). "Conservatism in Human Information Processing", in D. Kahneman, P. Slovic and A Tversky, Eds., *Judgment Under Uncertainty: Heuristics and Biases*, Cambridge, Cambridge University Press, 359–369; excepted from a paper that appeared in B. Kleinmuntz (Ed.), *Formal Representation of Human Judgment*, New York, John Wiley and Sons, Inc.

Fishburn, P. C. (1981). "Subjective Expected Utility: A Review of Normative Theories," *Theory and Decision*, **13**, 139–199.

Fishburn, P. C. (1986). "The Axioms of Subjective Probability," *Statistical Science*, **1** (3), 335–358.

Freudenburg, W. R. (1988). "Perceived Risk, Real Risk: Social Science and the Art of Probabilistic Risk Assessment," *Science*, **242** (70), 44–49.

Good, I. J. (1952). "Rational Decisions," *J. Royal Statist. Soc. (B)*, **14**, 107–114.

Good (1983). *Good Thinking: The Foundations of Probability and its Applications*, Minneapolis, The University of Minnesota Press.

Jaynes, E. T. & Bretthorst, G. Larry (2003), *Probability Theory: The Logic of Science*, Cambridge, Cambridge University Press.

Jeffrey, R. C. (Ed.) (1980). *Studies in Inductive Logic and Probability*, Vol. 2, Berkeley, The University of California Press.

Jeffreys, H. (1939, 1st edition; 1948, 2nd edition; 1961, 3rd edition). *Theory of Probability*, Oxford, Oxford University Press and Clarendon Press.

Keynes, J. M. (1921). *A Treatise on Probability*, London, Macmillan.

Kolmogorov, A. N. (1933). *Foundations of the Theory of Probability*, (translated from German, New York, Chelsey, 1950).

Laplace, P. S. (1814). *Essai Philosophique sur les Probabilites*, Paris. This book went through five editions (the fifth was in 1825) revised by Laplace. The sixth edition appeared in English translation by Dover Publications, New York, in 1951. While this philosophical essay appeared separately in 1814, it also appeared as a preface to his earlier work, *Theorie Analytique des Probabilites*.

Lindley, D. V. (1976). "Bayesian Statistics," in W. L. Harper and C. A. Hooker, Eds., *Foundations of Probability Theory, Statistical Inference, and Statistical Theories of Science*, Vol. II, Boston, Reidel, 353–363.

Lindley, D. V. (1982). "Scoring Rules and the Inevitability of Probability," *Int. Statist. Rev.* **50**, 1–26.

Popper, Sir K. (1959, 1968). *The Logic of Scientific Discovery*, New York, Harper Torchbooks, Harper and Row.

Pratt, J. W., Raiffa, H. and Schlaifer, R. (1965). *Introduction to Statistical Decision Theory*, New York, McGraw-Hill Book Co., Inc.

Ramsey, F. P. (1926). "Truth and Probability," in R. B. Braithwaite, Ed., *The Foundations of Mathematics and Other Logical Essays*, 1931, by permission of the Humanities Press,

New York, and Routledge and Kegan Paul Ltd., London, reprinted in Kyberg and Smokler, 1980, *Studies in Subjective Probability*, Melbourne, Florida, Krieger Publishing Co.

Rényi, A. (1970). *Probability Theory*, New York, American Elsevier.

Rice, J. A. (1994). *Mathematical Statistics and Data Analysis*, Wadsworth Publishing Co.

Roby, T. B. (1965). "Belief States and Uses of Evidence," *Behav. Sci.*, **10**, 255–270.

Rohatgi, V. K. (1976). *An Introduction to Probability Theory and Mathematical Statistics*, New York, John Wiley and Sons, Inc.

Savage, L. J. (1954). *The Foundations of Statistics*, New York, John Wiley and Sons, Inc.

Skyrms, B. (1984). *Pragmatics and Empiricism*, New Haven, CT, Yale University Press.

Stael von Holstein, C. A. S. (1970). *Assessment and Evaluation of Subjective Probability Distributions*, Stockholm, The Economic Research Institute, Stockholm School of Economics.

Toda, M. (1963). "Measurement of Subjective Probability Distributions," Report #3, Division of Mathematical Psychology, Institute for Research, State College Pennsylvania.

von Mises, R. (1928). *Probability, Statistics and Truth*, Translated from German, London, Allen and Unwin, 1957.

Von Neumann, J. and Morgenstern, O. (1st edition 1944; 2nd edition 1947; 3rd edition 1953; renewed copyright 1972), *Theory of Games and Economic Behavior*, Princeton, New Jersey, Princeton University Press.

Winkler, R. L. (1967). "The Quantification of Judgment: Some Methodological Suggestions," *J. Am. Statist. Assn.*, **62**, 1105–1120.

Winkler, R. L. and Murphy, A. H. (1968). "'Good' Probability Assessors," *J. Appl. Meteorology*, **7**, 751–758.

CHAPTER 3

The Likelihood Function

3.1 INTRODUCTION

This chapter focuses on the role of the likelihood function in Bayes' theorem. We will discuss the likelihood function itself, the likelihood principle as it applies to Bayesian statistics, the implications of the likelihood principle and the role of conditioning on earlier information in the likelihood function, the likelihood function as distinguished from the joint density or probability mass function in Bayesian inference, and finally, getting the Bayesian updating inference procedure started by developing the likelihood function based upon computer-generated graphical or other procedures.

3.2 LIKELIHOOD FUNCTION

The term "likelihood" appears to have been first used by Fisher (1921). A history of its use was given by Edwards (1972, 1974). A discussion of its use in maximum likelihood estimation was given by Aldrich (1997). For Fisher, it represented a rational belief about the unknown quantity. He said,

> The function of the θ's maximized is not, however, a probability and does not obey the laws of probability

(Fisher, 1930, p. 532). The likelihood function of a set of observations, $X_1, \ldots, X_n$, is their joint probability mass function (or joint probability density function) when viewed as a function of the unknown parameter θ, which indexes the distribution from which the X_is were generated. We denote the likelihood function by $L(\theta; x_1, \ldots, x_n)$. We denote it in this way to emphasize the fact that the likelihood function is viewed as a function of the parameter θ, and when we will be making inferences about θ using the likelihood function, such as finding maximum likelihood estimators, or making Bayesian inferences, the variable of interest is θ, not the X_is. We might denote the probability mass, or probability density, function of the data, conditional on θ, by $f(x_1, \ldots, x_n \mid \theta)$, but here, the expression is being viewed as a function of the data for a given value of θ. Mathematically, at least up to a multiplicative constant, we must have:

$$L(\theta; x_1, \ldots, x_n) = f(x_1, \ldots, x_n \mid \theta),$$

for all θ and the x_is. The likelihood function is uniquely defined only up to a multiplicative constant. We note that in finding a maximum likelihood estimator (MLE), a constant of proportionality plays no role. The MLE is the same regardless of the proportionality constant. (Also recall from Section 1.1 that the likelihood function is generally subjective because it depends upon subjectively observed and interpreted data points.)

To illustrate, suppose we are concerned with a binomial distribution for the number of successes, $r = \sum_{i=1}^{n} X_i$ in n independent Bernoulli trials with outcome X_i on the ith trial, where X_i is one or zero, with probabilities θ and $(1 - \theta)$. Then, the probability mass function is given by:

$$f(r \mid \theta) = \binom{n}{r} \theta^r (1 - \theta)^{n-r}.$$

However, by ignoring the binomial coefficient, $\binom{n}{r}$, the likelihood function for θ may be written as:

$$L(\theta; r) = \theta^r (1 - \theta)^{n-r},$$

to emphasize that interest focuses on θ, and that the binomial coefficient does not play a fundamental role in Bayesian inference. Note that $L(\theta; r)$ and $f(r \mid \theta)$ differ only by the proportionality constant $\binom{n}{r}$, and that the MLE for θ remains the same regardless of the value of the proportionality constant.

Fisher, (1922) used likelihood as a measure of plausibility. He suggested that the ratio of the likelihood function evaluated at the MLE to the likelihood function at various values of the unknown value of θ reflect the *plausibility* of a specific θ. He used the *plausibility ratio* $L(\hat{\theta})/L(\theta)$, where $\hat{\theta}$ denotes the MLE; see Reid (2000) for further discussion of this idea.

3.3 LIKELIHOOD PRINCIPLE

Let $x = (x_1, \ldots, x_n)'$ and $y = (y_1, \ldots, y_n)'$ denote two distinct sets of observations, and $f(x_1, \ldots, x_n \mid \theta)$ denotes the joint density of the x_is. Suppose $L(\theta; x) \propto L(\theta; y)$, that is, the likelihood functions for the two sets of observations are proportional (and the proportionality constant does not depend upon θ). Thus, their likelihood functions are the same, up to a multiplicative constant. But this means that MLEs of θ will be the same, and if we adopt the same prior for θ, posterior inferences about θ based upon x will be the same as those based upon y. This fact is called the *likelihood principle.*

An equivalent statement is that all the information about θ contained in the data from an experiment that is required for inference about this unknown quantity θ is available from the likelihood function for these data. According to the likelihood principle, information about how the experiment was carried out, or other background about the experiment is not relevant for inference since it does not appear in the likelihood function.

We note, along with Birnbaum (1962), that the evidence about θ arising from a given experiment and its observed outcome must be the same for two experiments whose associated likelihoods are proportional. We thus conclude, as a corollary to the likelihood principle, that all evidence about θ from an experiment and its observed outcome should be present in the likelihood function. If evidence about θ lies outside the likelihood function it is a violation of the likelihood principle. Birnbaum (1962) also pointed out that belief in sufficiency and conditionality lead to belief in the likelihood principle.

For example, suppose that in n independent Bernoulli trials, r successes and $(n - r)$ failures are obtained (we adopt here a more conventional notation used with binomial data rather than the generic notation used in Section 3.2). Let p denote the probability of success on a single trial. If the sampling takes place with n fixed, the resulting distribution for r, given p, is binomial. If, however, the sampling takes place with r held fixed, the resulting distribution for n, given p, is *negative binomial.* In either case, however, the likelihood function for p is proportional to $[p^r(1 - p)^x]$ (x denotes the number of failures before the rth success). Thus, the likelihood principle requires that all inferences about p be based upon the quantity in brackets, regardless of how the sampling was carried out. This idea that posterior inferences must be the same and that they must be based upon the likelihood function also follows from the ideas of coherence, and is implied by Bayes' theorem. Thus, the likelihood principle requires that statistical inference be based only upon data found in the likelihood function. Principles of unbiasedness and admissibility, which depend upon observations not yet taken (averages over the entire sample space), violate the likelihood principle. (For more extended discussions of the implications and foundations of the likelihood principle, see Berger and Wolpert, 1984.) For inferential inconsistencies arising from violations of the likelihood principle see Ghosh (1988, a collection of essays by D. Basu).

3.4 LIKELIHOOD PRINCIPLE AND CONDITIONING

Classical statistical inference is based upon long-run behavior over repeated experiments. The likelihood principle asserts that inference should be based only upon observations actually collected in this experiment (and not upon observations that might have been taken, but were not). Thus, we must condition upon the data from the current experiment only.

As an example (Berger and Wolpert, 1984), suppose a substance to be analysed can be sent either to a laboratory in New York or a laboratory in California. The two laboratories seem to be equally competent, so a fair coin is flipped to choose between them. The California laboratory is selected on the basis of the flip. Results of the analysis come back and a conclusion must be reached. Should the conclusion take into account that the coin could have fallen the other way and the New York laboratory could have been chosen? The likelihood principle says "no". Only the experiment actually performed should matter.

3.5 LIKELIHOOD AND BAYESIAN INFERENCE

It is revealing in terms of likelihood to examine the functional form of Bayes' theorem for a continuous unknown quantity θ, in terms of data $x_1, \ldots, x_n$. As we will see in Chapter 4, Bayes' theorem gives the posterior probability density function (pdf) for θ as:

$$h(\theta \mid x_1, \ldots, x_n) = \frac{f(x_1, \ldots, x_n \mid \theta)g(\theta)}{\int f(x_1, \ldots, x_n \mid \theta)g(\theta)d\theta}.$$

Here, f denotes the joint pdf of the data, g denotes the prior pdf of θ, and h denotes the posterior pdf. It is obvious, but important to note, that h depends upon the data only through the pdf f. Moreover, if we replace f by the likelihood function $L(\theta; x_1, \ldots, x_n)$, which is the same as f except for a proportionality constant, say c, as in:

$$h(\theta \mid x_1, \ldots, x_n) = \frac{cL(\theta; x_1, \ldots, X_n)g(\theta)}{\int cL(\theta; x_1, \ldots, x_n)g(\theta)d\theta},$$

the constant merely cancels out in the ratio. So the posterior pdf depends upon the data only through the likelihood function, L. Since all Bayesian inference is made from the posterior distribution, Bayesian inference depends on the data only through the likelihood function.

The same argument will apply to Bayes' theorem for discrete unknown quantities using probability mass functions (pmf).

It is clear from the above discussion that it is of fundamental importance to get the likelihood function right (the likelihood function should be a close approximation or representation of the data) if we are to depend upon Bayesian methods for decision making and statistical inference. But the likelihood is unknown at the outset of a problem involving Bayesian inference. It is only in large samples that the likelihood function can be closely approximated (see Sect. 3.6 below). In moderate or small samples there will be uncertainty about posterior inferences regarding an unknown θ attributed to both the likelihood function and the prior distribution.

3.6 DEVELOPMENT OF THE LIKELIHOOD FUNCTION USING HISTOGRAMS AND OTHER GRAPHICAL METHODS

In applications, we begin a Bayesian analysis by collecting a set of data $x_1, \ldots, x_n$. The data could of course be multidimensional, but for simplicity, let's confine our attention to the case in which the data are one-dimensional. Assume the data are observations from a continuous probability distribution. We must now begin to model the behavior of the data. It is usual to begin the modeling by studying the distribution of the data. A simple approach is to form a *histogram* of the data using almost any statistical computer program. For example, SAS, SPSS, STATA, and MINITAB can all generate histograms of the data with very little effort. The result is a graph and a printout of the empirical pdf for the data. Perhaps the histogram looks approximately unimodal suggesting an underlying normal density, or perhaps it looks one sided, like a gamma density function. In any case, the histogram could suggest a model for the data that could be used for generating the likelihood function you are going to use for Bayesian inference, such as normal, gamma, beta, and so on. If the histogram suggests a multimodal distribution, a mixture of two or more unimodal distributions might be appropriate.

Another approach would be to do *kernel density estimation*. STATA will do this quite painlessly, for example. With this approach, a smooth continuous density for the underlying data emerges. This form is very useful for applying the Bayesian paradigm swiftly, without making any further conjectures as would be required when histogram analysis is used to model the data.

It is also possible to form an empirical density from the histogram and to use this for Bayesian inference. You could retain the numerical values for the data frequencies and run the Bayesian analysis completely numerically, but this would not be the usual approach.

Sometimes, the model (data distribution) is revised in light of more careful study of the data, or because additional data are obtained so that the combined data set suggests a somewhat different model, or because of theoretical considerations. The Bayesian analysis must also then be revised.

If the data are discrete, a similar procedure can be followed. The data can be bar-graphed using one of the abovementioned computer programs, and again, an underlying discrete probability distribution should be suggested as a first measure in beginning the analysis. After further review, the assumed likelihood function may be altered to better fit the data.

If the data are multidimensional, they must be examined in each dimension to be sure that the assumed likelihood function will hold appropriately for each dimension. We assume that the data are jointly distributed so that the same family of distributions will be appropriate for all dimensions simultaneously even though there will probably be different parameterizations for each dimension. Mixed discrete/continuous data distributions (such as those encountered in some sample surveys) will present additional difficulties.

Note that for any of these methods to be meaningful, n must be reasonably large.

SUMMARY

This chapter has examined the part of Bayesian statistical science that involves the likelihood function. We saw that the likelihood function is not defined uniquely, but only up to a proportionality constant. It need not be the same as the joint pdf or pmf. We examined the likelihood principle and concluded that Bayesian statistical procedures should not violate the likelihood principle. Bayesian procedures should depend upon the data only through the likelihood function. Finally, we saw that great care should be exercised in selecting the likelihood function. Graphical methods for plotting the data in histogram form to suggest the underlying statistical models are generally available for plotting using computer programs such as SAS, SPSS, STATA, or MINITAB as long as large samples are involved.

EXERCISES

3.1* What is the likelihood principle and why is it important for Bayesian statistical science?

3.2 How would you decide whether to use a normal likelihood function or a Student t likelihood function for your data intended for a Bayesian analysis, assuming the data were likely to be roughly unimodal and symmetric? (*Hint*: see methods for comparing empirical distributions in other texts such as Rohatgi, 1976.)

3.3 Which computer program do you prefer for analysing data distributions? Why?

3.4 It is important to review your familiarity with histograms. The weights of 40 male students at the University of California, Riverside, were recorded to the nearest pound as shown below.

(a) Using a statistical computer program such as SAS, SPSS, STATA, or MINITAB, construct an appropriate histogram of the data.

(b) Is the data distribution best described approximately by a uniform distribution, a beta distribution, a gamma distribution, or a normal distribution? Why?

138	164	150	132	144	125	149	157
146	158	140	147	136	148	152	144
168	126	138	176	163	119	154	165
146	173	142	147	135	153	140	135
161	145	185	142	150	156	145	128

3.5 Give three examples of classical statistical procedures that depend upon data that are not taken in a given experiment, but might have been taken. Such procedures violate the likelihood principle. (*Hint*: often, such procedures involve averaging over all possible values of the observables.)

* Solutions for asterisked exercises may be found in Appendix 7.

3.6 Explain with three examples, the proposition: "The likelihood function need not be a probability."

3.7 Consider the average value of the function $G(\theta)$ with respect to the posterior distribution of the unobservable θ, based upon a set of continuous data: $x_1, \ldots, x_n$. The average G is given by:

$$\int G(\theta)h(\theta \mid x_1, \ldots, x_n)d\theta = \int G(\theta)\left[\frac{L(\theta; x_1, \ldots, x_n)g(\theta)}{\int L(\phi; x_1, \ldots, x_n)g(\phi)d\phi}\right]d\theta,$$

where h denotes the posterior pdf, L denotes the likelihood function, and g denotes the prior pdf. Suppose G denotes the squared difference between θ and $\hat{\theta}$, an estimate of it, based upon the data. Find the value of $\hat{\theta}$ that minimizes this $E[G(\theta)]$. (See Chapter 11 for a more general discussion of this problem.) Note that such integrals can be approximated by the Laplace approximation discussed in Section 7.3.2.

3.8* Explain why *unbiasedness* is a property that violates the spirit of the likelihood principle.

FURTHER READING

Aldrich, J. (1997). "R. A. Fisher and the Making of Maximum Likelihood 1912–1922," *Statistical Science*, **12**, 162–176.

Berger, J. O. and Wolpert, R. L. (1984). *The Likelihood Principle*, Lecture Notes Monograph Series, Vol. 6, Shanti S. Gupta (Ed.), Institute of Mathematical Statistics.

Birnbaum, A. (1962). "On the Foundations of Statistical Inference," with discussion, *J. Am. Statist. Assoc.*, **57**, 269–306.

Edwards, A. W. F. (1972). *Likelihood*, Cambridge, Cambridge University Press.

Edwards, A. W. F. (1974). "The History of Likelihood," *Int. Statist. Rev.*, **42**, 9–15.

Fisher, R. A. (1921). "On the 'Probable Error' of a Coefficient of Correlation Deduced From a Small Sample," *Metron*, **1**, 3–32.

Fisher, R. A. (1922). "On the Mathematical Foundations of Theoretical Statistics," *Philosophical Trans. Royal Soc. Ser. A*, **222**, 309–368.

Fisher, R. A. (1930). "Inverse Probability," *Proc. Camb. Phil. Soc.*, **26**, 528–535.

Ghosh, J. K. (1988). *Statistical Information and Likelihood*: A Collection of Critical Essays by Dr. D. Basu, New York, Springer-Verlag.

Reid, N. (2000). "Likelihood," *J. Am. Statist. Assoc.*, **95**(452), 1335–1340.

Rohatgi, V. K. (1976). *An Introduction to Probability Theory and Mathematical Statistics*, New York, John Wiley and Sons, Inc.

CHAPTER 4

Bayes' Theorem

4.1 INTRODUCTION

In this chapter we will present the formalism of the Bayesian approach to scientific analysis and decision making. We will introduce the notions of prior probabilities and prior probability distributions and posterior probabilities and posterior probability distributions, and we will examine cases in which the uncertain quantities might be events, models, or discrete or continuous random variables.

4.2 GENERAL FORM OF BAYES' THEOREM FOR EVENTS

Bayes' theorem is simply a statement of conditional probability. Suppose that $A_1, \ldots, A_k$ is any set of mutually exclusive and exhaustive events, and that events B and A_j are of special interest. Bayes' theorem for events provides a way to find the conditional probability of A_j given B in terms of the conditional probability of B given A_j. For this reason, Bayes' theorem is sometimes called a theorem about "inverse probability." Bayes' theorem for events is given by:

$$P\{A_j \mid B\} = \frac{P\{B \mid A_j\}P\{A_j\}}{\sum_{i=1}^{k} P\{B \mid A_i\}P\{A_i\}} \tag{4.1}$$

for $P\{B\} \neq 0$. Note that the interpretation of $P\{A_j\}$ in Equation (4.1) is personal. It is your personal *prior probability* of event A_j, in that it is your degree of belief about event A_j prior to your having any information about event B that may bear on A_j. Analogously, $P\{A_j \mid B\}$ in Equation (4.1) denotes your *posterior probability* of event A_j in that it is your degree of belief about event A_j posterior to you having the information about B. So one way to interpret the theorem is that it provides a means for updating your degree of belief about an event A_j in the light of new information B that bears on A_j. The updating that takes place is from your original degree of belief,

$P\{A_j\}$, to your updated belief, $P\{A_j \mid B\}$. The theorem might also be thought of as a rule for learning from previous experience.

4.2.1 Bayes' Theorem for Complementary Events

As a special case of Equation (4.1), let C denote any event, and $\bar{C}$ denote the complementary event to C (C and $\bar{C}$ are mutually exclusive and exhaustive). Then, for any other event B, $P\{B\} \neq 0$, Bayes' theorem for events becomes:

$$P\{C \mid B\} = \frac{P\{B \mid C\}P\{C\}}{P\{B \mid C\}P\{C\} + P\{B \mid \bar{C}\}P\{\bar{C}\}}. \tag{4.2}$$

4.2.2 Prior Probabilities

A prior probability, $P\{A_j\}$, is required in Bayes' theorem for events (Eq. (4.1)). Such prior probabilities are degrees of belief the analyst has prior to observing any data that may bear on the problem. In the event that there are no data available (such data may be too costly, too inconvenient, or too time consuming to obtain), we recommend that the prior probability be used for inference or decision making. Bayes' theorem is not used in such a case. In policy analysis, for example, perhaps most decisions are made without data; they are made merely on the basis of informed judgment. In such cases prior probabilities are all you have. This fact accentuates the importance of formulating prior probabilities very carefully. In science, business, law, engineering, and medicine, inferences and decisions about unknown quantities are generally made by learning from previous understanding of underlying theory and experience. If sample sizes of current data collected are large, the prior probabilities will usually disappear and the data will be left to "speak for themselves." However, if sample sizes are small, prior probabilities can weigh heavily in contrast to the small amount of observed data, and so could be extremely important. We will discuss the assessment of prior probabilities in Chapter 5.

4.2.3 Posterior Probabilities

Posterior probabilities are the probabilities that result from the application of Bayes' theorem. The posterior probabilities of mutually exclusive and exhaustive events must sum to one for them to be *bona fide* probabilities. If a new set of data becomes available, the posterior probabilities associated with the old data set should be used as the prior probabilities for the new data set, since those posterior probabilities represent the best information available.

4.2.4 Odds Ratios

An alternative representation of prior or posterior probabilities is in terms of odds ratios. If p denotes the probability of an event E, the odds *against* event E are given by $(1-p)/p$, or $(1-p):p$. The odds *in favor* of event E are given by $p/(1-p)$, or $p:(1-p)$.

Example 4.1 Bayes' Theorem for Events: DNA Fingerprinting

Fingerprinting based upon ridge patterns was developed by Sir Francis Galton in the 1890s. It rapidly became the standard method for distinguishing people, since fingerprints remain constant throughout life, and even identical twins have somewhat distinct fingerprints. As used, several different characteristics of two fingerprints had to be identical before a match could be declared. Accuracy and convenience became issues of importance, so new methods for distinguishing individuals were sought. A process for distinguishing people by the four blood types, A (possessing gene type A), B (possessing gene type B), AB (possessing both gene type A and gene type B), and O (possessing neither gene type A nor gene type B), was developed. Other characteristics of the blood were also used, and again, several different characteristics of two blood samples had to agree before a match would be declared. However, what if no blood samples were available? A more versatile method for distinguishing individuals was needed.

It was later noticed that people differ greatly in the DNA patterns (nucleotide sequences) found in every human cell. No two people are likely to have the same number of copies of repetitive DNA units at all of the places where the sequences occur. No longer would blood or fingerprints always be required for identification, but a sample of any cell of the individual would be sufficient in some situations.

DNA typing or fingerprinting (which does not depend upon fingerprints) began as a forensic process in which a sample of the DNA in one or more cells of a person is analyzed biologically to determine whether it matches the DNA of another sample of DNA obtained elsewhere. The procedure was developed in England, in 1984, by Alec Jeffreys, at the University of Leicester (Wambaugh, 1989). Several different characteristics of two DNA samples collected separately (using different biological *probes*) would have to agree before a match would be declared. Here is an application.

Someone has committed a crime (we will call that person "the perpetrator") and has inadvertently left some of his/her DNA at a crime scene (some blood, a strand of hair, some semen, some skin, etc.). Suppose that it is correctly assumed by the police that the person whose DNA was left at the crime scene has actually committed the crime. A person is accused of having committed the crime (we will call him/her "the suspect"). It is decided that the suspect's DNA should be compared with that of the perpetrator to see whether there is a match, which would provide evidence that the suspect is really the person who committed the crime. Bayes' theorem is applied. Let C denote the event that the suspect committed the crime. Suppose it turns out that a DNA match is declared. Does this match imply that the suspect really committed the crime? From Bayes' theorem in Equation (4.2), the *posterior probability* that the suspect committed the crime, given a DNA match, is

$$P\{C \mid \text{DNA match}\} = \frac{P\{\text{DNA match} \mid C\}P\{C\}}{P\{\text{DNA match} \mid C\}P\{C\} + P\{\text{DNA match} \mid \text{not } C\}P\{\text{not } C\}}.$$

Let M denote the event of a match between the perpetrator's DNA (left at the crime scene) and the suspect's DNA. Let p denote the prior probability, $P\{C\}$, that the suspect committed the crime. Then, if $\bar{C}$ denotes the complementary event "not C," the equation becomes

$$P\{C \mid M\} = \frac{P\{M \mid C\}p}{P\{M\}} = \frac{P\{M \mid C\}p}{P\{M \mid C\}p + P\{M \mid \bar{C}\}(1-p)}.$$

Since

$$P\{\bar{C} \mid M\} = \frac{P\{M \mid \bar{C}\}(1-p)}{P\{M \mid \bar{C}\}(1-p) + P\{M \mid C\}p},$$

the *posterior odds against* the suspect having committed the crime are

$$\frac{P\{\bar{C} \mid M\}}{P\{C \mid M\}} = \frac{P\{M \mid \bar{C}\}}{P\{M \mid C\}} \cdot \left[\frac{(1-p)}{p}\right].$$

where $[(1-p)/p]$ is called the *prior odds* against the suspect having committed the crime. We can see that an advantage of working with the odds ratio is that it is not necessary to compute the unconditional probability $P\{M\}$; that is, it is then unnecessary to compute the proportionality constant in Bayes' theorem.

Note that $P\{M \mid C\} = 1 - P\{\bar{M} \mid C\}$, and define

$$r = \frac{P\{M \mid C\}}{P\{M \mid \bar{C}\}} = \frac{P\{M \mid C\}}{1 - P\{\bar{M} \mid C\}}.$$

Then,

$$\frac{P\{C \mid M\}}{P\{\bar{C} \mid M\}} = r\left[\frac{P}{1-p}\right].$$

The ratio "r" is edited the *Bayes Factor.* The Bayes factor is the ratio of the posterior odds (in favor of the suspect having committed the crime) to the prior odds favoring the suspect. Thus, regardless of the value of p, the larger the Bayes factor, the greater are the odds favoring C. Note that r depends only upon the data. The Bayes factor will also be used in hypothesis testing problems (Chapter 8).

Suppose that the suspect was accused because he/she was seen near the scene of the crime at the approximate time the crime was committed. To keep the example simple, take the prior probability $p = 0.5$, so that prior odds ratio $= 1$. Then,

$$P\{C \mid M\} = \frac{P\{M \mid C\}}{P\{M \mid C\} + P\{M \mid \bar{C}\}} = \frac{1}{1 + 1/r}.$$

The two error probabilities, $P\{\bar{M} \mid C\}$, the probability of concluding that there is no match when there should be, and $P\{M \mid \bar{C}\}$, the probability of concluding that there is a match when there should not be (supposing the suspect to be a randomly selected person different from the perpetrator), are determined from the current state of the DNA technology. For simplicity, in this example, suppose that $P\{M \mid \bar{C}\} = 0.02$, and $P\{\bar{M} \mid C\} = 0.001$. The calculation of these probabilities is complicated and depends upon comparison of the DNA of the suspect with that found in a database of the DNA of people of similar ethnic background to that of the suspect. It also depends upon comparing several distinct characteristics of the two samples of DNA (evaluated by using different probes). Using these figures for this example, however, we find that the chance that the suspect committed the crime is approximately

$$P\{C \mid M = \text{match}\} = 0.9804, \qquad \text{or about 98 percent.}$$

The Bayes factor is $r = \dfrac{0.9804}{1 - 0.9804} = 50.02$. The posterior odds ratio against the suspect having committed the crime is $r(1-p)/p = 0.02002$, or the posterior odds favoring the suspect as the criminal are about 50 to 1. The DNA match in this case therefore provides very convincing evidence that the suspect committed the crime. If someone's prior odds ratio of the suspect having committed the crime is different from $|1|$ he merely needs to multiply it by r to get the new Bayes factor.

4.3 BAYES' THEOREM FOR DISCRETE DATA AND DISCRETE PARAMETER

Let $X_1, \ldots, X_n$ denote mutually independent and identically distributed observable random variables, each with probability mass function $f(x \mid \Theta = \theta)$, for Θ, an unknown and unobservable discrete random variable whose possible values $\theta_1, \ldots, \theta_p$ index the distribution of $(X \mid \Theta = \theta)$. You are interested in the chance that $\Theta = \theta$. Bayes' theorem in this situation asserts that

$$P\{\Theta = \theta \mid x_1, \ldots, x_n\} \equiv h(\theta \mid \text{data}) = \frac{f(x_1 \mid \theta) \cdots f(x_n \mid \theta) g(\theta)}{\sum_{i=1}^{p} f(x_1 \mid \theta_i) \cdots f(x_n \mid \theta_i) g(\theta_i)}, \tag{4.3}$$

where h denotes the posterior pmf of Θ, and g denotes the prior pmf of Θ. We assume, of course, that the denominator in Equation (4.3) is not zero.

4.3.1 Interpretation of Bayes' Theorem for Discrete Data and Discrete Parameter

We note first that both X and Θ have been assumed to have discrete probability distributions. The X_is are viewed as the data outcomes in a current experiment being performed to generate information that might bear on your beliefs about Θ. The denominator in Equation (4.3) is merely a constant that depends only upon the data, X, and not upon θ. So with respect to Θ, the denominator behaves as if it were a constant. For this reason, Equation (4.3) could be written equivalently as

$$h(\theta \mid x_1, \ldots, x_n) = cf(x_1 \mid \theta) \cdots f(x_n \mid \theta)g(\theta) \propto f(x_1 \mid \theta) \cdots f(x_n \mid \theta)g(\theta) \tag{4.4}$$

where c denotes a proportionality constant, and $\propto$ denotes proportionality. The constant c can always be reconstructed because it must be defined so that

$$\sum_{i=1}^{p} P\{\Theta = \theta_i \mid x_1, \ldots, x_n\} = \sum_{i=1}^{n} h(\theta_i \mid x_1, \ldots, x_n) = 1. \tag{4.5}$$

In Equation (4.4) the term $f(x_1 \mid \theta) \cdots f(x_n \mid \theta)$ is what is normally thought of as the likelihood function for the data (see Chapter 3). In this form, Bayes' theorem for discrete data and discrete parameters, as given in Equation (4.4), asserts that the posterior conditional pmf for Θ is proportional to the product of the likelihood function for the data and the prior pmf. Moreover, the data could be one dimensional or multidimensional and the analogous Bayesian result applies. Further, the data elements, X_i, need not be mutually independent; they could be correlated, as they would typically be in a stochastic process. However, Equation (4.4) would then be modified by changing the likelihood to the form $f(x_1, \ldots, x_n \mid \theta)$.

Example 4.2 Quality Control in Manufacturing: Discrete Data and Discrete Parameter (Inference About a Proportion)

The RAMM Company produces RAM (random access memory) printed circuits boards for personal computers. These modules are produced by the hundreds of thousands every month. But RAMM is deeply concerned about the quality of their product, as indeed they must be if they are to remain profitable.

Let θ denote the probability of a defective unit in a batch of n units. Suppose, for simplicity of illustration, that θ can assume just three possible values: 0.25 (good), 0.50 (fair), and 0.75 (poor), respectively.

The RAMM Company has kept excellent records and has found over time that it has produced good units, with $\theta = 0.25$, 60 percent of the time over the last five years, fair units, with $\theta = 0.5$, 30 percent of the time over the last five years, and poor units, with $\theta = 0.75$, 10 percent of the time. RAMM has decided to use the records as prior probabilities to update future questions about its degree of quality control. These prior probabilities are summarized in Table 4.1.

A new batch of 10,000 RAM memory circuit boards is produced and the company decides to check this batch to see whether its production process is under control. That is, does the current rate of defectives being produced in the manufacturing process correspond to its historical rate, or is the process now generating units at an unacceptably high rate of defectives?

Table 4.1 Prior Probabilities

	Quality of Product		
	Good	Fair	Poor
Probability of a defective (θ)	0.25	0.50	0.75
Probability mass function for θ ($P\{\theta\}$)	0.60	0.30	0.10

The company randomly samples a collection of three units and two defectives are found. Now what should RAMM believe about its current rate of defectives (equivalently, what is the posterior distribution for θ)?

In this problem the data are discrete and are assumed to follow a binomial distribution with parameter θ, with the number of successes $r = 2$ (a defective is defined as a success) out of $n = 3$ trials. So the likelihood is given by:

$$L(\theta) = \binom{3}{2}\theta^2(1 - \theta),$$

for $\theta = 0.25,\ 0.5,\ 0.75$. Equation (4.4) becomes in this case,

$$h(\theta \mid r, n) \propto L(r \mid \theta, n)g(\theta).$$

The posterior pmf h is given constructively in Table 4.2.

Note that the elements of the h column are found by dividing each element of the (prior) × (likelihood) column by the total of the elements in that column, in order to normalize to one. From Table 4.2 it is clear that the most likely value for θ in the recent batch of RAM memory circuit boards is $\theta = 0.5$, only a fair rate of defectives.

Before concluding that they should modify their manufacturing process, which would be very costly, the RAMM Company should probably take a much larger sample than $n = 3$, say, $n = 100$, 500, or 1000.

Table 4.2 Posterior Probabilities

θ	Prior for θ	Likelihood $L(\theta)$	(Prior) × (Likelihood)	h = Posterior pmf for θ
0.25	0.60	0.140625	0.0843750	0.35294
0.50	0.30	0.375000	0.1125000	0.47059
0.75	0.10	0.421875	0.0421875	0.17647
Totals			0.2390625	0.99999

4.3.2 Bayes' Theorem for Discrete Data and Discrete Models

Suppose you wish to compare several models, $M_1, M_2, \ldots, M_k$ to see which one you believe in most. The situation is analogous to the one in Example 4.2 where there were three possible values for the rate of defectives, θ. These possible θs might now be thought of as possible models.

The situation could be one in which there are several alternative models under consideration that might explain an observed phenomenon. You have some beliefs about each possible model as to whether it is the most appropriate model for the phenomenon, but you need to be more certain than you currently are. You take some data that bear on the phenomenon (carry out an experiment) and compare your posterior beliefs about each model when the different models are used to evaluate the data. Bayes' theorem for a discrete number of models is given by

$$P\{M_i \mid X = x\} = \frac{f(x \mid M_i)P\{M_i\}}{\sum_{j=1}^{k} f(x \mid M_j)P\{M_j\}}, \tag{4.6}$$

for $i = 1, \ldots, k$. Here, $f(x \mid M_i)$ denotes the likelihood of the discrete data under Model M_i, and the associated likelihood would be a pmf.

4.4 BAYES' THEOREM FOR CONTINUOUS DATA AND DISCRETE PARAMETER

Now suppose that X is an observable, absolutely continuous random variable with pdf $f(x \mid \theta)$, Θ denotes an unknown, unobservable, discrete parameter that indexes the distribution of X, and Θ has prior pmf $g(\theta)$. In this context, Bayes' theorem asserts that the posterior pmf for θ is given by

$$h(\theta \mid x_1, \ldots, x_n) = \frac{L(x_1, \ldots, x_n \mid \theta)g(\theta)}{\sum_{i=1}^{k} L(x_1, \ldots, x_n \mid \theta_i)g(\theta_i)}, \tag{4.7}$$

where $L(x_1, \ldots, x_n \mid \theta)$ denotes the likelihood function for θ. Equivalently,

$$h(\theta \mid x) \propto L(x_1, \ldots, x_n \mid \theta)g(\theta).$$

4.4.1 Interpretation of Bayes' Theorem for Continuous Data and Discrete Parameter

The data might actually be multidimensional, they might be mutually dependent, θ might be multidimensional, but the data are continuous, and the denominator in Equation (4.7) should be different from zero.

Example 4.3 Inferring the Section of a Class from which a Student was Selected: Continuous Data and Discrete Parameter (Choosing from a Discrete Set of Models)

There are three sections of a beginning language class. One student is selected at random from a list of names of all students in the class, including all three sections. The selected student is to be given an examination. The question will then be, "From which section was the student selected?"

To answer this question it is useful to bring prior information to bear. Assume that last year the instructors for the three sections were the same; the examination was the same; and the examination results of the students at the same point in time were as shown in Table 4.3. It may be noted from this table that student scores followed normal distributions with means m_i, and common variances $= 225$, $i = 1, 2, 3$, depending upon the section from which they came.

To help evaluate scores to be achieved this year, you decide to assign prior probabilities for this year's scores based upon last years results and upon your prior beliefs about student's abilities. The randomly selected student is named Maria. In fact, you know that Maria's grades have been pretty good so far, and that the students corresponding to the class whose score distribution follows model M_2 are better on average than those in the other two sections (they have been assigned to sections on that basis). So you decide to assign the prior probabilities to the three models as shown in the last column of Table 4.3.

The likelihood (probability density) functions are given by:

$$L(x \mid \theta_i) = \frac{1}{15\sqrt{2\pi}} \exp\left\{(-0.5)\left(\frac{x - \theta_i}{15}\right)^2\right\},$$

for $i = 1, 2, 3$. To evaluate the likelihood functions numerically we take $x = 0.76$ because Maria happens to have scored 76 percent on the exam this year. Standardize x by transforming to $y_i = (x - \theta_i)/15$. Now $y_i \sim N(0, 1)$, and if we define the standard normal density $\phi(y_i) \equiv (2\pi)^{-0.5} \exp\{(-0.5)y_i^2\}$, we can readily use the table in the complement to Chapter 4 to evaluate the likelihoods. Results are given in Table 4.4.

Examination of Table 4.4 shows that model M_2 has the highest posterior probability of the three models. We therefore conclude that the selected student came from Section 2.

Table 4.3 Models for Student Scores Last Year

Model Name = M	Mean = θ	Model	Prior Probability $P\{M\}$
M_1	$\theta_1 = 74$	N (74, 225)	0.25
M_2	$\theta_2 = 81$	N (81, 225)	0.50
M_3	$\theta_3 = 68$	N (68, 225)	0.25

Table 4.4 Posterior Probabilities for Student Score Models

Model	y_i	$\phi(y_i)$	Likelihood = $\phi(y_i)/15$	Prior	(Likelihood) × (Prior)	Posterior
M_1	0.133	0.395	0.02633	0.25	0.0065825	0.264
M_2	−0.333	0.377	0.02513	0.50	0.0125650	0.504
M_3	0.533	0.346	0.02307	0.25	0.0057675	0.231
Totals					0.0249150	0.999

Note that there is just one data point in this example. We might have asked about the section origins of several students who were taken from the same section, which would have increased the size of the sample data. Moreover, because the sample size was just one, the data (Maria's score) has little effect on the posterior relative to the prior, so the priors and posteriors are about the same. Had the data size been much larger, the prior beliefs would have played a much smaller role relative to the data.

4.5 BAYES' THEOREM FOR DISCRETE DATA AND CONTINUOUS PARAMETER

Bayes' theorem for discrete data and continuous parameter θ with pdf $g(\theta)$ takes the form:

$$h(\theta \mid x_1, \ldots, x_n) = \frac{L(x_1, \ldots, x_n \mid \theta)g(\theta)}{\int L(x_1, \ldots, x_n \mid \theta)g(\theta)d\theta}, \tag{4.8}$$

and the integral is taken over all values of θ. In this case, $L(x_1, \ldots, x_n \mid \theta)$, the likelihood function for the data is a pmf while the prior, $g(\theta)$, is a pdf. The posterior distribution is of course continuous so that h denotes a pdf.

Example 4.4 Quality Control in Manufacturing: Discrete Data and Continuous Parameter

We return to the problem of quality control in manufacturing that we examined in Example 4.2. Now, however, we permit θ, the probability of a defective RAM unit, to be a continuous parameter, with the clear constraint that $0 \leq \theta \leq 1$. In that example, three units were selected for testing and two defectives were found. The likelihood is still binomial, but now θ is no longer restricted to just the three values available in that example. We must now choose a prior distribution for θ.

Case 1: Uniform Prior. Suppose no records have been kept about earlier rates of defective units so that we have no idea whether our batch of units are all perfect, all defective, or somewhere in between. In such a case, we should consider taking the prior distribution of θ to be uniform over the unit interval. That is,

$$g(\theta) = \begin{cases} 1, & 0 < \theta < 1 \\ 0, & \text{otherwise.} \end{cases}$$

The posterior pdf for θ (with respect to Lebesgue measure) is given by:

$$h(\theta \mid r, n) = \frac{L(r \mid \theta, n)g(\theta)}{\int_0^1 L(r \mid \theta, n)g(\theta)d\theta}.$$

$$h(\theta \mid r = 2, n = 3) = \frac{\binom{3}{2}\theta^2(1-\theta)}{\int_0^1 \binom{3}{2}\theta^2(1-\theta)d\theta} = \frac{\theta^2(1-\theta)}{\int_0^1 \theta^2(1-\theta)d\theta}$$

$$= \frac{\theta^2(1-\theta)}{B(3,2)\int_0^1 \frac{1}{B(3,2)}\theta^{3-1}(1-\theta)^{2-1}d\theta} = \frac{1}{B(3,2)}\theta^{3-1}(1-\theta)^{2-1},$$

$$B(a, b) \equiv \int_0^1 y^{a-1}(1-y)^{b-1}dy, \tag{4.9}$$

and

$$B(a, b) = \frac{\Gamma(a)\Gamma(b)}{\Gamma(a+b)}, \tag{4.10}$$

and $\Gamma(a)$ denotes the gamma function of argument a. Moreover, a beta distribution with parameters a and b ($a > 0, b > 0$), for a random variable X, is defined to have density

$$f(x \mid a, b) = \begin{cases} \frac{1}{B(a,b)}x^{a-1}(1-x)^{b-1}, & 0 < x < 1 \\ 0, & \text{otherwise} \end{cases} \tag{4.11}$$

Thus, we see that the posterior density of θ is beta with parameters 3 and 2. The uniform prior density of θ and the beta posterior density of θ are both depicted in Figure 4.1.

Case 2: Beta Prior. Next consider the same problem examined in Case 1, above, except that now assume that you are more knowledgeable about θ than you were before; so now you can adopt a prior distribution for θ that reflects, and takes advantage of, that increased knowledge. The posterior density for θ is given in

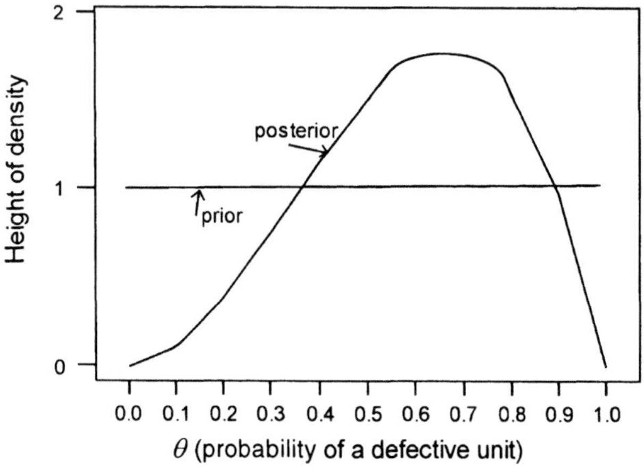

Figure 4.1 From uniform prior density to beta (3, 2) posterior density of θ

the same way as in Case 1, but now $g(\theta)$ is no longer a uniform density, but is a beta density.

Accordingly, adopt the beta prior density family:

$$g(\theta) = \begin{cases} \dfrac{1}{B(a, b)}\theta^{a-1}(1-\theta)^{b-1}, & 0 < \theta < 1 \\ 0, & \text{otherwise} \end{cases}$$

for appropriate a and b, to be selected within the beta family. Substituting gives the posterior density for θ:

$$h(\theta \mid r = 2, n = 3) = \frac{\left[\binom{3}{2}\theta^2(1-\theta)\right]\left[\dfrac{1}{B(a, b)}\theta^{a-1}(1-\theta)^{b-1}\right]}{\displaystyle\int_0^1 \left[\binom{3}{2}\theta^2(1-\theta)\right]\left[\dfrac{1}{B(a, b)}\theta^{a-1}(1-\theta)^{b-1}\right]d\theta}.$$

Simplifying,

$$\begin{aligned} h(\theta \mid r = 2, n = 3) &= \frac{\theta^{a+1}(1-\theta)^b}{\displaystyle\int_0^1 \theta^{a+1}(1-\theta)^b d\theta} \\ &= \frac{\theta^{(a+2)-1}(1-\theta)^{(b+1)-1}}{B(a+2, b+1)\displaystyle\int_0^1 \frac{\theta^{(a+2)-1}(1-\theta)^{(b+1)-1}}{B(a+2, b+1)}d\theta}, \end{aligned}$$

and since the second integral must be one, the posterior density for θ becomes

$$h(\theta \mid r = 2, n = 3) = \frac{1}{B(a+2, b+1)} \theta^{(a+2)-1}(1-\theta)^{(b+1)-1},$$

for $0 < \theta < 1$. Note that as in Case 1, the posterior density is again a member of the beta family, but the parameters of the posterior density are now different from what they were in Case 1. We must still assess a and b from prior knowledge.

It is well known that for the beta density in Equation (4.11),

$$E(\theta) = \frac{a}{a+b} \quad \text{and} \quad \text{var}(\theta) = \frac{ab}{(a+b)^2(a+b+1)}.$$

Suppose from past records of the RAMM Company we now find that $E(\theta) = 0.2$, and $\text{var}(\theta) = 0.01$. Substituting these values for the mean and variance and solving backwards for a and b gives $a = 3$, and $b = 12$. The prior density becomes

$$g(\theta) = \frac{1}{B(3, 12)} \theta^{3-1}(1-\theta)^{12-1}.$$

Substituting gives the final posterior density for θ:

$$h(\theta \mid r = 2, n = 3) = \frac{1}{B(5, 13)} \theta^{5-1}(1-\theta)^{13-1},$$

for $0 < \theta < 1$. Note that, not surprisingly, the prior and posterior densities, respectively, are quite similar. This is because the data added very little to your prior knowledge since the sample size was only $n = 3$; that is, only three units were sampled. The prior and posterior pdfs are depicted in Figure 4.2. It may be seen in that graph that the posterior mode is slightly to the right of the prior mode, which reflects the effect of the data that show that there were two defectives in three units sampled, increasing the probability of a higher defective rate than records have shown. The heights of the densities are about the same.

4.6 BAYES' THEOREM FOR CONTINUOUS DATA AND CONTINUOUS PARAMETER

Bayes' theorem for continuous data and continuous parameter is given by

$$h(\theta \mid x_1, \ldots, x_n) = \frac{L(x_1, \ldots, x_n \mid \theta)g(\theta)}{\int L(x_1, \ldots, x_n \mid \theta)g(\theta)d\theta}, \tag{4.12}$$

where $h(\theta \mid x_1, \ldots, x_n)$ denotes the posterior probability density function for θ, $L(x_1, \ldots, x_n \mid \theta)$ denotes the likelihood function for the continuous data (a

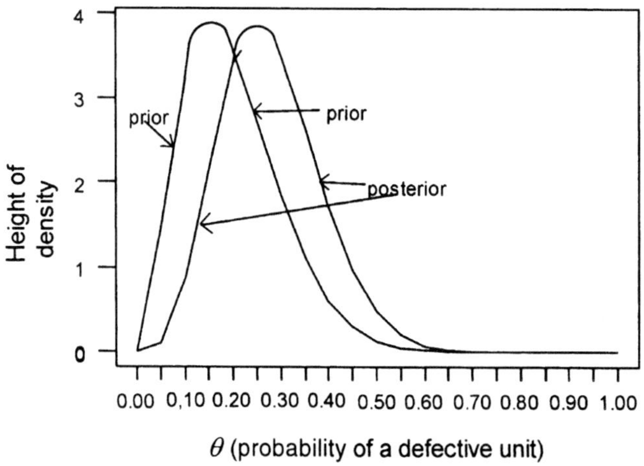

Figure 4.2 From beta (3,12) prior to beta (5,13) posterior pdf for θ

multiple of the joint pdf for the data), and $g(\theta)$ denotes the prior pdf for θ. The denominator integral is taken over all θ, and assumed to be nonzero. Again, the data may be multidimensional and correlated, and θ may be multidimensional.

Example 4.5 Normal Data: Unknown Mean, Known Variance

The Mutual of Phoenix insurance company requires a large sales force that is knowledgeable about the ins and outs of insurance. It decides to evaluate the efficacy of a salesmanship course it has been asked to consider. It chooses n of its salespeople to be given a test of their knowledge about insurance. Then they are given a three-week course about the subject, followed by another test of their knowledge about insurance. Let x_i denote the change in test score of salesperson i, $i = 1, \ldots, n$. We assume the x_is independently follow the distribution $N(\theta, 25)$. Note that the variance is assumed to be known. The company is of course interested in whether θ is positive, and if so, by how much? That is, how much improvement in knowledge can be expected of its sales force from this course? You decide to apply Bayes' theorem.

Case 1: Very Little Prior Information, Flat Prior. Suppose that you have very little information about the efficacy of this course, so that for you, over a rather large interval, all values of θ are equally likely. Accordingly, you decide to adopt a uniform prior density for θ over a large interval. In fact, since you really do not know how to define this large interval, as an approximation you decide to replace this proper uniform prior by an improper prior, one that is constant over the entire real line. Such a prior density cannot integrate to one, so it is called improper. But it is generally a good approximation to the uniform prior density over a large range, and it simplifies the mathematics of the computation. Such a prior density that is constant over the entire real line is often called a *flat, vague, noninformative, diffuse* or *default* prior density (see Chapter 5 for additional discussion of these priors). We express the flat prior density in the form

$$g(\theta) \propto \text{constant},$$

for all θ, $-\infty < \theta < +\infty$. Of course it can be thought of approximately as

$$g(\theta) = \lim_{a \to \infty} \left(\frac{1}{2a} \right), \qquad -a < \theta < a.$$

Now apply Bayes' theorem in the convenient form

$$h(\theta \mid x_1, \ldots, x_n) \propto L(x_1, \ldots, x_n \mid \theta) g(\theta),$$

and incorporating our flat prior gives:

$$h(\theta \mid x_1, \ldots, x_n) \propto L(x_1, \ldots, x_n \mid \theta).$$

Now write the likelihood function in terms of the sample mean, which is sufficient here. Take $n = 50$. Since $(\bar{x} \mid \theta) \sim N(\theta, 25/n) = N(\theta, 0.5)$, the posterior density becomes:

$$h(\theta \mid \bar{x}) \propto \exp\left\{ (-0.5) \left(\frac{\bar{x} - \theta}{0.5} \right)^2 \right\}.$$

It is clear that the posterior density of θ follows the distribution

$$(\theta \mid \bar{x}) \sim N(\bar{x}, 0.5).$$

Note that although the prior distribution was improper, the posterior distribution is proper, and is normal. Moreover, since you brought very little prior information to bear on this problem, not surprisingly, the posterior distribution is centered at the sample mean, which is also the maximum likelihood estimator, and it is the best information that you have. The variance of the posterior distribution has been reduced to 0.5 because of the large sample size taken for study. In this example we see the advantage to be gained by adopting the proportional form of Bayes' theorem; it was never necessary to carry out a formal integration since we could recognize the form of the posterior distribution family from the *kernel* of the posterior distribution (the form of the distribution without its normalizing constant that makes it integrate to one). Suppose $\bar{x} = 3$. The posterior becomes

$$(\theta \mid \bar{x}) \sim N(3, 0.5).$$

The posterior density is shown in Figure 4.3.

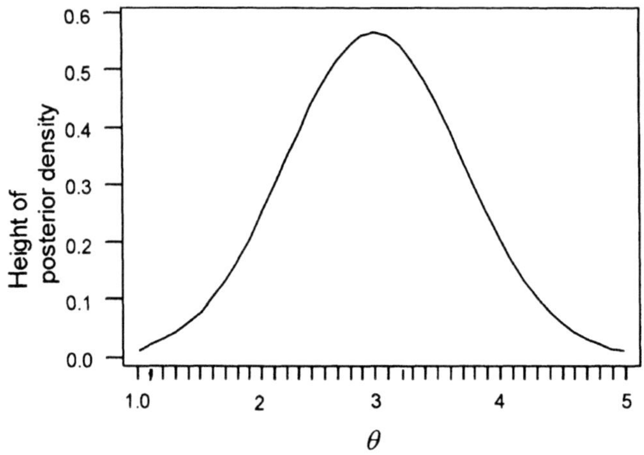

Figure 4.3 Posterior pdf N (3, 0.5) for change in test scores from salesmanship course, flat prior

It would seem from this result that the salesmanship course has made some of a difference in mean test scores with 99.7 percent = probability that the mean change in test scores is found in the interval (0.9, 5.1).

Case 2: Normal Distribution Prior. Next suppose that when Mutual of Phoenix was told about the salesmanship course it was advised that when the course was tried at a competitor company, changes in test scores θ followed the distribution $N(5, 2)$. To find your posterior distribution in this case, you decide to adopt this information as your prior distribution. More generally, if your prior distribution for θ is $N(m, \sigma^2)$, its density is given by

$$g(\theta) \propto \exp\left\{-0.5\left(\frac{\theta - m}{\sigma}\right)^2\right\}.$$

Substituting gives

$$h(\theta \mid \bar{x}) \propto \exp\left\{-0.5\left[\left(\frac{\theta - m}{\sigma}\right)^2 + \left(\frac{\bar{x} - \theta}{0.5}\right)^2\right]\right\}.$$

Completing the square in θ in the exponent gives:

$$h(\theta \mid \bar{x}) \propto \exp\left\{-0.5\left(\frac{\theta - \tilde{\theta}}{\tau}\right)^2\right\}$$

where

$$\frac{1}{\tau^2} = \frac{1}{\sigma^2} + \frac{1}{(0.5)^2},$$

and

$$\tilde{\theta} = \left(\frac{\sigma^{-2}}{\sigma^{-2} + (0.5)^{-2}}\right)m + \left(\frac{(0.5)^{-2}}{\sigma^{-2} + (0.5)^{-2}}\right)\bar{x};$$

that is, $(\theta \mid \bar{x}) \sim N(\tilde{\theta}, \tau^2)$. We may now see that when the sample mean, $\bar{x}$, follows a normal distribution with mean θ and known variance, and when θ also follows a normal distribution, the posterior distribution of θ is also normal. If we define *precision* as the reciprocal of variance, the posterior precision, τ^{-2}, is equal to the sum of the precision of the prior distribution, σ^{-2}, and the precision of the sampling distribution $(0.5)^{-2}$.

The posterior mean, $\tilde{\theta}$, is a convex combination or weighted average of the prior mean, m, and the sample mean, $\bar{x}$, and the weights are proportions of precision corresponding to the prior distribution and the sampling distribution, respectively. In this example, we therefore find that the posterior distribution is $N(3.22,\ 0.22)$. The prior and posterior distributions are depicted in Figure 4.4.

It may be seen from Figure 4.4 that as you modified your earlier prior beliefs about θ to your posterior beliefs, the most likely value of θ shifted from a value of 5.0 to a value of 3.22. Moreover, your posterior distribution tightened about its central value from a variance of 2.0 to a variance of about 0.2, showing that your belief about the mean change in test scores suggested by the purveyor of the test was

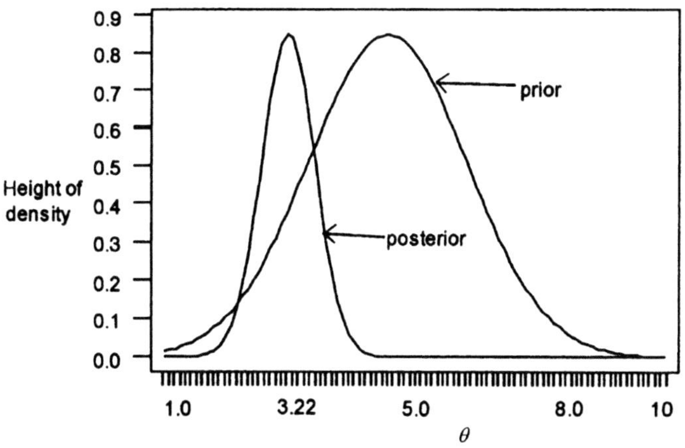

Figure 4.4 From prior density N (5.2) to posterior density N (3.22, 0.22) for change in test scores from salesmanship course, normal prior

reduced by your data; as you gained more knowledge you became more certain about your beliefs.

Example 4.6 Normal Data: Unknown Mean, Unknown Variance

Suppose the data in an experiment follow a normal distribution with general unknown mean and unknown variance. The special case in which the variance of the data is known was treated in Example 4.5 where we considered Case 1 in which the prior for the unknown mean θ was vague or flat, and Case 2 in which the prior for the unknown mean was taken to be another normal distribution. Now we address the more interesting and more realistic problem in which the data are normally distributed but not only is the mean unknown, the variance of the data is also unknown. It will be convenient to parameterize the data distribution in terms of its *precision*, that is, the reciprocal of its variance.

Likelihood Function. Suppose you have independent observations $X_1, \ldots, X_n$ that are identically distributed as $N(\theta, q^{-1})$, where q denotes the precision. The likelihood function for the data is given by

$$f(x_1, \ldots, x_n \mid \theta, q) \propto q^{0.5n} \exp\left\{\left(-\frac{q}{2}\right) \sum_{i=1}^{n} (x_i - \theta)^2\right\}.$$

Note that we can write:

$$\begin{aligned}\sum_{i=1}^{n} (x_i - \theta)^2 &= \sum_{i=1}^{n} \left[(x_i - \bar{x}) + (\bar{x} - \theta)\right]^2 \\ &= \sum_{i=1}^{n} (x_i - \bar{x})^2 + n(\theta - \bar{x})^2.\end{aligned}$$

Then, setting $(n-1)s^2 = \sum_{i=1}^{n} (x_i - \bar{x})^2$, the likelihood function may be written

$$f(x_1, \ldots, x_n \mid \theta, q) \equiv L(\bar{x}, s^2 \mid \theta, q) = q^{0.5n} \exp\left\{\left(-\frac{q}{2}\right)[(n-1)s^2 + n(\theta - \bar{x})^2]\right\}.$$

Bayes' theorem gives for the posterior density

$$h(\theta, q \mid \text{data}) \propto L(\bar{x}, s^2 \mid \theta, q) g(\theta, q),$$

where the prior density is denoted by $g(\theta, q)$.

Prior Density. We treat two very special types of prior distributions for (θ, q): first, the case in which very little prior information is available, where we use what is called the *vague* or *default prior*, and then, when there is substantial prior information available, we adopt what is called the *natural conjugate prior*, which is a mathematically convenient prior distribution. These prior distributions are discussed further in Chapter 5.

Case 1: Vague Prior Distribution. Adopt the vague prior densities for (θ, q):

$$g(\theta, q) = g_1(\theta)g_2(q), \tag{4.13}$$

where $g_1(\theta) \propto$ constant, and $g_2(q) \propto 1/q$. That is, θ and q are assumed to be *a priori* independent, we adopt a flat prior for θ, and we adopt a flat prior for $[\ln q]$, which implies that $g_2(q) \propto 1/q$ (see Section 5.3 for discussion of this prior). Therefore, a vague prior for (θ, q) corresponds to the (improper) density

$$g(\theta, q) \propto \frac{1}{q}. \tag{4.14}$$

Substituting gives the joint posterior density

$$h(\theta, q \mid \text{data}) \propto \left[\frac{1}{q}\right]\left[q^{0.5n} \exp\left\{\left(-\frac{q}{2}\right)\sum_{i=1}^{n}(x_i - \theta)^2\right\}\right].$$

Noting that $\sum_{i-1}^{n}(x_i - \theta)^2 = [(n-1)s^2 + n(\theta - \bar{x})^2]$, the joint posterior density may also be written

$$h(\theta, q \mid \text{data}) \propto \left[\frac{1}{q}\right] q^{0.5n} \exp\left\{\left(-\frac{q}{2}\right)[(n-1)s^2 + n(\theta - \bar{x})^2]\right\}.$$

Rewriting this equation in a more recognizable form gives:

$$\begin{aligned} h(\theta, q \mid \text{data}) &= h_1(\theta \mid q, \text{data})h_2(q \mid \text{data}) \\ &\propto \left[q^{0.5} \exp\left\{\left(-\frac{nq}{2}\right)(\theta - \bar{x})^2\right\}\right]\left[q^{0.5(n-3)} \exp\left\{\left(-\frac{q}{2}\right)(n-1)s^2\right\}\right]. \end{aligned} \tag{4.15}$$

Thus, by inspection, the conditional posterior distribution of $(\theta \mid q, \bar{x}, s^2) \sim N\left(\bar{x}, \frac{1}{qn}\right)$; and if $\sigma^2 \equiv 1/q$,

$$(\theta \mid \sigma^2, \bar{x}, s^2) \sim N\left(\bar{x}, \frac{\sigma^2}{n}\right);$$

and the posterior distribution of q is gamma $[(n-1)/2,\ 2/\{(n-1)s^2\}]$. That is,

$$r(q \mid \bar{x}, s^2) \propto q^{0.5(n-1)-1} \exp\left\{\left(-\frac{q}{2}\right)(n-1)s^2\right\}.$$

Recall that if Z follows a gamma (a, b) distribution, its density is given by

$$r(z) = \frac{1}{b^a\Gamma(a)} z^{a-1} \exp\{-z/b\}, \qquad 0 < z < \infty. \tag{4.16}$$

Moreover, $E(Z) = ab$, and $\text{var}(Z) = ab^2$. Next, find the marginal posterior distribution of θ by integrating out q. Thus,

$$p(\theta \mid \text{data}) = \int_0^\infty h(\theta, q)dq \propto \int_0^\infty q^{0.5(n-2)} \exp\{-qA\}dq,$$

where $A = \dfrac{[(n-1)s^2 + n(\theta - \bar{x})^2)]}{0.5}$ and A does not depend on q. Changing variable by letting $t = qA$ gives

$$p(\theta \mid \text{data}) \propto \frac{1}{A^{0.5n}} \int_0^\infty t^{0.5(n-2)} \exp(-t)dt.$$

Since the remaining integral is just a constant (it is a gamma function), we can incorporate it into the proportionality constant to get:

$$p(\theta \mid \text{data}) \propto \frac{1}{\{(n-1)s^2 + n(\theta - \bar{x})^2\}^{0.5n}}.$$

We recognize this kernel as that of a Student t-distribution. That is, marginally, the posterior distribution of θ is Student t, centered at the sample mean, with $(n - 1)$ degrees of freedom.

REMARK: The traditional Student t-distribution has density

$$f_0(t) = \frac{c}{(n + t^2)^{0.5(n+1)}}, \qquad -\infty < t < \infty,$$

where

$$c = \frac{n^{0.5n}}{B(0.5, 0.5n)},$$

and B denotes the beta function. A generalization of this distribution to one with arbitrary location and scale is the one with density

$$f(t) = \frac{c\sigma^{-1}}{\left[n + \{(t-\theta)/\sigma\}^2\right]^{0.5(n+1)}}.$$

Its first two moments are: $E(t) = \theta$, and $\text{var}(t) = \left\{\dfrac{n}{(n-2)}\right\}\sigma^2$.

Case 2: Natural Conjugate Prior Distribution. We adopt the natural conjugate prior distribution (see Section 5.3) for (θ, q):$(\theta \mid q) \sim N(\theta_0, Kq^{-1})$, and $q \sim \text{gamma}(a, b)$, where (θ_0, K, a, b) are hyperparameters (parameters of the prior distribution that must be assessed from prior information). In the gamma distribution, as, $a \to 0$ and $b \to \infty$, the density approaches the vague prior density adopted in Case 1. In terms of densities, the joint natural conjugate prior density is now given by:

$$\begin{aligned} g(\theta, q) &= g_1(\theta \mid \theta_0, K, q) g_2(q \mid a, b) \\ &\propto \left[q^{0.5} \exp\left\{\left(-\frac{q}{2K}\right)(\theta - \theta_0)^2\right\}\right]\left[q^{a-1} \exp\left\{-\left(\frac{q}{2b}\right)\right\}\right] \end{aligned}$$

or, after combining terms, the joint natural conjugate prior density is given by:

$$g(\theta, q) \propto q^{0.5(2a-1)} \exp\left\{(-q)\left(\frac{1}{2K}\right)(\theta - \theta_0)^2 + \frac{1}{2b}\right\}.$$

Note that, *a priori*, θ and q are not taken to be independent. Substituting the likelihood function and the natural conjugate prior into Bayes' theorem gives the joint posterior density:

$$\begin{aligned} h(\theta, q \mid data) &\propto [q^{0.5n} \exp\{(-0.5q)[n(\bar{x} - \theta)^2 + (n-1)s^2]\}] \\ &\quad \cdot \left[q^{0.5(2a-1)} \exp\left\{(-q)\left(\frac{1}{2K}\right)(\theta - \theta_0)^2 + \frac{1}{2b}\right\}\right]. \end{aligned}$$

Combining terms gives

$$\begin{aligned} h(\theta, q \mid \bar{x}, s^2) &\propto q^{0.5(2a+n-1)} \\ &\quad \exp\left\{(-0.5q)\left[n(\bar{x} - \theta)^2 + (n-1)s^2 + \frac{1}{K}(\theta - \theta_0)^2 + \frac{1}{b}\right]\right\}. \end{aligned}$$

Next, complete the square in θ in the exponent of the last equation and factor the expression, to obtain

$$\begin{aligned} h(\theta, q \mid \bar{x}, s^2) &\propto \left[q^{0.5} \exp\left\{(-0.5q)\left(n + \frac{1}{K}\right)(\theta - \tilde{\theta})^2\right\}\right] \\ &\quad \cdot [q^{0.5(2a+n-2)} \exp\{-0.5q\lambda\}], \end{aligned}$$

where

$$\tilde{\theta} \equiv \frac{n\bar{x} + (\theta_0/K)}{n + (1/K)},$$

and

$$\lambda = \frac{(n/K)}{n + (1/K)}(\bar{x} - \theta_0)^2 + (n - 1)s^2 + (1/b).$$

That is, if $\sigma^2 \equiv (1/q)$, the conditional posterior distribution of $\theta \mid q$ is:

$$(\theta \mid \sigma^2, \bar{x}, s^2) \sim N\left(\tilde{\theta}, \frac{\sigma^2}{n + (1/K)}\right);$$

and the posterior distribution of q is gamma $[0.5(2a + n), 2/\lambda]$. Note that the joint prior and the joint posterior densities are in the same family, namely the normal-gamma family. Such will always be the case when natural conjugate prior distributions are used. Note also that as in Example 4.5 with known variance of the data, the (conditional) posterior mean is again a convex combination of the prior mean and the sample mean. Now find the marginal posterior distribution for θ by integrating out q. We find that if

$$\tilde{A} = 0.5\left[n(\bar{x} - \theta)^2 + (n - 1)s^2 + \frac{1}{K}(\theta - \theta_0)^2 + \frac{1}{b}\right]$$

the marginal posterior density of θ becomes

$$\tilde{p}(\theta \mid \bar{x}, s^2) = \int_0^\infty q^{0.5(2a+n-1)} \exp\left(-q\tilde{A}\right) dq.$$

Making the same change of variable as above, and integrating, gives

$$\begin{aligned}\tilde{p}(\theta \mid \bar{x}, s^2) &\propto \frac{1}{\tilde{A}^{0.5(2a+n+1)}} \\ &= \frac{1}{\left\{n(\bar{x} - \theta)^2 + (n - 1)s^2 + \frac{1}{K}(\theta - \theta_0)^2 + \frac{1}{b}\right\}^{0.5(2a+n+1)}}.\end{aligned}$$

Completing the square in θ gives

$$\tilde{p}(\theta \mid \bar{x}, s^2) \propto \frac{1}{\{\alpha^2 + (\theta - \bar{\theta})^2\}^{0.5(2a+n+1)}},$$

where

$$\bar{\theta} \equiv \frac{n\bar{x} + (\theta_0/K)}{n + (1 + K)}, \qquad \alpha^2 \equiv \frac{(1/b) + (\theta_0^2/K) + \sum_{i=1}^n x_i^2}{n + (1 + K)}.$$

That is, the marginal posterior distribution of θ is Student t, centered at $\bar{\theta}$, a convex combination of the sample mean and the prior mean, with $(2a + n)$ degrees of freedom. The Student t-distribution was discussed in the Remark following Case 1.

SUMMARY

This chapter has introduced Bayes' theorem. The theorem was examined in various contexts. We saw it illustrated for a variety of situations:

1. First for events (the case of DNA comparisons);
2. Then for discrete data and discrete parameters (exemplified by quality control in manufacturing), as well as for discrete data and a discrete number of models being compared;
3. Then for discrete data and continuous parameters (again for quality control in manufacturing, except in this case the parameter was continuous);
4. Then for continuous data and discrete parameters (the example in which we inferred the section of a class from which a student was selected);
5. Then for continuous data and continuous parameters (where we evaluated the efficacy of a salesmanship course);
6. And finally, we examined the use of Bayes' theorem in the normal distribution with both known and unknown variance of the data. We examined both vague priors and natural conjugate priors in connection with Bayesian inference in the normal distribution.

We introduced the ideas of prior and posterior distributions, Bayes factors, odds ratios, kernels of distributions, flat prior densities, improper distributions, natural conjugate prior distributions, mixture priors, normalizing constants, and precision of a distribution. We discussed uniform, normal, gamma, Student t, beta, and binomial distributions.

EXERCISES

4.1* Distinguish between prior and posterior distributions.

4.2 You go to the doctor for a regular medical checkup and the doctor suggests a laboratory test for a particular form of cancer. Assume (unrealistically) that the doctor has available only the one test. We assume that she has not yet discovered any reason to believe, *a priori*, that you have the disease—it is not in the medical background you have told her.

Suppose that the medical test is a very sensitive one in that the chance that the laboratory test is positive when in fact you do not have the disease ("false

*Solutions for asterisked exercises may be found in Appendix 7.

positive") is 3 percent, and the chance that the laboratory test is negative when in fact you do have the disease ("false negative") is 1 percent.

Suppose also that the incidence of the disease in the population at large is one in a million (you adopt this value as the prior probability that you have the disease). You take the laboratory test, and to your surprise and horror, the laboratory result comes back positive! What is your posterior probability that you have the disease? (*Hint:* See Chapter 1, Section 1.1 for a discussion of this problem.)

4.3 In Exercise 4.1, suppose that when the doctor had asked you about your medical history, you did not know that there had been a strong incidence of cancer in your family. On further investigation, suppose that you found that three out of four of your grandparents had died early of the same type of cancer for which you are being tested and that your father had been misdiagnosed when he died. In fact, your father actually died from this form of cancer. Now you have to question whether the figure of one in a million that applies to the population at large applies to *you*. It probably does not apply to *you*. The chances of getting this type of cancer may be much higher for *you* than for others, because of something related to *your* genes. So the incidence of cancer for *your subpopulation* is not one in a million, but perhaps it is one in a thousand. Now what is *your* chance of having this type of cancer given your positive test result?

4.4 In Exercise 4.2, find the value of your prior probability that yields a posterior probability of 50 percent for having the disease.

4.5 The value of a second opinion: Consider the context of Exercise 4.1. Now suppose that because you feel that cancer is such a serious disease, you decide to seek a second opinion about your problem. You look for another physician, perhaps an oncologist, who will suggest either that a new independent laboratory test be obtained, or that a new pathologist evaluate the results of the same test. The purpose of either procedure is to see whether a new test or a new reading of the same results will change your opinion about your chance of your having the disease. (Because such tests can be quite invasive, and even dangerous, multiple readings of the same results are often taken rather than additional invasive procedures being performed.) In either case, the result will be either positive again (confirmatory), or perhaps this time it will be negative. Suppose the second test is positive again. How should that affect your belief about your having this cancer?

4.6 Tom Lee has been accused by the police of fathering Mary Jones' child. Samples are taken of Tom Lee's DNA and that of the child, and compared. Suppose that the probability of a DNA match given that Mr. Lee is not the true father is 1 percent, and the probability of a match given that Mr. Lee actually fathered the child is 99.6 percent. You reason that Mr. Lee would not have been accused if he were a randomly selected individual, so that there is probably a pretty good chance that he is the father of the child. You take this prior probability to be 75 percent. Suppose laboratory results show that there is a DNA match. What is your posterior probability that Tom Lee actually fathered the child?

4.7 As a particle physicist, you carry out an experiment involving a comparison between the physics of matter and the physics of antimatter. Is there a particular kind of symmetry or not? Suppose that your prior probability for finding this asymmetry is 50 percent. Moreover, suppose the experiment yields a "+1" if you measure a decay signature that corresponds to symmetry, and that the experiment yields a "−1" if you measure a decay signature that corresponds to asymmetry. The error in measuring the decay signature corresponding to symmetry is 0.001, whereas the error in measuring the decay signature corresponding to asymmetry is 0.01 in this experiment. Suppose you find a "+1" in your experiment. What is your posterior probability that you have found this type of asymmetry?

4.8 Example 4.2 concerned quality control in manufacturing. Suppose instead of the sample selected in that example, the RAMM Company had actually selected a sample for testing with $n = 100$, and that five defectives were found in the sample. Calculate your posterior distribution for θ assuming the same beta prior probability adopted in that example in Case 2.

4.9 Example 4.3 involved making an inference about which section of a large course a student came from. Adopt the same probabilities as in that example, but now suppose that 10 students were selected from a particular section (all from the same section) and that their average score was 72 percent. Find the section from which the 10 students were most likely selected.

4.10 In Example 4.5 we were concerned with the efficacy of a salesmanship course. Suppose in that example that your prior for θ followed the distribution N (4, 16) instead of the prior adopted in the example. For the same data as in that example, evaluate your posterior probability distribution for the change in test scores θ. Sketch the posterior density.

4.11 Find the Bayes factor in Exercise 4.10.

4.12 Find the posterior odds against Tom Lee being the father of Mary Jones' child in Exercise 4.6.

4.13 Find the mean and variance of the posterior distribution in Example 4.2.

4.14 Find the means and variances of the posterior distributions given in Equation (4.15) for the cases of normal data with unknown mean and unknown variance, for:

(a) A vague prior family of distributions;

(b) A natural conjugate prior family of distributions.

4.15 Let $X_1, \ldots, X_n$ denote independent and identically distributed observations from $N(\theta, \sigma^2)$. Suppose σ^2 is known, but you can assess a *mixture prior* for θ of the form: $g(\theta) = \alpha g_1(\theta) + (1 - \alpha) g_2(\theta)$, where for $0 \leq \alpha \leq 1$,

$$g_i(\theta) = \frac{1}{\tau_i \sqrt{2\pi}} \exp\left\{ \left(-\frac{1}{2\tau_i^2} \right) (\theta - \theta_i)^2 \right\}, \qquad i = 1, 2,$$

where $(\alpha, \theta_1, \theta_2, \tau_1, \tau_2)$ are found by assessment. Evaluate your posterior density for θ and place it in a simple recognizable form.

FURTHER READING

Wambaugh, J. (1989). *The Blooding*, New York, Morrow Publishing Co.

COMPLEMENT TO CHAPTER 4: HEIGHTS OF THE STANDARD NORMAL DENSITY

t	$\phi(t)$	t	$\phi(t)$	t	$\phi(t)$
0.00	0.39894	0.45	0.36053	0.90	0.26609
0.01	0.39892	0.46	0.35889	0.91	0.26369
0.02	0.39886	0.47	0.35723	0.92	0.26129
0.03	0.39876	0.48	0.35553	0.93	0.25888
0.04	0.39862	0.49	0.35381	0.94	0.25647
0.05	0.39844	0.50	0.35207	0.95	0.25406
0.06	0.39822	0.51	0.35029	0.96	0.25164
0.07	0.39797	0.52	0.34849	0.97	0.24923
0.08	0.39767	0.53	0.34667	0.98	0.24681
0.09	0.39733	0.54	0.34482	0.99	0.24439
0.10	0.39695	0.55	0.34294	1.00	0.24197
0.11	0.39654	0.56	0.34105	1.01	0.23955
0.12	0.39608	0.57	0.33912	1.02	0.23713
0.13	0.39559	0.58	0.33718	1.03	0.23471
0.14	0.39505	0.59	0.33521	1.04	0.23230
0.15	0.39448	0.60	0.33322	1.05	0.22988
0.16	0.39387	0.61	0.33121	1.06	0.22747
0.17	0.39322	0.62	0.32918	1.07	0.22506
0.18	0.39253	0.63	0.32713	1.08	0.22265
0.19	0.39181	0.64	0.32506	1.09	0.22025
0.20	0.39104	0.65	0.32297	1.10	0.21785
0.21	0.39024	0.66	0.32086	1.11	0.21546
0.22	0.38940	0.67	0.31874	1.12	0.21307
0.23	0.38853	0.68	0.31659	1.13	0.21069
0.24	0.38762	0.69	0.31443	1.14	0.20831
0.25	0.38667	0.70	0.31225	1.15	0.20594
0.26	0.38568	0.71	0.31006	1.16	0.20357
0.27	0.38466	0.72	0.30785	1.17	0.20121
0.28	0.38361	0.73	0.30563	1.18	0.19886
0.29	0.38251	0.74	0.30339	1.19	0.196520
0.30	0.38139	0.75	0.30114	1.20	0.19419
0.31	0.38023	0.76	0.29887	1.21	0.19186
0.32	0.37903	0.77	0.29659	1.22	0.18954

t	$\phi(t)$	t	$\phi(t)$	t	$\phi(t)$
0.33	0.37780	0.78	0.29431	1.23	0.18724
0.34	0.37654	0.79	0.29200	1.24	0.18494
0.35	0.37524	0.80	0.28969	1.25	0.18265
0.36	0.37391	0.81	0.28737	1.26	0.18037
0.37	0.37255	0.82	0.285W	1.27	0.17810
0.38	0.37115	0.83	0.28269	1.28	0.17585
0.39	0.36973	0.84	0.28034	1.29	0.17360
0.40	0.36827	0.85	0.27798	1.30	0.17137
0.41	0.36678	0.86	0.27562	1.31	0.16915
0.42	0.36526	0.87	0.27324	1.32	0.16694
0.43	0.36371	0.88	0.27086	1.33	0.16474
0.44	0.36213	0.89	0.26848	1.34	0.16256
1.35	0.16038	1.80	0.07895	2.25	0.03174
1.36	0.15822	1.81	0.07754	2.26	0.03103
1.37	0.15608	1.82	0.07614	2.27	0.03034
1.38	0.15395	1.83	0.07477	2.28	0.02965
1.39	0.15183	1.84	0.07341	2.29	0.02898
1.40	0.14973	1.85	0.07206	2.30	0.02833
1.41	0.14764	1.86	0.07074	2.31	0.02768
1.42	0.14556	1.87	0.06943	2.32	0.02705
1.43	0.14350	1.88	0.06814	2.33	0.02643
1.44	0.14146	1.89	0.06687	2.34	0.02582
1.45	0.13943	1.90	0.06562	2.35	0.02522
1.46	0.13742	1.91	0.06439	2.36	0.02463
1.47	0.13542	1.92	0.06316	2.37	0.02406
1.48	0.13344	1.93	0.06195	2.38	0.02349
1.49	0.13147	1.94	0.06077	2.39	0.02294
1.50	0.12952	1.95	0.05959	2.40	0.02239
1.51	0.12758	1.96	0.05844	2.41	0.02186
1.52	0.12566	1.97	0.05730	2.42	0.02134
1.53	0.12376	1.98	0.05618	2.43	0.02083
1.54	0.12188	1.99	0.05508	2.44	0.02033
1.55	0.12001	2.00	0.05399	2.45	0.01984
1.56	0.11816	2.01	0.05292	2.46	0.01936
1.57	0.11632	2.02	0.05186	2.47	0.01889
1.58	0.11450	2.03	0.05082	2.48	0.01842
1.59	0.11270	2.04	0.04980	2.49	0.01797
1.60	0.11092	2.05	0.04879	2.50	0.01753
1.61	0.10915	2.06	0.04780	2.51	0.01709
1.62	0.10741	2.07	0.04682	2.52	0.01667
1.63	0.10567	2.08	0.04586	2.53	0.01625
1.64	0.10396	2.09	0.04491	2.54	0.01585
1.65	0.10226	2.10	0.04398	2.55	0.01545
1.66	0.10059	2.11	0.04307	2.56	0.01506
1.67	0.09893	2.12	0.04217	2.57	0.01468
1.68	0.09728	2.13	0.04128	2.58	0.01431

(continued)

t	$\phi(t)$	t	$\phi(t)$	t	$\phi(t)$
1.69	0.09566	2.14	0.04041	2.59	0.01394
1.70	0.09405	2.15	0.03955	2.60	0.01358
1.71	0.09246	2.16	0.03871	2.61	0.01323
1.72	0.09089	2.17	0.03788	2.62	0.01289
1.73	0.08933	2.18	0.03706	2.63	0.01256
1.74	0.08780	2.19	0.03626	2.64	0.01223
1.75	0.08628	2.20	0.03547	2.65	0.01191
1.76	0.08478	2.21	0.03470	2.66	0.01160
1.77	0.08329	2.22	0.03394	2.67	0.01130
1.78	0.08183	2.23	0.03319	2.68	0.01100
1.79	0.08038	2.24	0.03246	2.69	0.01071
2.70	0.01042	3.15	0.00279	3.60	0.00061
2.71	0.01014	3.16	0.00271	3.61	0.00059
2.72	0.00987	3.17	0.00262	3.62	0.00057
2.73	0.00961	3.18	0.00254	3.63	0.00055
2.74	0.00935	3.19	0.00246	3.64	0.00053
2.75	0.00909	3.20	0.00238	3.65	0.00051
2.76	0.00885	3.21	0.00231	3.66	0.00049
2.77	0.00861	3.22	0.00224	3.67	0.00047
2.78	0.00837	3.23	0.00216	3.68	0.00046
2.79	0.00814	3.24	0.00210	3.69	0.00044
2.80	0.00792	3.25	0.00203	3.70	0.00042
2.81	0.00770	3.26	0.00196	3.71	0.00041
2.82	0.00748	3.27	0.00190	3.72	0.00039
2.83	0.00727	3.28	0.00184	3.73	0.00038
2.84	0.00707	3.29	0.00178	3.74	0.00037
2.85	0.00687	3.30	0.00172	3.75	0.00035
2.86	0.00668	3.31	0.00167	3.76	0.00034
2.87	0.00649	3.32	0.00161	3.77	0.00033
2.88	0.00631	3.33	0.00156	3.78	0.00031
2.89	0.00613	3.34	0.00151	3.79	0.00030
2.90	0.00595	3.35	0.00146	3.80	0.00029
2.91	0.00578	3.36	0.00141	3.81	0.00028
2.92	0.00562	3.37	0.00136	3.82	0.00027
2.93	0.00545	3.38	0.00132	3.83	0.00026
2.94	0.00530	3.39	0.00127	3.84	0.00025
2.95	0.00514	3.40	0.00123	3.85	0.00024
2.96	0.00499	3.41	0.00119	3.86	0.00023
2.97	0.00485	3.42	0.00115	3.87	0.00022
2.98	0.00471	3.43	0.00111	3.88	0.00021
2.99	0.00457	3.44	0.00107	3.89	0.00021
3.00	0.00443	3.45	0.00104	3.90	0.00020
3.01	0.00430	3.46	0.00100	3.91	0.00019
3.02	0.00417	3.47	0.00097	3.92	0.00018
3.03	0.00405	3.48	0.00094	3.93	0.00018
3.04	0.00393	3.49	0.00090	3.94	0.00017

t	$\phi(t)$	t	$\phi(t)$	t	$\phi(t)$
3.05	0.00381	3.50	0.00087	3.95	0.00016
3.06	0.00370	3.51	0.00084	3.96	0.00016
3.07	0.00358	3.52	0.00081	3.97	0.00015
3.08	0.00348	3.53	0.00079	3.98	0.00014
3.09	0.00337	3.54	0.00076	3.99	0.00014
3.10	0.00327	3.55	0.00073		
3.11	0.00317	3.56	0.00071		
3.12	0.00307	3.57	0.00068		
3.13	0.00298	3.58	0.00066		
3.14	0.00288	3.59	0.00063		

CHAPTER 5

Prior Distributions

5.1 INTRODUCTION

This chapter is concerned with the selection of a prior distribution for use in Bayes' theorem. How do you characterize the information that you have about the underlying process prior to taking data? The answers to this question are partly philosophical and partly pragmatic. Answers depend upon both the nature of the information that you possess, and your belief system regarding the appropriate compromise between being subjective or objective in the information you would like to introduce into the problem. We will discuss both subjective and objective prior distributions in both the univariate and multivariate cases.

5.2 OBJECTIVE AND SUBJECTIVE PRIOR DISTRIBUTIONS

The prior distribution to use in Bayes' theorem is not specified in the theorem. In fact, when the theorem was expounded by Thomas Bayes, he applied the theorem in a context in which the data followed a binomial distribution with unknown parameter p, the probability of success on a single trial, and his prior distribution for p was uniform (see Appendix 5). In other words, Bayes adopted a prior distribution that implied that all possible values of p were equally likely. This is one type of *objective* prior distribution. However, the theorem is not so limited. It has been generalized to include a wide variety of data distributions and a wide variety of prior distributions. When available information is minimal, we often begin the updating process with an objective prior distribution.

5.2.1 Objective Prior Distributions

Some of the prior distributions that have been developed attempt to reflect the notion of having very little prior information about the unknown parameters before any data are taken, subject to the prior distribution satisfying certain constraints. This idea translates to the class of *objective* prior distributions. But how should we define "very little"? There has not been general agreement about how that term should be defined, and as a result, there have been many proposals for how an objective prior distribution should be defined. Moreover, in many situations the result of a scientific experiment must be reported to the public as a general result (minimally specific to the beliefs of the researcher), and public policy may often have to be based upon this finding. For situations such as public policy judgments and decision making special efforts have been made to address this very-little-information issue in that context.

Public Policy Priors

We may wish to know what percentage change there will be in the number of high quality young people who will enlist in a volunteer army if they are offered a particular amount of money, say, a $10,000 bonus, for enlisting for at least four years. A particular military analyst may have a prior belief about this percentage, and an experiment may be carried out to study the problem, but in the final analysis, the army does not want posterior inferences made by different analysts to yield different answers (unless the different assumptions they use to develop their posteriors are clearly stated, and they are meaningful in terms of their policy implications).

The uniform prior distribution approach (used by Bayes) is quite reasonable for many problems. What should the statistician do in situations when he/she wants to choose a public policy prior (one that most observers will accept)? In such situations, the Bayesian statistician will either use a vague prior (see below) or will report a set of priors and their corresponding posteriors; that is, he/she will report a *mapping* from the priors to the implied posteriors (see below).

In some public policy situations we do not necessarily want a prior common to many researchers; we merely want to know the policy implications of particular prior beliefs. Dickey (1973) proposed presenting the posterior as a function of the prior (a mapping) to cover such situations. Although a single prior is not selected, the public is provided with the posterior inferences associated with a range of priors.

In public policy decision making where it is claimed by decision makers that they would like results to reflect the opinions of a group, such as the inhabitants of a city, objective prior distributions are recommended. Such claims are of course sometimes specious, and actually, sometimes decision makers' underlying desire is that their decisions reflect only the opinions of a few politicians.

Principle of Insufficient Reason (Laplace)

Pierre Simon Laplace (1749–1827) was also concerned with the public policy prior. He stated what we now call Bayes' theorem (Laplace, 1774, p. 623), 11 years after Bayes' essay appeared, in the following form (this translation is from Stigler, 1986):

If an event can be produced by a number n of different causes, then the probabilities of these causes given the event are to each other as the probabilities of the event given the causes, and the probability of the existence of each of these is equal to the probability of the event given the cause, divided by the sum of all the probabilities of the event given each of these causes.

Here, Laplace is taking the *a priori* probabilities to be equal, which was his early approach to dealing with prior beliefs. (He later generalized the prior.) Laplace is not likely to have seen Bayes' essay, which appeared originally in England in 1763 and probably did not circulate in France until about 1780 (Stigler, 1986). Thus, Laplace appears to have solved the problem of inverse probability independently of Bayes.

Laplace (1812) formally proposed that in situations where we need to use a public policy prior (he actually used words more appropriate to his time), and where we would therefore like to proceed by introducing as little external subjective information into the problem as possible (beyond the sample data obtained from an experiment), we should use the *principle of insufficient reason*. His principle suggested that in the absence of any reason to the contrary, all values of the unknown parameter indexing the occurrence of events should be taken to be equally likely, *a priori*. (In the second edition of his book, in 1814, he modified the principle to include unequal prior probabilities, if there were sufficient reason.)

Jeffreys (1961) (and in the earlier editions of his book, in 1939 and 1948) followed essentially the same insufficient reason principle, but he modified it slightly for parameters, with support only on the positive half of the real line (these ideas will be given mathematical form below). He also extended the principle to multiparameter families and he used invariance arguments to justify the procedures. Jaynes (1983) and Zellner (1971) recommended the same Laplace (or modified Laplace) procedure for choosing a public policy prior distribution. This prior can be thought of as objective, in that all observers will adopt the same prior as representing a prior state of indifference to one parameter value over another. Moreover, sometimes, use of such a prior distribution will yield statistical procedures analogous to those that are likely to be used by nonBayesian scientists, procedures developed using classical (frequentist) principles.

When the unknown parameter lies on a finite interval, the uniform distribution serves us well as the public policy, or objective, prior. When at least one end point of the domain of the parameter is not finite, however, the objective prior implied becomes improper (even though it is usually the case that the posterior remains a proper probability distribution). Additional research is needed to focus on the Rényi probability axiom system to legitimize these improper priors in those situations where they are appropriate.

5.2.2 Weighing the Use of Objective Prior Distributions

There are both advantages and disadvantages of adopting objective prior distributions.

Advantages

1. Objective prior distributions are sometimes used as a benchmark that will not reflect the particular biases of the analyst, so they are often used to provide an analytical *updating engine* that can be used by anyone regardless of his/her predilections about what is likely to be observed subsequent to carrying out an experiment.

2. Objective prior distributions reflect the idea of there being very little information available about the underlying process.
3. There are sometimes mathematically equivalent results obtained by Bayesian methods using objective prior distributions and results obtained by using frequentist methods on the same problem (but results in the two cases have different philosophical interpretations, and these differences are important).
4. Objective prior distributions are useful as public policy priors.

Disadvantages

As will be seen, it is difficult to specify a generally agreed upon *objective* prior distribution that expresses knowing little (see items 1, 4, 6, below), that will satisfy all the constraints that analysts would like to impose (see item 3, below), that will be sensible under a variety of conditions (see item 2, below), and that analysts are in reality likely to possess (see item 5, below). Disadvantages of objective prior distributions include:

1. Sometimes improper priors result from wanting to express complete ignorance, *a priori*, for parameters whose domain is the entire real line, or the entire half line, and the result can be improper posteriors.
2. In multiple parameter situations, the parameters are often taken to be independent in an objective prior distribution, but of course, multiple parameters can be correlated, and that causes its own problems.
3. Sometimes it is desirable that the posterior inference that will be made should be the same regardless of the particular parameterization of the problem that is used. Sometimes, invariance under certain groups of transformations seems appropriate (such as affine or rotational transformations), but the prior distributions that result are not necessarily sensible. Moreover, questions remain about whether to use left or right invariant Haar probability measures to determine the appropriate prior distributions. If left invariant Haar measure is used for determining a vague prior density for a normal distribution variance, σ^2, for example, right Haar measure will generate a different prior density. Both invariant measure groups yield $(1/\sigma^2)$ to a power, but the powers are different. See also the discussion about invariant priors in Section 5.3, following Figure 5.1, below; for details about Haar probability measures see, for example, Halmos, 1950.
4. Sometimes expressing ignorance, *a priori*, for a particular parameter using an objective prior results in undesirable inferences for some important function of the parameter.

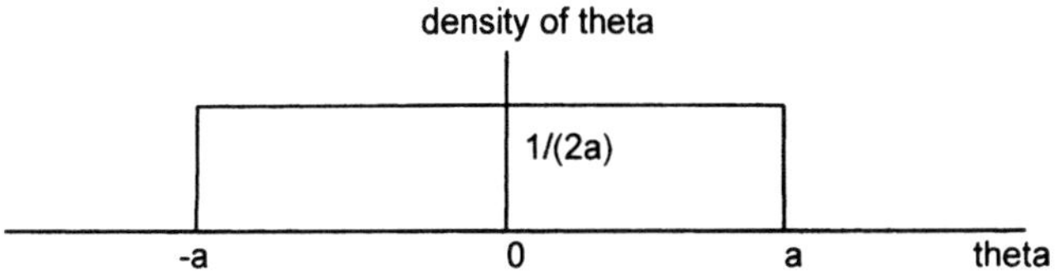

Figure 5.1 Vague prior density on $(-a, a)$

5. It is not always clear that any individual would seriously want to embrace an objective prior distribution to reflect his/her state of knowledge. That is, the prior does not really yield a meaningful probabilistic interpretation.
6. Improper objective prior distributions generate indeterminacies in their associated Bayes factors because it is not clear what constant to use in the Bayes factor ratio.
7. Stein (1956) pointed out that ordinary estimators are not even admissible (see Section 11.3) in dimensions higher than 2. From a Bayesian point of view, such estimators are often obtained as posterior means using vague (objective) objective prior distributions.
8. A paradox was found by Dawid, Stowe, and Zidik (1973) when they found results were different in Bayesian inference using vague prior distributions (objective Bayes) depending upon how a problem was parameterized, and which parameterization was used for marginalizing the joint posterior distribution to that of the parameter of interest.

5.2.3 Weighing the Use of Subjective Prior Distributions

Subjective prior distributions represent an attempt to bring prior knowledge about the phenomenon under study into the problem. They have their own advantages and disadvantages.

Advantages

1. Subjective prior distributions are proper (they integrate or sum to 1).
2. A subjective prior distribution is generally well behaved analytically. The effect of a subjective prior distribution on the posterior is as if there were additional replications of the data.
3. Subjective prior distributions may be used to introduce the informed understanding, beliefs, and experience of a scientist or decision maker into the Bayesian analysis of a problem to take advantage of this additional information about the phenomenon under study.
4. Often there is insufficient information in a problem to solve it by objective Bayesian or frequentist methods. The subjective Bayesian approach may be the only way to find an acceptable solution since it brings additional information to bear on the problem.

Example 5.1

As an example of a subjective prior distribution, suppose the problem contains an unknown parameter θ that is of fundamental interest. You may know that in an earlier study, θ was equal to 12. So the best information you have at this point is that $\theta = 12$, and you can use that information as your best *a priori* estimate of θ. You might also feel that your uncertainty about θ is probably symmetric about $\theta = 12$, so it might be reasonable to adopt a normal prior distribution for θ centered about a prior mean of 12. In this way, you are introducing *subjective* prior information into the current problem. You might also be able to reason that it is extremely unlikely (much less than a 1 percent chance) that $\theta < 9$, and extremely unlikely that $\theta > 15$. If you take these bounds as three standard deviations on either side of a normal mean (assuming that $9 < \theta < 12$ with 99.7 percent probability) you readily find that your prior distribution has a standard deviation of 1. So you adopt a subjective prior distribution for that θ that is $N(12, 1)$.

Example 5.2

As an example of advantage 3 above, suppose that the data follow a binomial distribution with unknown parameter θ, for r successes in n independent trials. Then the likelihood may be written (ignoring any constants) $L(r \mid \theta, n) = \theta^r(1-\theta)^{n-r}$. You adopt a beta distribution prior on θ of the form beta (a, b), with density proportional to $\theta^{a-1}(1-\theta)^{b-1}$, $a > 0, b > 0$, for known (a, b). By Bayes' theorem, the posterior density is proportional to $\theta^{(a+r)-1}(1-\theta)^{(b+n-r)-1}$, another beta distribution, but now the distribution is beta $(a + r, b + n - r)$. Had you taken a prior that expressed little knowledge (vague prior) *a priori*, with $a = 0$ and $b = 0$, the posterior density would have been beta $(r, n - r)$. By bringing subjective prior information to bear, the posterior behaves as if there were effectively $(a + r)$ and $(n - r + b)$ degrees of freedom.

Scientists have used subjectivity in their methodology throughout history (see Chapter 1, and Press and Tanur, 2001). Their inferences from data have benefited from informed guesses, hunches, intuition, and earlier experiments that they or others carried out. They introduced such inferences informally, without the use of Bayes' theorem, but their subjectivity influenced their methodology nevertheless.

Disadvantages

1. It is not always easy to assess subjective prior distributions (those based upon earlier knowledge) because it is not always easy to translate the prior knowledge into a meaningful probability distribution. The parameters that index subjective prior distributions are called *hyperparameters*. They must be assessed, and the assessment problem is sometimes difficult, especially in multiparameter problems.
2. Results of a Bayesian analysis that used a subjective prior distribution are meaningful to the particular analyst whose prior distribution was used, but not necessarily to other researchers, because their priors might have been different. So a subjective prior distribution would not always be appropriate in a public policy situation, for example. In contexts in which we would like to take advantage of the prior knowledge of the scientist it is very meaningful to adopt a subjective prior distribution.
3. A subjective prior distribution might not be very tractable mathematically (particularly if it is assessed numerically, and it is not a distribution from a simple mathematical family such as a natural conjugate family of distributions). Fortunately, the current computer environment makes it straightforward

to compute posterior and predictive distributions even when the prior distribution is not mathematically tractable.

5.3 (UNIVARIATE) PRIOR DISTRIBUTIONS FOR A SINGLE PARAMETER

In this section we discuss the formation of prior distributions for problems involving data distributions indexed by just one parameter. We will generalize these ideas to parameters of multivariate data distributions in Section 5.4.

5.3.1 Vague (Indifference, Default, Objective) Priors

You can develop a prior distribution representing your degree of belief for any quantity unknown to you.

Suppose you are asked to assess a prior distribution for the density (weight per unit volume) of potassium in your bloodstream, θ. Unless you have a strong background in biology, at first you may throw up your hands and say, "I am totally uninformed about θ," and then you may take all values of θ as equally likely. However, when you think about it you realize that there are really some things you know about θ.

For example, suppose θ denotes a mean potassium density (this use of the term *density* refers to a physical property of a substance). You know that you must have $\theta > 0$. You might take the prior probability density

$$g(\theta) = \begin{cases} \dfrac{1}{a}, & 0 < \theta < a \\ 0, & \text{otherwise.} \end{cases} \tag{5.1}$$

However, what should you take for a? And what would this imply about your prior knowledge of, say, $[\log \theta]$, and $[\theta^2]$? It will be seen later in this section that for $\theta > 0$, if we want to display "knowing little" about θ, or ignorance, or vagueness, we will take $[\log \theta]$ to be uniformly distributed, which implies that $g(\theta) \propto \theta^{-1}$; this represents vagueness about θ.

Suppose, for illustration, we consider the problem in which $X \mid \theta \sim N(\theta, 1)$, $-\infty < \theta < \infty$, and we wish to estimate θ. A feeling of being totally uninformed about the true value of θ might lead some of us to an improper prior distribution of θ over the entire real line. If we demand the use of only proper priors, we are led to confront an impossible situation, as we will see below. It will be seen that if we are uninformed about θ, we necessarily become informed about some functions of θ, and conversely.

Suppose, for example, we try to argue that to be indifferent to all values of θ should imply that if we transform θ by projecting it in a simple, monotonically smooth way into the unit interval and then put a uniform (proper) prior distribution on all values in the unit interval, we will have circumvented our problem. Not so, as we shall see below.

Our problem is to develop a prior that will reflect indifference to all values of θ. Let

$$\phi \equiv F(\theta) \tag{5.2}$$

where $F(\theta)$ denotes any monotone, nondecreasing, differentiable transformation of θ such that

$$0 \leq F(\theta) \leq 1. \tag{5.3}$$

For example, $F(\cdot)$ can be any cumulative distribution function (cdf). Let the prior distribution for θ be uniform; that is, its density is

$$p(\phi) = \begin{cases} 1, & 0 < \phi < 1 \\ 0, & \text{otherwise.} \end{cases} \tag{5.4}$$

Clearly, $p(\phi)$ reflects indifference to all values of ϕ. But the inverse transformation $\theta = F^{-1}(\phi)$ induces a prior density on θ,

$$g(\theta) = \left| \frac{dF(\theta)}{d\theta} \right|, \tag{5.5}$$

by the basic theorem on transformation of densities. We are free to choose any $F(\theta)$. Suppose we choose the normal cdf:

$$\phi \equiv F(\theta) = \int_{-\infty}^{\theta} \frac{1}{\sqrt{2\pi}} e^{-0.5t^2}\, dt. \tag{5.6}$$

Then, the density of θ is

$$g(\theta) = \frac{1}{\sqrt{2\pi}} e^{-0.5\theta^2}, \tag{5.7}$$

or $\theta \sim N(0, 1)$. The posterior density is

$$p(\theta \mid X) \propto e^{-0.5(x-\theta)^2 - 0.5\theta^2}, \tag{5.8}$$

$$\theta \mid x \sim N(0.5x, 0.5). \tag{5.9}$$

However, note that we are being extremely informative about θ, in spite of assigning equal probability to all possible intervals of ϕ of equal length, and all we did was try to be indifferent (Laplace principle of insufficient reason) to all values of a simple monotone transformation of θ. In order for the prior distribution of θ to express indifference toward specific values of θ, we would need to have

$$g(\theta) \propto \text{ constant}, \tag{5.10}$$

which, in turn, implies that $\frac{dF(\theta)}{d\theta}$ is constant, or $F(\theta)$ is a linear function of θ. However, it is impossible (in the usual probability calculus) for θ on the real line (or even on the positive real line) to have $0 \leq F(\theta) \leq 1$, and have $F(\theta)$ be a straight line for $-\infty < \theta < \infty$.

Placing uniform distributions on simple monotone transformations of θ will not solve our problem, nor will more complicated transformation procedures. The problem is more basic, and it exhorts us to develop some sort of probability algebra in which it is reasonable to adopt improper prior distributions on θ over the entire real line.

Vague Prior Density for Parameter on $(-\infty, \infty)$

In the remainder of this volume, whenever we want to express the belief that, *a priori*, no value of θ is any more likely than any other value, or that you are indifferent to one value over any other, we will adopt a vague prior, which is explained below.

To express vagueness, or indifference about our knowledge of θ, when θ is a parameter lying on the real line, $-\infty < \theta < \infty$, we use an improper prior distribution, called *vague*, one that is of the form given in Equation (5.10), which is a limiting form of $g_a(\theta)$, where

$$g(\theta) = \lim_{a \to \infty} g_a(\theta) \tag{5.11}$$

and (Fig. 5.1) where

$$g_a(\theta) = \begin{cases} \frac{1}{2a}, & -a < \theta < a \\ 0, & \text{otherwise.} \end{cases} \tag{5.12}$$

(*Note:* The limiting process used to arrive at a vague prior distribution is not unique.)

Vague Prior Density for Parameter on $(\mathbf{0}, \infty)$

For a positive parameter, say σ, we take $\log \sigma$ to be uniformly distributed on the entire real line, as we did previously with θ. By transforming variables we find that (using the same g, generically, to mean density function, but not the same one that was used for θ), the prior density becomes

$$g(\sigma) \propto \frac{1}{\sigma},$$

and equivalently

$$g(\sigma^2) \propto \frac{1}{\sigma^2}. \tag{5.13}$$

These vague prior densities have useful properties of invariance of inferences under various simple group structures. That is, if the parameter θ or σ is transformed (using simple, common, transformations), posterior inferences based upon the new parameter will be consistent with those based upon the old parameter. Thus, if $g(\theta)$ is the prior density for θ, and if we transform θ to ϕ by $\phi = F(\theta)$, and if $h(\phi)$ is the prior density for ϕ, we should have $g(\theta)d\theta = h(\phi)d\phi$, and posterior inferences in either case should reflect this consistency.

There have been various proposals for how to express the notion of indifference to one parameter value over another by using group invariance (e.g., Hartigan, 1964; Jeffreys, 1961, pp. 179–181; Villegas, 1969). Other principles for selecting prior distributions that reflect indifference have been based upon information theory (Bernardo, 1979, reference posterior distributions; Berger and Bernardo, 1992; Jaynes, 1983, pp. 116–130 "prior probabilities," plus other related articles; Zellner, 1971, pp. 50–51, minimal information priors; Zellner, 1977, pp. 211–232, maximal data information prior distributions). The reason for the diversity of views on the subject is that the method for expressing the notion of *knowing little* differs depending upon the context. For this reason, various authors propose different priors for different groups of transformations appropriate in different settings. Catalogues of objective prior distributions may be found in Kass and Wasserman, 1996 and Yang and Berger, 1996.

5.3.2 Families of Subjective Prior Distributions

Suppose you are not vague about θ but have some (nonequally likely) prior beliefs. Such is the case generally with subjective prior distributions. You would like to bring these prior beliefs into the analysis in a formal way. It will often suffice to permit your prior beliefs to be represented by some smooth distribution that is a specific member of a family of subjective prior distributions.

One such family is called a *natural conjugate family* (Raiffa and Schlaifer, 1961, Chapter 3). This family is chosen largely on the basis of mathematical convenience. A member of such a family is sometimes also called a *convenience prior*. This family of prior distributions is that class that is closed under transformation from prior to posterior. That is, the implied posterior distribution with respect to a natural conjugate prior distribution is in the same family as the prior distribution.

Another family of mathematically convenient subjective prior distributions is called the family of *exponential power distributions* (Box and Tiao, 1973, pp. 157ff.). This class was originally suggested as a family of distributions to represent the data distribution. However, we propose the class as a useful family of prior distributions as well. The family of exponential power distributions is proposed as a prior family because its richness of hyperparameters within the class provides great flexibility of expression of prior views (the family of exponential power distributions is symmetric after centering). It was proposed as a family of data distributions because it conveniently provided a way of fitting data distributions that are normal-like, but not necessarily normal.

Another class of subjective prior distributions is the class of *mixture prior distributions*.

These three classes of subjective prior distributions are discussed below.

A. Natural Conjugate Families of Prior Distributions

The recipe for developing an appropriate family of natural conjugate prior distributions in general is to first form the likelihood function, and then:

1. Interchange the roles of the random variable and the parameter in the likelihood function.
2. "enrich" the parameters of the resulting density kernel of a distribution, that is, make their values perfectly general and not dependent upon the current data set.
3. Identify the distribution corresponding to the resulting kernel and adjoin the appropriate normalizing constant to make the density integrate to one.

Example 5.3 A Natural Conjugate Prior: Binomial Data

Suppose the likelihood function is binomial, so that the likelihood may be written:

$$L(\theta) = \theta^{y}(1-\theta)^{n-y} \tag{5.14}$$

Recall that a likelihood function is unique only up to a multiplicative constant. In the case of the binomial likelihood in Equation (5.14), suppose that θ were the random variable and y and n were just unknown parameters. $L(\theta)$ would then look like the kernel of a *beta distribution*. But you do not want the beta prior family to depend upon sample data (y and n), so you use arbitrary parameters α and β, and you norm the density to make it proper, to get the beta prior density family

$$g(\theta) = \begin{cases} \dfrac{1}{B(\alpha, \beta)}\theta^{\alpha-1}(1-\theta)^{\beta-1}, & 0 < \theta < 1 \\ 0, & \text{otherwise} \end{cases} \tag{5.15}$$

for $0 < \alpha$, $0 < \beta$. Now use your prior beliefs to assess the hyperparameters α and β, that is, having fixed the family of priors as the beta distribution family, only (α, β) remain unknown (that is, you do not have to assess the entire prior distribution, but just the two parameters that index the beta family).

An additional mathematical convenience arises in computing the posterior for a natural conjugate prior distribution family. Since by Bayes' theorem, the posterior

Table 5.1 Some Natural Conjugate Prior Distribution Families

Sampling Distribution	Natural Conjugate Prior Distribution
1 Binomial	Success probability follows a beta distribution
2 Negative binomial	Success probability follows a beta distribution
3 Poisson	Mean follows a gamma distribution
4 Exponential with mean λ	λ follows a gamma distribution
5 Normal with known variance but unknown mean	Mean follows a normal distribution
6 Normal with known mean but unknown variance	Variance follows an inverted gamma distribution

density is proportional to the likelihood times the prior density, in this case a beta prior density, the posterior density is given by

$$\begin{aligned} h(\theta \mid y) &\propto [\theta^{y}(1-\theta)^{n-y}][\theta^{\alpha-1}(1-\theta)^{\beta-1}] \\ h(\theta \mid y) &\propto \theta^{y+\alpha-1}(1-\theta)^{n-y+\beta-1}, \end{aligned} \tag{5.16}$$

a beta posterior density kernel. Natural conjugate prior distributions have the advantage that the induced posterior distributions have the same form (distributional family) as the prior distributions. This "closure property" for distributions of the natural conjugate family holds for all sampling distribution members of the exponential family, that is, the family of distributions with sufficient statistics (subject to some simple regularity conditions).

Table 5.1 provides some sampling distributions in the exponential family, along with their corresponding natural conjugate priors. The reader should try to establish these relations by following the example of the binomial, below.

B. Exponential Power Family (EPF) of Prior Distributions

The class of exponential power distributions for a parameter θ has the density

$$g(\theta) = \frac{k}{\sigma} \exp\left\{ -0.5 \left| \frac{\theta - \theta_0}{\sigma} \right|^{(2/1+\beta)} \right\}, \tag{5.17}$$

where $-\infty < \theta < \infty$, $-1 < \beta \le 1, 0 < \sigma, k^{-1} = \Gamma(1 + 0.5(1+\beta))2^{1+0.5(1+\beta)}$. It can be found that

$$E(\theta) = \theta_0,$$

and

$$\text{var}(\theta) = 2^{1+\beta}\sigma^2 \frac{\Gamma[(1+\beta)(3/2)]}{\Gamma[(1+\beta)(1/2)]}. \tag{5.18}$$

The parameter β is a measure of kurtosis that reflects the degree of nonnormality of the distribution. When $\beta = 0$ the distribution is normal. When $\beta = 1$ the distribution is the double exponential with density

$$g(\theta) = \frac{1}{\sigma\sqrt{2}} \exp\left\{ -\sqrt{2}\left|\frac{\theta - \theta_0}{\sigma}\right| \right\}, \qquad -\infty < \theta < \infty. \tag{5.19}$$

When $\beta \to -1$, the distribution tends to the uniform distribution on $(\theta_0 - \sigma\sqrt{3}, \theta_0 + \sigma\sqrt{3})$. Thus, taking the limit in Equation (5.17) gives

$$\lim_{\beta \to -1} g(\theta) = \frac{1}{2\sigma\sqrt{3}}, \qquad (\theta_0 - \sigma\sqrt{3}, \theta_0 + \sigma\sqrt{3}). \tag{5.20}$$

EPF as a Data Distribution

The exponential power family of distributions may also be used as a *data distribution*. The exponential power class has a density for a continuous observable random variable X of the form:

$$f(x \mid \theta, \tau, \alpha) = \frac{K}{\tau} \exp\left\{ -0.5\left|\frac{x - \theta}{\tau}\right|^{2/1+\alpha} \right\}, \tag{5.21}$$

where $K^{-1} = 2^{1+0.5(1+\alpha)}\Gamma(1 + 0.5(1 + \alpha))$, $-\infty < x < \infty$, $-1 < \alpha \leq 1$, $0 < \tau < \infty$, $-\infty < \theta < \infty$. Suppose that α is known, θ and τ are *a priori* independent, and that you adopt a vague prior density for (θ, τ) of the form $g(\theta, \tau) \propto 1/\tau$. The posterior density for (θ, τ) was studied for this case by Box and Tiao (1973, p. 160 ff.).

C. Mixture Prior Distribution Families

We can easily generate very general classes of prior distributions that contain many hyperparameters that can be used to adjust the prior to wide varieties of prior information by using mixtures of natural conjugate prior density families. When the prior information available suggests that data might cluster around more than one mode, a mixture family of prior distributions comes to mind to translate the prior information into a mathematical form.

Example 5.4 (Binomial)

Suppose the data distribution is binomial with success probability θ, and n trials, with the likelihood function given in Equation (5.14). Adopt the prior density for θ that is a mixture of beta densities:

$$g(\theta) = \sum_{i=1}^{m} \frac{\pi_i}{B(\alpha_i, \beta_i)} \theta^{\alpha_i - 1}(1 - \theta)^{\beta_i - 1}, \tag{5.22}$$

for $0 \leq \pi_i \leq 1$, $\sum_{i=1}^{m} \pi_i = 1$, $0 < \alpha_i$, $0 < \beta_i$, and $0 < \theta < 1$. The $\pi_i s$ are the mixing parameters, and there are m beta densities in the mixture. By Bayes' theorem, the posterior density for θ is also a mixture of beta densities, and it takes the form

$$h(\theta \mid r) = \sum_{i=1}^{m} \frac{\pi_i^*}{B(\alpha_i^*, \beta_i^*)} \theta^{\alpha_i^* - 1}(1 - \theta)^{\beta_i^* - 1}, \tag{5.23}$$

where

$$\pi_i^* = \frac{\pi_i B(\alpha_i^*, \beta_i^*)}{\sum_{i=1}^{m} \pi_i B(\alpha_i^*, \beta_i^*)}, \qquad \alpha_i^* = \alpha_i + r, \qquad \beta_i^* = \beta_i + (n - r), \tag{5.24}$$

and r denotes the number of successes in n trials. To be specific, suppose $m = 3$, $(\pi_1, \pi_2, \pi_3) \equiv (0.5, 0.2, 0.3)$, $(\alpha_1, \alpha_2, \alpha_3 \equiv (10, 15, 20)$, and $(\beta_1, \beta_2, \beta_3) \equiv (20, 15, 10)$. If $r = 3$ and $n = 10$, $(\pi_1^*, \pi_2^*, \pi_3^*) = (0.77, 0.16, 0.07)$, $(\alpha_1^*, \alpha_2^*, \alpha_3^*) = (13, 18, 23)$, and $(\beta_1^*, \beta_2^*, \beta_3^*) = (27, 22, 17)$. This distribution is multimodal.

Exponential Families

A general result appropriate for exponential families is the following. Let $x \equiv (x_1, \ldots, x_n)$ denote a random sample from a regular exponential family with density:

$$p(x \mid \theta) = \left\{ \prod_{j=1}^{n} f(x_j) \right\} [g(\theta)]^n \exp\left\{ \left[\sum_{i=1}^{m} c_i \phi_i(\theta) \right] \left[\sum_{j=1}^{n} h_i(x_j) \right] \right\}. \tag{5.25}$$

Adopt the mixture prior density

$$q(\theta \mid \tau_1, \ldots, \tau_m) = \sum_{i=1}^{m} \pi_i q_i(\theta \mid \tau_i), \qquad 0 \leq \pi_i \leq 1, \qquad \sum_{i=1}^{m} \pi_i = 1, \tag{5.26}$$

where for $i = 1, \ldots, m$,

$$q_i(\theta \mid \tau_i) = \frac{[g(\theta)]^{\tau_{i0}}}{K(\tau_i)} \exp\left\{ \sum_{l=1}^{k} c_l \phi_l(\theta) \tau_{il} \right\} \tag{5.27}$$

are elements of the natural conjugate family of distributions, and τ_i denotes a vector of parameters $\tau_i \equiv (\tau_{il})$, for each i. Applying Bayes' theorem, the posterior density becomes

$$p^*(\theta \mid x) = \sum_{i=1}^{m} \pi_i^* q_i(\theta \mid \tau_i^*), \tag{5.28}$$

and

$$\tau_i^* = \tau_i + t_n(x), \qquad \pi_i^* \propto (\pi_i) \prod_{j=i}^{n} f(x_j) \frac{K(\tau_i^*)}{K(\tau_i)}, \tag{5.29}$$

$$t_n(x) = \left\{ n, \sum_{j=1}^{n} h_1(x_j), \ldots, \sum_{j=1}^{n} h_k(x_j) \right\}. \tag{5.30}$$

Thus, the mixture posterior density is in the same family as the mixture prior density. Generally, any prior density for an exponential family parameter can be approximated arbitrarily closely by such a mixture (see also Dalal and Hall, 1983; Diaconis and Ylvisaker, 1985).

5.3.3 Data-Based Prior Distributions

A. Historical Priors

Suppose $x_1, \ldots, x_n$ are i.i.d. from the exponential distribution with density $f(x \mid \theta) = \theta \exp\{-\theta x\}$. You obtained this data set last month. Because you had very little information about θ, you decided to adopt the vague prior $g(\theta) \propto 1/\theta$. The implied posterior density for θ is:

$$h(\theta \mid \bar{x}) \propto \theta^{n-1} \exp\{-n\bar{x}\theta\}. \tag{5.31}$$

Next suppose that this month you obtained a new data set $y_1, \ldots, y_m$, that is, an i.i.d. data set from $f(x \mid \theta)$. What should you use as a prior to update your new belief about θ. The best information that you have about θ is given in Equation (5.31). So you decide to adopt $h(\theta \mid \bar{x})$ given in Equation (5.31) as your prior. Note that it is a natural conjugate prior with hyperparameters based upon the earlier data set (it is a data-based prior). Your implied updated posterior becomes

$$h^*(\theta \mid \bar{x}, \bar{y}) \propto \theta^{m+n-1} \exp\{-\theta(n\bar{x} + m\bar{y})\}, \tag{5.32}$$

a density in the same distribution family (gamma), but now it is based upon an equivalent set of $(m + n)$ observations. So the effect of your earlier experience is to base your posterior belief on the totality of data from last month and this month, and thereby effectively increase the sample size.

B. Sample Splitting Priors

In the absence of historical data and when a subjective prior distribution is sought, it is sometimes convenient to split the sample data into two parts: one part is used to form a prior distribution; the other part is used to form the likelihood function. The advantage of this approach is that a subjective prior distribution is obtained relatively easily in the absence of the availability of historical data. The disadvantages are that: (1) the sample size available to form a likelihood function is thereby reduced; (2)

Bayes' theorem is violated because the prior distribution is based upon the current data set. (The same problem arises when *empirical Bayes estimation* approaches, discussed later in connection with Bayes estimation procedures, are used.) It is also not clear how to partition the data set so as to make the partitioning optimal. However, the procedure works and yields useful results in many situations.

The splitting of the sample into two parts is also suggested in a different context, namely, when very little is known *a priori* and the analyst is searching for an objective prior distribution. These two situations are quite different and should not be confused. In this latter situation, the analyst will compute an *intrinsic Bayes factor* for use in comparing models (discussed in our section on *Bayes factors*); see also, Berger and Pericchi, 1993. In connection with intrinsic Bayes factors, the sample is split into two parts one of which is used for a training sample, to obtain proper posterior densities for the parameters under each model, after starting with an improper vague prior. A certain minimal size subset of the data will be required to achieve such proper posterior densities. The remaining part of the sample is used for discriminating among models. That is, by using the proper posteriors obtained, and the remaining data, to compute Bayes factors, you can use the resulting Bayes factor to determine which model corresponds to the maximum posterior probability.

5.3.4 *g*-Prior Distributions

A variant of the natural conjugate prior family that is sometimes used in regression problems is the "g-prior" distribution family (see Zellner, 1986). This family takes into account the form of the experiment used to generate the observational data. Suppose

$$(\mathbf{y} \mid \mathbf{X}) = \mathbf{X}\boldsymbol{\beta} + \mathbf{u} \tag{5.33}$$

denotes a univariate regression, where $\mathbf{y}$: (n $\times$ 1) denotes the vector of dependent variable observations, $\boldsymbol{\beta}$: $(k \times 1)$ denotes an unknown regression coefficient vector, $\mathbf{u}$: $(n \times 1)$ denotes an n-vector of disturbance terms, with $\mathrm{E}(\mathbf{u}) = 0$, $\mathrm{var}(\mathbf{u}) = \sigma^2 \boldsymbol{I}_n$, $\boldsymbol{I}_n$ denotes the identity matrix of order n, and $\mathbf{X}$: $(n \times k)$ denotes a matrix of explanatory (independent) variables fixed by the design of the experiment generating the $\mathbf{y}$ vector. We adopt the g-prior distribution family

$$g(\beta, \sigma) = g_1(\beta \mid \sigma) g_2(\sigma), \tag{5.34}$$

where $(\beta \mid \sigma)$ is taken *a priori* to be normally distributed, and a vague prior is adopted for σ. That is,

$$g_1(\beta \mid \sigma) \propto \frac{1}{\sigma^k} \exp\left\{\left(-\frac{g}{2\sigma^2}\right)(\beta - \bar{\beta})(\mathbf{X}'\mathbf{X})(\beta - \bar{\beta})\right\}, \tag{5.35}$$

and

$$g_2(\sigma) \propto \frac{1}{\sigma}, \tag{5.36}$$

where $(g, \bar{\beta})$ denote hyperparameters that must be assessed, and $g > 0$.

Information about the experiment is incorporated in $g_1(\beta \mid \sigma)$ through $(\mathbf{X}'\mathbf{X})$. Inferences about β or σ in the above regression may be made from their marginal posterior densities based upon this prior. A procedure for assessing g based upon a "conceptual sample" may be found in Zellner (1986).

5.3.5 Stable Estimation Prior Distributions

Savage (1962) noted that by Bayes' theorem, the posterior density depends only upon the product of the likelihood function and the prior. Therefore, he suggested that to be vague about θ we should take the prior on θ to be uniform over the range where the likelihood function is non-negligible. Outside of this range, we can take the prior to be any convenient smooth function. It will not matter which function because the posterior density will be close to zero outside of the non-negligible interval, regardless.

5.3.6 Assessing Fractiles of Your Subjective Prior Probability Distribution

Suppose I want to assess individual fractiles of your subjective prior distribution for an unobservable parameter θ, a scalar. In this case I want to assess more than just the hyperparameters of a pre-assigned family of distributions that might be mathematically convenient. Below we suggest a four-step procedure that is useful in simple situations for assessing individual fractiles of your subjective prior distribution.

Assessment Steps

1. I ask you to give me a value $\theta_{0.5}$ such that, in your view, it is equally likely that $\theta > \theta_{0.5}$, or that $\theta < \theta_{0.5}$. Note that $\theta_{0.5}$ is the median of your prior distribution.
2. Now suppose that I tell you that $\theta > \theta_{0.5}$. Then, I ask you to specify a value $\theta_{0.75} > \theta_{0.5}$ such that it is equally likely that $\theta > \theta_{0.75}$, or that $\theta < \theta_{0.75}$. It is easy to check that $\theta_{0.75}$ is the 75th percentile of your prior distribution.
3. We can repeat Step 2 for other fractiles greater than 50 percent. Then suppose I tell you that $\theta < \theta_{0.5}$, and I ask you to specify a value $\theta_{0.25}$, and so on. In this manner we can assess any number of fractiles of your prior distribution.

Combine the assessed fractiles by connecting them on a graph to form a smooth cdf curve.

This procedure provides an assessment of your subjective prior probability distribution for θ, without assuming that your prior distribution belongs to any specific family.

5.4 PRIOR DISTRIBUTIONS FOR VECTOR AND MATRIX PARAMETERS

5.4.1 Vague Prior Distributions for Parameters on $(-\infty, \infty)$

In this section, we develop the notion of *knowing little* for unknown vector and matrix parameters, just as it was developed for one-dimensional parameters in Section 5.3.

Suppose that θ denotes a $p \times 1$ column vector of unknown parameters that index a data distribution. Suppose that for all i, $-\infty < \theta_i < \infty$. We will call $g(\theta)$ the density of a *vague prior distribution* for $\theta \equiv (\theta_i)$ if:

$$g(\theta) = \prod_{i=1}^{p} g_i(\theta_i), \qquad g_i(\theta_i) \propto \text{constant},$$

and

$$g(\theta) \propto \text{constant}. \tag{5.37}$$

The result in Equation (5.37) generalizes the single parameter (univariate) case given in Equation (5.10) to the vector or matrix (multivariate) case, whose components are defined on the entire real line. For example, θ might correspond to the mean vector of a multivariate normal distribution.

5.4.2 Vague Prior Distributions for Parameters on $(0, \infty)$

Next suppose that $\mathbf{X} \equiv (X_i)$ denotes a p-dimensional random variable whose components are independent, and whose variances are given by $\text{var}(X_i) = \sigma_i^2$, $0 < \sigma_i^2 < \infty$. Then, if $\mathbf{D} \equiv \text{diag}(\sigma_1^2, \ldots, \sigma_p^2)$, denotes a $p \times p$ diagonal matrix whose ith diagonal element is given by σ_i^2, and if the variances are mutually independent, a *vague prior density* for the parameter D is given by:

$$g(\mathbf{D}) = g(\sigma_1^2, \ldots, \sigma_p^2) = \prod_{i=1}^{p} g_i(\sigma_i^2) \propto \frac{1}{(\sigma_1^2) \cdots (\sigma_p^2)}, \tag{5.38}$$

or

$$g(\mathbf{D}) \propto \frac{1}{|\mathbf{D}|}, \qquad \mathbf{D} > 0. \tag{5.39}$$

In Equation (5.39), $\mathbf{D} > 0$ means that the matrix $\mathbf{D}$ is positive definite (its latent roots are positive), and $|\mathbf{D}|$ denotes the determinant of the matrix $\mathbf{D}$. This result generalizes that given for the single parameter (univariate) case in Equation (5.13) to the vector or matrix (multivariate) case, whose components are positive, defined on the semi-infinite line $(0, \infty)$.

The notion of a vague prior distribution is now developed for a full covariance matrix, $\Sigma : (p \times p)$. First note that since Σ is symmetric ($\Sigma = \Sigma'$), there are p distinct elements in the first row, only $(p - 1)$ distinct elements remaining in the second row, only $(p - 2)$ distinct elements remaining in the third row, and so on. Hence, by addition over the rows, there are a total of $p(p+1)/2$ distinct elements in the matrix, or $(p+1)/2$ groups of p distinct elements, and the result for each group of p elements is given by Equation (5.39). The generalization of Equation (5.39) for a full covariance matrix Σ is given in terms of the *generalized variance*, $| \Sigma |$. Knowing little (vagueness) about the covariance matrix Σ is expressed in terms of the vague prior density for Σ:

$$g(\Sigma) \propto \frac{1}{| \Sigma |^{(p+1)/2}}, \qquad \Sigma > 0. \tag{5.40}$$

For example, we will adopt this prior density for the covariance matrix in a multivariate normal distribution where $\Sigma > 0$. The exponent of $|\Sigma|$ given here was first suggested by Geisser and Cornfield, 1963.

5.4.3 Jeffreys' Invariant Prior Distribution: Objective Bayesian Inference in the Normal

Jeffreys (1961) suggested that a reasonable criterion for developing prior distributions should imply that probabilities made about the observable random variables should remain invariant under changes in the parameterization of the problem. To satisfy this criterion, Jeffreys showed that a random vector parameter $\boldsymbol{\delta}$: $(p \times 1)$, or a random matrix parameter δ: $(p \times p)$ must be given a prior density of the form:

$$g(\boldsymbol{\delta}) \propto | \mathbf{J} |^{0.5}, \tag{5.41}$$

where $\mathbf{J} \equiv (J_{ij})$ denotes the square *Fisher Information Matrix* associated with the likelihood function for the data. That is, if $f(\mathbf{x}_1, \ldots, \mathbf{x}_n \mid \boldsymbol{\delta})$ denotes a likelihood function for the possibly p-dimensional data $\mathbf{x}_1, \ldots, \mathbf{x}_n$, where $\mathbf{x}_i$ denotes a possibly p-dimensional vector of data for each i (p might be equal to 1), the components of J are defined by:

$$J_{ij} \equiv -E\left[\frac{\partial^2 \log f(\mathbf{x}_1, \ldots, \mathbf{x}_n \mid \boldsymbol{\delta})}{\partial\delta_i \partial\delta_j}\right], \qquad i, j = 1, \ldots, p. \tag{5.42}$$

For example, if the data are a sample from a multivariate normal distribution with mean vector $\boldsymbol{\theta}$ and covariance matrix Σ, and $\boldsymbol{\theta}$ and Σ are *a priori* independent, with $\delta \equiv (\boldsymbol{\theta}, \Sigma)$, it is straightforward to find that the Jeffreys' invariant prior is given by the vague prior density given in Equations (5.37) and (5.40) (see, for example, Press, 1982, pp. 79–80), so that

$$g(\boldsymbol{\delta}) \equiv g(\boldsymbol{\theta}, \Sigma) \propto \frac{1}{| \Sigma |^{(p+1)/2}}. \tag{5.43}$$

In this example, we have formed the information matrix separately, first for $\boldsymbol{\theta}$, and then Σ, and then multiplied the (assumed) independent prior densities.

When the information matrix is not diagonal, Jeffreys suggests applying the invariance criterion to each scalar parameter separately, one at a time, assuming the others are constant, in order to obtain a reasonable prior density result. (Recall the discussion in Section 2.2.3 about the inconsistencies in Jeffreys' logical probability approach).

REMARK 1: It is interesting to note that the Jeffreys' invariant prior distribution leads to the same results we obtain for a vague prior distribution, the latter of which is the prior distribution that corresponds to the notion of knowing little. But the Jeffreys' prior was not derived using the notion of knowing little. It was derived merely requiring invariance of probabilistic inferences made about the data. The Jeffreys' prior corresponds to an objective prior distribution for both univariate and multivariate distributions.

REMARK 2: When there are no nuisance parameters, and regularity conditions hold, the *reference prior distribution* suggested by Bernardo (1979), and Berger and Bernardo (1989) reduce to the Jeffreys' prior.

Example 5.5 Univariate Normal Data (Both Parameters Unknown)

A. Vague Prior Density. In Example 4.5 we examined a situation in which the sample data were normally distributed, but with known common variance so that there was just one unknown parameter. Now examine the more common two-parameter case in which the data are again normally distributed, but both the mean and variance are unknown. Accordingly, suppose that $x_1, \ldots, x_n$ denotes a sample of data that are i.i.d. $N(\theta, \sigma^2)$. Find the posterior distribution corresponding to a vague prior distribution.

The joint density function of the data is given by

$$f(x_1, \ldots, x_n \mid \theta, \sigma^2) = \prod_{i=1}^{n} f(x_i \mid \theta^2) = \prod_{i=1}^{n} \left[\frac{1}{(2\pi\sigma^2)^{0.5}} \exp\left\{ \left(-\frac{1}{2\sigma^2} \right) (x_i - \theta)^2 \right\} \right], \tag{5.44}$$

so the likelihood function is given by

$$L(\theta, \sigma^2) \propto \frac{1}{(\sigma^2)^{0.5n}} \exp\left\{ \left(-\frac{1}{2\sigma^2} \right) \sum_{i=1}^{n} (x_i - \theta)^2 \right\}. \tag{5.45}$$

From Equations (5.10) and (5.13), and *a priori* independence of the parameters, a vague prior density for (θ, σ^2) is given by

$$g(\theta, \sigma^2) \propto \frac{1}{\sigma^2}. \tag{5.46}$$

By Bayes' theorem, the joint posterior density of (θ, σ^2) is given by

$$h(\theta, \sigma^2 \mid \text{data}) \propto \frac{1}{(\sigma^2)^{0.5(n+2)}} \exp\left\{ \left(-\frac{1}{2\sigma^2} \right) \sum_{i=1}^{n} (x_i - \theta)^2 \right\}. \tag{5.47}$$

Usually, interest focuses on inferences about θ. We can make marginal posterior inferences about θ by integrating Equation (5.47) with respect to σ^2. That is, the marginal posterior density of θ is given by:

$$h_1(\theta \mid \text{data}) \equiv \int_0^\infty h(\theta, \sigma^2 \mid \text{data}) d\sigma^2 \tag{5.48}$$

and substituting Equation (5.47) into Equation (5.48) and integrating by substituting

$$t = \frac{1}{\sigma^2} \sum_{i=1}^n (x_i - \theta)^2,$$

and noting that the remaining integral is just a constant that does not depend on θ (absorbing it into the proportionality constant), gives

$$h_1(\theta \mid \text{data}) \propto \frac{1}{\left\{ \sum_{i=1}^n (x_i - \theta)^2 \right\}^{0.5n}}. \tag{5.49}$$

Now note that $\bar{x}$ denotes the sample mean, and $s^2 = \frac{1}{n} \sum_{i=1}^n (x_i - \bar{x})^2$ denotes the sample variance. Since

$$\sum_{i=1}^n (x_i - \theta)^2 = \sum_{i=1}^n [(x_i - \bar{x}) + (\bar{x} - \theta)]^2 = \sum_{i=1}^n (x_i - \bar{x})^2 + n(\theta - \bar{x})^2, \tag{5.50}$$

and since the cross-product term vanishes, the marginal posterior density of θ may be written as

$$h_1(\theta \mid \text{data}) \propto \frac{1}{\left\{ 1 + \left(\frac{\theta - \bar{x}}{s} \right)^2 \right\}^{0.5\{(n-1)+1\}}}, \tag{5.51}$$

which is the kernel of a Student t-density, centered at $\bar{x}$, with $(n - 1)$ degrees of freedom. So marginal posterior inferences about θ may be made from the Student t-distribution in Equation (5.51).

By integrating Equation (5.47) with respect to θ (a simple normal density integral), we obtain the marginal posterior density of σ^2. The result is an inverted gamma distribution. That is, $(1/\sigma^2)$, *a posteriori*, follows a gamma distribution. That result may be used to make marginal posterior inferences about σ^2. The marginal posterior density of σ^2 is given by

$$h_2(\sigma^2 \mid \text{data}) \propto \frac{\exp\{-(ns^2)/(2\sigma^2)\}}{(\sigma^2)^{0.5(n+1)}}. \tag{5.52}$$

B. Jeffreys' Prior Density. The prior density used in this example to find the implied posterior density for normally distributed data was based in Section A upon using a vague belief, as expressed as a limit of a uniform prior density over a finite range for θ as the finite limits went off to infinity, and for similar considerations for $\log \sigma$. However, we examine in the sequel what the Jeffreys' prior might be in this situation, and thereby, also see how to evaluate Jeffreys' prior distributions.

Recall that if $\boldsymbol{\delta} \equiv (\theta, \sigma^2)$ the Jeffreys' prior density is given by $g(\boldsymbol{\delta}) = g(\theta, \sigma^2) \propto \mid \mathbf{J} \mid^{0.5}$, where $\mathbf{J} = (J_{ij})$ denotes the 2×2, symmetric Fisher Information Matrix, and

$$\begin{aligned} J_{11} &= -E\left[\frac{\partial^2}{\partial \theta^2} \log f(x_1, \ldots, x_n \mid \theta, \sigma^2)\right], \\ J_{12} &= -E\left[\frac{\partial^2 \log f(x_1, \ldots, x_n \mid \theta, \sigma^2)}{\partial \theta \partial(\sigma^2)}\right], \\ J_{22} &= -E\left[\frac{\partial^2 \log f(x_1, \ldots, x_n \mid \theta, \sigma^2)}{\partial(\sigma^2)^2}\right]. \end{aligned} \tag{5.53}$$

Note that $J_{12} = J_{21}$. Since in this example,

$$f(x_1, \ldots, x_n \mid \theta, \sigma^2) = \frac{1}{(\sigma^2)^{0.5n}} \exp\left\{-\left(\frac{1}{2\sigma^2}\right) \sum_{i=1}^{n} (x_i - \theta)^2\right\} \tag{5.54}$$

$$\log f(x_1, \ldots, x_n \mid \theta, \sigma^2) = (-0.5n) \log \sigma^2 - \left(\frac{1}{2\sigma^2}\right) \sum_{i=1}^{n} (x_i - \theta)^2. \tag{5.55}$$

Therefore, we may readily find that $J_{11} = \frac{1}{\sigma^2}, J_{12} = 0, J_{21} = 0, J_{22} = \frac{n}{2\sigma^4}$. So $\mid \mathbf{J} \mid = \frac{n}{2\sigma^6}$, and

$$g(\theta, \sigma^2) \propto \frac{1}{\sigma^3}. \tag{5.56}$$

This is the result obtained for Jeffreys' invariant prior density without treating the two parameters separately (which Jeffreys himself warns against doing; that is, Jeffreys warns against treating the parameters jointly, even though we just did that obtaining the result in Equation (5.56) to see what we would get, for illustrative purposes).

Now treat the parameters separately. Suppose that σ^2 is a known constant, although θ is actually unknown. In this case, J is a scalar and $J = J_{11} =$ constant. Then Jeffreys' prior density becomes:

$$g(\theta) \propto \text{constant.} \tag{5.57}$$

Next suppose that θ is a known constant, but σ^2 is unknown. Then, $J = J_{22} = 1/(2\sigma^4)$, and the associated Jeffreys' prior becomes:

$$g(\sigma^2) \propto \frac{1}{\sigma^2}. \tag{5.58}$$

Combining terms for the parameter by parameter case, assuming θ and σ^2 are independent, gives:

$$g(\theta, \sigma^2) \propto \frac{1}{\sigma^2}. \tag{5.59}$$

So we see that at least for the general univariate normal distribution (both parameters unknown), *evaluating the Jeffreys' prior density for one parameter at a time gives the same result as does the vague prior argument* (Equation 5.46).

Example 5.6 Multivariate Normal Data (Both Parameters Unknown)

In this example we seek the posterior distribution for the case of a multivariate normal data distribution under a vague prior distribution. Suppose that $(\mathbf{X}_1, \ldots, \mathbf{X}_n)$ are an i.i.d. sample of p-dimensional vectors that follow the p-dimensional normal distribution: $N(\boldsymbol{\theta}, \Sigma)$. The data density function is:

$$f(\mathbf{x}_1 \ldots, \mathbf{x}_n \mid \boldsymbol{\theta}, \Sigma) = \frac{1}{(2\pi)^{0.5np} \mid \Sigma \mid^{0.5n}} \exp\left\{(-0.5) \sum_{i=1}^{n} (\mathbf{x}_i - \theta)' \Sigma^{-1} (\mathbf{x}_i - \boldsymbol{\theta})\right\}. \tag{5.60}$$

The log density is therefore

$$\begin{aligned} L &\equiv \log f(\mathbf{x}_1, \ldots, \mathbf{x}_n \mid \boldsymbol{\theta}, \Sigma) \\ &= \log(2\pi)^{0.5np} - (0.5n) \log \mid \Sigma \mid - \left\{0.5 \sum_{i=1}^{n} (\mathbf{x}_i - \boldsymbol{\theta})' \Sigma^{-1} (\mathbf{x}_i - \boldsymbol{\theta})\right\}. \end{aligned} \tag{5.61}$$

We will develop the Jeffreys' invariant prior density by applying the procedure to one matrix parameter at a time.

First assume that Σ is a known constant matrix. Then, using vector differentiation (see for example, Press, 1982, Chapter 2) and noting that the derivative of a scalar L

with respect to a vector $\boldsymbol{\theta}$ is the vector $\partial L/\partial \boldsymbol{\theta}$ of derivatives of L with respect to each element of the vector $\boldsymbol{\theta} = (\boldsymbol{\theta}_i)$, $i = 1, \ldots, p$:

$$\frac{\partial L}{\partial \boldsymbol{\theta}} \equiv \left(\frac{\partial L}{\partial \theta_i}\right) = -n\Sigma^{-1}(\boldsymbol{\theta} - \bar{\mathbf{x}}). \tag{5.62}$$

Similarly,

$$-E\left\{\frac{\partial^2 L}{\partial \boldsymbol{\theta} \partial \boldsymbol{\theta}'}\right\} = n\Sigma^{-1} = \text{constant}. \tag{5.63}$$

So the information matrix is constant, and the Jeffreys' invariant prior for $\boldsymbol{\theta}$ is given by:

$$g(\boldsymbol{\theta}) \propto \text{constant}. \tag{5.64}$$

Now assume $\boldsymbol{\theta}$ is a known constant vector. Write L in the more convenient form:

$$\begin{aligned} L &\equiv \log f(\mathbf{x}_1, \ldots, \mathbf{x}_n \mid \boldsymbol{\theta}, \Sigma) \\ &= \log(2\pi)^{-0.5np} + (0.5n) \log \mid \Lambda \mid -0.5 \operatorname{tr} \Lambda \left\{ \sum_{i=1}^{n} (\mathbf{x}_i - \boldsymbol{\theta})(\mathbf{x}_i - \boldsymbol{\theta})' \right\}, \end{aligned} \tag{5.65}$$

where "tr" denotes the trace of a square matrix (the sum of its diagonal elements), and we have reparameterized in terms of the precision matrix $\Lambda \equiv \Sigma^{-1}$. Forming the first two derivatives of L with respect to $\Lambda \equiv (\lambda_{ij})$ gives (see, for example, Press, 1982, p. 79 for details)

$$\begin{aligned} \frac{\partial L}{\partial \Lambda} &= n\Lambda^{-1} - 0.5\text{diag}(\Lambda^{-1}) - \left\{ 0.5 \sum_{i=1}^{n} (\mathbf{x}_i - \boldsymbol{\theta})(\mathbf{x}_i - \boldsymbol{\theta})' \right\}, \\ \frac{\partial^2 L}{\partial \lambda_{ij} \partial \lambda_{kl}} &\propto \frac{1}{\mid \Lambda \mid^{(p+1)}}. \end{aligned} \tag{5.66}$$

Taking expectations in the second derivatives with respect to the data vectors leaves the result unchanged. So the Jeffreys' invariant prior density for Λ is given by:

$$g(\Lambda) \propto \frac{1}{\mid \Lambda \mid^{0.5(p+1)}}. \tag{5.67}$$

Since the Jacobian from Λ to Σ is $\mid \Sigma \mid^{-(p+1)}$, the implied invariant prior density for Σ is:

$$g(\Sigma) \propto \frac{1}{\mid \Sigma \mid^{0.5(p+1)}}. \tag{5.68}$$

Finally, if θ and Σ are, *a priori*, independent, the invariant prior density implied by the Jeffreys' argument is:

$$g(\theta, \Sigma) \propto \frac{1}{| \Sigma |^{0.5(p+1)}}. \tag{5.69}$$

Equation (5.69) gives the multivariate generalization of the vague (and identical Jeffreys' prior density developed for the univariate case in Equation (5.59). It is the objective prior density most frequently used in the multivariate case to express knowing little, or vagueness. There are no hyperparameters to assess.

It would be interesting to examine the case of natural conjugate families of prior distributions for $(\boldsymbol{\theta}, \Sigma)$ in the multivariate situation. Such discussion is somewhat beyond the scope of this volume so the reader is referred to sources such as Press (1982).

5.4.4 Assessment of a Subjective Prior Distribution for a Group

You are a decision maker who would like to be helped in making a meaningful judgment. There are several correlated observable variables as well as several correlated unobservables. How might you generate multivariate subjective probability prior distributions for the unknown unobservables? We suggest that you use the combined judgments of an informed group to assess a prior distribution for the unobservables and then use Bayes' theorem to make posterior inferences conditional on the observables. Some references to other approaches to combining opinions of a group (usually, a group of experts) may be found at the end of this chapter, in the section on Further Reading.

In this section:

1. We propose a model for grouping opinions;
2. We describe elicitation procedures, based upon the model, for assessing a multivariate prior density;
3. We show how to merge the points assessed into a density;
4. We report the results of an empirical application involving assessing a prior probability density for the event of nuclear war;
5. Finally, we summarize the biases introduced into the subjective probability assessments by psychological factors.

A more detailed discussion of this problem, including other approaches, can be found in Press (1985a).

Multvariate Subjective Assessment for a Group

People cannot easily think in many dimensions simultaneously. For example, they are overwhelmed by being asked for their subjective probability that a random variable X_1 lies in a given interval (a_1, b_1), and simultaneously, a second random variable X_2 lies in another interval (a_2, b_2). It is much harder, of course, if we ask for their joint probability for p variables, and require:

$$P\{a_1 < X_1 < b_1, a_2 < X_2 < b_2, \ldots, a_p < X_p < b_p\}.$$

One of the most important cutting-edge problems of Bayesian inference is how to improve our methods of assessing prior distributions, especially multivariate prior distributions. While the approach we propose here will not be applicable in all situations, it should provide a useful solution in many problems. The approach we propose here is applicable to assessment of both univariate and multivariate prior distributions.

Assessment Overview for a Group

Instead of assessing the fractiles of your prior distribution, we propose to combine the opinions of N individuals, treating each opinion as a sample point from some common underlying distribution, and substitute the combined judgments of the group for your prior. This is just a formalization of the old idea of using a consensus view of informed individuals to express a judgment about an important proposition.

In the multivariate case, every opinion of a subject in the group is a p-dimensional vector. Thus, each subject can be asked for his/her p responses (correlated) to each of p questions. (A questionnaire with p items could be used.)

Model for a Group

Suppose that $\boldsymbol{\theta}$: $(p \times 1)$ is a continuous, unobservable, random vector denoting the mean of some p-dimensional sampling distribution with density $f(\mathbf{z} \mid \boldsymbol{\theta})$. You would like to assess your prior probability density for $\boldsymbol{\theta}$, and you would like your prior density to reflect the combined judgments of some informed population. We assume the members of the informed population are "experts," in that they possess contextual knowledge. We take a sample from the population of experts and ask each subject for a "best guess" opinion about $\boldsymbol{\theta}$. Let $\boldsymbol{\theta}_j$: $(p \times 1)$ denote the vector of opinions about $\boldsymbol{\theta}$ given by the jth subject, for $j = 1, \ldots, N$.

Note: We must be careful to ask each expert for his/her opinion vector privately and independently of all other experts. Then we can guarantee that the $\boldsymbol{\theta}_j$s are mutually independent (but the assessments within a $\boldsymbol{\theta}_j$ vector are all correlated, because they are responses to related questions by the same individual). This approach was generalized to group assessment in sample surveys involving factual recall, where the θ_j vectors were not assumed to be identically distributed, but followed a hierarchical model (see Chapter 13). There, respondents are asked for upper and lower bounds on their assessments as well as for basic "best guess" responses for the values they are attempting to recall. The approach is called respondent-generated intervals (RGI); see Press (2002).

Multivariate Density Assessment for a Group

Assume that there are N assessments of the expert panel, $\boldsymbol{\theta}_1, \ldots, \boldsymbol{\theta}_N$, all $(p \times 1)$, and they are all independent and identically distributed, all with density $g(\boldsymbol{\theta})$. To synthesize these N opinions, we adopt the nonparametric kernel density estimation approach of Parzen (1962) and Cacoullos (1966) (for discussion of general methodology of this subject, see, for example Taipia and Thompson (1982) or Silverman (1986); for discussion relevant to this application, see Press (1983) and the references therein). We permit the density of the combined opinions to assume the form

$$g_N(\boldsymbol{\theta}) = \frac{1}{N} \sum_{j=1}^{N} \phi_j(\boldsymbol{\theta}), \tag{5.70}$$

where

$$\phi_j(\boldsymbol{\theta}) = \frac{1}{\delta^p(N)} K\left(\frac{\boldsymbol{\theta} - \boldsymbol{\theta}_j}{\delta(N)}\right). \tag{5.71}$$

$K(\boldsymbol{\theta})$ is a kernel chosen to satisfy suitable regularity conditions, and $\delta(N)$ is a sequence of positive constants satisfying:

$$\lim_{N\to\infty} \delta(N) = 0; \qquad \lim_{N\to\infty} N\delta^p(N) = \infty. \tag{5.72}$$

So far, the particular kernel is not specified. Moreover, this approach to density estimation should be fairly insensitive to the specific kernel selected. In the sequel, for specificity, we adopt the normal density kernel.

Normal Density Kernel

We adopt the normal density kernel

$$K(\boldsymbol{\theta}) = \frac{1}{(2\pi)^{0.5p}} \exp(-0.5\boldsymbol{\theta}'\boldsymbol{\theta}), \tag{5.73}$$

and we take

$$\delta(N) = \frac{C}{N^{1/(p+1)}}, \tag{5.74}$$

where C is any preassigned constant (fixed in any given sample to smooth the density). Thus, if $\mathbf{X}_j$ has density $\phi_j(\mathbf{x})$, we have

$$\mathbf{X}_j \sim N[\boldsymbol{\theta}_j, \delta^2(N)\mathbf{I}_p] \tag{5.75}$$

Moreover,

$$g_N(\boldsymbol{\theta}) = \frac{1}{C^p N^{1/(p+1)}(2\pi)^{0.5p}} \sum_{j=1}^{N} \exp\left\{-\frac{N^{2/(p+1)}}{2C^2}(\boldsymbol{\theta} - \boldsymbol{\theta}_j)'(\boldsymbol{\theta} - \boldsymbol{\theta}_j)\right\}. \tag{5.76}$$

Also, $g_N(\boldsymbol{\theta})$ is asymptotically unbiased, and

$$\lim_{N\to\infty} E[g_N(\boldsymbol{\theta}) - g(\boldsymbol{\theta})]^2 = 0, \tag{5.77}$$

at every point of continuity $\boldsymbol{\theta}$ of $g(\boldsymbol{\theta})$.

Summary of Group Assessment Approach

1. *Consensus of a group of experts is not required, as in some group procedures.* It is the diversity of opinions of the experts that is interesting. When a group of experts is given some battery of questions, and they are asked to respond independently, their degree of disagreement reflects the amount of inherent uncertainty there is in the underlying issue. Often the opinions in the tails of the distribution are the most interesting in terms of creative ideas.
2. *No need for an assumed functional form for the prior.* We do not use natural conjugate prior structures, or any other family of priors, and then assess hyperparameters. We merely let the actual empirical distribution of expert opinion determine the prior.
3. *Assessment ease.* It is not necessary for subjects to think about the likelihood of certain events occurring simultaneously, in order to assess a p-vector. We need only to administer a battery of p-questions to N people, and each person's p-responses are correlated. Subjects need to think only about one-dimensional (marginal) events. It becomes as easy to assess a p-dimensional prior as it is to assess a one-dimensional prior.
4. *Additional fractiles.* It is usually of interest to assess several fractiles of $g(\boldsymbol{\theta})$ from each subject (such as medians and quartile points, or means and variances, and so on). This second moment information reflects how certain each respondent is about his/her belief regarding $\boldsymbol{\theta}$. Such information could be used in a variety of ways. The "best guess" assessments of $\boldsymbol{\theta}$ could be weighted by precisions (reciprocal variances) provided by the respondents; high-variance assessments of $\boldsymbol{\theta}$ could be evaluated separately, or even ignored; and so on. In our empirical application of group assessment prior distribution procedures discussed below (in an application involving experts' opinions about the chances of nuclear war in the 1980s), we used 90 percent assessed fractiles with 5 percent assessments in each tail.
5. *Convergence to "truth".* If there is indeed "truth," in some absolute sense, increasing the size of the group will not necessarily cause convergence to truth. In groups with individuals whose opinions follow substantially different distributions, such convergence is unlikely. In groups with identically distributed opinions, convergence is guaranteed. In group situations "in between," it is hard to say.

Empirical Application of Group Assessment: Probability of Nuclear War in the 1980s

It was of interest in 1980 to know how experts in the United States felt about the chances of nuclear war between the United States and the Soviet Union during the 1980s. We decided that the group judgment of these experts could be developed by using the approach proposed in this section. A list was prepared containing the names of experts in strategic policy who were employed at The Rand Corporation in Santa Monica, California, or who were members of the California Arms Control

Seminar, a collection of experts in strategic arms control, military defense, and weapon systems analysis. A random sample of 149 experts was selected, and they were asked to complete written questionnaires. Each subject was asked in early 1981:

> Please express your subjective assessment that the United States or the Soviet Union will detonate a nuclear weapon in the other's homeland during the year 1981.

Denote the event of this occurrence by E, and its probability by $P\{E\}$. It was of concern that although all of the subjects in the group were substantive experts in strategic arms control, they might not all have been sufficiently quantitatively oriented to be able to express their views equally well from a subjective probability standpoint. We decided to ask them to express their views on the same question on several scales, and then to evaluate their consistency.

A. Ordinal Scale

First, subjects were asked to respond on an *ordinal scale* (by circling one + point on the scale below).

+ The probability is greater than the probability of getting a "head," by flipping a fair coin once. (The chance of getting a "head" is 0.5.)
+ The probability is less than the probability of getting a "head" by flipping a fair coin once, but is greater than the probability that a randomly selected state in the United States would be the state in which you live. (The actual chance that a randomly selected state would be yours is 2 percent.)
+ The probability is less than the probability that a randomly selected state would be the one in which you live but is greater than the probability of throwing boxcars (double sixes) in two successive throws of the dice. (The actual chance of throwing double boxcars is 1 in 1296.)
+ The probability is less than the probability of throwing double boxcars in two successive throws of the dice, but is greater than the probability that a randomly selected person in Los Angeles would be you if you lived in Los Angeles. (There are about 2 million people living in Los Angeles.)
+ The probability is less than the chance that a randomly selected person in Los Angeles would be you if you lived in Los Angeles, but is greater than the probability that a randomly chosen American will be you. (There are about 220 million Americans.)
+ The probability is less than the probability that a randomly chosen American will be you.

B. Linear Scale

Subjects were then asked the same question, but they were asked to respond on a *linear scale* (by circling one "+" point on the scale shown in Table 5.2.

Table 5.2 Linear Scale

Percentage Probability (%)	Circle One	Absolute Probability	Word Prompt
100	+	1.00	Certainty
95	+	0.95	
90	+	0.90	Very likely
85	+	0.85	
80	+	0.80	
75	+	0.75	Somewhat likely
70	+	0.70	
65	+	0.65	
60	+	0.60	
55	+	0.55	
50	+	0.50	Chances about even
45	+	0.45	
40	+	0.40	
35	+	0.35	
30	+	0.30	
25	+	0.25	Somewhat unlikely
20	+	0.20	
15	+	0.15	
10	+	0.10	Very unlikely
5	+	0.05	
0	+	0.00	Impossible

C. Log Scale

Subjects were asked the same question again, but they were asked to respond on a *log-scale* (by circling one "+" point on the scale shown in Table 5.3.

Finally, subjects were asked to respond to the same question, and to answer on a log scale, but for *all of the years separately* in the 1980s (see Table 5.4).

Consistency of Response

Subject's responses on the three different scales were compared for consistency. When the subject was found to be inconsistent on at least one scale (that is, the subject's answers on that scale differed from his/her answers on the other scales), that subject was eliminated from the panel. This was done to avoid confounding the effects of inadequate quantitative training with substantive degree of belief. Of the subjects, 107 remained; these subjects were experts who responded consistently on all three scales. In Figure 5.2 we see the mean assessed probability as a function of year in the next decade. We also see the mean 90 percent credibility bands (assessed).

Implications

1. The average (mean) assessed probability never exceeds about 3 percent.
2. The median assessed probability, in sharp contrast, never exceeds about 0.01 percent.

Table 5.3 Log Scale

Word Prompt	Probability	Circle One	The Probability of:
A sure thing	1.0	+	
		+	
	1/3	+	
		+	
A 10% chance	1/10	+	A randomly chosen American being Black
		+	
	1/30	+	
		+	
A 1% chance	1/100	+	
		+	
	1/300	+	A randomly chosen American male age 35–44 years dying during a one-year period
		+	
One chance in a thousand	1/1000	+	
		+	
	1/3000	+	
		+	A randomly chosen American being killed in a motor vehicle accident during a one-year period
	1/10,000	+	
	1/30,000	+	
		+	
	1/100,000	+	Getting all "heads" while flipping a fair coin 17 times
		+	
	1/300,000	+	
		+	
One chance in a million	1/1,000,000	+	
		+	Being killed in an accident during a single scheduled domestic airline flight
	1/3,000,000	+	
		+	
	1/10,000,000	+	
		+	Selecting a given Californian from among all Californians
	1/30,000,000	+	
		+	
	1/100,000,000	+	
		+	A randomly chosen American will be you
	1/300,000,000	+	
		+	
One chance in a billion	1/1,000,000,000	+	
		+	
Less than one chance in a billion	< 1/1,000,000,000	+	

Table 5.4 Log Scale For All Years Separately

	Year									
Probability	1981	1982	1983	1984	1985	1986	1987	1988	1989	1990
1	+	+	+	+	+	+	+	+	+	+
	+	+	+	+	+	+	+	+	+	+
1/3	+	+	+	+	+	+	+	+	+	+
	+	+	+	+	+	+	+	+	+	+
1/10	+	+	+	+	+	+	+	+	+	+
	+	+	+	+	+	+	+	+	+	+
1/30	+	+	+	+	+	+	+	+	+	+
	+	+	+	+	+	+	+	+	+	+
1/100	+	+	+	+	+	+	+	+	+	+
	+	+	+	+	+	+	+	+	+	+
1/300	+	+	+	+	+	+	+	+	+	+
	+	+	+	+	+	+	+	+	+	+
1/1000	+	+	+	+	+	+	+	+	+	+
	+	+	+	+	+	+	+	+	+	+
1/3000	+	+	+	+	+	+	+	+	+	+
	+	+	+	+	+	+	+	+	+	+
1/10,000	+	+	+	+	+	+	+	+	+	+
	+	+	+	+	+	+	+	+	+	+
1/30,000	+	+	+	+	+	+	+	+	+	+
	+	+	+	+	+	+	+	+	+	+
1/100,000	+	+	+	+	+	+	+	+	+	+
	+	+	+	+	+	+	+	+	+	+
1/300,000	+	+	+	+	+	+	+	+	+	+
	+	+	+	+	+	+	+	+	+	+
1/1,000,000	+	+	+	+	+	+	+	+	+	+
	+	+	+	+	+	+	+	+	+	+
1/3,000,000	+	+	+	+	+	+	+	+	+	+
	+	+	+	+	+	+	+	+	+	+
1/10,000,000	+	+	+	+	+	+	+	+	+	+
	+	+	+	+	+	+	+	+	+	+
1/30,000,000	+	+	+	+	+	+	+	+	+	+
	+	+	+	+	+	+	+	+	+	+
1/100,000,000	+	+	+	+	+	+	+	+	+	+
1/300,000,000	+	+	+	+	+	+	+	+	+	+
	+	+	+	+	+	+	+	+	+	+
1/1,000,000,000	+	+	+	+	+	+	+	+	+	+
less than 1/1,000,000,000	+	+	+	+	+	+	+	+	+	+

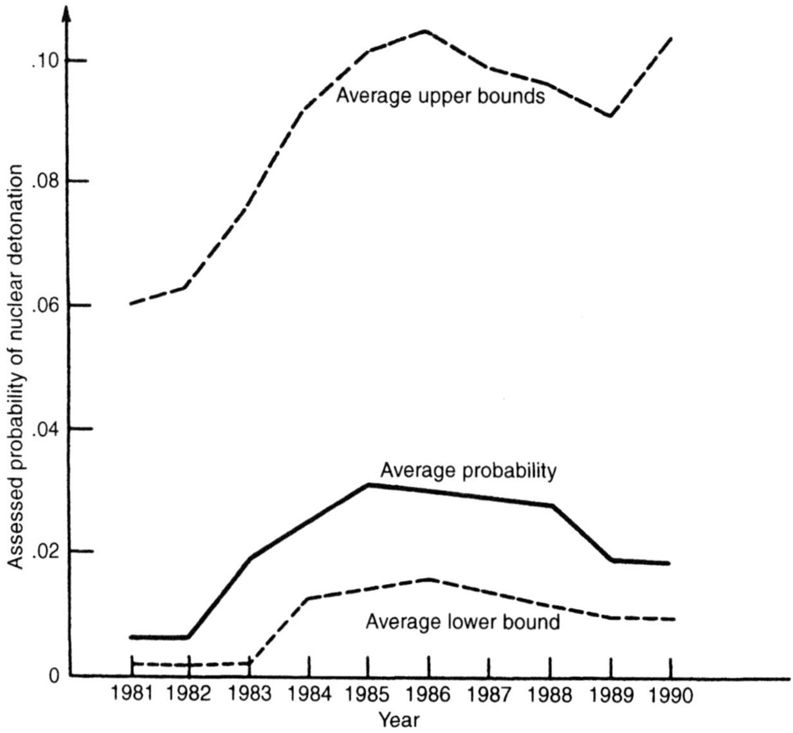

Figure 5.2

3. If the estimates of the probabilities for each year were independent (which they are not), a probability of 0.03 (3 percent) per year, for each year in the decade, would imply, using the binomial theorem, that the probability of war in the decade (i.e., the probability of exactly one success in 10 trials is about 22.8 percent).

Histogram

In Figure 5.3 we see a histogram for the year 1984.

Smoothed Prior Density (Fitted)

Define $\boldsymbol{\theta}_j$: (10×1) as the response vector for panelist j; each component of $\boldsymbol{\theta}_j$ is the assessed probability for each of the 10 years in the decade, from 1981 through 1990. We assume $\boldsymbol{\theta}_j$ is an observation of $\boldsymbol{\theta}$, a continuous random vector with density $g(\boldsymbol{\theta})$; $j = 1, \ldots, 107$. The empirical density of responses (the prior density) is given by

$$g_{107}(\boldsymbol{\theta}) = \frac{1}{C^{10}(107)^{1/11}} \sum_{j=1}^{107} \frac{\exp\left\{\left[-(107)^{2/11}/2C^2\right](\boldsymbol{\theta} - \boldsymbol{\theta}_j)'(\boldsymbol{\theta} - \boldsymbol{\theta}_j)\right\}}{(2\pi)^5}.$$

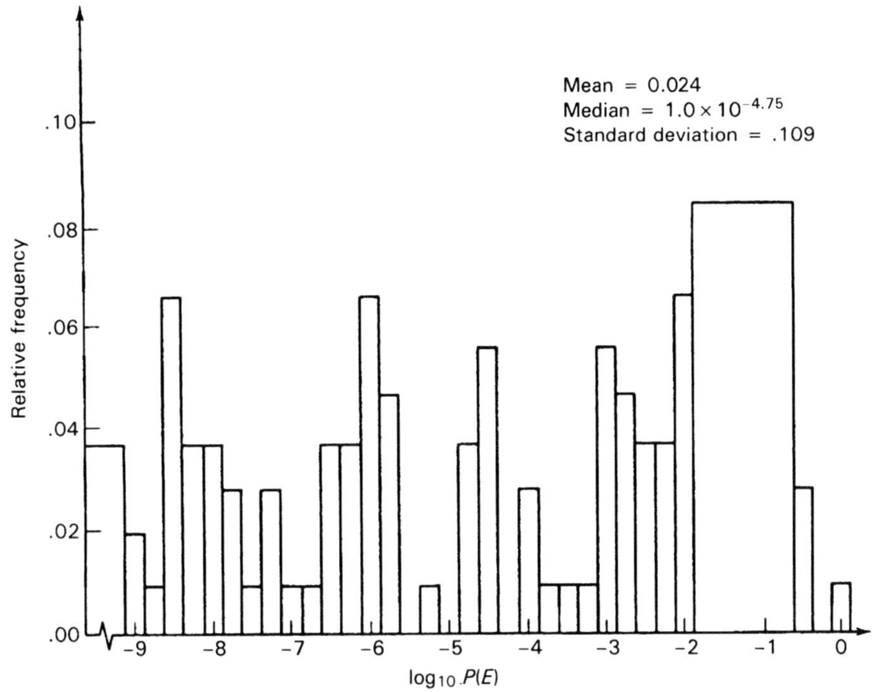

Figure 5.3 Histogram for assessed probability of nuclear detonation for 1984 ($N = 107$)

This is a 10-dimensional density. The one-dimensional marginal density for the ith component is given by

$$g_{107}(\theta^{(i)}) = \frac{1}{C(107)^{10/11}} \sum_{j=1}^{107} \frac{\exp\left\{\left[-(107)^{2/11}/2C^2\right]\left(\theta^{(i)} - \theta_j^{(i)}\right)^2\right\}}{\sqrt{2\pi}}.$$

A plot of this density is given in Figure 5.4. Note that the smoothing constant $C = 0.1$. This choice yielded a smooth well-behaved density on a scale that was readily interpretable. (The density became extremely noisy for smaller C, and it was overfitted for larger C.)

Qualitative Data Provided by Expert Panelists (Qualitative Controlled Feedback: Content Analysis, Ethnography)

Panelists were requested to give not only their opinions about the probability of nuclear war during each of the years of the decade, but also their *reasons* for the responses they gave. Table 5.5 lists the three most frequently given reasons for the event of nuclear war.

The qualitative reasons provided by the experts are an extremely important aspect of the subjective probability elicitation process. In some senses the qualitative

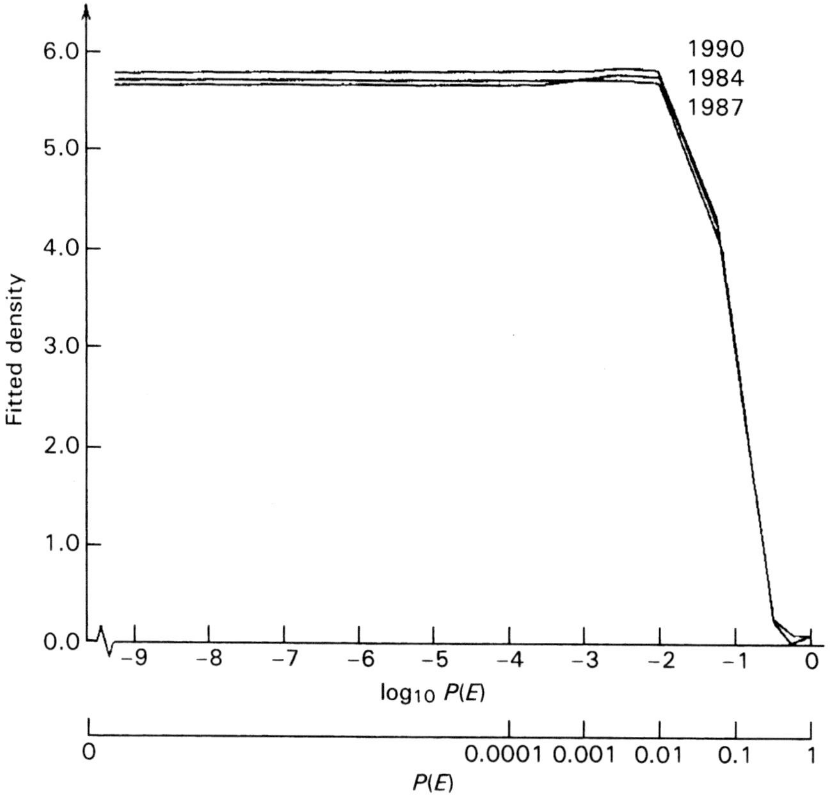

Figure 5.4 Fitted density for assessed probability of nuclear detonation in 1984, 1987 and 1990, $N = 107$ (log scale)

information processes provide more information than the quantitative probability assessments because they provide each expert's rationale, and thereby afford the analyst the opportunity to understand the entire process. An elicitation process called QCF, for *qualitative controlled feedback* (see Press 1978, 1980, 1985a; Press et al. 1979) was superimposed upon the elicitation procedure, to refine it. In the QCF process the questions are asked repeatedly over several stages.

At Stage one of the process, panelists are asked independently (by mail) for both their subjective probabilities for event E, and for their reasons for giving that probability. The independence of response afforded by a mail questionnaire permitted the respondents to answer without the social pressure of knowing who else was in the group. In each case their reasons are given in a paragraph of prose. Their subjective probabilities are combined into a group distribution, as described above. The next question is, how to combine their collections of reasons? This is an issue in content analysis. What is the content of each paragraph of prose? How should we merge the contents of them? The problems here are really problems in how best to quantify the content of qualitative ethnographic information. We really do not know how to do this very well yet. However, what we did was to review each paragraph of prose and

Table 5.5 Most Frequently Given Reasons for the Event of Nuclear War

No.	Reason	Response Frequency (%)	Response Frequency Rank
1	Increasing world nuclear proliferation increases the chances of nuclear war among other third-party nations facing critical economic, political, and demographic problems, and it makes third parties much more likely to get into direct nuclear conflict than the United States and the Soviet Union.	14	1
2	The United States and the Soviet Union both view Middle East oil as vital to their economies and will find their oil contention peaking over the next decade, causing them to become more directly involved in unstable Third World countries, perhaps using military force to secure supplies.	13	2
3	Probability of accident (computer, radar, or human failure) causing a nuclear strike is greater than the probability of the United States or the Soviet Union intentionally striking the other side.	12.9	3

subdivide it into separate enumerated reasons. Then, using scissors to cut up each paragraph of prose into what we felt were distinct reasons, we created piles of reasons that we felt were really the same reason, but phrased differently by different panelists. We ended up with a handful of distinct reasons provided by the group and called this collection of distinct reasons *the composite*. We then moved on to Stage 2.

At Stage two we asked the panelists a second time, independently, for their subjective probabilities for event E. But this time, when we questioned them, we did so after having provided them the composite of reasons given by the group. We did not advise them of the frequencies with which each reason was given—just the composite of reasons. At this juncture the panelists could respond after taking into account reasons they might not have thought of at Stage one. So they might modify their first-stage response. Many did modify their first-stage responses accordingly. Again we formed a composite and repeated the process.

By the end of the third stage, the QCF process had stabilized, since the number of new reasons created by panelists had ended, and the number of panelists who were still modifying their subjective probabilities for event E had been reduced to zero. We then brought the QCF process to an end. We used the third stage QCF results to form the prior distribution for the group.

Psychological Factors Relating to Subjective Probability Assessments for Group Members (or Individuals)

There are many biases in the subjective assessments made by individuals (for additional discussion of this topic see Tversky and Kahneman (1974), Hogarth (1980), Kahneman et al. (1982), Yates (1990). A few major biases are listed below.

Biases

1. *Desirable/Undesirable Events.* Subjects tend to under-assess the probability of undesirable events (nuclear war).
2. *Availability Heuristic.* Subjects tend to underassess the probability of events for which they cannot quickly construct previous similar occurrences or scenarios, and certainly for events that have never occurred.
3. *Low/High Probability Events.* Subjects tend to overassess extremely low probability events because they have only incomplete understanding of how to distinguish among such infrequently occurring events and do not have real appreciation of their rarity (such as the rare event of nuclear war).
4. *Anchoring Heuristic.* Subjects are unlikely to substantially change an intellectual position they have espoused. This affects their ability to modify their positions from one stage to the next in a sequence of QCF rounds in which they are effectively asked if their opinions have changed.

Conclusions Regarding Psychological Factors

Psychologically-related sources of potential bias in the subjective assessments sometimes leads to overassessment, but other times appears to lead to underassessment. There has been very little research on how to evaluate the effects of competing heuristics. We conclude that the above assessments are probably somewhat too low, and we view the results as conservative. Much more research remains to be carried out in this area.

Summary of Group Prior Distribution Assessment

In this section we have focused on the problem of how to assess multivariate prior distributions. We have suggested that in some contexts it may be reasonable to substitute the distribution of the p-point assessments of a random vector for N individuals for the p-dimensional distribution of an unknown vector for a single individual. In such cases a normally very hard problem is very easily solved. We have applied this procedure to the problem of finding a prior distribution for the probability of nuclear war between the United States and the Soviet Union over the decade of the 1980s.

Posterior Distribution for Probability of Nuclear War

It is now reasonable to develop a posterior distribution for the probability of nuclear war using the prior distribution assessed by the group of experts jointly. We merely combine it via Bayes' theorem with an appropriate likelihood function based upon observable events that could trigger a nuclear exchange.

5.4.5 Assessing Hyperparameters of Multiparameter Subjective Prior Distributions

Subjective prior distributions always have hyperparameters to assess, and the assessment task is not always easy. The problem is exacerbated in the case in which the prior distribution contains vector or matrix parameters.

For example, suppose the data distribution for a scalar random variable X follows the distribution $N(\theta, \sigma^2)$, for both θ and σ^2 unknown. For the prior distribution we might, for simplicity of interpretation, reparameterize in terms of the precision $\lambda \equiv 1/\sigma^2$, and adopt the (normal-gamma) prior density structure:

$$g(\theta, \lambda) = g_1(\theta \mid \lambda) g_2(\lambda), \tag{5.78}$$

with $(\theta \mid \lambda)$ following a normal distribution with density:

$$g_1(\theta \mid \lambda) \propto \lambda^{0.5} \exp\{-(0.5k\lambda)(\theta - \bar{\theta})^2 \tag{5.79}$$

for some constant k, and λ following a gamma distribution with density:

$$g_2(\lambda) \propto \lambda^{a-1} \exp\{-b\lambda\}, \qquad a > 0, b > 0. \tag{5.80}$$

In Equations (5.79) and (5.80), the hyperparameters $(k, \tilde{\theta}, a, b)$ remain to be assessed. Following a procedure suggested by Arnold et al. (1999, Section 13.6), we first evaluate the conditional moments (noting that $(\lambda \mid \theta)$ also follows a gamma distribution):

$$\begin{aligned} &E(\theta \mid \lambda) = \theta_0, \qquad \text{var}(\theta \mid \lambda) = k/\lambda, \\ &E(\theta \mid \lambda) = \frac{a + 0.5}{b + 0.5(\theta - \bar{\theta})^2}, \qquad \text{var}(\lambda \mid \theta) = \frac{a + 0.5}{[b + 0.5(\theta - \bar{\theta})^2]^2}. \end{aligned} \tag{5.81}$$

Such moments are sometimes computed in connection with desired Markov Chain Monte Carlo calculations. Next, using elicitation, we ask the person who has substantive information about the four unknown quantities $(k, \tilde{\theta}, a, b)$ for assessments of these moments at particular values of the variables on which we conditioned. Thus, we can find:

$$\begin{aligned} &E(\theta \mid \lambda_i) \approx \tilde{\theta}_{A_i}, \qquad \text{var}(\theta \mid \lambda_i) \approx k_{A_i}/\lambda_{A_i}, \qquad i = 1, \ldots, m, \\ &E(\lambda \mid \theta_j) \approx \frac{a_{A_j} + 0.5}{b_{A_j} + 0.5(\theta_j - \tilde{\theta}_{A_j})^2}, \\ &\text{var}(\lambda \mid \theta_j) \approx \frac{a_{A_j} + 0.5}{[b_{A_j} + 0.5(\theta_j - \tilde{\theta}_{A_j})^2]^2}, \qquad j = 1, \ldots, l, \end{aligned} \tag{5.82}$$

where the A subscript denotes "assessed value" m and l are the numbers of values on which we have conditioned. All quantities with A subscripts are assumed known from assessment. Moreover, $\theta_1, \ldots, \theta_l$ and $\lambda_l, \ldots, \lambda_m$ are set before the assessment procedure takes place. So all quantities on the right-hand sides of Equation (5.82) are known. Forming the squared differences between the assessed values in Equation (5.81) and the corresponding quantities in Equation (5.82), and simultaneously minimizing the sums of squared differences in the four relationships, gives least squares estimators of the hyperparameters.

The procedure just outlined provides estimates of the hyperparameters for the normal-gamma form of prior densities proposed. Arnold et al. (1999, Section 13.6) suggest that if the form of the prior information available does not fit the prior family proposed very well, prior families are available that are richer in parameters so that there would be more flexibility to more closely approximate more sophisticated forms of prior information. They call such families *conditionally specified distributions*. The procedure also generalizes to non-normal distributions. The principal advantage of the approach just outlined is that subjective probability assessments may be made from one-dimensional conditional distributions, even when the joint prior distributions are multidimensional. In this sense, the approach reduces the difficulty of assessing hyperparameters of multidimensional distributions.

Maximum Entropy (Maxent) Prior Distributions (Minimum Information Priors)
When maximum entropy is the criterion used to generate the prior distribution, the resulting prior is sometimes called a *maxent* (minimum information) prior distribution.

For example, the *maximal data information priors* proposed by Zellner, (1977, using the Shannon notion of information) select the proper prior $g(\theta)$ by forming the average prior information in the data, $\bar{I}$, with $\bar{I} = \int_a^b I(\theta)g(\theta)d(\theta)$, $I(\theta) = \int_R p(y \mid \theta) \log p(y \mid \theta)dy, a \leq \theta \leq b$, a and b finite, and $p(y \mid \theta) > 0$ in R denoting the data density, and subtracting from it the information in the prior, $\int_a^b g(\theta) \log g(\theta)d(\theta)$, and maximizing the difference with respect to $g(\theta)$. We discuss such prior distributions in Chapter 10 (see Section 10.5), along with predictive distributions.

It is sometimes useful to develop prior distributions that have been derived from maximum entropy considerations when exchangeability of the observable variables in the data distribution can be assumed. It will be seen (Chapter 10) that under exchangeability, there is a one-to-one correspondence between the prior distribution and the predictive distribution. Such considerations give rise to the *de Finetti Transform*.

5.5 DATA-MINING PRIORS

Data mining involves various (usually) computer-intensive methods for searching for relationships among variables. Often, the relationships discovered are unexpected. The methods typically involve analyzing very large (massive) quantities of multidimensional data. The sizes of the data sets are generally in the thousands, hundreds of thousands, or even in the millions. The number of dimensions (variables per observation vector) might also be in the thousands or millions. In addition, the observation vectors are frequently available over many time points or space points, or both, so that the data occupy a three- or four-dimensional hyper rectangle—with missing data as well. Such data collections are usually referred to in the data mining literature as *the data cube* (which of course is a misnomer since the arrayed data generally have different sizes). The data cube may be represented by many *flat files* (which is the way statisticians refer to two-dimensional data (rows of cases, and columns of attributes), or it may be represented in a *relational database* (which is the way computer scientists refer to it; see, for example, http://163.238.182.99/chi/715/theory/html), and computer scientists use Structured Query Language, or SQL (see, for example, http://www.sqlcourse.com/) to manipulate the data). In the relational database format, there are many tables (rows of cases or records, by many columns of variables, attributes, or characteristics) related by at least one characteristic common to all the tables (for example, a number that identifies the case).

The data in a data cube are all discrete; those data that started out as continuous are discretized. The data typically violate many of the usual assumptions of traditional statistics, assumptions such as the variables being independent and identically distributed. The data are just collections of multidimensional observations without regard to how they were collected; traditional experimental design does not enter the problem at this point. Sample estimates are used to approximate cell probabilities, and other approximations might be used as well. The data cube is searched for relationships by proceeding through the cube along both its principal axes, and diagonally, in steps. The search of the data cube is open to any and all kinds of exploratory data analysis from clustering, classification, and regression, and other methods, to visual displays of all kinds, to try to see what relationships there might be among the variables.

The development of subjective prior distributions for unknown quantities in a model typically involves considerable introspection on the part of the analyst. The analyst might search for underlying theory to bring to bear some real information about the process generating the outcome variables, and thereby be able to assess prior distributions for the underlying variables of interest. Many other methods of assessing hyperparameters have been described above in this chapter. Yet another method that might be useful in appropriate circumstances is to use data-mining methods to establish the relationships required for assessing desired prior distributions. Assessment of a prior distribution does not require that the methods used to generate it be rigorous mathematically or statistically, only that the distribution be the analyst's best guess, judgment, or intuition about the underlying process. Data mining fits the requirement very well.

Suppose, for example, there is a vector of unknown parameters, β, and the analyst would like to assess a prior distribution for it. We have available in the data cube input variables for person or case $i, x_i = (x_{i1}, \ldots, x_{ip}), i = 1, \ldots, n$, that we believe are comparable to the variables in our model, and at least one corresponding output variable, y_i, which we think might depend upon these inputs. We might form a score (objective) function, $f(\beta) = \sum_i (y_i - x_i\beta)^2$, and select β to minimize $f(\beta)$. This value of β could be used as the mean of a family of prior distributions for β. Moreover, typically in data mining, there are large numbers of observations; that fact opens many possible courses of further action. The data set could, for example, be subdivided into k groups, and the same objective function could be calculated for each group. The values of β obtained by minimizing those objective functions could be used to form a sample cdf for β, and that sample cdf could then be used as the prior distribution for β. A kernel density for β could also be estimated (see Section 5.4.4 for methodology for doing this). For additional discussion and references on data mining, see, e.g. Press, 2003.

5.6 "WRONG" PRIORS

Can your prior distribution ever be wrong? The answer is both yes and no. In fact, your prior for some unknown quantity θ cannot be wrong, in the sense that the prior represents your degree of belief about this unknown quantity. But your beliefs could be wrong! This distinction becomes important in scientific research in which you may begin a study of some phenomenon and believe that the phenomenon is driven in a particular way, but your belief may be quite wrong. For this reason, it is important that your prior be constructed in such a way that no possible values of θ be assigned zero probability *a priori*. If certain values of θ are assigned probability zero, *a priori*, no amount of data will ever be able to bring those values back to consideration.

For example, if θ^* were given prior probability zero, since the posterior probability for θ^* must be proportional to the (prior probability) $\times$ (the likelihood of the data), and if the prior probability is taken to be zero, it would make the posterior for θ^* also equal to zero.

This issue is also of importance in decision making. A crisis arises in some organization. The suggestions of the decision maker's closest advisors are in conflict. There is little agreement as to what to do. Such disagreements can envelop the decision maker in confusion. In such critical situations the decision maker must make important decisions with only a fraction of the necessary information available, and in spite of the fact that all the initial reports may turn out to be wrong. The prior information supplied to the decision maker may be correct with respect to the sincere beliefs of the advisors, but it may be wrong with respect to reality, and with respect to the consequences and implications of the decision being made. The decision maker in such situations must recognize that because all the prior beliefs of advisors may be wrong, he/she must make decisions based upon his/her best-guess judgment, intuition, and common sense. In this case, the decision maker should allot some probability to the case that all his/her advisors are wrong and something else entirely is the case.

Sometimes an individual has a very strong belief about some phenomenon, prior to his/her collecting contradictory data. That individual may continue to insist on that belief in spite of even substantial amounts of data to the contrary. Such dogmatic priors can lead to scientific fraud (see Chapter 1, and Press and Tanur, 2001), or even worse. Seife (2000) points out (p. 38, op. cit.) that Hippasus of ancient Greece (6th century BC) was probably tossed overboard at sea by his fellow Pythagoreans "for ruining a beautiful theory with harsh facts." His crime had been to permit the outside scientific world to learn that irrational numbers exist (a fact known to him because it was discovered earlier by his secret society of Pythagoreans). His being condemned to death by drowning was the penalty the Pythagoreans imposed on Hippasus for revealing to Greek mathematicians that their priors (relative to mathematical evidence) about all numbers being rational, were at odds with the truth. The fact that their priors were diametrically opposed to the facts constrained the further development of Greek mathematics for centuries.

SUMMARY

This chapter has discussed the advantages and disadvantages of both objective and subjective prior distributions for both discrete and continuous variables (unknown quantities), and for both one-dimensional unknowns, and for unknown vectors and matrices. We have suggested many ways to assess prior distributions for these variables, both by individuals and by groups, using assessment of historical priors, and for assessment of the hyperparameters of families of prior distributions. We discussed vague priors, Jeffreys' invariant priors, g-priors, natural conjugate priors, exponential power priors, stable estimation, public policy priors, the principle of insufficient reason, sample-splitting priors, mixture priors, assessing the fractiles of your complete subjective prior, and priors obtained by data mining.

EXERCISES

5.1 Suppose X denotes an observation that follows the exponential distribution with density given by: $f(x \mid \theta) = \theta \exp\{-\theta x\},\ 0 < x,\ \theta < \infty$. Give a vague prior density for θ.

5.2 Suppose that $x_1, \ldots, x_n$ denotes i.i.d. observations from the gamma distribution with density: $f(x \mid \theta, \tau) \propto x^{\theta-1} \exp\{-\tau x\}, \infty < \theta < \infty,\ 0 < \tau < \infty$. Give a vague prior density for (θ, τ).

5.3 For Exercise 5.1, give a natural conjugate prior density for θ.

5.4 Suppose $\mathbf{X}_1, \ldots, \mathbf{X}_n$ denotes an i.i.d. sample of p-vectors from a $N(\theta, \mathbf{A})$ distribution, where $\mathbf{A}$ denotes a known $p \times p$ covariance matrix. Give a vague prior density for $(\theta, \mathbf{A})$.

5.5 Give a natural conjugate prior density for the parameters in Exercise 5.4.

5.6 Give the Jeffreys' prior density for the parameter in Exercise 5.1.

5.7 Give the Jeffreys' prior density for the parameters in Exercise 5.2.

5.8 Distinguish between objective and subjective prior distributions.

5.9* What is meant by a public policy prior distribution?

*Solutions for asterisked exercises may be found in Appendix 7.

5.10 What is a "reference prior" distribution?

5.11 How would you handle a problem in which the posterior distribution for an unknown quantity turned out to be improper?

5.12 Suppose the probability mass function for a random variable r is given by:

$$f(r \mid p, n) = f(r/p, n) = \binom{n}{r} p^r (1-p)^{n-r}, \quad 0 < p < 1, \quad r = 0, 1, \ldots, n.$$

(a) Find a vague prior density for p.

(b) Find the Jeffreys' prior density for p.

5.13 For the distribution in Exercise 5.12, find a natural conjugate prior density for p.

5.14 What is a *g-prior* distribution and when might you want to use one?

5.15 What is a *mixture prior* distribution and when might you want to use one?

5.16* What is Laplace's *Principle of Insufficient Reason*? When might you want to use such a principle?

5.17 **(a)** What are the advantages of adopting objective prior distributions?

(b) What are the advantages of adopting subjective prior distributions?

5.18 What is an exponential power distribution family of prior densities?

5.19 What is *stable estimation*?

5.20 **(a)** What is a *data-based prior* distribution?

(b) What is meant by *sample splitting* in developing prior distributions?

5.21 Assume I am a geneticist. What sequence of questions would you ask me to assess fractiles of my prior distribution for θ, the number of years from now when I believe it will be possible through genetic engineering to affect the intelligence of an unborn child?

5.22 Assume that I am a male sociologist. What sequence of questions would you ask me to assess fractiles of my prior distribution for θ, the proportion of time that husbands of working couples spend delivering and picking up their children from day care?

5.23 Assume that I am a female sociologist. What sequence of questions would you ask me to assess fractiles of my prior distribution for θ, the proportion of time that wives of working couples spend delivering and picking up their children from day care?

5.24 Assume I am a geologist. What sequence of questions would you ask me to assess fractiles of my prior distribution for (r, θ), the polar coordinates (r = radial distance from here, θ = angular direction from here) of the location that I believe is most likely to contain large reserves of natural gas?

5.25 Assume that I am a macro economist. What sequence of questions would you ask me to assess fractiles of my prior distribution for θ, my probability that the gross domestic product will decline in both of the next two quarters?

*Solutions for asterisked exercises may be found in Appendix 7.

5.26 Suppose that a data distribution for a random variable X follows the law $N(\theta, \sigma^2)$, and that $X_1, \ldots, X_n$ denotes a random sample from this distribution. Suppose further that for a prior distribution family you assume that $(\theta \mid \sigma^2)$ follows the natural conjugate distribution $N(\tilde{\theta}, K\sigma^2)$ for some hyperparameters $(\tilde{\theta}, K)$, where K denotes some pre assigned constant.

(a) Adopt an appropriate natural conjugate prior distribution family for σ^2 and evaluate the induced posterior distribution for $(\theta, \sigma^2 \mid x_1, \ldots, x_n)$.

(b) Find the marginal posterior distribution for $(\theta \mid x_1, \ldots, x_n)$.

(c) Find the marginal posterior distribution for $(\sigma^2 \mid x_1, \ldots, x_n)$.

5.27* In assessing a multivariate prior distribution by assessing a p-dimensional vector from N subjects and merging them by density estimation, we used a kernel, $K(\theta)$, which we took to be normal. Suppose instead that we choose $K(\theta)$ to be multivariate Student t, so that

$$K(\theta) \propto \{n + \theta'\theta\}^{-0.5(n+p)},$$

for some suitable n. How would this change be likely to affect the usefulness of the fitted density of opinions? (*Hint:* see literature on density estimation, such as Taipia and Thompson, 1982.)

5.28* How should an expert be defined for the purposes of group prior distribution assessment?

5.29* Suppose you are trying to decide whether to buy shares in the XYZ corporation. You seek an opinion from several of the well-known stock brokerage firms in the closest major financial center. One decision rule is to follow the advice of the majority of opinions. Another might involve finding out how confident each firm is in its opinion. How would you generalize this concept to the multivariate prior distribution assessment context?

5.30* Compare the mean assessed probability with the median assessed probability in the nuclear war application in Section 5.4. How do you interpret the substantial discrepancy?

5.31* By using the availability heuristic, explain why many people do not perceive the advent of a nuclear war in the near future as a likely event.

5.32* Let θ denote the expected number of live chickens in California on a given day. Explain how you would formulate your prior distribution for θ.

5.33* Explain the "closure property" for natural conjugate prior distributions.

5.34* If X given λ follows a Poisson distribution, give a natural conjugate prior for λ.

5.35* Give another method that has been proposed for assessing multivariate prior distributions. (*Hint:* see Kadane et al., 1980.)

5.36* Suppose $\mathbf{X}$: $p \times 1$ and $(\mathbf{X} \mid \boldsymbol{\theta}) \sim N(\boldsymbol{\theta}, \mathbf{I})$. Give a natural conjugate prior family for θ.

5.37* Suppose $\mathbf{X}$: $p \times 1$ and $\mathscr{L}(\mathbf{X} \mid \Lambda) = N(\mathbf{0}, \Sigma)$, for $\Lambda = \Sigma^{-1}$. Give a natural conjugate prior family for Λ. (Assume Σ is a positive definite matrix.)

5.38* Suppose **X**: $p \times 1$, and $\mathscr{L}(\mathbf{X} \mid \theta, \Lambda) = N(\theta, \Lambda^{-1})$; $\Sigma \equiv \Lambda^{-1}$ denotes a positive definite matrix. Find a natural conjugate prior family for (θ, Λ) by noting that $g(\theta, \Lambda) = g_1(\theta \mid \Lambda)_2(\Lambda)$, and assuming that *a priori*, var $(\theta \mid \Sigma) \propto \Sigma$.

FURTHER READING

Abramson, J. H. and Abramson, Z. H. (1999). *Survey Methods in Community Medicine*, 5th Edition, Churchill Livingston.

Arnold, B. C., Castillo, E. and Sarabia, J. M. (1999). *Conditional Specification of Statistical Models*, New York, Springer-Verlag.

Berger, J. O. and Pericchi, L. R. (1993). "The Intrinsic Bayes Factor for Model Selection and Prediction," Technical Report. #93-43C, Department of Statistics, Purdue University.

Berger J. O. and Pericchi L. R. (1997). "Objective Bayesian Methods for Model Selection: Introduction and Comparison," presented at the Bayesian Model Selection Workshop, Cagliari, Italy, Technical Report. #00-09, Duke University, Institute of Statistics and Decision Sciences.

Berger, J. O. and Bernardo, J. M. (1989). "Estimating a Product of Means: Bayesian Analysis with Reference Priors," *J. Am. Statist. Assoc.*, **84**, 200–207.

Berger J. O. and Bernardo J. M. (1992). "On the Development of the Reference Prior Method," in J. M. Bernardo et al., Eds., *Bayesian Statistics IV*, London, Oxford University Press, 35–60.

Bernardo, J. M. (1979). "Reference Posterior Distributions for Bayesian Inference," *J. Royal Statis. Soc. (B)*, **41**(2), 113–147.

Box, G. E. P. and Tiao, G. C. (1973). *Bayesian Inference in Statistical Analysis*, Reading MA, Addison-Wesley Publishing Co.

Cacoullos, T. (1966). "Estimation of a Multivariate Density," *Ann. Inst. Statist. Math.*, **2**(2), 179–189.

Chatterjee, S. and S. (1987). "On Combining Expert Opinions." *Am. J. Math. Mgmnt Sci.*, 271–295.

Dalal, S. and Hall, W. J. (1983). "Approximating Priors by Mixtures of Natural Conjugate Priors," *J. Royal Stat. Soc. (B)*, **45**, 278–286.

Dawid, A. P., Stone, M., and Zidik, J. V. (1973). "Marginalization Paradoxes in Bayesian and Structural Inference," *J. Royal Statist. Soc. B*, **35**, 189–233.

Diaconis, P. and Ylvisaker, D. (1985). "Quantifying Prior Opinion," in J. M. Bernardo, M. H. DeGroot, D. V. Lindley, and A. F. M. Smith, Eds., *Bayesian Statistics 2*, Amsterdam, North Holland Pub. Co., 133–156.

Dickey, J. M. (1973). "Scientific Reporting and Personal Problems: Student's Hypothesis," *J. Royal Statist. Soc. (B)*, **35**, 285–305.

Garthwaite, P. H. and Dickey, J. M. (1988). "Quantifying Expert Opinion in Linear Regression Problems," *J. Royal Statist. Soc. B*, **50**(3), 462–474.

Geisser, S. and Cornfield, J. (1963). "Posterior Distributions for Multivariate Normal Parameters," *J. Royal Statist. Soc. B*, **25**, 368–376.

Gelfand, A. Mallick, B. and Dey, D. (1995). "Modeling Expert Opinion Arising as a Partial Probabilistic Specification," *Jr. Am. Statist. Assoc.* 598–604.

Genest, C. and Schervish, M. (1985). "Modeling Expert Judgements for Bayesian Updating." *Annals Statist.*, 1198–1212.

Halmos, P. A. (1950). *Measure Theory*, Princeton, New Jersey, D. Van Nostrand Co., Inc.

Hartigan, J. A. (1964). "Invariant Prior Distributions," *Ann. Math. Stat.*, **35**, 836–845.

Hogarth, R. (1980). *Judgment and Choice*, New York, John Wiley and Sons, Inc.

Jaynes, E. T. (1983). *Papers on Probability, Statistics and Statistical Physics of E. T. Jaynes*, R. D. Rosenkranz (Ed.), Dordrecht, Holland, D. Reidel Publishing Co.

Jeffreys, H. (1939, 1st edition; 1948, 2nd edition; 1961, 3rd edition). *Theory of Probability*, Oxford, Oxford University Press and Clarendon Press.

Kahneman, D., Slovic, P. and Tversky, A. (1982). *Judgment Under Uncertainty: Heuristics and Biases*, Cambridge, Cambridge University Press.

Kass, R. and Wasserman, L. (1996). "The Selection of Prior Distributions by Formal Rules," *J. Am. Statist. Assoc.*, **91**, 1343–1370.

Laplace, P. S. (1774, 1814). *Essai Philosophique sur les Probabilites*, Paris. This book went through five editions (the fifth was in 1825) revised by Laplace. The sixth edition appeared in English translation by Dover Publications, New York, in 1951. While this philosophical essay appeared separately in 1814, it also appeared as a preface to his earlier work, *Theorie Analytique des Probabilites*.

Lindley, D. V. (1983). "Reconciliation of Probability Distributions," *Op. Res.*, 866–880.

Morris, P. (1977). "Combining Expert Judgments: A Bayesian Approach," *Mngmnt. Sci.*, 679–693.

Parzen, E. (1962). "On Estimation of a Probability Density Function and Mode," *Ann. Math. Statist.*, **33**, 1065–1076.

Press, S. J. (1978). "Qualitative Controlled Feedback for Forming Group Judgments and Making Decisions," *J. Am. Statist. Assoc.*, **73**(363), 526–535.

Press, S. J. (1980). "Bayesian Inference in Group Judgment Formulation and Decision Making Using Qualitative Controlled Feedback," in J. M. Bernardo, M. H. De Groot, D. V. Lindley, A. F. M. Smith, (Eds.), *Bayesian Statistics*, 383–430.

Press, S. J. (1980). "Multivariate Group Judgments by Qualitative Controlled Feedback," in P. R. Krishnaiah, Ed., *Multivariate Analysis V*, New York, North Holland Publishing Co., 581–591.

Press, S. J. (1982). *Applied Multivariate Analysis: Using Bayesian and Frequentist Methods of Inference*, Malabar, FL, Krieger Publishing Co.

Press, S. J. (1983). "Group Assessment of Multivariate Prior Distributions," *Technological Forecasting and Social Change*, **23**, 247–259.

Press, S. J. (1985a). "Multivariate Group Assessment of Probabilities of Nuclear War," in J. M. Bernardo, M. H. De Groot, D. V. Lindley and A. F. M. Smith, Eds. *Bayesian Statistics 2*, Amsterdam, North-Holland, 425–462.

Press, S. J. (1985b). "Multivariate Analysis (Bayesian)," in S. Kotz and N. L. Johnson, Eds., *Encyclopedia of Statistical Sciences*, New York, John Wiley & Sons, Inc., Vol. 6, 16–20.

Press, S. J. (2002). "Respondent-Generated Intervals for Recall in Sample Surveys," submitted manuscript.

Press, S. J. (2003). "The Role of Bayesian and Frequentist Multivariate Modeling In Statitical Data Mining," in Proceedings of the C. Warren Neel Conference on the New Frontiers of

Statistical Data Mining (DM), Knowledge Discovery (KD), and e-Business, New York: John Wiley and Sons, Inc. (in press, to be published on-line).

Press, S. J., Ali, M. W. and Yang, E. (1979). "An Empirical Study of a New Method for Forming Group Judgments: Qualitative Controlled Feedback," *Technological Forecasting & Social Change*, **15**, 171–189.

Press, S. J. and Tanur, J. M. (2001). *The Subjectivity of Scientists and the Bayesian Approach*, New York, John Wiley and Sons, Inc.

Raiffa, H. and Schlaifer, R. (1961). *Applied Statistical Decision Theory*, Boston, MA, Harvard University, Graduate School of Business Administration.

Savage, L. J. (1962). *The Foundations of Statistical Inference*, G. A. Barnard and D. R. Cox (Eds.) London, Methuen.

Seife, C. (2000). *Zero: The Biography of a Dangerous Idea*, New York: Penguin.

Silverman, B. W. (1986). *Density Estimation for Statistics and Data Analysis*, New York: Chapman and Hall.

Stein, C. (1956). "Inadmissibility of the Usual Estimator for the Mean of a Multivariate Normal Distribution," in Proceedings of the Third Berkely Symposium on Mathematical Statistics and Probability, University of California Press, Berkeley, 197–206.

Stigler, S. M. (1986). *The History of Statistics*, Cambridge, MA, The Belknap Press of Harvard University Press.

Taipia, R. A. and Thompson, J. R. (1982). *Nonparametric Probability Density Estimation*, Baltimore, MD, The Johns Hopkins University Press.

Tversky, A. and Kahneman, D. (1974). "Judgment Under Uncertainty: Heuristics and Biases," *Science*, **185**, 1124–1131.

Villegas, C. (1969). "On the A Priori Distribution of the Covariance Matrix," *Ann. Math. Statist.*, **40**, 1098–1099.

West, M. (1988), "Modelling Expert Opinion," in *Bayesian Statistics 3*, 493–508.

Yang, R. and Berger, J. O. (1996). "A Catalog of Noninformative Priors," Technical Report. #97-42, Duke University, Institute of Statistics and Decision Sciences.

Yates, J. F. (1990). *Judgment and Decision Making*, Englewood Cliffs, New Jersey, Prentice Hall, Inc.

Zellner, A. (1971). An Introduction to Bayesian Inference in Econometrics, New York, John Wiley and Sons, Inc.

Zellner, A. (1977). "Maximal Data Information Prior Distributions," in A. Aykac and C. Brumat (Eds.), *New Developments in the Application of Bayesian Methods*, Amsterdam, North Holland Publishing Co., 211–232.

Zellner, A. (1986). "On Assessing Prior Distributions and Bayesian Regression Analysis with g-Prior Distributions," in P. Goel and A. Zellner, Eds., *Bayesian Inference and Decision Techniques: Essays in Honor of Bruno de Finetti*, New York, North Holland Publishing Co., 233–243.

PART II

Numerical Implementation of the Bayesian Paradigm

CHAPTER 6

Markov Chain Monte Carlo Methods

*Siddhartha Chib**

6.1 INTRODUCTION

In this chapter we describe a general simulation approach for numerically calculating the quantities (such as the moments and quantiles of posterior and predictive densities) that arise in the Bayesian prior–posterior analysis. Suppose that we are interested in finding the mean of the posterior density

$$E(\boldsymbol{\theta} \mid \mathbf{y}) = \int \boldsymbol{\theta} h(\boldsymbol{\theta} \mid \mathbf{y}) d\boldsymbol{\theta}$$

where $\boldsymbol{\theta}$ is a vector-valued parameter vector and $\mathbf{y}$ is a vector of observations. Now suppose that this integral cannot be computed analytically and that the dimension of the integration exceeds three or four (which essentially rules out the use of standard quadrature-based methods). In such cases one can compute the integral by Monte Carlo sampling methods. The general idea is to abandon the immediate task at hand (which is the computation of the above integral) and to ask how the posterior density $h(\boldsymbol{\theta} \mid \mathbf{y})$ may be sampled. The reason for changing our focus is that if we were to have the draws

$$\boldsymbol{\theta}^{(1)}, \ldots, \boldsymbol{\theta}^{(M)} \sim h(\boldsymbol{\theta} \mid \mathbf{y}),$$

from the posterior density, then provided the sample is large enough, we can estimate not just the above integral but also other features of the posterior density by taking those draws and forming the relevant sample-based estimates. For example, the

* John M. Olin School of Business, Washington University, CB 1133, 1 Brookings Dr., St. Louis, MO 63130; E-mail: chib@olin.wustl.edu.

sample average of the sampled draws would be our simulation-based estimate of the posterior mean, while the quantiles of the sampled output would be estimates of the posterior quantiles, with other summaries obtained in a similar manner. Under suitable laws of large numbers these estimates would converge to the posterior quantities as the simulation size becomes large. In short, the problem of computing an intractable integral is reduced to the problem of sampling the posterior density.

To implement the above tactic we need general methods for drawing variates from the sort of posterior densities that arise in Bayesian inference. If a given model admits a natural conjugate prior density, then under that prior, the posterior density is also in the same family and can be sampled by known methods. When such simplifications are not possible, the sampling of the posterior density can become more complicated.

Fortunately, it is possible to sample complex and high-dimensional posterior densities by a set of methods that are called Markov chain Monte Carlo methods, or MCMC methods for short. These methods involve the simulation of a suitably constructed Markov chain that converges to the target density of interest (usually the posterior density).

The defining feature of Markov chains is the property that the conditional density of $\boldsymbol{\theta}^{(j)}$ (the jth element of the sequence) conditioned on the entire preceding history of the chain depends only on the previous value $\boldsymbol{\theta}^{(j-1)}$. We denote this conditional density (called the transition density) as $T(\boldsymbol{\theta}^{(j-1)}, \cdot \,|\, \mathbf{y})$.

The idea behind MCMC simulations is to construct a transition density that converges to the posterior density from any starting point $\boldsymbol{\theta}_0$ (in the sense that for any measurable set A under h, $\Pr(\boldsymbol{\theta}^{(j)} \in A \,|\, \mathbf{y}, \boldsymbol{\theta}_0)$ converges to $\int_A h(\boldsymbol{\theta} \,|\, \mathbf{y}) d\boldsymbol{\theta}$ as j becomes large). Then, under this approach, we obtain draws from the posterior density by recursively sampling the transition density as

$$\begin{aligned}
\boldsymbol{\theta}^{(1)} &\sim T(\boldsymbol{\theta}^{(0)}, \cdot \,|\, \mathbf{y}) \\
\boldsymbol{\theta}^{(2)} &\sim T(\boldsymbol{\theta}^{(1)}, \cdot \,|\, \mathbf{y}) \\
&\;\vdots \\
\boldsymbol{\theta}^{(j)} &\sim T(\boldsymbol{\theta}^{(j-1)}, \cdot \,|\, \mathbf{y}) \\
&\;\vdots
\end{aligned}$$

Because the chain was constructed to converge to the posterior density, the values beyond the first n_0 iterations,

$$\boldsymbol{\theta}^{(n_0+1)}, \boldsymbol{\theta}^{(n_0+2)}, \ldots, \boldsymbol{\theta}^{(n_0+M)}$$

can be taken as draws from $h(\boldsymbol{\theta} \,|\, \mathbf{y})$. The initial draws up to n_0 are typically discarded to allow the effect of the starting value to wear off; this period is referred to as the "burn-in" period. It is important to recognize that the draws obtained by a Markov chain sampling procedure are a correlated sample from the posterior density. Nonetheless, these draws can still be used to form sample averages and sample quantiles. From laws of large numbers for Markov sequences we can show that under general conditions these estimates will converge to the posterior quantities as the number of drawings becomes large. For example, for any integrable function of $\boldsymbol{\theta}$, say $g(\boldsymbol{\theta})$, the sample average of the values $\{g(\boldsymbol{\theta}^{(j)})\}$ converges, under weak regularity conditions, to its expectation:

$$M^{-1} = \sum_{j=1}^{M} g(\boldsymbol{\theta}^{(j)}) \to \int g(\boldsymbol{\theta})h(\boldsymbol{\theta}\,|\,\mathbf{y})d\boldsymbol{\theta}, \quad \text{as } M \uparrow \infty. \tag{6.1}$$

The usefulness of Markov chain sampling derives from the fact that it is usually possible to construct a transition density $T(\boldsymbol{\theta}^{(j-1)}, \cdot\,|\,\mathbf{y})$ that converges to the target density (implied by the model at hand) (Gamerman, 1997; Gilks et al., 1996). One way to construct the appropriate Markov chain is by a method called the Metropolis–Hastings (M–H) algorithm, which was introduced by Metropolis et al. (1953) and Hastings (1970). Two important references for this method are Tierney (1994), where the Markov chain underpinnings of the method are elucidated, and Chib and Greenberg (1995), which contains a detailed discussion of the method along with a derivation of the algorithm from the logic of reversibility. A second technique for constructing Markov chain samplers is by the Gibbs sampling algorithm. This method, which was introduced by Geman and Geman (1984), Tanner and Wong (1987) and Gelfand and Smith (1990), was the impetus for the current interest in Markov chain sampling methods. The Gibbs sampling algorithm is actually a special case of the M–H algorithm. In this chapter both methods are explained; details on the use of the WinBUGS software program (Spieglehalter et al., 2000) for implementing the Gibbs algorithm are provided in Complement A to this chapter.

6.2 METROPOLIS–HASTINGS (M–H) ALGORITHM

Suppose that we are interested in sampling the target density $h(\boldsymbol{\theta}\,|\,\mathbf{y})$, where $\boldsymbol{\theta}$ is a vector-valued parameter and $h(\boldsymbol{\theta}\,|\,\mathbf{y})$ is a continuous density. Let us also suppose that we are sampling $\boldsymbol{\theta}$ in one block. Suppose that we specify a Markov chain through the transition density $q(\boldsymbol{\theta}', \boldsymbol{\theta}\,|\,\mathbf{y})$, where $(\boldsymbol{\theta}', \boldsymbol{\theta})$ are any two points. Also suppose that this transition density is specified without any reference to the target density and is, therefore, not convergent to the target density. For example, the transition density $q(\boldsymbol{\theta}', \boldsymbol{\theta}\,|\,\mathbf{y})$ may be taken as multivariate normal with mean vector $\boldsymbol{\theta}'$ and some variance matrix $\mathbf{V}$. With $q(\boldsymbol{\theta}', \boldsymbol{\theta}\,|\,\mathbf{y})$ as the input, the objective is to construct a Markov chain that converges to $h(\boldsymbol{\theta}\,|\,\mathbf{y})$. This is achieved by the Metropolis–Hastings method. This method essentially provides a scheme to modify or alter the $q(\boldsymbol{\theta}', \boldsymbol{\theta}\,|\,\mathbf{y})$ Markov chain to ensure that the transition density of the modified chain "converges" to the target density.

To define the M–H algorithm, let $\boldsymbol{\theta}^{(0)}$ be the starting value of the chain and suppose that the algorithm has been run to obtain the values $\boldsymbol{\theta}^{(0)}, \boldsymbol{\theta}^{(1)}, \ldots, \boldsymbol{\theta}^{(j-1)}$.

Now, the next item of the chain $\boldsymbol{\theta}^{(j)}$ is produced by a two-step process consisting of a "proposal step" and a "move step."

Proposal step: Sample a proposal value $\boldsymbol{\theta}$ from $q(\boldsymbol{\theta}^{(j-1)}, \boldsymbol{\theta}\,|\,\mathbf{y})$ and calculate the quantity

$$\alpha(\boldsymbol{\theta}^{(j-1)}, \boldsymbol{\theta}\,|\,\mathbf{y}) = \min\left\{1, \frac{h(\boldsymbol{\theta}\,|\,\mathbf{y})}{h(\boldsymbol{\theta}^{(j-1)}\,|\,\mathbf{y})}\frac{q(\boldsymbol{\theta}, \boldsymbol{\theta}^{(j-1)}\,|\,\mathbf{y})}{q(\boldsymbol{\theta}^{(j-1)}, \boldsymbol{\theta}\,|\,\mathbf{y})}\right\}. \tag{6.2}$$

Move step: Set $\boldsymbol{\theta}^{(j)}$ to equal $\boldsymbol{\theta}$ with probability $\alpha(\boldsymbol{\theta}^{(j-1)}, \boldsymbol{\theta}\,|\,\mathbf{y})$; set $\boldsymbol{\theta}^{(j)}$ to equal $\boldsymbol{\theta}^{(j-1)}$ with probability $1 - \alpha(\boldsymbol{\theta}^{(j-1)}, \boldsymbol{\theta}\,|\,\mathbf{y})$.

We should mention that because $q(\boldsymbol{\theta}', \boldsymbol{\theta}\,|\,\mathbf{y})$ is the source density for this Markov chain simulation, it is called the candidate generating density or proposal density. General ways in which it can be specified are explained below. Also notice that the function $\alpha(\boldsymbol{\theta}^{(j-1)}, \boldsymbol{\theta}\,|\,\mathbf{y})$ in this algorithm can be computed without knowledge of the norming constant of the posterior density $h(\boldsymbol{\theta}\,|\,\mathbf{y})$. We refer to $\alpha(\boldsymbol{\theta}^{(j-1)}, \boldsymbol{\theta}\,|\,\mathbf{y})$ as the *acceptance probability* or, more descriptively, as the *probability of move*.

The theoretical properties of this algorithm depend crucially on the nature of the proposal density. It is typically required that the proposal density be positive everywhere in the support of the posterior density. This implies that the M–H chain can make a transition to any point in the support in one step. In addition, due to the randomization in the move step, the transition density $T(\boldsymbol{\theta}^{(j-1)}, \cdot\,|\,\mathbf{y})$ of this Markov chain has a particularly interesting form, being a mixture of a continuous density (for moves away from $\boldsymbol{\theta}^{(j-1)}$) and a discrete part (for the probability of staying at $\boldsymbol{\theta}^{(j-1)}$). It is given by

$$T(\boldsymbol{\theta}^{(j-1)}, \boldsymbol{\theta}\,|\,\mathbf{y}) = q(\boldsymbol{\theta}^{(j-1)}, \boldsymbol{\theta}\,|\,\mathbf{y})\alpha(\boldsymbol{\theta}^{(j-1)}, \boldsymbol{\theta}\,|\,\mathbf{y}) + r(\boldsymbol{\theta}^{(j-1)}\,|\,\mathbf{y})\delta_{\boldsymbol{\theta}^{(j-1)}} \tag{6.3}$$

where

$$r(\boldsymbol{\theta}^{(j-1)}\,|\,\mathbf{y}) = \int q(\boldsymbol{\theta}^{(j-1)}, \boldsymbol{\theta}\,|\,\mathbf{y})\alpha(\boldsymbol{\theta}^{(j-1)}, \boldsymbol{\theta}\,|\,\mathbf{y})d\boldsymbol{\theta},$$

and $\delta_{\boldsymbol{\theta}^{(j-1)}}$ is the Dirac function at $\boldsymbol{\theta}^{(j-1)}$ defined as $\delta_{\boldsymbol{\theta}}(\boldsymbol{\theta}') = 0$ for $\boldsymbol{\theta}' \neq \boldsymbol{\theta}$ and $\int \delta_{\boldsymbol{\theta}}(\boldsymbol{\theta}')d\boldsymbol{\theta}' = 1$. It is easy to check that the integral of the transition density over all possible values of $\boldsymbol{\theta}$ is one, as required. Note that the functions $T(\boldsymbol{\theta}^{(j-1)}, \boldsymbol{\theta}\,|\,\mathbf{y}$ and $r(\boldsymbol{\theta}^{(j-1)}\,|\,\mathbf{y})$ are not computed while implementing the two steps of the M–H algorithm defined above.

Also because of the way in which $\boldsymbol{\theta}^{(j)}$ is defined, and made explicit in the form of the transition density, the M–H Markov chain can repeat the value $\boldsymbol{\theta}^{(j-1)}$. It follows that in order to move efficiently over the support of the target density the chain should not frequently stay at the same point for many iterations. Such behavior can be avoided by choice of the proposal density $q(\boldsymbol{\theta}^{(j-1)}, \boldsymbol{\theta}\,|\,\mathbf{y})$. Further discussion of this important issue is provided by Tierney (1994) and Chib and Greenberg (1995), and in the examples below.

REMARK: An effective way to monitor the empirical behavior of the M–H output is by *autocorrelation time* or *inefficiency factor* of each component θ_k of $\boldsymbol{\theta}$ defined as

$$a_k = \left\{1 + 2\sum_{s=1}^{M}\left(1 - \frac{s}{M}\right)\rho_{ks}\right\},$$

where ρ_{ks} is the sample autocorrelation at lag s from the M sampled draws

$$\theta_k^{(n_0+1)}, \ldots, \theta_k^{(n_0+M)}.$$

One way to interpret this quantity is in terms of the *effective sample size*, or ESS, defined for the kth component of $\boldsymbol{\theta}$ as

$$\text{ESS}_k = \frac{M}{a_k}. \tag{6.4}$$

With independent sampling the autocorrelation times are theoretically equal to one, and the effective sample size is M. When the inefficiency factors are high, the effective sample size is much smaller than M.

6.2.1 Example: Binary Response Data

Consider the data in Table 6.1, taken from Fahrmeir and Tutz (1994), which is concerned with the occurrence or nonoccurrence of infection following birth by Caesarian section. The response variable y is one if the caesarian birth resulted in an infection, and zero if not. The available covariates are three indicator variables: x_1 is an indicator for whether the Caesarian was nonplanned; x_2 is an indicator for whether risk factors were present at the time of birth; and x_3 is an indicator for whether antibiotics were given as a prophylaxis. The data in the table contains information from 251 births. Under the column of the response, an entry such as

Table 6.1 Caesarian Infection Data

y (1/10)	x_1	x_2	x_3
11/87	1	1	1
1/17	0	1	1
0/2	0	0	1
23/3	1	1	0
28/30	0	1	0
0/9	1	0	0
8/32	0	0	0

11/87 means that there were 98 deliveries with covariates (1,1,1) of whom 11 developed an infection and 87 did not.

Let us model the binary response by a *probit* model, letting the probability of infection for the ith birth be given as

$$\Pr(y_i = 1 \mid \mathbf{x}_i, \boldsymbol{\beta}) = \Phi(\mathbf{x}_i'\boldsymbol{\beta}), \tag{6.5}$$

where $\mathbf{x}_i = (1, x_{i1}, x_{i2}, x_{i3})'$ is the covariate vector, $\boldsymbol{\beta} = (\beta_0, \beta_1, \beta_2, \beta_3)$ is the vector of unknown coefficients and Φ is the cdf of the standard normal random variable. Under the assumption that the outcomes $\mathbf{y} = (y_1, y_2, \ldots, y_{251})$ are conditionally independent the likelihood function is given by

$$L(\mathbf{y} \mid \mathbf{X}, \boldsymbol{\beta}) = \prod_{i=1}^{251} \Phi(\mathbf{x}_i'\boldsymbol{\beta})^{y_i} \left\{1 - \Phi(\mathbf{x}_i'\boldsymbol{\beta})\right\}^{(1-y_i)}.$$

The likelihood function does not admit a natural conjugate prior density. Let us assume that our prior information about $\boldsymbol{\beta}$ can be represented by a multivariate normal density with mean centered at zero for each parameter, and variance given by $5\mathbf{I}_4$, where $\mathbf{I}_4$ is the four-dimensional identity matrix. Therefore, the prior density of $\boldsymbol{\beta}$ is

$$g(\boldsymbol{\beta}) \propto \exp\left\{-0.5\boldsymbol{\beta}'(5\mathbf{I}_4)^{-1}\boldsymbol{\beta}\right\},$$

and the posterior density is

$$\begin{aligned} h(\boldsymbol{\beta} \mid \mathbf{y}, \mathbf{X}) &\propto g(\boldsymbol{\beta})L(\mathbf{y} \mid \mathbf{X}, \boldsymbol{\beta}) \\ &\propto \exp\left\{-0.5\boldsymbol{\beta}'(5\mathbf{I}_4)^{-1}\boldsymbol{\beta}\right\} \prod_{i=1}^{251} \Phi(\mathbf{X}_i'\boldsymbol{\beta})^{y_i} \left\{1 - \Phi(\mathbf{X}_i'\boldsymbol{\beta})\right\}^{(1-y_i)}. \end{aligned}$$

Clearly this posterior density does not belong to a known family of distributions. If we want to estimate $\boldsymbol{\beta}$ by the posterior mean of this distribution then we need to evaluate the mean vector (and probably the covariance matrix as well) of this distribution. As argued above, the most simple way to achieve this is by producing draws from the posterior density and then using those draws to estimate the posterior summaries of interest. In Section 6.4 we will show that this posterior density can be sampled quite easily by the Markov chain Monte procedure of Albert and Chib (1993), which relies on modifying the target distribution by the introduction of latent data. For now we employ the M–H algorithm without latent data.

Random Walk Proposal Density
To define the proposal density let us gather some information from the likelihood function. Let

$$\begin{aligned}\hat{\boldsymbol{\beta}} &= \arg\max \ln L(\mathbf{y} \mid \mathbf{X}, \boldsymbol{\beta}) \\ &= (-1.093022 \quad 0.607643 \quad 1.197543 \quad -1.904739)'\end{aligned}$$

be the MLE found using the Newton–Raphson algorithm and let

$$\mathbf{V} = \begin{pmatrix} 0.040745 & -0.007038 & -0.039399 & 0.004829 \\ & 0.073101 & -0.006940 & -0.050162 \\ & & 0.062292 & -0.016803 \\ & & & 0.080788 \end{pmatrix}$$

be the symmetric matrix obtained by inverting the negative of the Hessian matrix (the matrix of second derivatives) of the log-likelihood function evaluated at $\hat{\boldsymbol{\beta}}$. Now suppose that proposal values are generated according to a random walk as

$$\begin{aligned}\boldsymbol{\beta} &= \boldsymbol{\beta}^{(j-1)} + \boldsymbol{\varepsilon}^{(j)} \\ \boldsymbol{\varepsilon}^{(j)} &\sim N_4(\mathbf{0}, \tau\mathbf{V}),\end{aligned} \tag{6.6}$$

where the scaler parameter τ is a tuning parameter adjusted at the start of the iterations to produce competitive proposal values (more about this shortly). In this case, the proposal density has the form

$$q(\boldsymbol{\beta}^{(j-1)}, \boldsymbol{\beta} \mid \mathbf{y}) \propto \exp\left\{-0.5\left(\boldsymbol{\beta} - \boldsymbol{\beta}^{(j-1)}\right)'(\tau\mathbf{V})^{-1}\left(\boldsymbol{\beta} - \boldsymbol{\beta}^{(j-1)}\right)\right\} \tag{6.7}$$

where we have suppressed the normalizing constant because it is irrelevant for the calculations.

To obtain draws from the multivariate normal density, we recall a fact from matrix algebra that any positive definite matrix $\mathbf{A}$ can be decomposed in a unique way into the form $\mathbf{LL}'$ where $\mathbf{L}$ is a lower triangular matrix with positive entries on the diagonal. Let us suppose that a matrix $\mathbf{L}$ has been found such that $\tau\mathbf{V} = \mathbf{LL}'$. Now the vector $\boldsymbol{\varepsilon}^{(j)}$ can be obtained from the $N_4(\mathbf{0}, \tau\mathbf{V})$ distribution as

$$\boldsymbol{\varepsilon}^{(j)} = \mathbf{L}\mathbf{z}^{(j)}$$

where $\mathbf{z}^{(j)}$ is distributed as $N_4(\mathbf{0}, \mathbf{I}_4)$.

Suppose we have completed $(j-1)$ steps of the algorithm and are deciding on the next simulated value. Then, we proceed to draw a proposal value $\boldsymbol{\beta}$ from Equation (6.6) and calculate the probability of move

$$\alpha(\boldsymbol{\beta}^{(j-1)}, \boldsymbol{\beta} \mid \mathbf{y}) = \min\left\{1, \frac{h(\boldsymbol{\beta} \mid \mathbf{y})}{h(\boldsymbol{\beta}^{(j-1)} \mid \mathbf{y})} \frac{q(\boldsymbol{\beta}, \boldsymbol{\beta}^{(j-1)} \mid \mathbf{y})}{q(\boldsymbol{\beta}^{(j-1)}, \boldsymbol{\beta} \mid \mathbf{y})}\right\}. \tag{6.8}$$

Note that the terms involving the proposal density cancel, a consequence of the fact that the random-walk proposal density is symmetric in its arguments. If we let $U \sim$ Uniform(0,1), then the next iterate is obtained as

$$\boldsymbol{\beta}^{(j)} = \begin{cases} \boldsymbol{\beta}, & \text{if } U < \alpha(\boldsymbol{\beta}^{(j-1)}, \boldsymbol{\beta} \mid \mathbf{y}) \\ \boldsymbol{\beta}^{(j-1)}, & \text{otherwise.} \end{cases}.$$

We can repeat this process to produce the sequence of values $\{\boldsymbol{\beta}^{(1)}, \ldots, \boldsymbol{\beta}^{(n_0+M)}\}$.

It is important to understand the role of the tuning parameter τ. From the expression of the probability of move we see that if τ is too large then the proposal values will be far from the current value and will likely get rejected; if τ is too small the proposal value will likely be accepted but successive values will be close together and hence the chain will only slowly explore the posterior distribution. In both cases, the sampled values of $\boldsymbol{\beta}$ from the posterior distribution will display strong serial correlations. To avoid this problem, it may be necessary to try some trial runs with different values of τ. Chib and Greenberg (1995) recommend adjusting τ to obtain an acceptance percentage of proposal values of between 30 and 50 percent. Using such considerations, we settle on the value $\tau = 1$ for the data at hand and run the M–H sampler for 5000 iterations beyond a burn-in of 100 iterations.

There are various ways, some graphical and some tabular, to summarize the simulation output. In Table 6.2, for example, we provide the prior and posterior first two moments, and the 2.5th (lower) and 97.5th (upper) percentiles, of the marginal densities of $\boldsymbol{\beta}$. Each of the quantities is computed from the posterior draws in an obvious way, the posterior standard deviation in the table is the standard

Table 6.2 Caesarian Data: Prior–Posterior Summary Based on 5000 Draws (Beyond a Burn-in of 100 Cycles) from the Random-walk M–H Algorithm

	Prior		Posterior			
	Mean	s.d.	Mean	s.d.	Lower	Upper
β_0	0.000	3.162	−1.110	0.224	−1.553	−0.677
β_1	0.000	3.162	0.612	0.254	0.116	1.127
β_2	0.000	3.162	1.198	0.263	0.689	1.725
β_3	0.000	3.162	−1.901	0.275	−2.477	−1.354

deviation of the sampled variates and the posterior percentiles are just the percentiles of the sampled draws.

As expected, both the first and second covariates increase the probability of infection while the third covariate (the antibiotics prophylaxis) reduces the probability of infection.

To get an idea of the form of the posterior density we plot in Figure 6.1 the four marginal posterior densities. The density plots are obtained by smoothing the histogram of the simulated values with a Gaussian kernel. In the same plot we also report the autocorrelation functions (correlation against lag) for each of the sampled parameter values. The autocorrelation plots provide information of the extent of serial dependence in the sampled values. Here we see that the serial correlations start out high but decline to almost zero by lag 20.

Tailored Proposal Density

Proposal values can be drawn according to another scheme, as discussed by Chib and Greenberg (1995). In this scheme, one utilizes both $\hat{\boldsymbol{\beta}}$ and $\mathbf{V}$, the two quantities defined above. If we assume a normal distribution, then the form of the proposal density is given by

$$q(\boldsymbol{\beta} \mid \mathbf{y}) \propto \exp\left\{-0.5\left(\boldsymbol{\beta} - \hat{\boldsymbol{\beta}}\right)' (\tau \mathbf{V})^{-1}\left(\boldsymbol{\beta} - \hat{\boldsymbol{\beta}}\right)\right\}.$$

This proposal density is similar to the random-walk proposal except that the distribution is centered at the fixed point $\hat{\boldsymbol{\beta}}$, not the previous item of the chain. We can refer to this proposal density as a *tailored* proposal density because it is matched to the target. The proposal values are now generated as

$$\boldsymbol{\beta} = \hat{\boldsymbol{\beta}} + \boldsymbol{\varepsilon}^{(j)},$$
$$\boldsymbol{\varepsilon}^{(j)} \sim N_4(\mathbf{0}, \tau \mathbf{V}),$$

and the probability of move is given by

$$\alpha(\boldsymbol{\beta}^{(j-1)}, \boldsymbol{\beta} \mid \mathbf{y}) = \min\left\{1, \frac{w(\boldsymbol{\beta} \mid \mathbf{y})}{w(\boldsymbol{\beta}^{(j-1)} \mid \mathbf{y})}\right\}, \tag{6.9}$$

where

$$w(\mathbf{t} \mid \mathbf{y}) = \frac{h(\mathbf{t} \mid \mathbf{y}, \mathbf{X})}{q(\mathbf{t} \mid \mathbf{y})}.$$

For this tailored proposal to work, it is necessary that the function $w(\mathbf{t} \mid \mathbf{y})$ be bounded. For a unimodal, absolutely continuous $h(\mathbf{t} \mid \mathbf{y}, \mathbf{X})$, this condition is satisfied if the tails of the proposal density are thicker than the tails of the target. This

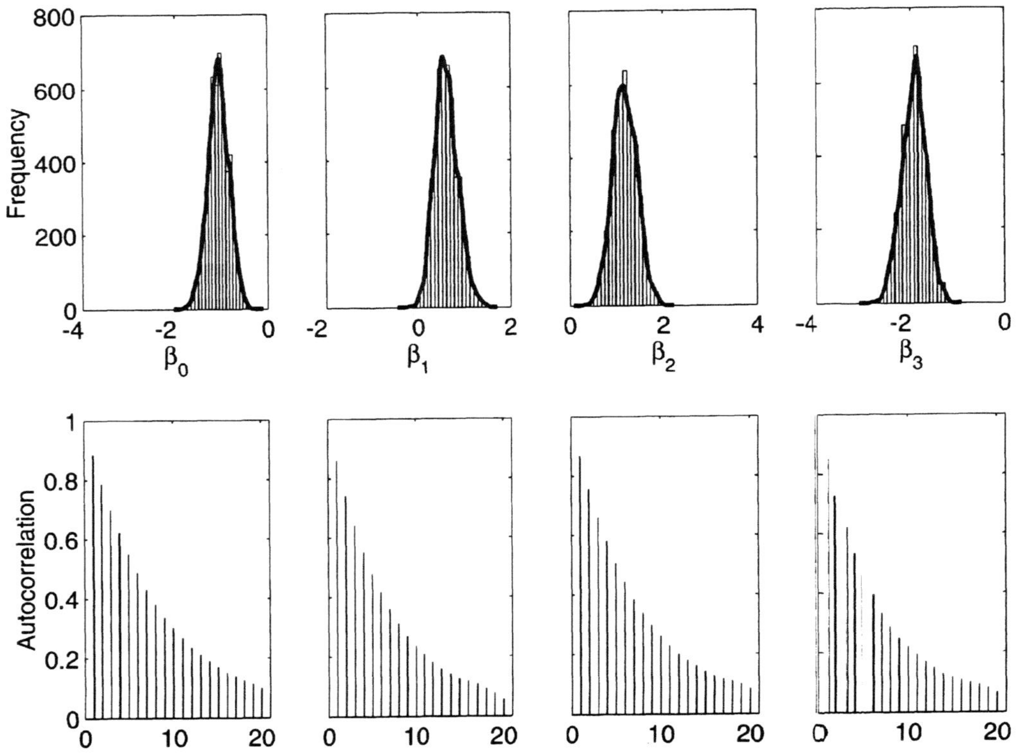

Figure 6.1 Caesarian data with random-walk M–H algorithm: Marginal posterior densities (top panels) and autocorrelation plot (bottom panels).

Table 6.3 Caesarian Data: Prior–posterior Summary Based on 5000 draws (Beyond a Burn-in of 100 cycles) from the Tailored M–H Algorithm

	Prior		Posterior			
	Mean	s.d.	Mean	s.d.	Lower	Upper
β_0	0.000	3.162	−1.080	0.220	−1.526	−0.670
β_1	0.000	3.162	0.593	0.249	0.116	1.095
β_2	0.000	3.162	1.181	0.254	0.680	1.694
β_3	0.000	3.162	−1.889	0.266	−2.421	−1.385

condition needs to be checked analytically although it is usually true when the proposal density is multivariate-t or when $\tau > 1$.

The prior–posterior summary based on 5000 draws of this M–H algorithm is given in Table 6.3. The proposal values are drawn as above from a t distribution with $\nu = 10$ degrees of freedom. We see that the marginal posterior moments are quite similar to those in Table 6.1. The marginal posterior densities are reported in the top panel of Figure 6.2. These are virtually identical to those computed using the random-walk M–H algorithm. The most notable difference is in the serial correlation

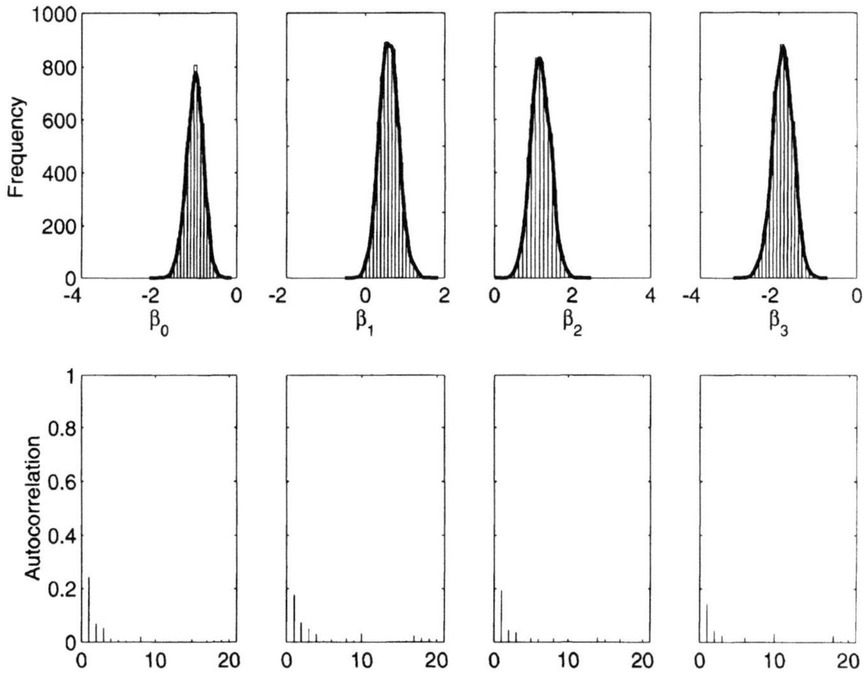

Figure 6.2 Caesarian data with tailored M–H algorithm: Marginal posterior densities (top panels) and autocorrelation plot (bottom panels)

plots, which now decline much more quickly to zero indicating that the algorithm is mixing quite well. The same information is revealed by the inefficiency factors, which are much closer to one than those from the previous algorithm.

The message from this analysis is that the two proposal densities produce similar results, with the differences appearing only in the autocorrelation plots (and inefficiency factors) of the sampled draws.

6.3 MULTIPLE-BLOCK M–H ALGORITHM

In applications where the dimension of $\boldsymbol{\theta}$ is large, it can be difficult to construct a single block M–H algorithm that is rapidly converging to the posterior density. In such cases, it is helpful to break up the parameter space into smaller blocks and then to construct a Markov chain in terms of these smaller blocks. To describe how such a chain can be constructed, suppose that $\boldsymbol{\theta}$ is split into two vector blocks $(\boldsymbol{\theta}_1, \boldsymbol{\theta}_2)$. For example, in a regression model, one block may consist of the regression coefficients and the other block may consist of the error variance. Next we specify two proposal densities,

$$q_1(\boldsymbol{\theta}_1, \boldsymbol{\theta}_1' \,|\, \mathbf{y}, \boldsymbol{\theta}_2); \; q_2(\boldsymbol{\theta}_2, \boldsymbol{\theta}_2' \,|\, \mathbf{y}, \boldsymbol{\theta}_1),$$

one for each block. Note that the proposal density q_k is allowed to depend on the data and the current value of the remaining block. Also define (by analogy with the single-block case)

$$\alpha(\boldsymbol{\theta}_1, \boldsymbol{\theta}_1' \,|\, \mathbf{y}, \boldsymbol{\theta}_2) = \min\left\{1, \frac{h(\boldsymbol{\theta}_1' \,|\, \mathbf{y}, \boldsymbol{\theta}_2) q_1(\boldsymbol{\theta}_1', \boldsymbol{\theta}_1 \,|\, \mathbf{y}, \boldsymbol{\theta}_2)}{h(\boldsymbol{\theta}_1 \,|\, \mathbf{y}, \boldsymbol{\theta}_2) q_1(\boldsymbol{\theta}_1, \boldsymbol{\theta}_1' \,|\, \mathbf{y}, \boldsymbol{\theta}_2)}\right\}, \tag{6.10}$$

and

$$\alpha(\boldsymbol{\theta}_2, \boldsymbol{\theta}_2' \,|\, \mathbf{y}, \boldsymbol{\theta}_1) = \min\left\{1, \frac{h(\boldsymbol{\theta}_2' \,|\, \mathbf{y}, \boldsymbol{\theta}_1) q_2(\boldsymbol{\theta}_2', \boldsymbol{\theta}_2 \,|\, \mathbf{y}, \boldsymbol{\theta}_1)}{h(\boldsymbol{\theta}_2 \,|\, \mathbf{y}, \boldsymbol{\theta}_1) q_2(\boldsymbol{\theta}_2, \boldsymbol{\theta}_2' \,|\, \mathbf{y}, \boldsymbol{\theta}_1)}\right\}, \tag{6.11}$$

as the probability of move for block $\boldsymbol{\theta}_k (k = 1, 2)$ conditioned on the other block. The conditional posterior densities

$$h(\boldsymbol{\theta}_1 \,|\, \mathbf{y}, \boldsymbol{\theta}_2) \text{ and } h(\boldsymbol{\theta}_2 \,|\, \mathbf{y}, \boldsymbol{\theta}_1)$$

that appear in these functions are called the *full conditional densities* and by Bayes' theorem each is just proportional to the joint posterior density. For example,

$$h(\boldsymbol{\theta}_1 \,|\, \mathbf{y}, \boldsymbol{\theta}_2) \propto h(\boldsymbol{\theta}_1, \boldsymbol{\theta}_2 \,|\, \mathbf{y}),$$

and, therefore, the probabilities of move in Equations (6.10) and (6.11) can be expressed equivalently in terms of the kernel of the joint posterior density

$h(\boldsymbol{\theta}_1, \boldsymbol{\theta}_2 \mid \mathbf{y})$ because the normalizing constant of the full conditional density (the norming constant in the latter expression) cancels in forming the ratio.

Then, in the multiple-block M–H algorithm, one sweep of the algorithm is completed by updating each block, say sequentially in fixed order, using a M–H step with the above probabilities of move, given the most current value of the other block. The algorithm is summarized as follows.

Algorithm: Multiple-Block Metropolis–Hastings

1. Specify an initial value $\boldsymbol{\theta}^{(0)} = (\boldsymbol{\theta}_1^{(0)}, \boldsymbol{\theta}_2^{(0)})$:

2. Repeat for $j = 1, 2, \ldots, n_0 + M$.

Repeat for $k = 1, 2$

a. Propose a value for the kth block, conditioned on the previous value of the kth block, and the current value of the other block $\boldsymbol{\theta}_{-k}$:

$$\boldsymbol{\theta}_k' \sim q_k(\boldsymbol{\theta}_k^{(j-1)}, \cdot \mid \mathbf{y}, \boldsymbol{\theta}_{-k}).$$

b. Calculate the probability of move

$$\alpha_k(\boldsymbol{\theta}_k^{(j-1)}, \boldsymbol{\theta}_k' \mid \mathbf{y}, \boldsymbol{\theta}_{-k}) = \min\left\{1, \frac{h(\boldsymbol{\theta}_k' \mid \mathbf{y}, \boldsymbol{\theta}_{-k}) q_k(\boldsymbol{\theta}_k', \boldsymbol{\theta}_k^{(j-1)} \mid \mathbf{y}, \boldsymbol{\theta}_{-k})}{h(\boldsymbol{\theta}_k^{(j-1)} \mid \mathbf{y}, \boldsymbol{\theta}_{-k}) q_k(\boldsymbol{\theta}_k^{(j-1)}, \boldsymbol{\theta}_k' \mid \mathbf{y}, \boldsymbol{\theta}_{-k})}\right\}.$$

c. Update the kth block as

$$\boldsymbol{\theta}_k^{(j)} = \begin{cases} \boldsymbol{\theta}_k' & \text{with Pr}\left[\alpha_k(\boldsymbol{\theta}_k^{(j-1)}, \boldsymbol{\theta}_k' \mid \mathbf{y}, \boldsymbol{\theta}_{-k})\right] \\ \boldsymbol{\theta}_k^{(j-1)} & \text{with Pr}\left[1 - \alpha_k(\boldsymbol{\theta}_k^{(j-1)}, \boldsymbol{\theta}_k' \mid \mathbf{y}, \boldsymbol{\theta}_{-k})\right] \end{cases}.$$

3. Return the values $\{\boldsymbol{\theta}^{(n_0+1)}, \boldsymbol{\theta}^{(n_0+2)}, \ldots, \boldsymbol{\theta}^{(n_0+M)}\}$.

The extension of this method to more than two blocks is straightforward.

REMARK: Versions of either random-walk or tailored proposal densities can be used in this algorithm, analogous to the single-block case. An important special case occurs if each proposal density is taken to be the full conditional density of that block. Specifically, if we set

$$q_1(\boldsymbol{\theta}_1^{(j-1)}, \boldsymbol{\theta}_1' \mid \mathbf{y}, \boldsymbol{\theta}_2) = h(\boldsymbol{\theta}_1' \mid \mathbf{y}, \boldsymbol{\theta}_2),$$

and

$$q_2(\boldsymbol{\theta}_2^{(j-1)}, \boldsymbol{\theta}_2' \mid \mathbf{y}, \boldsymbol{\theta}_1) = h(\boldsymbol{\theta}_2' \mid \mathbf{y}, \boldsymbol{\theta}_1),$$

then an interesting simplification occurs. The probability of move (for the first block) becomes

$$\alpha_1(\theta_1^{(j-1)}, \theta_1' \,|\, \mathbf{y}, \theta_2) = \min\left\{1, \frac{h(\theta_1' \,|\, \mathbf{y}, \theta_2)h(\theta_1^{(j-1)} \,|\, \mathbf{y}, \theta_2)}{h(\theta_1^{(j-1)} \,|\, \mathbf{y}, \theta_2)h(\theta_1' \,|\, \mathbf{y}, \theta_2)}\right\}$$
$$= 1,$$

and similarly for the second block, implying that if proposal values are drawn from their full conditional densities then the proposal values are accepted with probability one. This special case of the multiple-block M–H algorithm (in which *each* block is proposed using its full conditional distribution) is called the Gibbs sampling algorithm.

6.3.1 Gibbs Sampling Algorithm

The Gibbs sampling algorithm is one of the most simple Markov chain Monte Carlo algorithms. It was introduced by Geman and Geman (1984) in the context of image processing and then discussed in the context of missing data problems by Tanner and Wong (1987). The paper by Gelfand and Smith (1990) helped to demonstrate the value of the Gibbs algorithm for a range of problems in Bayesian analysis.

In the Gibbs sampling algorithm, the Markov chain is constructed by sampling the set of full conditional densities

$$\{h(\theta_1 \,|\, \mathbf{y}, \theta_2, \ldots, \theta_p); \quad h(\theta_2 \,|\, \mathbf{y}, \theta_1, \theta_3, \ldots, \theta_p); \quad \ldots, h(\theta_p \,|\, \mathbf{y}, \theta_1, \ldots, \theta_{p-1})\}, \tag{6.12}$$

where the blocks are chosen to ensure that the full conditional densities are all tractable. It turns out that in many Bayesian problems a Gibbs sampling scheme is suggested by the model structure itself.

As a simple example, consider data $\mathbf{y} = (y_1, \ldots, y_n)$ that is modeled by the linear regression

$$y_i \,|\, \boldsymbol{\beta}, \sigma^2 \sim N(\mathbf{x}_i'\boldsymbol{\beta}, \sigma^2), \qquad i = 1, 2, \ldots, n$$
$$\boldsymbol{\beta} \sim N_k(\boldsymbol{\beta}_0, \mathbf{B}_0)$$
$$\sigma^2 \sim \mathrm{IG}(0.5\nu_0, 0.5\delta_0),$$

where IG denotes the inverted-gamma distribution with density

$$p(\sigma^2) \propto \left(\frac{1}{\sigma^2}\right)^{0.5\nu_0+1} \exp\left(-\frac{\delta_0}{2\sigma^2}\right), \qquad \sigma^2 > 0.$$

Then, in this case, if one takes $\boldsymbol{\beta}$ and σ^2 as the two blocks, the full conditional density of $\boldsymbol{\beta}$ is the density $h(\boldsymbol{\beta} \mid \mathbf{y}, \sigma^2)$, which can be derived through some algebra as

$$\begin{aligned} h(\boldsymbol{\beta} \mid \mathbf{y}, \sigma^2) &\propto f(\mathbf{y}, \boldsymbol{\beta}, \sigma^2) \\ &\propto g(\boldsymbol{\beta}) L(\mathbf{y} \mid \boldsymbol{\beta}, \sigma^2) \\ &\propto \exp\{-0.5(\boldsymbol{\beta} - \hat{\boldsymbol{\beta}})' \mathbf{B}_n^{-1} (\boldsymbol{\beta} - \hat{\boldsymbol{\beta}})\}, \end{aligned}$$

where

$$\hat{\boldsymbol{\beta}} = \mathbf{B}_n \left(\mathbf{B}_0^{-1} \boldsymbol{\beta}_0 + \sigma^{-2} \sum_{i=1}^{n} \mathbf{x}_i y_i \right),$$

and

$$\mathbf{B}_n = \left(\mathbf{B}_0^{-1} + \sigma^{-2} \sum_{i=1}^{n} \mathbf{x}_i \mathbf{x}_i' \right)^{-1}.$$

Thus, the full conditional density of $\boldsymbol{\beta}$ is multivariate normal. The full conditional density of σ^2 is the density $h(\sigma^2 \mid \mathbf{y}, \boldsymbol{\beta})$ and can be shown to be an updated inverted-gamma distribution:

$$\sigma^2 \mid \mathbf{y}, \boldsymbol{\beta} \sim \mathrm{IG}\left\{ \frac{v_0 + n}{2}, \frac{\delta_0 + \sum_{i=1}^{n} (y_i - \mathbf{x}_i' \boldsymbol{\beta})^2}{2} \right\}.$$

Both full conditional densities are thus tractable. Note that the device of *data augmentation*, which is discussed below, is often helpful in obtaining a set of tractable full conditional distributions.

One cycle of the Gibbs sampling algorithm is completed by simulating $\{\boldsymbol{\theta}_k\}_{k=1}^{p}$ from each full conditional distribution, recursively updating the conditioning variables as one moves through the set of distributions. The Gibbs sampler in which each block is revised in fixed order is defined as follows.

Algorithm: Gibbs Sampling

1. Specify an initial value $\boldsymbol{\theta}^{(0)} = (\boldsymbol{\theta}_1^{(0)}, \ldots, \boldsymbol{\theta}_p^{(0)})$:

2. Repeat for $j = 1, 2, \ldots, n_0 + M$.

$$\begin{aligned}
&\text{Generate } \theta_1^{(j)} \text{ from } h(\theta_1 \,|\, \mathbf{y}, \theta_2^{(j-1)}, \theta_3^{(j-1)}, \ldots, \theta_p^{(j-1)})\\
&\text{Generate } \theta_2^{(j)} \text{ from } h(\theta_2 \,|\, \mathbf{y}, \theta_1^{(j)}, \theta_3^{(j-1)}, \ldots, \theta_p^{(j-1)})\\
&\vdots\\
&\text{Generate } \theta_p^{(j)} \text{ from } h(\theta_p \,|\, \mathbf{y}, \theta_1^{(j)}, \ldots, \theta_{p-1}^{(j)}).
\end{aligned}$$

3. Return the values $\{\theta^{(n_0+1)}, \theta^{(n_0+2)}, \ldots, \theta^{(n_0+M)}\}$.

In this algorithm, block θ_k is generated from the full conditional distribution

$$h(\theta_k \,|\, \mathbf{y}, \theta_1^{(j)}, \ldots, \theta_{k-1}^{(j)}, \theta_{k+1}^{(j-1)}, \ldots, \theta_p^{(j-1)}),$$

where the conditioning elements reflect the fact that when the kth block is reached, the previous $(k-1)$ blocks have already been updated. The transition density of the chain, again under the maintained assumption that h is absolutely continuous is, therefore, given by the product of transition kernels for each block:

$$T(\theta^{(j-1)}, \theta^{(j)} \,|\, \mathbf{y}) = \prod_{k=1}^{p} h(\theta_k \,|\, \mathbf{y}, \theta_1^{(j)}, \ldots, \theta_{k-1}^{(j)}, \theta_{k+1}^{(j-1)}, \ldots, \theta_p^{(j-1)}). \tag{6.13}$$

To illustrate the manner in which the blocks are revised, we consider a two-block case, each with a single component, and trace out in Figure 6.3 a possible trajectory of the sampling algorithm. The contours in the plot represent the joint distribution of θ and the labels (0), (1) etc., denote the simulated values. Note that one iteration of the algorithm is completed after both components are revised. Also notice that each component is revised along the direction of the coordinate axes. This feature can be a source of problems if the two components are highly correlated because then the contours become compressed and movements along the coordinate axes tend to produce only small moves. We return to this issue below.

Use of the Gibbs sampling algorithm is predicated on having tractable full conditional distributions. If one or more full conditional distributions are not of recognizable form then sampling of that block of parameters can be done by utilizing a suitable proposal density and applying the multiple-block M–H algorithm. Some authors have referred to such an algorithm as "Metropolis-within-Gibbs," but both Besag et al. (1995) and Chib and Greenberg (1995) disavow this terminology since it is a just a particular implementation of the multiple-block M–H algorithm.

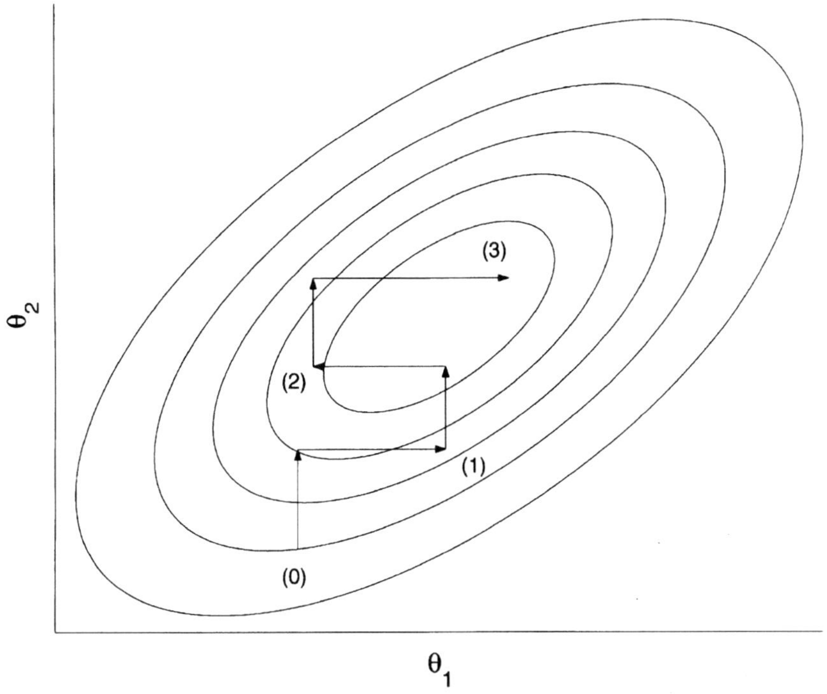

Figure 6.3 Gibbs algorithm with two blocks: three complete iterations of the algorithm

6.4 SOME TECHNIQUES USEFUL IN MCMC SAMPLING

6.4.1 Data Augmentation

In designing MCMC simulations, it is sometimes helpful to modify the target distribution by introducing *latent variables* or *auxiliary variables* into the sampling. This idea was called data augmentation by Tanner and Wong (1987) in the context of missing data problems. *Slice sampling*, which we do not discuss in this chapter, is a particular way of introducing auxiliary variables into the sampling; for example see Damien et al. (1999).

To fix notations, suppose that the target distribution is

$$h(\boldsymbol{\theta} \mid \mathbf{y}) \propto g(\boldsymbol{\theta}) L(\mathbf{y} \mid \boldsymbol{\theta}),$$

and that either because of the prior or because of the likelihood, the target distribution is computationally difficult. Now let $\mathbf{z}$ denote a vector of latent variables and let the modified target distribution be

$$h(\boldsymbol{\theta}, \mathbf{z} \mid \mathbf{y}) \propto g(\boldsymbol{\theta}, \mathbf{z}) L(\mathbf{y}, \mathbf{z} \mid \boldsymbol{\theta}).$$

If the latent variables were introduced tactically it may be the case that either $L(\mathbf{y}, \mathbf{z} \mid \boldsymbol{\theta})$ becomes tractable or that the conditional distribution of $\boldsymbol{\theta}$ given $(\mathbf{y}, \mathbf{z})$ is easy to derive. In the latter case, a multiple-block M–H simulation may be conducted with the blocks $\boldsymbol{\theta}$ and $\mathbf{z}$. The MCMC simulations would then produce the draws

$$\left(\boldsymbol{\theta}^{(n_0+1)}, \mathbf{z}^{(n_0+1)}\right), \ldots, \left(\boldsymbol{\theta}^{(n_0+M)}, \mathbf{z}^{(n_0+M)}\right) \sim h(\boldsymbol{\theta}, \mathbf{z} \mid \mathbf{y}),$$

where the draws on $\boldsymbol{\theta}$, ignoring those on the latent data, are from $h(\boldsymbol{\theta} \mid \mathbf{y})$, as required. Thus, the sampling of the desired target density is achieved in this indirect manner.

To demonstrate this technique, consider a univariate regression model defined by the specification

$$\begin{aligned} L(y_i \mid \boldsymbol{\beta}, \sigma^2) &\propto \sigma^{-2}\left\{1 + \frac{1}{v\sigma^2}(y_i - \mathbf{x}_i'\boldsymbol{\beta})^2\right\}^{-0.5(1+v)} \\ \boldsymbol{\beta} \mid \sigma^2 &\sim N_k(\boldsymbol{\beta}_0, \mathbf{B}_0) \\ \sigma^2 &\sim \mathrm{IG}(0.5v_0, 0.5\delta_0), \end{aligned}$$

in which the observations conditionally follow a Student t-density. If we assume that the degrees of freedom v of the Student t-density is known, and $\mathbf{B}_0$ does not depend upon σ^2 (so that $\boldsymbol{\beta}$ and σ^2 are independent), the target density becomes:

$$h(\boldsymbol{\beta}, \sigma^2 \mid \mathbf{y}) \propto g(\boldsymbol{\beta}) g(\sigma^2) \prod_{i=1}^{n} L(y_i \mid \boldsymbol{\beta}, \sigma^2).$$

This posterior density does not give rise to tractable full conditional densities $h(\boldsymbol{\beta} \mid \mathbf{y}, \sigma^2)$ and $h(\sigma^2 \mid \mathbf{y}, \boldsymbol{\beta})$.

However, if we note that the Student t-distribution can be expressed in the form (Carlin and Polson, 1991)

$$y_i \mid \boldsymbol{\beta}, \sigma^2, z_i \sim N(\mathbf{x}_i'\boldsymbol{\beta}, z_i^{-1}\sigma^2),$$

where

$$z_i \sim G(0.5v, 0.5v),$$

then a tractable full conditional distribution structure is available from the "augmented" or modified target density

$$\begin{aligned} h(\boldsymbol{\beta}, \sigma^2, \mathbf{z} \mid \mathbf{y}) \propto{} & \exp\{-0.5(\boldsymbol{\beta} - \boldsymbol{\beta}_0)'\mathbf{B}_0^{-1}(\boldsymbol{\beta} - \boldsymbol{\beta}_0)\}\left(\frac{1}{\sigma^2}\right)^{0.5v_0+1} \exp\left(-\frac{2}{\delta_0\sigma^2}\right) \\ & \times (\sigma^2)^{-0.5n} \prod_{i=1}^{n} z_i^{0.5} \exp\left(-\frac{z_i}{2\sigma^2}(y_i - \mathbf{x}_i'\boldsymbol{\beta})^2\right) z_i^{0.5v-1} \exp(-0.5z_i v), \end{aligned}$$

where $\mathbf{z} = (z_1, z_2, \ldots, z_n)$. One can now see that the full conditional posterior distribution $\boldsymbol{\beta} \,|\, \mathbf{y}, \sigma^2, \mathbf{z}$ is normal, $\sigma^2 \,|\, \mathbf{y}, \boldsymbol{\beta}, \mathbf{z}$ is inverse-gamma and $\mathbf{z} \,|\, \mathbf{y}, \boldsymbol{\beta}, \sigma^2$ factors into a set of independent gamma distributions. In particular, the three full conditional distributions are

$$N_k\left(\hat{\boldsymbol{\beta}}_{\mathbf{z}}, \mathbf{B}_{n,\mathbf{z}}\right),$$

$$\mathrm{IG}\left(\frac{\upsilon_0 + n}{2}, \frac{\delta_0 + \sum_{i=1}^{n} z_i(y_i - \mathbf{x}_i'\boldsymbol{\beta})^2}{2}\right),$$

and

$$G\left(\frac{\nu + 1}{2}, \frac{\nu + \sigma^{-2}(y_i - \mathbf{x}_i\boldsymbol{\beta})^2}{2}\right), \qquad i \leq n,$$

respectively, where $\hat{\boldsymbol{\beta}}_{\mathrm{z}} = \mathbf{B}_{n,\mathbf{z}}(\mathbf{B}_0^{-1}\boldsymbol{\beta}_0 + \sigma^{-2}\sum_{i=1}^{n} z_i \mathbf{x}_i y_i)$ and $\mathbf{B}_{n,\mathrm{z}} = (\mathbf{B}_0^{-1} + \sigma^{-2}\sum_{i=1}^{n} z_i \mathbf{x}_i \mathbf{x}_i')^{-1}$. These are easily sampled. Thus, by a judicious choice of augmenting variables $\mathbf{z}$, an intractable problem is turned into one that is fully tractable.

6.4.2 Method of Composition

The method of composition is a simple technique for sampling a joint density. It is based on the observation that if some joint density $\pi(\mathbf{z}, \boldsymbol{\theta})$ is expressed as

$$\pi(\mathbf{z}, \boldsymbol{\theta}) = \pi(\mathbf{z} \,|\, \boldsymbol{\theta})\pi(\boldsymbol{\theta})$$

and each density on the right-hand side is easily sampled, then a draw from the joint distribution may be obtained by

1. Drawing $\boldsymbol{\theta}^{(j)}$ from $\pi(\boldsymbol{\theta})$ and then
2. Drawing $\mathbf{z}^{(j)}$ from $\pi(\mathbf{z} \,|\, \boldsymbol{\theta}^{(j)})$

The method of composition can also be stated another way. If we are interested in obtaining a draw from the marginal distribution

$$\pi(\mathbf{z}) = \int \pi(\mathbf{z} \,|\, \boldsymbol{\theta})\pi(\boldsymbol{\theta})\, d\boldsymbol{\theta},$$

then the draw $\mathbf{z}^{(j)}$ obtained from the method of composition is a draw from the marginal distribution $\pi(\mathbf{z})$. This is just an implication of the fact that the draw $(\mathbf{z}^{(j)}, \boldsymbol{\theta}^{(j)})$ is a draw from the joint distribution. Note that in some applications,

$\pi(\boldsymbol{\theta})$ is a discrete mass distribution with support $\{\boldsymbol{\theta}_1, \ldots, \boldsymbol{\theta}_J\}$ and probabilities $\{p_1, \ldots, p_J\}$. In that case, $\pi(\mathbf{z})$ is a discrete mixture distribution

$$\pi(\mathbf{z}) = \sum_{j=1}^{J} \pi(\mathbf{z} \mid \boldsymbol{\theta}_j) p_j,$$

and a variate from $\pi(\mathbf{z})$ can again be drawn by the method of composition.

The method of composition finds many uses in simulation. One important use is in the context of drawing samples from the Bayes prediction density for a set of future observations $\mathbf{z}$. By definition, the prediction density (see Chapter 10) is defined as

$$h(\mathbf{z} \mid \mathbf{y}) = \int L(\mathbf{z} \mid \mathbf{y}, \theta) h(\boldsymbol{\theta} \mid \mathbf{y})\, d\boldsymbol{\theta},$$

where $L(\mathbf{z} \mid \mathbf{y}, \boldsymbol{\theta})$ is the density of the future observations conditioned on the data $\mathbf{y}$, and the parameters $\boldsymbol{\theta}$. To obtain a sample of draws from the predictive density we need to take each draw $\boldsymbol{\theta}^{(j)} \sim h(\boldsymbol{\theta} \mid \mathbf{y})$ from the MCMC simulation and produce the draw $\mathbf{z}^{(j)}$ from $L(\mathbf{z} \mid \mathbf{y}, \boldsymbol{\theta}^{(j)})$. The set of such draws constitute a sample from the predictive density, which can be summarized in the usual ways.

6.4.3 Reduced Blocking

As a general rule, sets of parameters that are highly correlated need to be sampled in one block when applying the multiple-block M–H algorithm. The importance of such coarse blocking or *reduced blocking* has been highlighted in a number of different problems, for example, the state space model, hidden Markov model, and clustered data models with random effects. In each of these models, auxiliary variables are required in the sampling and the parameter space is quite large. The problem arises from the fact that the latent variables tend to be highly correlated, either amongst themselves, or with the other parameters in the model.

One way to implement reduced blocking is by the method of composition. For example, suppose that $\boldsymbol{\theta}_1$, $\boldsymbol{\theta}_2$ and $\boldsymbol{\theta}_3$ are three blocks and that the distribution $\boldsymbol{\theta}_1 \mid \boldsymbol{\theta}_3$ is tractable (i.e., can be sampled directly). Then, the blocks $(\boldsymbol{\theta}_1, \boldsymbol{\theta}_2)$ can be collapsed by first sampling $\boldsymbol{\theta}_1$ from $\boldsymbol{\theta}_1 \mid \boldsymbol{\theta}_3$ followed by $\boldsymbol{\theta}_2$ from $\boldsymbol{\theta}_2 \mid \boldsymbol{\theta}_1, \boldsymbol{\theta}_3$. This amounts to a two block MCMC algorithm. In addition, if it is possible to sample $(\boldsymbol{\theta}_1, \boldsymbol{\theta}_2)$ marginalized over $\boldsymbol{\theta}_3$ then the number of blocks is reduced to one. Liu (1994) and Liu et al. (1994) discuss the value of these strategies in the context of a three-block Gibbs MCMC chain. Roberts and Sahu (1997) provide further discussion of the role of blocking in the context of Gibbs Markov chains used to sample multivariate normal target distributions.

6.4.4 Rao–Blackwellization

The MCMC simulations can be used to estimate marginal and conditional densities of the parameters. Suppose that interest is in the marginal density of $\boldsymbol{\theta}_k$ at the point $\boldsymbol{\theta}_k^*$

$$h(\boldsymbol{\theta}_k^* \mid \mathbf{y}) = \int h(\boldsymbol{\theta}_k^* \mid \mathbf{y}, \boldsymbol{\theta}_{-k})\, h(\boldsymbol{\theta}_{-k} \mid \mathbf{y})\, d\boldsymbol{\theta}_{-k}, \tag{6.14}$$

where $\boldsymbol{\theta}_{-k}$ as before (Section 6.2), denotes the set of blocks excluding $\boldsymbol{\theta}_k$. If we know the normalizing constant of $h(\boldsymbol{\theta}_k^* \mid \mathbf{y}, \boldsymbol{\theta}_{-k})$, which is often the case, then we can estimate the marginal density as an average of the full conditional density over the simulated values of $\boldsymbol{\theta}_{-k}$:

$$\hat{h}(\boldsymbol{\theta}_k^* \mid \mathbf{y}) = M^{-1} \sum_{j=1}^{M} h(\boldsymbol{\theta}_k^* \mid \mathbf{y}, \boldsymbol{\theta}_{-k}^{(j)}).$$

Gelfand and Smith (1990) refer to this approach as "Rao–Blackwellization" because of the connection with the Rao–Blackwell theorem in classical statistics. That connection is more clearly seen in the context of estimating (say) the mean of $\boldsymbol{\theta}_k$, $E(\boldsymbol{\theta}_k \mid \mathbf{y}) = \int \boldsymbol{\theta}_k h(\boldsymbol{\theta}_k \mid \mathbf{y}) d\boldsymbol{\theta}_k$. By the law of the iterated expectation,

$$E(\boldsymbol{\theta}_k) = E\{E(\boldsymbol{\theta}_k \mid \boldsymbol{\theta}_{-k})\},$$

and therefore the estimates

$$M^{-1} \sum_{j=1}^{M} \boldsymbol{\theta}_k^j,$$

and

$$M^{-1} \sum_{j=1}^{M} E(\boldsymbol{\theta}_k \mid \boldsymbol{\theta}_{-k}^{(j)}),$$

both converge to $E(\boldsymbol{\theta}_k \mid \mathbf{y})$ as $M \to \infty$. Under i.i.d. sampling, and under Markov sampling provided some conditions are satisfied (Liu et al., 1994; Casella and Robert 1996), the variance of the latter estimate is less than that of the former. Thus, if possible, one should average the conditional mean $E(\boldsymbol{\theta}_k \mid \mathbf{y}, \boldsymbol{\theta}_{-\mathbf{k}})$. Gelfand and Smith appeal to this analogy to argue that the Rao-Blackwellized estimate of the density is preferable to that based on the method of kernel smoothing. Chib (1995) extends the Rao–Blackwellization approach to estimate "reduced conditional ordinates," defined as the density of $\boldsymbol{\theta}_k$ conditioned on one or more of the remaining blocks. More discussion of this is provided below in Section 6.5.

6.5 EXAMPLES

6.5.1 Binary Response Data (Continued)

We now return to the binary data model discussed in Section 6.1.1 to show how a MCMC sampler can be developed without use of the Metropolis algorithm. Recall that the posterior density of $\boldsymbol{\beta}$ in the binary response model is proportional to

$$h(\boldsymbol{\beta} \mid \mathbf{y}) \propto g(\boldsymbol{\beta}) \prod_{i=1}^{n} p_i^{y_i}(1 - p_i)^{(1-y_i)},$$

where $p_i = \Phi(\mathbf{x}_i'\boldsymbol{\beta})$. To deal with this model, and others involving binary and ordinal data, Albert and Chib (1993) introduce a technique that has been applied to many problems. The Albert–Chib algorithm capitalizes on the simplifications afforded by introducing latent or auxiliary data into the sampling.

Instead of the specification above, the model is specified as

$$\begin{aligned} z_i \mid \boldsymbol{\beta} &\sim N(\mathbf{x}_i'\boldsymbol{\beta}, 1), \\ y_i &= I[z_i > 0], \qquad i \leq n, \\ \boldsymbol{\beta} &\sim N_k(\boldsymbol{\beta}_0, \mathbf{B}_0). \end{aligned} \tag{6.15}$$

This specification is equivalent to the binary probit regression model since

$$\Pr(y_i = 1 \mid \mathbf{x}_i, \boldsymbol{\beta}) = \Pr(z_i > 0 \mid \mathbf{x}_i, \boldsymbol{\beta}) = \Phi(\mathbf{x}_i'\boldsymbol{\beta}),$$

as required.

Now the algorithm proceeds with the sampling of the full conditional distributions

$$\boldsymbol{\beta} \mid \mathbf{y}, \{z_i\}; \qquad \{z_i\} \mid \mathbf{y}, \boldsymbol{\beta},$$

where both these distributions are tractable (i.e., requiring no M–H steps). Specifically, the distribution of $\boldsymbol{\beta}$ conditioned on the latent data becomes independent of the observed data and has the same form as in the Gaussian linear regression model with the response data given by $\{z_i\}$:

$$\begin{aligned} h(\boldsymbol{\beta} \mid \mathbf{y}, \{z_i\}) &= h(\boldsymbol{\beta} \mid \{z_i\}) \\ &\propto \exp\{-0.5(\boldsymbol{\beta} - \boldsymbol{\beta}_0)'\mathbf{B}_0^{-1}(\boldsymbol{\beta} - \boldsymbol{\beta}_0)\} \\ &\times \prod_{i=1}^{n} \exp(-0.5(z_i - \mathbf{x}_i'\boldsymbol{\beta})^2), \end{aligned}$$

which by the standard calculations is a multivariate normal density with mean $\hat{\boldsymbol{\beta}} = \mathbf{B}(\mathbf{B}_0^{-1}\boldsymbol{\beta}_0 + \sum_{i=1}^{n} \mathbf{x}_i z_i)$ and variance matrix $\mathbf{B} = (\mathbf{B}_0^{-1} + \sum_{i=1}^{n} \mathbf{x}_i\mathbf{x}_i')^{-1}$.

Next, the distribution of the latent data conditioned on the data and the parameters factor into a set of n independent distributions with each depending on the data only through y_i:

$$\{z_i\} \,|\, \mathbf{y}, \boldsymbol{\beta} \overset{d}{=} \prod_{i=1}^{n} z_i \,|\, y_i, \boldsymbol{\beta}.$$

The distributions $z_i \,|\, y_i, \boldsymbol{\beta}$ are obtained by reasoning as follows. Suppose that $y_i = 0$; then from Bayes' theorem

$$\begin{aligned} L(z_i \,|\, y_i = 0, \boldsymbol{\beta}) &\propto N(z_i \,|\, \mathbf{x}_i'\boldsymbol{\beta}, 1) L(y_i = 0 \,|\, z_i, \boldsymbol{\beta}) \\ &\propto N(z_i \,|\, \mathbf{x}_i'\boldsymbol{\beta}, 1) I[z_i \leq 0], \end{aligned} \tag{6.16}$$

because $L(y_i = 0 \,|\, z_i, \boldsymbol{\beta})$ is equal to one if z_i is negative, and equal to zero otherwise, which is the definition of $I[z_i \leq 0]$. Hence, the information $y_i = 0$ simply serves to truncate the support of z_i. By a similar argument, the support of z_i is $(0, \infty)$ when conditioned on the event $y_i = 1$:

$$L(z_i \,|\, y_i = 1, \boldsymbol{\beta}) \propto N(z_i \,|\, \mathbf{x}_i'\boldsymbol{\beta}, 1) I[z > 0]. \tag{6.17}$$

We note that each of the truncated normal distributions is simulated by applying the *inverse cdf* method (Ripley, 1987). Specifically, it can be shown that the draw

$$\mu + \sigma\Phi^{-1}\left[\Phi\left(\frac{a-\mu}{\sigma}\right) + U\left(\Phi\left(\frac{b-\mu}{\sigma}\right) - \Phi\left(\frac{a-\mu}{\sigma}\right)\right)\right], \tag{6.18}$$

where Φ^{-1} is the inverse cdf of the $N(0, 1)$ distribution and $U \sim \text{Unif}(0, 1)$ is a draw from a $N(\mu, \sigma^2)$ distribution truncated to the interval (a, b).

The algorithm may now be summarized as follow.

Algorithm: Binary Probit

1. Sample

$$\boldsymbol{\beta} \sim N_k\left(\mathbf{B}(\mathbf{B}_0^{-1}\boldsymbol{\beta}_0 + \sum_{i=1}^{n} \mathbf{x}_i z_i), \mathbf{B} = (\mathbf{B}_0^{-1} + \sum_{i=1}^{n} \mathbf{x}_i \mathbf{x}_i')^{-1}\right)$$

2. Sample

$$z_i \sim \begin{cases} N(z_i \,|\, \mathbf{x}_i'\boldsymbol{\beta}, 1) I[z_i < 0], & \text{if } y_i = 0, \quad i \leq n \\ N(z_i \,|\, \mathbf{x}_i'\boldsymbol{\beta}, 1) I[z > 0], & \text{if } y_i = 1, \quad i \leq n \end{cases}$$

3. Go to 1.

Table 6.4 Caesarian Data: Prior–Posterior Summary Based on 5000 Draws (Beyond a Burn-in of 100 Cycles) from the Albert–Chib Algorithm

	Prior		Posterior			
	Mean	s.d.	Mean	s.d.	Lower	Upper
β_0	0.000	3.162	−1.100	0.210	−1.523	−0.698
β_1	0.000	3.162	0.609	0.249	0.126	1.096
β_2	0.000	3.162	1.202	0.250	0.712	1.703
β_3	0.000	3.162	−1.903	0.266	−2.427	−1.393

The results, based on 5000 MCMC draws beyond a burn-in of 100 iterations, are reported in Table 6.4 and Figure 6.4. We see that the results correspond closely to those reported above, especially to those from the tailored M–H chain.

6.5.2 Hierarchical Model for Clustered Data

Markov chain Monte Carlo methods have proved useful in all areas of Bayesian applications. As another illustration, consider a hierarchical Bayesian model for

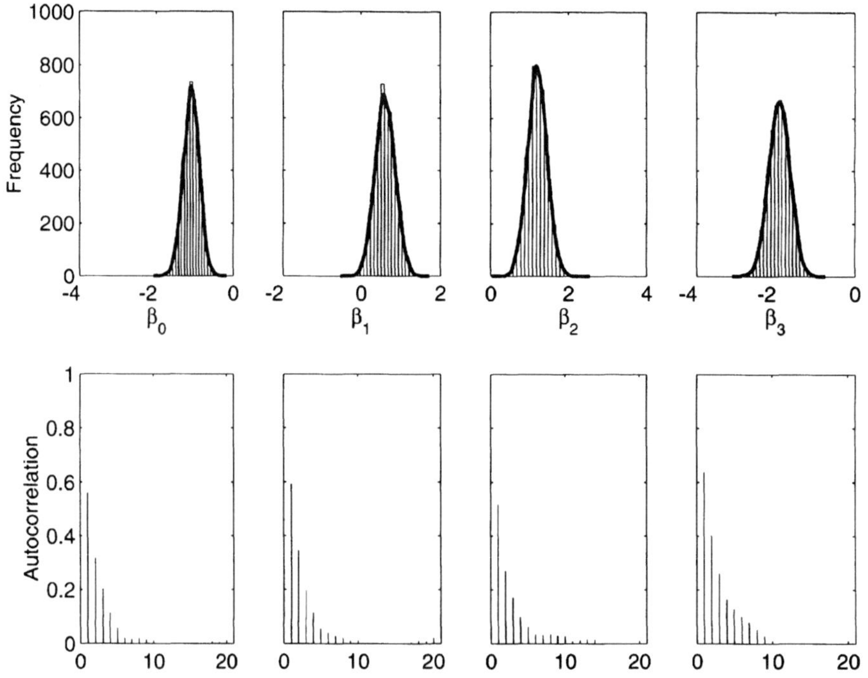

Figure 6.4 Caesarian data with Albert–Chib algorithm: marginal posterior densities (top panels) and autocorrelation plot (bottom panels)

clustered data (hierarchical models are discussed in detail in Chapter 14). The data are from a clinical trial on the effectiveness of two antiretroviral drugs (didanosine or (ddI) and zalcitabine or (ddC)) in 467 persons with advanced HIV infection. The response variable y_{ij} for patient i at time j is the square root of the patient's CD4 count, a serological measure of immune system health and prognostic factor for AIDS-related illness and mortality. The data set records patient CD4 counts at study entry and again at 2, 6, 12, and 18 months after entry, for the ddI and ddC groups, respectively.

The model is formulated as follows. If we let y_i denote a n_i vector of responses across time for the ith patient, then following the discussion in Carlin and Louis (2000), suppose

$$\begin{aligned} \mathbf{y}_i \mid \boldsymbol{\beta}, \mathbf{b}_i \sigma^2 &\sim N_{n_i}(\mathbf{X}_i \boldsymbol{\beta} + \mathbf{W}_i \mathbf{b}_i, \sigma^2 \mathbf{I}_{n_i}) \\ \mathbf{b}_i \mid \mathbf{D} &\sim N_2(\mathbf{0}, \mathbf{D}), \qquad i \leq 467, \end{aligned} \tag{6.19}$$

where the jth row of the patient i's design matrix $\mathbf{W}_i$ takes the form $\mathbf{w}_{ij} = (1, t_{ij})$, t_{ij} belongs to the set $\{0, 2, 6, 12, 18\}$, and the fixed design matrix $\mathbf{X}_i$ is obtained by horizontal concatenation of $\mathbf{W}_i$, $d_i \mathbf{W}_i$ and $a_i \mathbf{W}_i$, where d_i is a binary variable indicating whether patient i received ddI ($d_i = 1$) or ddC ($d_i = 0$), and a_i is a binary variable indicating if the patient was diagnosed as having AIDS at baseline ($a_i = 1$) or not ($a_i = 0$).

The prior distribution of $\boldsymbol{\beta}: 6 \times 1$ is assumed to be $N_6(\boldsymbol{\beta}_0, \mathbf{B}_0)$ with

$$\boldsymbol{\beta}_0 = (10, 0, 0, 0, -3, 0),$$

and

$$\mathbf{B}_0 = \mathrm{diag}(2^2, 1^2, (0.1)^2, 1^2, 1^2, 1^2),$$

while that on $\mathbf{D}^{-1}$ is taken to be Wishart $W(\mathbf{R}_0/\rho_0, 2, \rho_0)$ with $\rho_0 = 24$ and $\mathbf{R}_0 = \mathrm{diag}(0.25, 16)$. Finally, σ^2 is *a priori* assumed to followed the inverse-gamma distribution

$$\sigma^2 \sim \mathrm{IG}(0.5\nu_0, 0.5\delta_0),$$

with $\nu_0 = 6$ and $\delta_0 = 120$ (which imply a prior mean and standard deviation both equal to 30).

To deal with this model, it is helpful to include the $\{\mathbf{b}_i\}$ in the sampling. The joint distribution of the data and the parameters (under the assumption that the observations across clusters are conditionally independent) is given by

$$\begin{aligned} L(\mathbf{y}, \boldsymbol{\beta}, \{\mathbf{b}_i\}, \mathbf{D}, \sigma^2 &= g(\boldsymbol{\beta}, \{\mathbf{b}_i\}, \mathbf{D}, \sigma^2) L(\mathbf{y} \mid \boldsymbol{\beta}, \{\mathbf{b}_i\}, \mathbf{D}, \sigma^2) \\ &= g(\boldsymbol{\beta}) g(\mathbf{D}) g(\sigma^2) \prod_{i=1}^{n} L(\mathbf{y}_i \mid \boldsymbol{\beta}, \mathbf{b}_i, \sigma^2) g(\mathbf{b}_i \mid \mathbf{D}). \end{aligned} \tag{6.20}$$

Wakefield et al. (1994) propose a Gibbs MCMC approach for joint distribution that is based on full blocking (i.e., sampling each block of parameters from their full conditional distribution). This blocking scheme is not very desirable because the random effects and the fixed effects $\boldsymbol{\beta}$ tend to be highly correlated and treating them as separate blocks creates problems with mixing (Gelfand et al., (1995). To deal with this problem, Chib and Carlin (1999) suggest a number of reduced blocking schemes. One of the most simple proceeds by noting that $\boldsymbol{\beta}$ and $\{\mathbf{b}_i\}$ can be sampled in one block by the method of composition, first sampling $\boldsymbol{\beta}$ marginalized over $\{\mathbf{b}_i\}$ and then sampling $\{\mathbf{b}_i\}$ conditioned on $\boldsymbol{\beta}$.

What makes reduced blocking possible is the fact that the $\mathbf{b}_i$ in Equation (6.20) can be marginalized out leaving a normal distribution that can be combined with the assumed normal prior on $\boldsymbol{\beta}$. In particular, we can derive that

$$\begin{aligned} L(\mathbf{y}_i \mid \boldsymbol{\beta}, \mathbf{D}, \sigma^2) &= \int L(\mathbf{y}_i \mid \boldsymbol{\beta}, \mathbf{b}_i, \sigma^2) g(\mathbf{b}_i \mid \mathbf{D})\, d\mathbf{b}_i \\ &\propto |\mathbf{V}_i|^{-0.5} \exp\{-0.5(\mathbf{y}_i - \mathbf{X}_i\boldsymbol{\beta})' \mathbf{V}_i^{-1} (\mathbf{y}_i - \mathbf{X}_i\boldsymbol{\beta})\}, \end{aligned}$$

where

$$\mathbf{V}_i = \sigma^2 \mathbf{I}_{n_i} + \mathbf{W}_i \mathbf{D} \mathbf{W}_i'$$

and hence the reduced conditional posterior of $\boldsymbol{\beta}$ is obtained as multivariate normal by the usual calculations:

$$\begin{aligned} h(\boldsymbol{\beta} \mid \mathbf{y}, \mathbf{D}, \sigma^2) &\propto g(\boldsymbol{\beta}) \prod_{i=1}^{n} |\mathbf{V}_i|^{-0.5} \exp\{-0.5(\mathbf{y}_i - \mathbf{X}_i\boldsymbol{\beta})' \mathbf{V}_i^{-1} (\mathbf{y}_i - \mathbf{X}_i\beta)\} \\ &\propto \exp\{-0.5(\boldsymbol{\beta} - \hat{\boldsymbol{\beta}})' \mathbf{B}^{-1} (\boldsymbol{\beta} - \hat{\boldsymbol{\beta}})\}, \end{aligned}$$

where

$$\hat{\boldsymbol{\beta}} = \mathbf{B}\left(\mathbf{B}_0^{-1} \boldsymbol{\beta}_0 + \sum_{i=1}^{n} \mathbf{X}_i' \mathbf{V}_i^{-1} \mathbf{y}_i\right),$$

and

$$\mathbf{B} = \left(\mathbf{B}_0^{-1} + \sum_{i=1}^{n} \mathbf{X}_i' \mathbf{V}_i^{-1} \mathbf{X}_i\right)^{-1}.$$

The rest of the algorithm follows the steps of Wakefield et al., (1994). In particular, the sampling of the random effects is from independent normal distributions that are derived by treating $(\mathbf{y}_i - \mathbf{X}_i\boldsymbol{\beta})$ as the data, $\mathbf{b}_i$ as the regression coefficient, and $\mathbf{b}_i \sim N_q(\mathbf{0}, \mathbf{D})$ as the prior. The sampling of $\mathbf{D}^{-1}$ is from a Wishart distribution and that of σ^2 from an inverse gamma distribution.

Algorithm: Gaussian Panel

1. Sample

$$\boldsymbol{\beta} \sim N_6(\hat{\boldsymbol{\beta}}, \mathbf{B})$$

2. Sample

$$\mathbf{b}_i \sim N_2\left[\mathbf{D}_i\sigma^{-2}\mathbf{W}_i'(\mathbf{y}_i - \mathbf{X}_i\boldsymbol{\beta}), \mathbf{D}_i = (\mathbf{D}^{-1} + \sigma^{-2}\mathbf{W}_i'\mathbf{W}_i)^{-1}\right], \qquad i \leq n$$

3. Sample

$$\mathbf{D}^{-1} \sim W_2\left\{\rho_0 + n, \left(\mathbf{R}_0^{-1} + \sum_{i=1}^{n} \mathbf{b}_i\mathbf{b}_i'\right)^{-1}\right\}$$

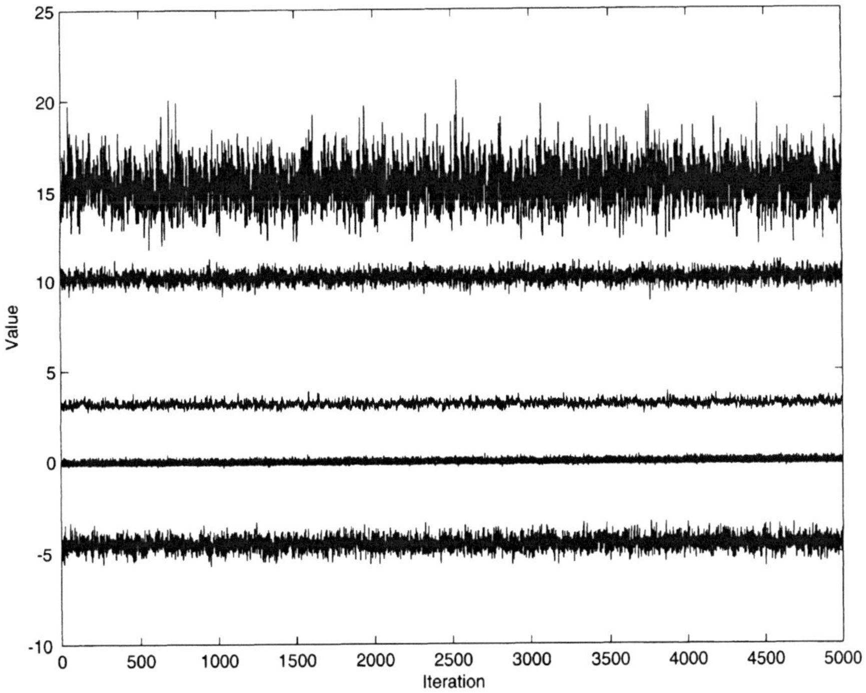

Figure 6.5 AIDS clustered data: simulated values by iteration for each of ten parameters

4. Sample

$$\sigma^2 \sim \text{IG}\left(\frac{\nu_0 + \sum n_i}{2}, \frac{\delta_0 + \sum_{i=1}^{n} \| \mathbf{y}_i - \mathbf{X}_i \boldsymbol{\beta} - \mathbf{W}_i \mathbf{b}_i \|^2}{2} \right)$$

5. Goto 1

The MCMC simulation is run for 5000 cycles beyond a burn-in of 100 cycles. The simulated values by iteration for each of the ten parameters are given in Figure 6.5. Except for the parameters that are approximately the same, the sampled paths of the parameters are clearly visible and display little correlation.

These draws from the posterior distribution, as before, can be used to produce different summaries of the posterior distribution. In Figure 6.6 we report the marginal posterior distributions in the form of histogram plots. We see that three of the regression parameters are centered at zero, that D_{11} is large and D_{22} (which is the variance of the time-trend random effect) is small. As for the mixing of the sampler, the inefficiency factors for all parameters are low: 1 for each of the regres-

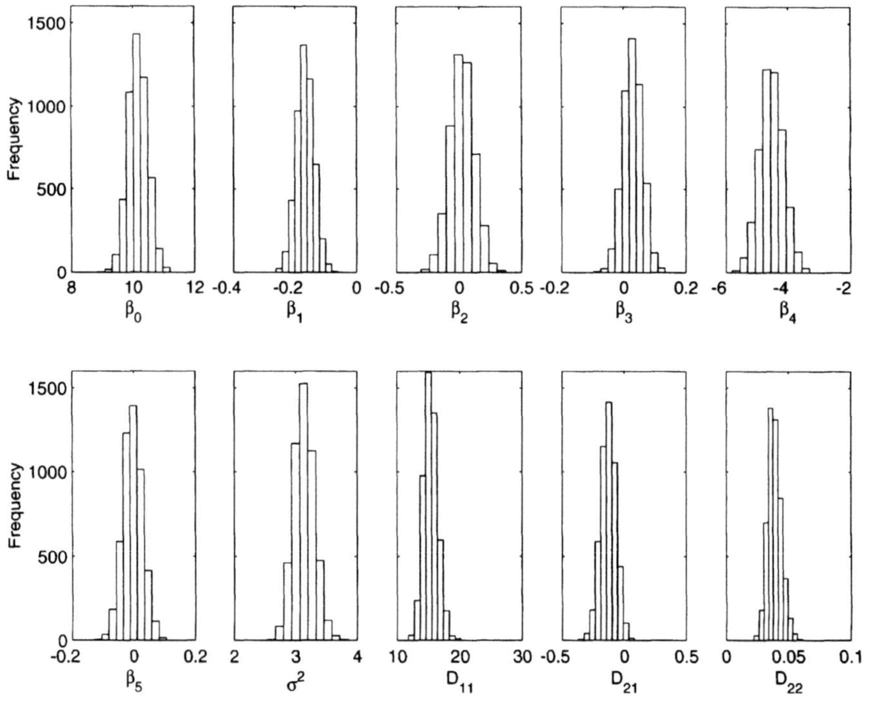

Figure 6.6 AIDS clustered data: marginal posterior distributions of parameters on 5000 MCMC draws

sion parameters, 2.78 for σ^2, and 1.54, 4.83, and 9.20 for the three elements of **D**.

6.6 COMPARING MODELS USING MCMC METHODS

Posterior simulation by MCMC methods does not require knowledge of the normalizing constant of the posterior density. Nonetheless, if we are interested in comparing alternative models, then knowledge of the normalizing constant is essential. This is because the standard and formal Bayesian approach for comparing models is via *Bayes' factors*, or ratios of *marginal likelihoods*. The marginal likelihood of a particular model is the normalizing constant of the posterior density and is defined as

$$m(\mathbf{y}) = \int L(\mathbf{y} \mid \boldsymbol{\theta}) g(\boldsymbol{\theta})\, d\boldsymbol{\theta}, \tag{6.21}$$

the integral of the likelihood function with respect to the prior density. If we have two models M_k and M_l, then the Bayes' factor is the ratio

$$\mathrm{B_{kl}} = \frac{\mathrm{m}(\mathbf{y} \mid \mathcal{M}_\mathbf{k})}{\mathrm{m}(\mathbf{y} \mid \mathcal{M}_\mathbf{l})} \tag{6.22}$$

Computation of the marginal likelihood is, therefore, of some importance in Bayesian statistics (DiCiccio, et al., 1997 Chen and Shao 1997). Unfortunately, because MCMC methods deliver draws from the posterior density, and the marginal likelihood is the integral with respect to the prior, the MCMC output cannot be used directly to average the likelihood. To deal with this problem, a number of methods have appeared in the literature. One quite simple and widely applicable method is due to Chib (1995) and Chib and Jeliazkov (2001), which we briefly explain as follows.

Begin by noting that $m(\mathbf{y})$, by virtue of being the normalizing constant of the posterior density, can be expressed as

$$m(\mathbf{y}) = \frac{L(\mathbf{y} \mid \boldsymbol{\theta}^*) g(\boldsymbol{\theta}^*)}{h(\boldsymbol{\theta}^* \mid \mathbf{y})}, \tag{6.23}$$

for any given point $\boldsymbol{\theta}^*$ (generally taken to be a high density point such as the posterior mean). Thus, provided we have an estimate $\hat{h}(\boldsymbol{\theta}^* \mid \mathbf{y})$ of the posterior ordinate, the marginal likelihood can be estimated on the log scale as

$$\log m(\mathbf{y}) = \log L(\mathbf{y} \mid \boldsymbol{\theta}^*) + \log g(\boldsymbol{\theta}^*) - \log \hat{h}(\boldsymbol{\theta}^* \mid \mathbf{y}). \tag{6.24}$$

In the context of both single and multiple block M—H chains, good estimates of the posterior ordinate are available. For example, in the latter situation with k blocks, to estimate the posterior ordinate we can employ the marginal-conditional decomposition

$$h(\theta^* \mid \mathbf{y}) = h(\theta_1^* \mid \mathbf{y}) h(\theta_2^* \mid \mathbf{y}, \theta_1^*) \ldots h(\theta_k^* \mid \mathbf{y}, \theta_1^*, \theta_2^*, \ldots, \theta_{k-1}^*), \tag{6.25}$$

where the first ordinate can be estimated by draws from the full MCMC runs, and the second and subsequent ordinates from the output of *reduced MCMC* runs. These reduced runs are defined by fixing the appropriate parameter block and conducting the MCMC simulations with the blocks that are free.

As an example of this method, the posterior ordinate for the clustered data model in Section 6.4.2 can be written as

$$h(\mathbf{D}^{-1*}, \sigma^{2*}, \boldsymbol{\beta}^* \mid \mathbf{y}) = h(\mathbf{D}^{-1*} \mid \mathbf{y}) h(\sigma^{2*} \mid \mathbf{y}, \mathbf{D}^*) h(\boldsymbol{\beta}^* \mid \mathbf{y}, \mathbf{D}^*, \sigma^{2*}),$$

where the first term is obtained by averaging the Wishart density over draws on $\{\mathbf{b}_i\}$ from the full run. To estimate the second ordinate, which is conditioned on $\mathbf{D}^*$, we run a reduced MCMC simulation with the full conditional densities

$$h(\boldsymbol{\beta} \mid \mathbf{y}, \mathbf{D}^*, \sigma^2);\ h(\sigma^2 \mid \mathbf{y}, \boldsymbol{\beta}, \mathbf{D}^*, \{\mathbf{b}_i\});\ h(\{\mathbf{b}_i\} \mid \mathbf{y}, \boldsymbol{\beta}, \mathbf{D}^*, \sigma^2),$$

where each conditional utilizes the fixed value of $\mathbf{D}$. The second ordinate can now be estimated by averaging the inverse-gamma full conditional density of σ^2 at σ^{2*} over the draws on $(\boldsymbol{\beta}, \{b_i\})$ from this reduced run. The third ordinate is multivariate normal as given above and available directly.

An alternative approach for finding the normalizing constant is via a simulation over both model space and parameter space, for example, as in the work of Carlin and Chib (1995) and Green (1995). Dellaportas et al. (2000) and Godsill (1999) compare and contrast the latter two methods. The model space approach is often applied when the model space is large, as for example in a variable-selection problem. The use of these methods when the model space is large is not without difficulties, however, because one cannot be sure that all models supported by the data have been visited according to their posterior probabilities. If the model space is diminished to ensure better coverage of the various models then it may happen that direct computation of the marginal likelihoods is more simple and faster. This tension in the choice between direct computation and model space algorithms is real and cannot be adjudicated in the absence of a specific problem.

SUMMARY

In this chapter we have presented Markov chain Monte Carlo simulation methods and given examples of how these powerful methods are used in Bayesian statistics. We have described the Metropolis–Hastings algorithm and the special case of the Gibbs sampling algorithm. We also discussed related data augmentation methods (using latent variables), and Rao–Blackwellization. We showed how to compare models using MCMC methods. We discussed the matter of convergence in these computationally intensive methods, and suggested software to implement the algorithms (see Complement A to this chapter).

EXERCISES

6.1 Suppose that the posterior distribution of θ is $N(0, 1)$ with density $h(\theta \mid y) \propto \exp(-0.5\theta^2)$.

- **(a)** Use a standard normal generator to obtain $M = 5000$ draws from the posterior density. Use these draws to estimate the posterior mean and posterior standard deviation of θ.
- **(b)** Set up a Metropolis–Hastings sampler to obtain $M = 5000$ values from the target density using the random-walk proposal generator $\theta \mid y, \theta^{(j-1)} \sim N(\theta^{(j-1)}, \tau V)$ with $V = 1$. Adjust the value of τ in trial runs to ensure that approximately 35–50 percent of the proposed values are accepted.
- **(c)** Use the posterior sample in (b) to find the posterior mean and posterior standard deviation of θ. How do your estimates compare with those in (a)? Are 5000 MCMC draws enough to estimate accurately these moments?

6.2 Repeat the preceding exercise with the tailored proposal generator $\theta \mid y \sim N(0, 1.5)$.

6.3 Suppose that the target distribution is $N(0.5, 1)$ truncated to the interval $(0, \infty)$ with density

$$h(\theta \mid y) \propto \exp\big(-0.5(\theta - 0.5)^2\big) I_{(0,\infty)}.$$

- **(a)** What is the normalizing constant of this target density?
- **(b)** Use the expression given in Equation (6.18) to obtain $M = 5000$ draws from the target density. Use this sample to find the posterior mean and variance.
- **(c)** Set up a random-walk Metropolis–Hastings sampler to obtain $M = 5000$ values from the target density, adjusting the value of τ as in problem 6.1(b) to ensure that approximately 35–50 percent of the proposed values are accepted. Did you need the normalizing constant of the target density to implement this algorithm?
- **(d)** Find the posterior mean and standard deviation from the MCMC output and compare your results with those in part (b).

6.4 Suppose that the posterior density $\theta \mid y$ is $N(0, 1)$, as in Problem 6.1, and that

$$z \mid y, \theta \sim N(\theta, 0.5).$$

*Solutions for asterisked exercises may be found in Appendix 7.

The goal is to sample the target density $z \mid y$.

(a) Derive the target density $h(z \mid y) = \int h(z \mid y, \theta)h(\theta \mid y)\, d\theta$ by direct calculation. (Hint: the marginalized density is normal).

(b) Use your result in (a) to obtain $M = 5000$ draws from $h(z \mid y)$ with the help of a standard normal generator. Estimate the mean and standard deviation of $z \mid y$ from this sample of draws.

(c) Use the method of composition to obtain $M = 5000$ draws from $h(z \mid y)$, drawing θ from $\theta \mid y$ using the algorithm you developed in problem 6.2 followed by the drawing of z from the distribution $z \mid y, \theta$. Find the mean and standard deviation of the sampled zs and compare these estimates with those you found in part (b).

6.5* Consider the standard Gaussian model for univariate observations y_i:

$$\begin{aligned} y_i \mid \boldsymbol{\beta}, \sigma^2 &\sim N(\boldsymbol{\beta}, \sigma^2), \quad i = 1, 2, \ldots, n \\ \beta &\sim N(\boldsymbol{\beta}_0, B_0) \\ \sigma^2 &\sim \mathrm{IG}(0.5\nu_0, 0.5\delta_0). \end{aligned}$$

(a) Derive the full conditional distribution $\beta \mid y, \sigma^2$.

(b) Derive the full conditional distribution $\sigma^2 \mid y, \beta$.

(c) Explain how these full conditional distributions could be used to set up a Gibbs sampler to sample the posterior density $h(\beta, \sigma^2 \mid, mskip 1 y)$.

6.6 Montgomery, et al. (2000, p. 75) give $n = 25$ observations on the amount of time taken to service vending machines by the route driver. The amount of time y (measured in minutes) is associated with the number of cases of product stocked (x_1) and the distance in feet walked by the route driver (x_2). The full data is given in Table 6.5.

(a) Suppose that we ignore the covariates for the time being and model y with the model given in Problem 6.5. Fix the hyperparameters of the normal and inverse-gamma prior distributions.

(b) Use your results from Problem 6.5 to sample the posterior density of (β, σ^2) given the data on $y = (y_1, y_2, \ldots, y_{25})$.

(c) Now consider the Bayesian multiple regression model

$$\begin{aligned} y_i \mid \boldsymbol{\beta}, \sigma^2 &\sim N(\mathbf{x}_i'\boldsymbol{\beta}, \sigma^2), \qquad i = 1, 2, \ldots, 25 \\ \boldsymbol{\beta} &\sim N_3(\boldsymbol{\beta}_0, \mathbf{B}_0) \\ \sigma^2 &\sim \mathrm{IG}(0.5\nu_0, 0.5\delta_0), \end{aligned}$$

where $\mathbf{x}_i' = (1, x_{1i}, x_{2i})$. You can fix the hyperparameter values $(\boldsymbol{\beta}_0, \mathbf{B}_0)$ and (ν_0, δ_0) as you see fit. Use the full conditional densities given in the chapter and the Gibbs sampler to obtain $M = 5000$ draws from the posterior density $h(\boldsymbol{\beta}, \sigma^2 \mid \mathbf{y})$. Use the sampled draws to estimate the posterior mean and standard deviations of $\boldsymbol{\beta}$ and σ^2.

Table 6.5 Delivery Time Data

	Delivery Time(y)	Cases(x_1)	Distance (x_2)
1	16.68	7	560
2	11.50	3	220
3	12.03	3	340
4	14.88	4	80
5	13.75	6	150
6	18.11	7	330
7	8.00	2	110
8	17.83	7	210
9	79.24	30	1460
10	21.50	5	605
11	40.33	16	688
12	21.00	10	215
13	13.50	4	255
14	19.75	6	462
15	24.00	9	448
16	29.00	10	776
17	15.35	6	200
18	19.00	7	132
19	9.50	3	36
20	35.10	17	770
21	17.90	10	140
22	52.32	26	810
23	18.75	9	450
24	19.83	8	635
25	10.75	4	150

(d) Given your sampling scheme in part (c), use the method of composition to obtain $M = 5000$ draws from the predictive density

$$f(y_{26} \mid \mathbf{y}) = \int N(y_{26} \mid \mathbf{x}'_{26}\boldsymbol{\beta}, \sigma^2) h(\boldsymbol{\beta}, \sigma^2 \mid \mathbf{y}) \, d\boldsymbol{\beta} d\sigma^2$$

under the assumption that $\mathbf{x}'_{26} = (1, 6, 500)$. Use the sampled draws to estimate the predictive mean and standard deviation.

FURTHER READING

Albert, J. and Chib, S. (1993). "Bayesian analysis of binary and polychotomous response data," *J. Am. Statist. Assoc.*, **88**, 669–679.

Besag, J., Green, E. Higdon, D. and Mengersen, K. L. (1995). "Bayesian Computation and Stochastic Systems (with discussion)," *Statist. Sci.*, **10**, 3–66.

Carlin, B. P., and Polson, N. G. (1991). "Inference for Non-Conjugate Bayesian Models Using the Gibbs Sampler," *Canadian J. Statist.*, **19**, 399–405.

Carlin, B. P. and Chib, S. (1995). "Bayesian model choice via Markov Chain Monte Carlo methods," *J. Royal Statist. Soc. B*, **57**, 473–484.

Carlin, B. P. and Louis, T. (2000). *Bayes and Empirical Bayes Methods for Data Analysis*, 2nd edition, Chapman and Hall, London.

Casella, G. and Robert, C. P. (1996). "Rao-Blackwellization of sampling schemes," *Biometrika*, **83**, 81–94.

Chen, M-H. and Shao, Q-M. (1997). "On Monte Carlo Methods for Estimating Ratios of Normalizing Constants," *Ann. Statist.*, **25**, 1563–1594.

Chib, S. (1995). "Marginal likelihood from the Gibbs Output," *J. Am. Statist. Assoc.*, **90**, 1313–1321.

Chib, S. and Greenberg, E. (1995). "Understanding the Metropolis–Hastings Algorithm," *Am. Statist.*, **49**, 327–335.

Chib, S. and Carlin, B. P. (1999). "On MCMC Sampling in Hierarchical Longitudinal Models," *Statist. Comput.*, **9**, 17–26.

Chib, S. and Jeliazkov, I. (2001). "Marginal Likelihood from the Metropolis–Hastings Output," *J. Am. Statist. Assoc.*, **96**, 270–281.

Damien, P., Wakefield, J. and Walker, S. (1999). "Gibbs Sampling for Bayesian Nonconjugate and Hierarchical Models Using Auxiliary Variables," *J. Royal Statist. Soc. B*, **61**, 331–344.

Dellaportas, P., Forster, J. J. and Ntzoufras, I. (2000). "On Bayesian Model and Variable Selection Using MCMC," *Statist. Comput.*, in press.

DiCiccio, T. J., Kass, R. E., Raftery, A. E., and Wasserman, L. (1997). "Computing Bayes factors by combining simulation and asymptotic approximations," *J. Am. Statist. Assoc.*, **92**, 903–915.

Fahrmeir, L and Lutz, G. T. (1994). *Multivariate Statistical Modelling Based on Generalized Linear Models*, Springer-Verlag, New York.

Gamerman, D. (1997). *Markov Chain Monte Carlo: Stochastic Simulation for Bayesian Interface*, London, Chapman and Hall.

Gelfand, A. E. and Smith, A. F. M. (1990). "Sampling-Based approaches to Calculating Marginal Densities," *J. Am. Statist. Assoc.*, **85**, 398–409.

Gelfand, A. E., Sahu, S. K. and Carlin, B. P. (1995). "Efficient Parameterizations for Normal Linear Mixed Models," *Biometrika*, **82**, 479–488.

Geman, S. and Geman, D. (1984). "Stochastic Relaxation, Gibbs Distributions and the Bayesian Restoration of Images," *IEEE Trans. Pattern Analysis and Machine Intelligence*, **12**, 609–628.

Gilks, W. R., Richardson, S., and Spieglehalter, D. J. (Eds.), (1996). *Markov Chain Monte Carlo in Practice*, London, Chapman and Hall.

Godsill, S. J. (1999). "On the Relationship Between Model Uncertainty Methods," *Technical Report*, Signal Processing Group, Cambridge University.

Green, P. E. (1995). "Reversible Jump Markov Chain Monte Carlo Computation and Bayesian Model Determination," *Biometrika*, **82**, 711–732.

Hastings, W. K. (1970). "Monte Carlo Sampling Methods Using Markov Chains and Their Applications," *Biometrika*, **57**, 97–109.

Liu, J. S. (1994). "The Collapsed Gibbs sampler in Bayesian Computations with Applications to a Gene Regulation Problem," *J. Am. Statist. Assoc.*, **89**, 958–966.

Liu, J. S., Wong, W. H. and Kong, A. (1994). "Covariance Structure of the Gibbs Sampler with Applications to the Comparisons of Estimators and Data Augmentation Schemes," *Biometrika*, **81**, 27–40.

Metropolis, N., Rosenbluth, A. W., Rosenbluth, M. N., Teller, A. H., and Teller, E. (1953). "Equation of State Calculations by Fast Computing Machine," *J. Chem. Phys.*, **21**, 1087–1091.

Montgomery, D. C., Peck E. A. and Vining G. G. (2000). *Introduction to Linear Regression Analysis*, 3rd ed., New York, John Wiley and Sons.

Ripley, B. D. (1987). *Stochastic Simulation*, New York, John Wiley and Sons, Inc.

Roberts, G. O. and Sahu, S. K. (1997). "Updating Schemes, Correlation Structure, Blocking, and Parametization for the Gibbs Sampler," *J. Royal Statistist. Soc. B*, **59**, 291–317.

Spiegelhalter, D., Thomas, A, and Best, N. (2000). WinBUGS Version 1.3 User Manual. London, MRC Biostatistics Unit, Institute of Public Health.

Tanner, M. A. and Wong W. H. (1987). "The Calculation of Posterior Distributions by Data Augmentation," *J. Am. Statist. Assoc.*, **82**, 528–549.

Tierney, L. (1994). "Markov Chains for Exploring Posterior Distributions (with discussion)," *Ann. Statist.*, **22**, 1701–1762.

Wakefield, J. C., Smith, A. F. M., Racine Poon, A. and Gelfand, A. E. (1994). "Bayesian Analysis of Linear and Non-Linear Population Models by Using the Gibbs Sampler," *App. Statist.*, **43**, 201–221.

Complement A to Chapter 6: The WinBugs Computer Program by George Woodworth*

Introduction

WinBUGS© (the MS Windows operating system version of Bayesian Analysis Using Gibbs Sampling) is a versatile package that has been designed to carry out Markov chain Monte Carlo computations for a wide variety of Bayesian models. The software is currently distributed electronically from the BUGS Project web site. The address is http://www.mrcbsu.cam.ac.uk/bugs/overview/contents.shtml (click the WinBUGS link). If this address fails, try the search phrase "WinBUGS Gibbs" in a web browser. The downloaded software is restricted to fairly small models, but can be made fully functional by acquiring a license, currently for no fee, from the BUGS Project web site. WinBUGS runs under the Microsoft Windows© operating system. Versions of BUGS for other operating systems can be found on the BUGS project web site.

This complement is a tutorial on computing Bayesian analyses via WinBUGS; it assumes that the reader has basic knowledge of Bayesian methods, basic knowledge

*Professor George Woodworth, Department of Statistics and Actuarial Science, University of Iowa, Iowa City, Iowa 52242, E-mail: george-woodworth@uiowa.edu

of Markov chain Monte Carlo, as discussed in Chapter 6, and access to a licensed copy of WinBUGS.

The WinBUGS installation contains an extensive user manual (Spiegelhalter, et al., 2000) and 35 completely worked examples. The manual and examples are under the "help" pulldown menu on the main WinBUGS screen (Fig. 6.7). The user manual is a detailed and helpful programming and syntax reference; however, we have found that the quickest way to become familiar with WinBUGS programming and syntax is to work through a few of the examples.

WinBUGS requires that the Bayesian model be expressible as a directed graph. There is no requirement that the user literally make a drawing of the model in the form of a directed graph; however, an understanding that the model must be a directed graph helps explain why some models cannot be handled by WinBUGS. The authors of WinBUGS strongly recommend drawing the directed graph as the first step in any analysis. To encourage this practice, WinBUGS includes a "doodle editor" (under the "doodle" pulldown menu), which permits the user to draw a directed graph. However, the authors of WinBUGS note, "...there are many features in the *BUGS* language that cannot be expressed with doodles. If you wish to proceed to serious, non-educational use, you may want to dispense with *DoodleBUGS* entirely, or just use it for initially setting up a simplified model that can be elaborated later using the *BUGS* language." We agree, and for that reason we do not discuss the "doodle" facility (directed graph model specification) in detail here. Users wishing to learn to use the doodle facility should consult the WinBUGS user manual and "Doodle help" under the "help" pulldown menu.

WinBUGS implements various Markov chain Monte Carlo algorithms to generate observations from a Markov chain on the parameter space. The posterior distribution of the parameters is the stationary (marginal) distribution of the Markov chain; consequently, posterior moments, distributions, and event probabilities can be estimated as sample analogs computed on the output of the Markov chain. WinBUGS uses direct sampling for conjugate prior–posterior models, adaptive rejection Gibbs sampling for log-concave posterior densities, and the Metropolis–Hastings algorithm for nonconjugate priors and nonlog-concave posterior densities. See references in the user manual and on the BUGS Project web page for more details.

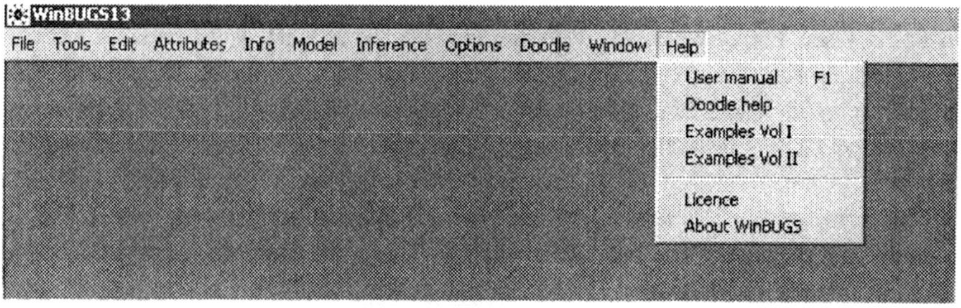

Figure 6.7 Main WinBUGS screen showing pulldown menus

The WinBUGS Programming Environment

A WinBUGS analysis—model specification, data, initial values, and output—is contained in a single "compound document." Analytic tools are available as drop-down menus and dialog boxes. Data tables are embedded as subdocuments. Components of the document, such as data tables or output tables, can be "folded" out of sight to make the document easier to work with. Data can be expressed in S-Plus list structures or as rectangular tables in ASCII format; however, WinBUGS cannot read data from an external file.

Specifying the Model

WinBUGS requires that the Bayesian model be a directed graph. A directed graph consists of nodes connected by downward links. A node is either a stochastic variable, in which case its distribution is conditioned by the values of its parent nodes (either parameters or other variables), or it is a logical node—a deterministic function of its parents.

Example 6.1 Inference on a Single Proportion

The instructor in a statistics class tossed a thumbtack 25 times in a standardized way. The tack landed "point up" 11 times. Obtain the posterior distribution of p, the probability that this tack tossed this way will land point up.

The likelihood function is $f(\text{data} \mid p) \propto p^x(1-p)^{n-x}, x = 11, n = 25$. We will specify a uniform (0,1) prior for p, that is a beta distribution with parameters $a = 1$ and $b = 1$. Figure 6.8 depicts the model as a directed graph. Each parameter or variable in the model, including the hyperparameters of the binomial distribution, is represented by a node in the graph. Each node has none, one, or more parents and none, one, or more children. Constants are in rectangles; variable nodes, which

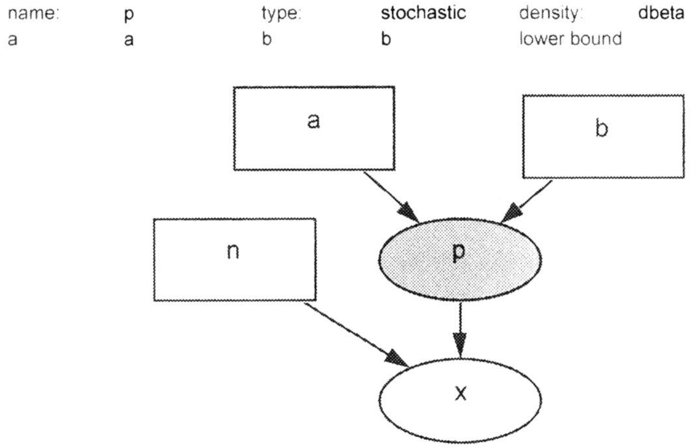

Figure 6.8 Directed graph representation of the Bayesian analysis of the proportion of successes in n tosses of a thumbtack

Table 6.6 Some Distributions Available in WinBUGS

Name	Usage	pdf or pmf	Notes
Normal	z ~ dnorm (mu, tau)	$\frac{\tau}{\sqrt{2\pi}}\exp(0.5\tau(z-\mu)^2)$	τ = precision $= \frac{1}{\text{variance}}$
Beta	u ~ dbeta (a,b)	$\frac{\Gamma(a+b)}{\Gamma(a)\Gamma(b)}u^{a-1}(1-u)^{b-1}$	
Gamma	v ~ dgamma (a,b)	$\frac{b^{a-1}}{\Gamma(a)}v^{a-1}\exp(-bv)$	Mean $= \frac{a}{b}$ Variance $= \frac{a}{b^2}$
Binomial	x ~ dbin (p,n)	$\binom{n}{x}p^x(1-p)^{n-x}$	

depend functionally or stochastically on their parents, are in ovals. When a node is selected, its properties are displayed at the top of the doodle window. For example, node p has a beta distribution with parameters a and b, which are constants.

Distributions available in WinBUGS are listed in Table I of the WinBUGS user manual; some of these are listed in Table 6.6 below. Pay attention to the nonstandard notations for the normal (precision, not variance), binomial (p before n), and gamma distributions.

It is possible (and generally easier) to describe rather than draw the model. Figure 6.9 shows the model in WinBUGS model specification language. The analysis is a self-contained compound document containing the model specification, the data, and the values of any fixed hyperparameters.

The words "MODEL," "DATA," and "INITIAL VALUES" are not required. WinBUGS treats everything between the opening and closing brackets { } as a description of the model. Lists of data and parameter values are written in S-plus format (see the WinBUGS user manual for details). But S-plus format is not necessary; simple ASCII files are sufficient.

```
MODEL {
  p ~ dbeta(a,b)
  x ~ dbin(p,n)
}

DATA list(a=1,b=1,x=11,n=25)

INITIAL VALUES list(p=.5)
```

Figure 6.9 A beta-binomial model

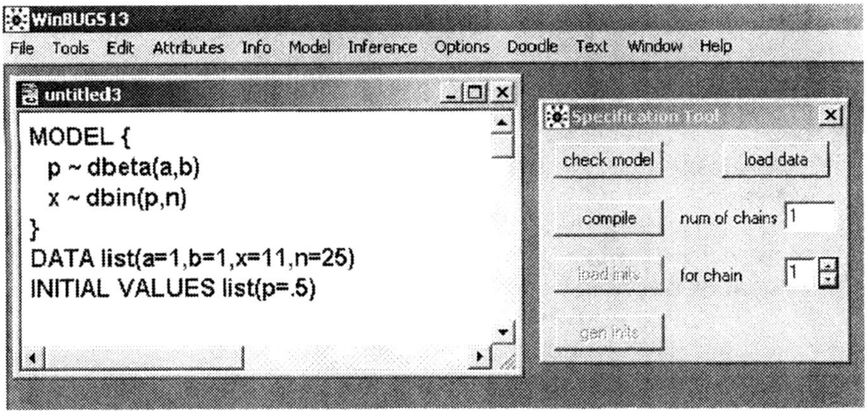

Figure 6.10 The specification tool: click the "check model" button for a syntax check

How to Run this Example

Launch WinBUGS. (The logo, which resembles a spider, is in the directory where you installed WinBUGS—you can drag it to the desktop for convenience). Read and then close the license agreement window and open a blank window (pull down: file/new).

Specify the Model. Type the contents of Figure 6.9 in the blank window you just opened.

Check the Syntax. Pull down model/specification and observe the "Specification Tool" (Fig. 6.10). Click the "check model" button. Look in the gray message bar along the bottom of the screen. You should see "model is syntactically correct." Figure 6.11 shows what happens when there is a syntax error (in this

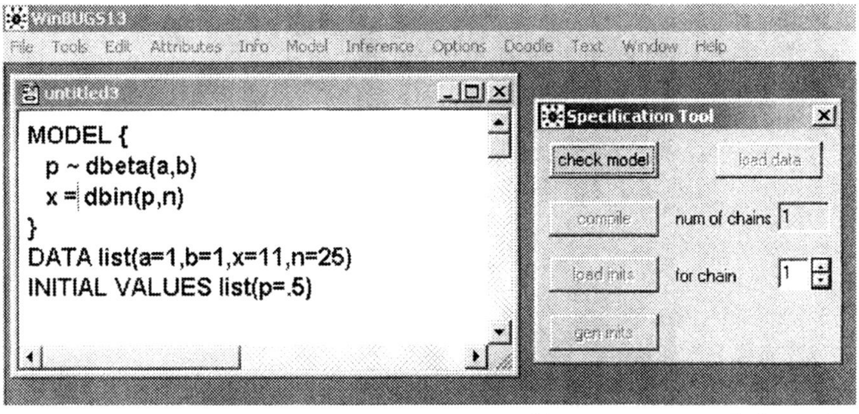

Figure 6.11 A syntax error: the syntax checker expected a tilde, not an equal sign (before the dotted cursor)

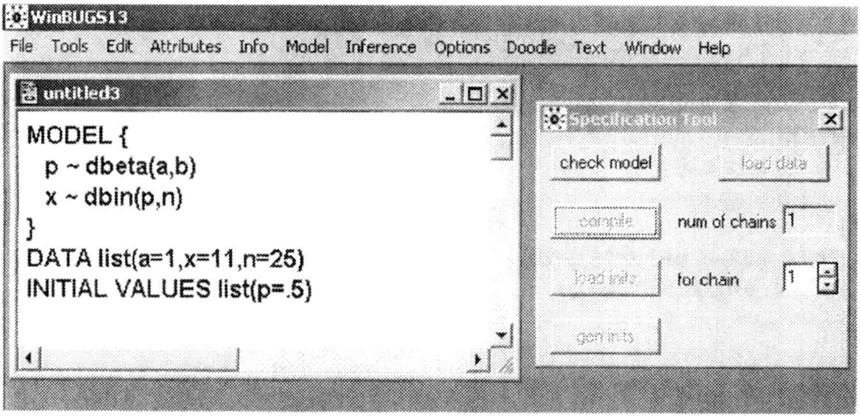

Figure 6.12 A compiler error message (a value is needed for b)

case " = " instead of " ~ "). Note that there is a (barely visible) vertical dotted line cursor showing the location of the error.

Enter Data and Constants. Highlight the word "list" after "Data" and click "load data." Highlight carefully to avoid an error message.

Compile the Model. Click "compile" and look for the words "model compiled" in the message bar across the bottom of the screen. Figure 6.12 shows a compilation error caused by not providing a value for parameter b. Another common compilation error is misspelling a variable name—WinBUGS is case-sensitive.

Enter or Generate Initial Values. Highlight the word "list" after "Initial Values" and click "load inits." WinBUGS can generate initial values using the prior distributions (click "gen inits" to do this); however, we recommend providing initial values for precision parameters (See Example 6.2 below) to avoid an error message. When some initial values are listed and some are generated, first click "load inits" then "gen inits."

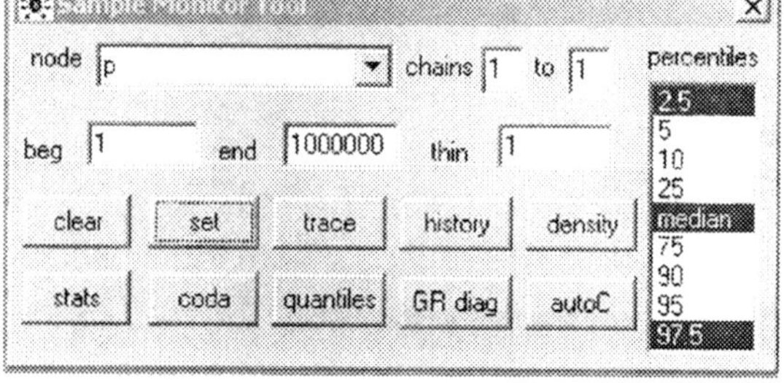

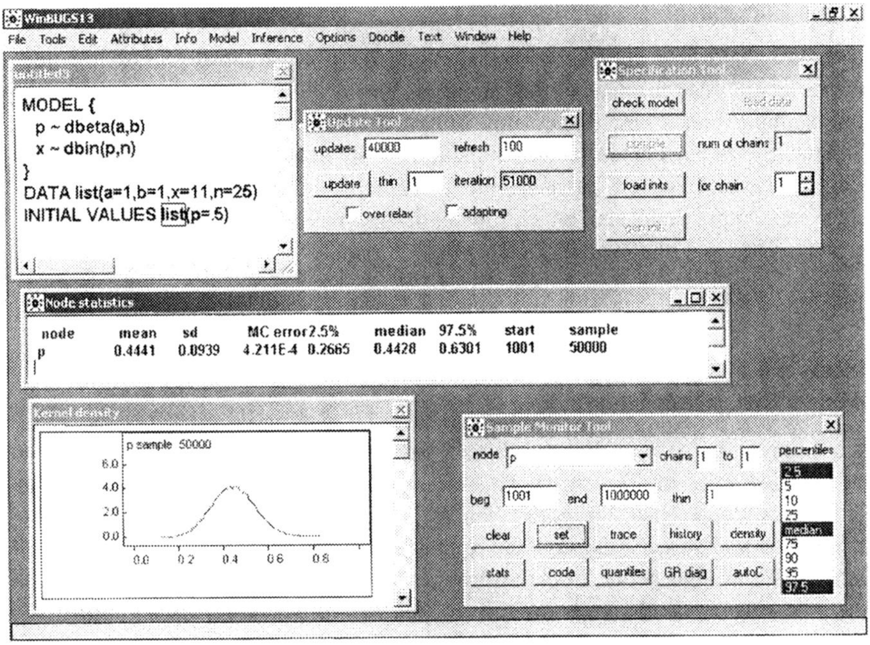

Figure 6.13 Complete WinBUGS session showing model, tool windows, and output windows

Select the Nodes to be Monitored. Monitoring a node means asking WinBUGS to keep a file of the values of that node generated by the Markov chain. In this case we need to monitor the "p" node. Under the "inference" pulldown menu select "samples." The "Sample Monitor Tool" will appear. In the "node" field type "p" (without quotes) and click "set."

Run the Markov Chain. Under the "model" pulldown menu, select "update." The "update tool" will appear. In the "updates" field enter the desired number of updates of the chain (for example 51,000) as shown in Figure 6.13 and click "update." The chain can be stopped and restarted by clicking the "update" button.

Explore the Posterior Distribution. In the "Sample Monitor Tool" enter a number in the "beg" field—this instructs WinBUGS to discard the initial, transient part of the Markov chain. The number to discard is determined by examining convergence diagnostics discussed below. Enter the node to be examined in the "node" field and click "distribution" and then "stats". In this case the posterior distribution has posterior mean 0.4441, posterior standard deviation 0.0939, and the 95 percent equal-tail posterior credible interval is 0.27 to 0.63. Note that the exact posterior distribution is beta (12,15) which has mean 0.4444 and standard deviation 0.0939.

Simple Convergence Diagnostics

A Markov chain that approaches its stationary distribution slowly or exhibits high autocorrelation can produce an inaccurate picture of the posterior distribution. The next example shows how to use some simple convergence diagnostics available in WinBUGS: traces, autocorrelation functions, and the Gelman–Rubin diagnostic. Additional convergence diagnostics are available in the independent package BOA (Bayesian Output Analysis), which runs under S-plus or R. For information on how to obtain BOA see the WinBUGS Project web page.

Example 6.2 Comparing Two Proportions: Difference, Relative Risk, Odds Ratio

In a comparison of radiation therapy vs. surgery, cancer of the larynx remained uncontrolled in 2 of 23 surgery patients and 3 of 18 radiation patients. Figure 6.14 shows the analysis. The parameters p_{rad} and p_{srg} are the probabilities of failure of radiation and surgery, respectively. x_{rad}, n_{rad}, x_{srg} and n_{srg}, are the data. The prior distributions are independent beta distributions with parameters (0.5, 0.5). Note that text following a pound sign is interpreted as a comment.

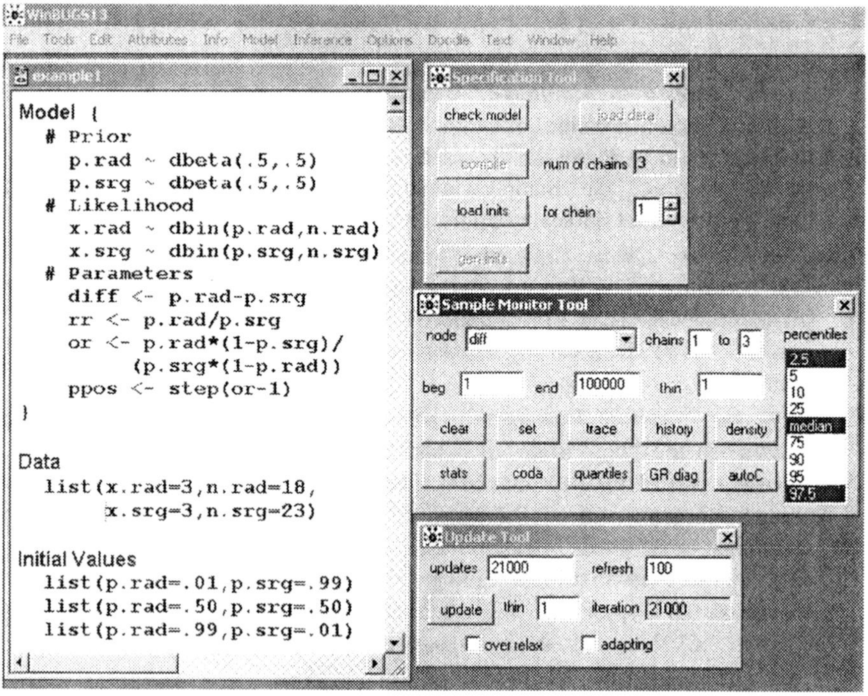

Figure 6.14 Comparing two proportions: Illustrating computed parameters and convergence diagnostics.

Table 6.7 Some WinBUGS Scalar Functions

Name	Value
step(x)	1 if $x \geq 0$, otherwise 0
log(x)	ln(x)
logit(p)	$\ln(p/(1-p))$
exp(x)	exp(x)
abs(x)	$\lvert x \rvert$
pow(x,c)	x^c

This example illustrates the "arrow" syntax for "logical" nodes that are deterministic, not stochastic, functions of their parent nodes. Here the logical nodes are the difference of the two failure probabilities, the relative risk, the odds ratio, and an indicator variable (ppos) for the event that radiation has a higher failure rate than surgery ($p_{rad} > p_{srg}$).

Stochastic relationships are indicated by a tilde or "twiddle" and deterministic relationships are indicated by an arrow "<–" constructed of a "less than" symbol and a minus sign. Inadvertent use of an " = " sign instead of a twiddle or an arrow will produce an error message. The equal sign is never used in WinBUGS model specification language.

This example uses the step( ) function to compute the indicator function of the event $(p_{rad} \geq p_{srg})$. The posterior mean of this function is the posterior probability of that event. Table 6.7 is a list of some other functions available in WinBUGS; consult Table II of the user manual for the complete list.

Convergence diagnostics are based on running two or more chains in parallel starting from widely dispersed starting values. If the multiple chains are statistically indistinguishable and "mix" rapidly, the chains are likely to have converged to the stationary distribution.

How to Compute Convergence Diagnostics

Specify Multiple Chains. In the model specification tool, check the model and load the data list. Before compiling the model enter the number of chains in the "num of chains" field. Compile the model. Highlight the first initial value list and click "load inits"; do the same for the second and third initial value lists.

Select Parameters to be Monitored. Values of monitored parameters generated by the Markov chain are retained in a data file (in effect a sample from the posterior distribution). Posterior marginal distributions, event probabilities, and moments can be approximated using these values.

To open the sample monitor tool, pull down the "Inference" menu to "samples." Type the name of the first parameter to be monitored in the "node" field and

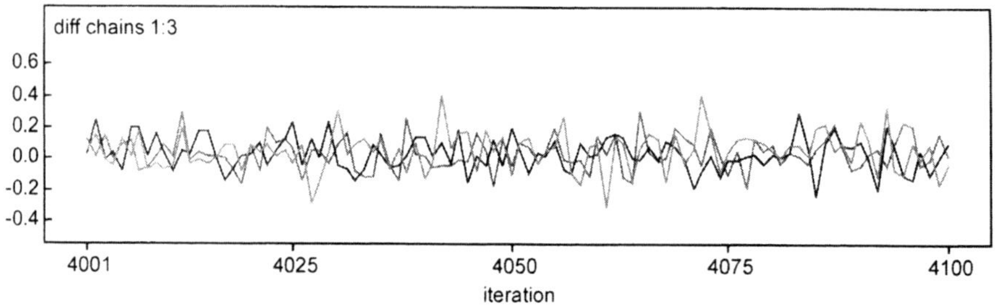

Figure 6.15 Short segment of the history of the difference parameter for three parallel chains

click "set." Repeat for the other parameters to be monitored. (Fixed quantities—including fixed hyperparameters, observed data values, and functions of data—cannot be monitored.)

Run the Chains. Pull down the "model" menu and select "update." Enter the desired number of updates (for a complex, slow-running chain select a more frequent refresh rate such as 10 or 5). Selecting "over relax" may reduce the autocorrelation between successive updates of the chain—see the user manual for details. Click the "update" button to start the chain. The chain can be stopped at the next refresh by clicking the update button a second time. For models that require WinBUGS to use the Metropolis–Hastings algorithm, the sampler "adapts" for 4000 updates. During this phase the "adapting" box is checked and posterior moments and percentiles cannot be computed.

Compute Convergence Diagnostics. To observe the values of a parameter as the chain updates, enter the parameter name in the "node" field of the sample monitor tool (or enter * to monitor all parameters). Click "trace" for a dynamic view or "history" for a static view. Figure 6.15 is a short segment of a longer history. After a short initial "settling in" period, the multiple chains should cover the same range and should not show trends or long cycles.

Click "GR diag" to view a cumulative graph of the Gelman–Rubin convergence diagnostic. Convergence is indicated when the ratio (the red trace) of the between and within chain variances rapidly approaches 1.

Click "quantiles" for a cumulative graph of the selected quantiles. Convergence is indicated when the quantiles of the parallel chains rapidly coincide.

Click "autoC" for the autocorrelation function of the selected parameter. Although this is not directly a convergence diagnostic, a long-tailed autocorrelation graph suggests that the model is ill conditioned and that the chains will converge more slowly.

Examine the Posterior Distribution. In the sample monitor tool, enter the parameter name in the "node" field (or enter * to examine all parameters).

Table 6.8 Marginal Moments and Quantiles for Difference, Odds Ratio, Relative Risk, and the Event "Surgery has a Lower Failure Rate Than Radiation"

Node	Mean	s.d.	2.5%	Median	97.5%
diff	0.03859	0.1115	−0.1767	0.03548	0.2676
or	1.977	2.298	0.2364	1.331	7.533
ppos	0.6326	0.4821	0.0	1.0	1.0
rr	1.694	1.627	0.288	1.269	5.601

In the "beg" field enter the number of updates to be discarded (this is the initial transient phase of the chain, as determined by examining history traces and convergence diagnostics). Click "stats" to compute marginal moments and quantiles for the selected parameters. Table 6.8 indicates that the posterior mean difference between the failure rates of radiation and surgery is about four percentage points and that there is about a 63 percent chance that the failure rate is higher for radiation.

Click "density" in the sample monitor tool to examine marginal density functions of the selected parameters (Fig. 6.13).

Advanced Tools: Loops, Matrices, Imbedded Documents, Folds

The purpose of the WinBUGS programming language is to specify a model—prior and likelihood. It is not a programming language. It does not specify a series of commands to be executed in sequence. In fact, model specification statements can be written in almost any order without changing the meaning of the model. Repetitive model components, such as random coefficients, can be specified using "for loops" but conditional branching structures such as "if ... then ... else" are not available and, indeed, have no meaning in model specification.

Always keep in mind that a model must be a directed graph. That means that each node should be on the right side of only one equation (twiddle or arrow) and no node can directly or indirectly be its own parent or its own child. The purpose of the WinBUGS model specification language is to "paint a word picture" of the directed graph.

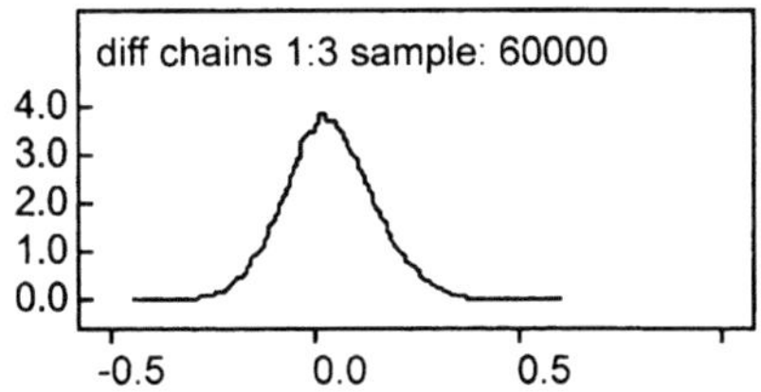

Figure 6.16 Posterior marginal distribution of the difference between the failure rates

One consequence of this restriction is that some legitimate Bayesian models cannot be implemented in WinBUGS. An example is the three-parameter Box–Cox model, $y_i^{\lambda} \sim N(\mu, \tau), i = 1, \ldots, n$. This equation is illegal in WinBUGS because the left side contains two nodes y_i and λ. One strategy might be to define an auxiliary node $z_i \leftarrow y_i^{\lambda}$, *and* $z_i \sim N(\mu, \tau)$, but this is not permitted because z_i would appear on the left of two equations.

WinBUGS allows one minor exception to the "one node, one equation" restriction to permit fixed transformations of data. For example, the model, $\sqrt{y_i} \sim N(\mu, \tau), i = 1, \ldots, n$, can be implemented in WinBUGS via two equations, $z_i \leftarrow pow(y_i, 0.5)$, and $z_i \sim N(\mu, \tau)$.

Example 6.3 Multiple Logistic Regression

From a study (Baldus et al., 1989) of convicted murderers, five variables were selected: sentence (life or death), race of victim (white or other), race of defendant (black or other), number of aggravating circumstances, and number of mitigating circumstances. The goal is to study the influence of race, controlling for the number of aggravating and mitigating circumstances. Let $\mathbf{x}$ be the vector of case characteristics (race of victim, defendant, numbers of aggravating and mitigating circumstances) and let f_x and n_x be the numbers of death sentences and number of cases, respectively, with characteristics x. The likelihood is $f_x \sim \text{binomial}(p_x, n_x)$, where $\text{logit}(p_x) = \ln(p_x/(1-p_x)) = \alpha + x'\beta$. The intercept and regression coefficients are given diffuse but proper normal distributions with means 0 and precisions 0.01. The data (Table 6.9) has 101 rows (observed combinations of circumstances) and 6 columns.

A directed graph depicting this model would have about seven hundred nodes. The model specification (Fig. 6.17) uses "for loops" to specify repetitive substructures in the model. The syntax of a "for loop" is

Table 6.9 First 7 of 101 Rows of the Data Matrix

Black Defendant	White Victim	Aggravating Circumstances	Mitigating Circumstances	No. of Death Sentences	No. of Cases
0	0	0	0	0	1
0	0	0	3	0	3
0	0	0	5	0	1
0	0	1	0	0	2
0	0	1	2	0	5
0	0	1	3	0	1
0	0	1	5	0	3
etc.					

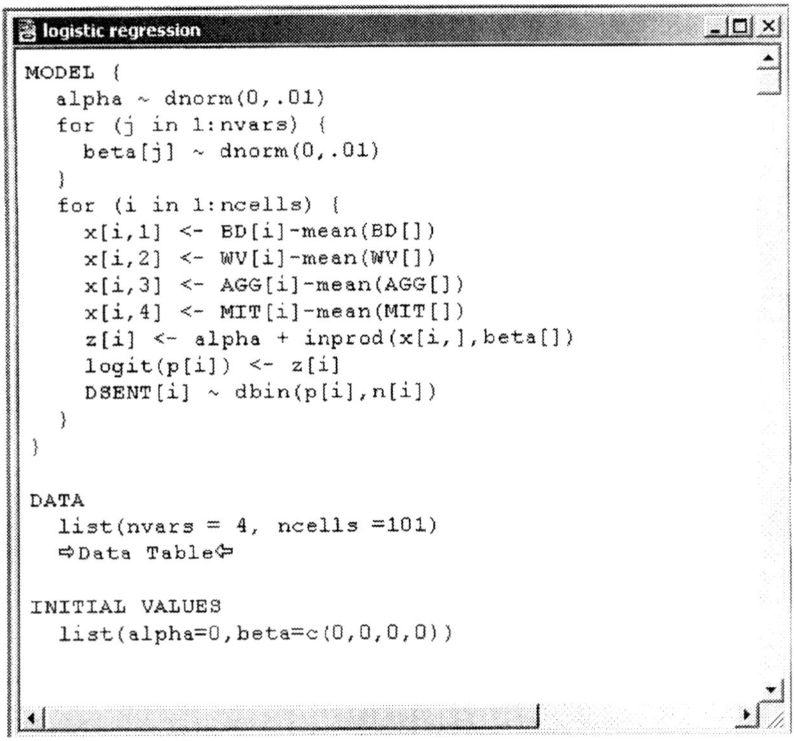


```
MODEL {
  alpha ~ dnorm(0,.01)
  for (j in 1:nvars) {
    beta[j] ~ dnorm(0,.01)
  }
  for (i in 1:ncells) {
    x[i,1] <- BD[i]-mean(BD[])
    x[i,2] <- WV[i]-mean(WV[])
    x[i,3] <- AGG[i]-mean(AGG[])
    x[i,4] <- MIT[i]-mean(MIT[])
    z[i] <- alpha + inprod(x[i,],beta[])
    logit(p[i]) <- z[i]
    DSENT[i] ~ dbin(p[i],n[i])
  }
}

DATA
  list(nvars = 4, ncells =101)
  ⇨Data Table⇦

INITIAL VALUES
  list(alpha=0,beta=c(0,0,0,0))
```

Figure 6.17 Multiple logistic regression. A 101-line data table is "folded" out of sight between the white arrows

FOR (index IN first:last) {body of loop}

We used indentation in Figure 6.17 to visually clarify the logic of the model; however, it is not required. The "for" loop structure does not specify a sequence of computations, nor does it have anything to do with the actual order in which nodes are updated in the Markov chain Monte Carlo algorithm. The statements inside a "for loop" are "expanded" to create the actual model specification. For example, the first for loop "expands" to:

beta[1] ~ dnorm(0, .01)

beta[2] ~ dnorm(0, .01)

beta[3] ~ dnorm(0, .01)

beta[4] ~ dnorm(0, .01)

and the second "for" loop expands to 707 lines.

Table 6.10 Some Scalar Functions of Vector Arguments

Name	Value
inprod(u[ ], v[ ])	$u_l v_l + \cdots + u_k v_k$
inprod(m[i,],v[ ])	$m_{il} v_l + \cdots + m_{ik} v_k$
mean(v[ ])	Average of elements of **v**
rank(v[ ], s)	Number of elements of **v** $\leq s$
ranked(v[ ], s)	The sth smallest element of **v**

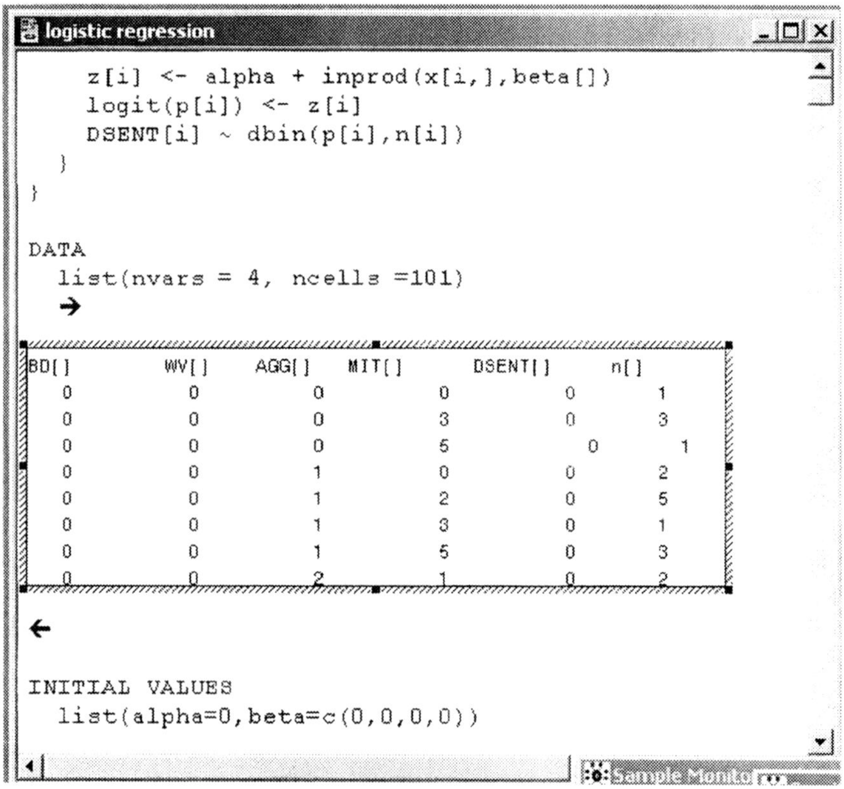

Figure 6.18 The "unfolded" view of Figure 6.17; the 101-line data table is an imbedded document that can be "folded" out of sight by clicking on either black arrow

The "expanded" model specification must form a directed graph. As a consequence, "for" loops that would be legal in a programming language may be illegal in WinBUGS. An example is

mean = 0

for (i in 1:n){mean <−mean + $y[i]$}

This would not be permitted in WinBUGS model specification language because the node labeled "mean" occurs on the left of n + 1 equations and is its own parent.

"for" loops operate on subscripted variables (vectors and matrices), with one or two subscripts enclosed in square brackets. WinBUGS model specification language has several functions that take vectors as arguments. This model specification uses the mean(v[ ]) function to center the design variables at their means and it uses the inner product function inprod(u[ ], v[ ]) to compute $x'\beta$. Table 6.8 is a partial list; see Table II of the WinBUGS user manual for a complete list.

Data Tables, Imbedded Documents, Folds

The data for this problem are a 107 by 6 matrix (Table 6.9). It is more convenient to enter this table directly rather than attempting to convert it to an S-plus list structure. In Figure 6.17 the data table is hidden in a "fold" labeled "⇒ Data Table ⇐". Clicking on either white arrow will "unfold" the table as in Figure 6.18. To refold click either of the black arrows above and below the table. The table itself was created in a separate WinBUGS document and the entire document was pasted into the "fold" using the following procedure.

How to Embed a Data Table or Other Document in a Fold

- Open a new blank document window (pull down: file/new).
- Paste in the data in as ASCII text or open an existing ASCII file in the window.
- Type a subscripted variable name at the head of each column as shown in Figure 6.18.
- Copy the document (pull down: edit/select document, then edit/copy).
- Go to the main document, and create a "fold" where the data table is to be placed (pull down: tools/create fold).
- Position the cursor inside the fold (white arrows) and type the data title.
- Click either white arrow to open the fold and position the cursor between the black arrows.
- Paste in the document containing the data table (pull down: edit/paste).
- Click either black arrow to close the fold.

Loading a Data Table

After checking the model, load the data list in the usual way, then open the fold containing the data table. Click anywhere in the data window, but do not highlight any text, as that will produce an error message. The data table should be surrounded

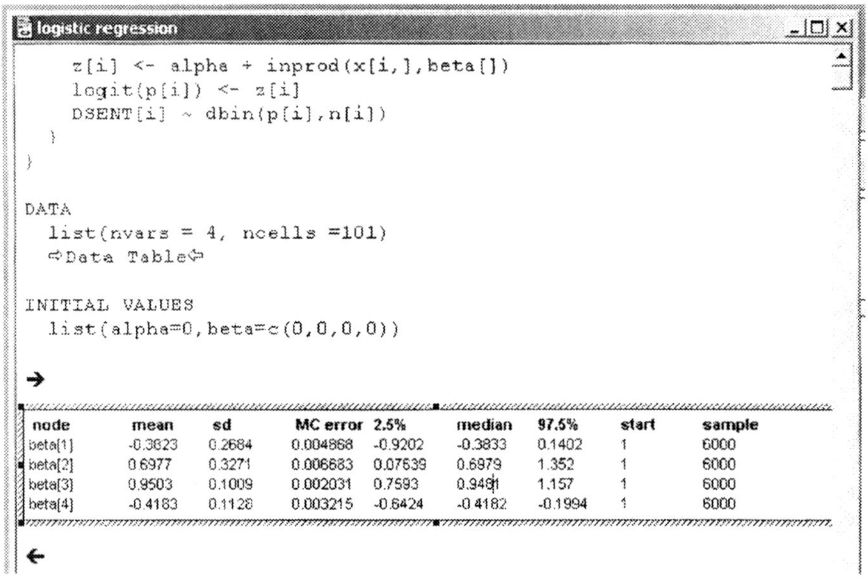

node	mean	sd	MC error	2.5%	median	97.5%	start	sample
beta[1]	-0.3823	0.2684	0.004868	-0.9202	-0.3833	0.1402	1	6000
beta[2]	0.6977	0.3271	0.006683	0.07639	0.6979	1.352	1	6000
beta[3]	0.9503	0.1009	0.002031	0.7593	0.9491	1.157	1	6000
beta[4]	-0.4183	0.1128	0.003215	-0.6424	-0.4182	-0.1994	1	6000

Figure 6.19 The "stats" document has been pasted into a fold. The 95 percent posterior credible interval for β_2 (the white victim "effect") includes only positive values; however, the race of defendant effect (β_1) is neither clearly positive nor clearly negative

by a "hairy border." Click "load data," close the fold and proceed with the compilation. Note that there must be a carriage return at the end of the last line in the embedded data document. If this is missing, WinBUGS will report that there is an incomplete data line.

Placing Output in a Fold

We recommend that output tables and graphs be pasted into the "compound document" containing the model specification. This is easy to do and creates a complete record of the analysis as one document, which can be saved and if desired reopened for modification or alternate analyses.

For example, the table of posterior moments and quantiles is initially reported in a separate document. Click anywhere in this document, select it and copy it (pull down: edit/select document, then edit/copy). Note that you must "select document," not "select all." Insert and label a fold in the main document. Open the fold and paste in the "stats" document (Fig. 6.19).

Additional Resources

Readers are urged to look at the examples distributed with WinBUGS (look under the "Help" pulldown). An efficient way to set up a data analysis is to imitate an example similar to the analysis that you want to do. The examples include hierarchical models (parametric empirical Bayes), survival analysis and Cox Regression, random effects logistic regression, nonparametric smoothing, spatial models, and multivariate normal models.

There is an active WinBUGS electronic mail list where users can post modeling and debugging questions. To join the mail list follow the instructions on this web page: http://www.mrc-bsu.cam.ac.uk/bugs/overview/list.shtml.

FURTHER READING

Baldus, David C., George Woodworth, and Charles A. Pulaski, Jr. (1989). "Charging and Sentencing of Murder and Voluntary Manslaughter Cases in Georgia 1973–1979". Data and documentation: http://www.icpsr.umich.edu/cgi/archive.prl?path = NACJD&format = tb&num = 9264.

Spiegelhalter, D., Thomas, A, and Best, N. (2000). WinBUGS Version 1.3 User Manual. MRC Biostatistics Unit, Institute of Public Health, London.

Complement B to Chapter 6: Bayesian Software

It can be extremely useful for those interested in applying the Bayesian paradigm to be able to draw upon an armamentarium of readily available computer software—without the consumer necessarily having to have the ability to personally program some problem of interest. Of course software development is usually very much in flux because it is so frequently modified, updated, and augmented. It is therefore generally out of date very quickly. An Internet search by Google of "Bayesian software" generated 56,600 references (in 0.37 seconds). Nevertheless, for the convenience of the consumer, we list below a few of the currently most popular references for Bayesian software. Many more references to Bayesian software may be found at the web site: http://www.cs.berkeley.edu/~murphyk/Bayes/economist.html.

1. AutoClass, a Bayesian classification system (http://ic-www.arc.nasa.gov/ic/projects/bayes-group/group/autoclass/).
2. BACC: Bayesian Analysis, Computation and Communication, http://www.citano.fc.ca/mccaus/index.html, by John Geweke, University of Iowa, bacc@ecumn.edu. These programs use .
3. BATS, designed for Bayesian time series analysis (http://www.stat.duke.edu/-mw/bats.html).
4. Bayesian biopolymer sequencing software (http://www-stat.stanford.edu/~jliu/).

5. Bayesian regression and classification software based on neural networks, Gaussian processes, and Bayesian mixture models (http://www.cs.utoronto.ca/~radford/fbm.software.html).
6. BAYESPACK, etc., numerical integration algorithms (http://www.math.wsu.edu/math/faculty/genz/homepage).
7. B/D, a linear subjective Bayesian system (http://fourier.dur.ac.uk:8000/stats/bd/).
8. Belief Networks—BN PowerConstructor: An efficient system that learns Bayesian belief network structures and parameters from data (www.cs.ualberta.ca/~jcheng/bnsoft.htm).
9. Belief Networks—Examples for use with belief network software. Norsys maintains a Library of Networks. Microsoft has proposed a Bayesian Network Standard. (www.stat.washington.edu/almond/belief.html).
10. BIPS: Bayesian Inference for the Physical Sciences—Tom Loredo's Index Page. (astrosun.tn.cornell.edu/staff/loredo/bayes/).
11. BMA, software for Bayesian model averaging for predictive and other purposes (http://www.research.att.com/~volinsky/bma.html).
12. BRCAPRO, which implements a Bayesian analysis for genetic counseling of women at high risk for hereditary breast and ovarian cancer (http://www.stat.duke.edu/~isds-info/soft).
13. BUGS, designed to analyse general hierarchical models via MCMC (http://www.mrc-bsu.cam.ac.uk/bugs/). See also Complement A to Chapter 6.
14. Darren's www links: Bayesian software. Stats: Software/Statistical computing: Bayesian software (www.mas.ncl.ac.uk/~ndjw1/bookmarks/Stats/Software-Statistical_computing/Bayesian_software/).
15. FIRST BAYES—For Bayesian teaching (http://www.shef.ac.uk/~stlao/lb.html).
16. Flexible Bayesian modeling and Markov chain sampling. (http://www.cs.toronto.edu/~radford/fbm.software.html).
17. Graphical models/Bayesian networks. Last updated 20 September 2001 (www.cs.berkeley.edu/~murphyk/Bayes/bnsoft.html).
18. Hierarchical Modeling (see Chapter 14, Sect. 14.7) SAS (PROC MIXED), S plus/R, Stata (gllamm), MLWEN, HLM, MIXOR/MIXREG, MLA.
19. Lawrence Joseph's homepage—Bayesian sample size, related papers. Normal means, program files, instructions/examples (www.epi.mcgill.ca/Joseph/software.html).
20. LISP-STAT (various Bayesian capabilities). Tierney 1991; LISP-STAT, *An object-oriented environment for statistical computing and dynamic graphics*. New York: John Wiley and Sons, Inc.

21. Matlab and Minitab Bayesian computational algorithms for introductory Bayes and ordinal data (http://www-math.bgsu,edu/~albert/).
22. Nuclear magnetic resonance Bayesian software; this is the manual (http://www.bayes.wustl.edu/glb/manual.pdf).
23. StatLib, a repository for statistics software, mostly Bayesian (http://lib.stat.cmu.edu/).
24. Time series software for nonstationary time series and analysis with autoregressive component models (http://www.stat.duke.edu/~mw/books_software_data.html).

CHAPTER 7

Large Sample Posterior Distributions and Approximations

7.1 INTRODUCTION

The Bayesian paradigm is conceptually simple, intuitively plausible, and probabilistically elegant. As we saw in Chapter 6, however, its numerical implementation sometimes requires computer intensive calculation. Posterior distributions are often expressible only in terms of complicated analytical functions. We often know only the kernel of the posterior density (and not the normalizing constant), and we often cannot readily compute the marginal densities and moments of the posterior distribution in terms of exact closed form explicit expressions. It is also sometimes difficult to find numerical fractiles of the posterior cdf. We quote Dempster (1980, p. 273):

> The application of inference techniques is held back by conceptual factors and computational factors. I believe that Bayesian inference is conceptually much more straightforward than non-Bayesian inference, one reason being that Bayesian inference has a unified methodology for coping with nuisance parameters, whereas non-Bayesian inference has only a multiplicity of ad hoc rules.

It is therefore sometimes useful to study the simple, approximate, and large sample behavior of posterior distributions. In Chapter 6 we saw how to implement many types of Bayesian procedures using Markov chain Monte Carlo methods. However, it is not always easy to apply the MCMC sampling approach (the conditionals may not be available); nor is it always easy to interpret and extrapolate the MCMC result. In this chapter we therefore present a summary of other major methods that have been developed to attack the computational problems of implementing Bayesian inference. We begin with the large sample (normal) approximation to the posterior distribution (Section 7.2). In Section 7.3 we move on to the Lindley approximation to the large sample posterior distribution (which involves the third derivative of the log-likelihood function); the Tierney–Kadane/Laplace approximation, and the Naylor/Smith approximations that provide analytical forms for the large sample posterior distribution and improve the rate of convergence. In Section 7.4 we discuss importance sampling, which is useful for high-dimensional problems.

7.2 LARGE-SAMPLE POSTERIOR DISTRIBUTIONS

One method of dealing with the practical problems of implementing the Bayesian paradigm, which can be useful when moderate or large samples are available, is to adopt the limiting large-sample normal distribution for posterior inferences. (It is often a surprisingly good approximation even in small samples.) This limiting distribution is straightforward to determine, and the regularity conditions required for its existence are usually satisfied. The result is given in the theorem below.

Theorem 7.1 Let $\mathbf{x}_1, \ldots, \mathbf{x}_n$ denote p-variate observations from a sampling distribution with joint density $f(\mathbf{x}_1, \ldots, \mathbf{x}_n \mid \boldsymbol{\theta})$, $\boldsymbol{\theta} : (k \times 1)$, and denote the prior density for $\boldsymbol{\theta}$ by $g(\boldsymbol{\theta}) > 0$. Under suitable regularity conditions, the limiting posterior distribution of $\boldsymbol{\theta}$ conditional on the data is given in terms of its probability law $\mathcal{L}$, by:

$$\lim_{n\to\infty} \mathcal{L}\{\hat{\Sigma}^{-0.5}(\theta - \hat{\boldsymbol{\theta}}) \mid \mathbf{x}_1, \ldots, \mathbf{x}_n\} = N(\mathbf{0}, \mathbf{I}_k),$$

where $\hat{\boldsymbol{\theta}}$ denotes the maximum likelihood estimator of $\boldsymbol{\theta}$, $\hat{\Sigma}^{-1} \equiv \hat{\Lambda} \equiv (\hat{\lambda}_{ij})$, and

$$\hat{\boldsymbol{\lambda}}_{ij} = \left.\frac{\partial^2 \log f(\mathbf{x}_1, \ldots, \mathbf{x}_n \mid \boldsymbol{\theta})}{\partial\theta_i \partial\theta_j}\right|_{\theta=\hat{\theta}},$$

$\mathbf{I}_k$ denotes the identity matrix of order k, and N denotes the normal distribution.

REMARK 1: An equivalent statement of the result in the theorem is that if

$$\boldsymbol{\phi} \equiv \hat{\Sigma}^{-0.5}(\boldsymbol{\theta} - \hat{\boldsymbol{\theta}}) \equiv (\phi_i),$$

then

$$\lim_{n\to\infty} P\{\phi_1 \le t_1, \ldots, \phi_k \le t_k \mid \mathbf{x}_1, \ldots, \mathbf{x}_n\} = \int_{-\infty}^{t_1} \cdots \int_{-\infty}^{t_k} \frac{1}{(2\pi)^{k/2}} e^{-\boldsymbol{\phi}'\boldsymbol{\phi}/2} d\boldsymbol{\phi}.$$

REMARK 2: Note that for the theorem to hold, $g(\boldsymbol{\theta})$ must be strictly positive (and continuous) at all interior points of the parameter space (otherwise, no amount of sample data could revise the prior belief that $g(\boldsymbol{\theta}_0) = 0$ for some $\boldsymbol{\theta}_0$). Moreover, the likelihood function $f(\mathbf{x}_1, \ldots, \mathbf{x}_n \mid \boldsymbol{\theta})$ must be twice differentiable at $\hat{\boldsymbol{\theta}}$, so that $\hat{\Sigma}$ will exist; of course we must also have $\hat{\Sigma}$ nonsingular, so that $\hat{\Lambda}$, the precision matrix, will also exist. We assume also that $\boldsymbol{\theta}$ is not one of the bounds of the support of $\mathbf{x}_i$.

REMARK 3: Note that the large sample posterior distribution of $\boldsymbol{\theta}$ does not depend upon the prior; that is, in large samples, the data totally dominate the prior beliefs.

Proof

See Le Cam (1956, p. 308). It may also be shown that when the sample data is i.i.d., the required regularity conditions are the same as those required for proving asymptotic normality of the MLE (Hyde and Johnstone, 1979). When the observations are correlated (such as in time series data, or in spatially correlated data arising in geophysics), the required regularity conditions are more complicated. We note that the result in this theorem arises from a Taylor series expansion of the log-likelihood function about $\hat{\boldsymbol{\theta}}$. For a convenient heuristic proof, see Lindley (1965, Volume 2).

Example 7.1

Let $x_1, \ldots, x_n$ denote the numbers of small aircraft arriving at the municipal airport in Riverside, California, during time intervals $t_1, \ldots, t_n$. Assume the x_is are mutually independent and suppose the probability mass function of X_j is Poisson and is given by

$$\ell(x_j \mid \theta) = \frac{e^{-\theta}\theta^{x_j}}{x_j!}, \theta > 0,$$

so that

$$f(x_1, \ldots, x_n \mid \theta) = \prod_1^n \frac{e^{-\theta}\theta^{x_j}}{x_j!} = \frac{e^{-n\theta}\theta^{\sum_1^n x_j}}{\prod_1^n x_j!}.$$

The maximum likelihood estimator of θ is found as follows. The log-likelihood of θ is

$$\begin{aligned} L(\theta) &\equiv \log f(x_1, \ldots, x_n \mid \theta) \\ &= -n\theta + (\log\theta)\sum_1^n x_j - \log\left(\prod_1^n x_j!\right). \end{aligned}$$

$$\frac{dL(\theta)}{d\theta} = -n + \frac{\sum_1^n x_j}{\theta}.$$

$$\frac{dL(\theta)}{d\theta} = 0, \quad \text{implies} \quad \hat{\theta} = \bar{x} = \frac{1}{n}\sum_1^n x_j.$$

The second derivative of $L(\theta)$ with respect to θ, evaluated at $\hat{\theta}$, is

$$-\hat{\Lambda} \equiv \left.\frac{d^2L(\theta)}{d\theta^2}\right|_{\theta=\hat{\theta}} = -\left.\frac{1}{\theta^2}\sum_1^n x_j\right|_{\theta=\hat{\theta}}$$

or if we define, as in the theorem,

$$\hat{\Sigma}^{-1} = \hat{\Lambda} = \frac{1}{\hat{\theta}^2}\Sigma_1^n x_j = \frac{n}{\bar{x}}.$$

Suppose the prior distribution of θ is log-normal with density

$$g(\theta) = \frac{1}{\theta\sigma_0\sqrt{2\pi}} e^{-(0.5\sigma_0^2)(\log\theta-\mu)^2}, \quad \theta > 0,$$

where (μ, σ_0^2) are known parameters that are assessed. By Bayes' theorem, the posterior density then becomes

$$h(\theta \mid x_1, \ldots, x_n) \propto \left[\frac{1}{\theta} e^{-(1/2\sigma_0^2)(\log\theta-\mu)^2}\right][e^{-n\theta}\theta^{\Sigma_1^n x_j}]$$

$$h(\theta \mid x_1, \ldots, x_n) \propto \theta^{\Sigma_1^n x_j - 1} e^{-n\theta-(2\sigma_0^2)^{-1}(\log\theta-\mu)^2}, \quad \theta > 0.$$

This posterior distribution is quite complicated; we cannot readily evaluate the moments, and we do not even know the proportionality constant. If n is sufficiently large, however, we can rely on the large-sample result in Theorem 7.1. Since for this example, $\hat{\Sigma}^{-1} = n/\bar{x}$, we find that for large n, we have the large sample distributional result

$$\mathcal{L}\left\{\sqrt{\frac{n}{\bar{x}}}(\theta - \bar{x}) \mid x_1, \ldots, x_n\right\} \simeq N(0, 1).$$

Equivalently, for large n, we have

$$\mathcal{L}\{\theta \mid x_1, \ldots, x_n\} \simeq N\left(\bar{x}, \frac{\bar{x}}{n}\right),$$

a much simpler result from which to make inferences than the exact posterior distribution.

7.3 APPROXIMATE EVALUATION OF BAYESIAN INTEGRALS

To implement the Bayesian paradigm it is necessary to be able to evaluate ratios of integrals of the form

$$I(\mathbf{x}_1, \ldots, \mathbf{x}_n) \equiv \frac{\int u(\boldsymbol{\theta}) e^{L(\boldsymbol{\theta})+\rho(\boldsymbol{\theta})} d\boldsymbol{\theta}}{\int e^{L(\boldsymbol{\theta})+\rho(\boldsymbol{\theta})} d\boldsymbol{\theta}}, \tag{7.1}$$

where $L(\boldsymbol{\theta}) \equiv \log f(\mathbf{x}_1, \ldots, \mathbf{x}_n \mid \theta) = \log \prod_1^n \ell(\mathbf{x}_j \mid \theta)$ denotes the log of the likelihood function, $\rho(\boldsymbol{\theta}) \equiv \log g(\boldsymbol{\theta})$ denotes the log of the prior density, and $u(\boldsymbol{\theta})$ is an arbitrary function of $\boldsymbol{\theta}$.

For example, if θ is one-dimensional, $u(\theta) = \theta$ is the mean of the posterior distribution (the Bayesian estimator of θ for quadratic loss; see Chapter 11); $u(\theta) = \theta^k$ more generally yields the kth moment of the posterior distribution. (Note that θ may be multidimensional.) Since $I(x_1, \ldots, x_n) \equiv E[u(\theta) \mid x_1, \ldots, x_n]$, $I(\cdot)$ provides the posterior mean of the arbitrary function $u(\theta)$. Evaluating the denominator integral in Equation (7.1) yields the normalizing constant in the posterior density.

Marginal posterior densities may be obtained from a variant of the ratio of integrals in Equation (7.1). Partition $\boldsymbol{\theta}$, letting $\boldsymbol{\theta} \equiv (\dot{\boldsymbol{\theta}}', \ddot{\boldsymbol{\theta}}')'$, let $h(\cdot)$ denote the posterior density, and define

$$\begin{aligned} I^*(\dot{\boldsymbol{\theta}}; x_1, \ldots, x_n) &\equiv \int h(\dot{\boldsymbol{\theta}}, \ddot{\boldsymbol{\theta}} \mid x_1, \ldots, x_n)\, d\ddot{\boldsymbol{\theta}} \\ &= \int h(\boldsymbol{\theta} \mid x_1, \ldots, x_n)\, d\ddot{\boldsymbol{\theta}} \\ &= \frac{\int e^{L(\boldsymbol{\theta})+\rho(\boldsymbol{\theta})} d\ddot{\boldsymbol{\theta}}}{\int e^{L(\boldsymbol{\theta})+\rho(\boldsymbol{\theta})} d\boldsymbol{\theta}}. \end{aligned} \tag{7.2}$$

Note that the ratio of integrals in Equation (7.2) is the marginal posterior density of $\dot{\boldsymbol{\theta}}$, given the data.

In Section 7.3.1 below, we present three of the analytical approximations that have been proposed for evaluating $I(\cdot)$ and $I^*(\cdot)$ in Equations (7.1) and (7.2). They differ from one another in their scope of applicability and in their accuracy. They all improve on the large-sample normal approximation presented in Section 7.2 by not requiring the sample size to be so large.

7.3.1 The Lindley Approximation

The first approximation, developed by Lindley (1980), is given in the following theorem. It is most useful when the dimension (number of parameters, p) is low (say ≤ 5).

Theorem 7.2 For n sufficiently large, so that $L(\boldsymbol{\theta})$ defined in Equation (7.1) concentrates around a unique maximum likelihood estimator $\hat{\boldsymbol{\theta}} \equiv \hat{\theta}(\mathbf{x}_1, \ldots, \mathbf{x}_n)$, for $\boldsymbol{\theta} \equiv (\theta_i) : p \times 1, \hat{\boldsymbol{\theta}} = (\hat{\boldsymbol{\theta}}_i)$, it follows that $I(\cdot)$, defined in Equation (7.1), is expressible approximately as

$$I(\mathbf{x}_1, \ldots, \mathbf{x}_n) \simeq u(\hat{\boldsymbol{\theta}}) + 0.5 \sum_{i=1}^{p} \sum_{j=1}^{p} \left[\left. \frac{\partial^2 u(\boldsymbol{\theta})}{\partial \theta_i \partial \theta_j} \right|_{\theta=\hat{\theta}} \cdot \left\{ 2 \left. \frac{\partial u(\boldsymbol{\theta})}{\partial \theta_i} \right|_{\theta=\hat{\theta}} \right\} \left\{ \left. \frac{\partial \rho(\theta)}{\partial \theta_j} \right|_{\theta=\hat{\theta}} \right\} \right] \hat{\sigma}_{ij}$$
$$+ 0.5 \sum_{i=1}^{p} \sum_{j=1}^{p} \sum_{k=1}^{p} \sum_{l=1}^{p} \left[\left. \frac{\partial^3 L(\theta)}{\partial \theta_i \partial \theta_j \partial_l} \right|_{\theta=\hat{\theta}} \cdot \left\{ \left. \frac{\partial u(\boldsymbol{\theta})}{\partial \theta_k} \right|_{\theta=\hat{\theta}} \right\} \hat{\sigma}_{ij} \hat{\sigma}_{kl} \right], \tag{7.3}$$

where $\hat{\sigma}_{ij}$ denotes the (i, j) element $\hat{\boldsymbol{\Sigma}}^{-1} = \hat{\boldsymbol{\Gamma}} = (\hat{\lambda}_{ij})$, and

$$\hat{\lambda}_{ij} = - \left. \frac{\partial^2 L(\boldsymbol{\theta})}{\partial \theta_i \partial \boldsymbol{\theta}_j} \right|_{\theta=\hat{\theta}}.$$

REMARK 1: The first term in Equation (7.3) is $O(1)$; the other terms are all $O(1/n)$ and are called *correction terms*. The overall approximation in the theorem is $O(1/n)$, so the first term neglected is $O(1/n^2)$.

REMARK 2: The approximation in Equation (7.3) involves the third derivative of $L(\theta)$. The last term, however, does not involve the prior at all, and only the first derivative of the prior is involved. Interestingly, by contrast with the large-sample approximation in Section 7.2, which depended strictly on the form of the likelihood function, this approximation depends upon the form of the prior as well.

Example 7.2

We reconsider Example 7.1, but now we focus attention on the Bayesian estimator for a quadratic loss function, the posterior mean (see Chapter 11). In that example, we saw that the large-sample (first-order) approximation to the posterior mean is the MLE, namely $\hat{\theta} = \bar{x}$. Now we use Theorem 7.2 to improve upon the approximation and to see the effect of the prior on the posterior mean.

First simplify Equation (7.3) for the case of a single one-dimensional parameter of interest and take $p = 1$. This gives

$$I(x_1, \ldots, x_n) \simeq u(\hat{\theta}) + 0.5 \left\{ \left. \frac{\partial^2 u(\theta)}{\partial \theta^2} \right|_{\theta=\hat{\theta}} + 2 \left. \frac{\partial u(\theta)}{\partial \theta} \frac{\partial \rho(\theta)}{\partial \theta} \right|_{\theta=\hat{\theta}} \right\} \hat{\sigma}^2$$
$$+ \left\{ 0.5 \left. \frac{\partial^3 L(\theta)}{\partial \theta^3} \frac{\partial u(\theta)}{\partial \theta} \right|_{\theta=\hat{\theta}} \right\} \hat{\sigma}^4. \tag{7.4}$$

Next take $u(\theta) = \theta$ (so that $\partial u(\theta)/\partial\theta = 1$ and $\partial^2 u(\theta)/\partial\theta^2 = 0$), and recall that $\hat{\theta} = \bar{x}$. Substituting into Equation (7.4) gives

$$I(x_1, \ldots, x_n) \simeq \bar{x} + \left\{ \frac{\partial\rho(\theta)}{\partial\theta}\bigg|_{\theta=\bar{x}} \right\} \hat{\sigma}^2 + 0.5 \frac{\partial^3 L(\theta)}{\partial\theta^3}\bigg|_{\theta=\bar{x}} \hat{\sigma}^4. \tag{7.5}$$

Now recall from Example 7.1 that $\hat{\Sigma} = \hat{\sigma}^2 = \bar{x}/n$ and that $L(\theta) = -n\theta + (\log\theta)\sum x_j - \log(\prod_j x_j!)$. Then

$$\frac{\partial L(\theta)}{\partial\theta}\bigg|_{\theta=\bar{x}} = 0; \quad \frac{\partial^2 L(\theta)}{\partial\theta^2}\bigg| = -\frac{n}{\bar{x}}; \quad \frac{\partial^3 L(\theta)}{\partial\theta^2}\bigg|_{\theta=\bar{x}} = \frac{2n}{\bar{x}^2}.$$

Substituting into Equation (7.5) gives

$$\begin{aligned} I(x_1, \ldots, x_n) &\simeq \bar{x} + \left[\frac{\partial\rho(\theta)}{\partial\theta}\bigg|_{\theta=\bar{x}} \right] \left(\frac{\bar{x}}{n}\right) + \left(\frac{n}{\bar{x}^2}\right)\left(\frac{\bar{x}^2}{n^2}\right) \\ &= \bar{x} + \frac{1}{n}\left[1 + \bar{x} \frac{\partial\rho(\theta)}{\partial\theta}\bigg|_{\theta=\bar{x}} \right]. \end{aligned}$$

Note also that since in this case, $I(x_1, \ldots, x_n) = E(\theta \mid x_1, \ldots, x_n)$, it follows that

$$E(\theta \mid x_1, \ldots, x_n) \simeq \bar{x} + \frac{1}{n}\left[1 + \bar{x} \frac{\partial\rho(\theta)}{\partial\theta}\bigg|_{\theta=\bar{x}} \right]. \tag{7.6}$$

Since in this example the prior distribution is log-normal, we have

$$\rho(\theta) = \log g(\theta) = -\log\sigma_0\sqrt{2\pi} - \log\theta - /12\sigma_0^2(\log\theta - \mu)^2.$$

So

$$\frac{\partial\rho(\theta)}{\partial\theta}\bigg|_{\theta=\bar{x}} = -\frac{1}{x}\left[1 + \frac{\log\bar{x} - \mu}{\sigma_0^2} \right].$$

Substituting into Equation (7.6) gives

$$E(\theta \mid x_1, \ldots, x_n) \simeq \bar{x} + \frac{1}{n}\left[1 - \left(1 + \frac{\log\bar{x} - \mu}{\sigma_0^2} \right) \right],$$

or

$$E(\theta \mid x_1, \ldots, x_n) \simeq \bar{x} - \frac{1}{n}\left(\frac{\log\bar{x} - \mu}{\sigma_0^2} \right). \tag{7.7}$$

Thus, the parameters (μ, σ_0^2) of the prior distribution enter into the second-order approximation for the posterior mean. Note that the terms neglected in the approximation in Equation (7.7) are $O(1/n^2)$.

7.3.2 Tierney–Kadane–Laplace Approximation

Another analytical approximation result that is quite useful for evaluation of Bayesian integrals is due to Tierney and Kadane (1986). which is given in Theorem 7.3. (The proof is based upon the steepest descent algorithm used for evaluation of Laplace integrals; see, for example, Erdelyi, 1956, p. 29)

Theorem 7.3 For n sufficiently large, if the posterior distribution of $u(\theta)$ (given the data) is concentrated on the positive (or negative) half-line, and if $[L(\theta) + \rho(\theta)]$, defined in Equation (7.1), concentrates around a unique maximum, under suitable regularity conditions the ratio of integrals in Equation (7.1) is given approximately by

$$I(x_1, \ldots, x_n) \simeq \frac{\sigma^*}{\sigma} \exp\{n[\mathcal{L}^*(\tilde{\theta}^*) - \mathcal{L}(\tilde{\theta})]\}, \tag{7.8}$$

where $n\mathcal{L}(\theta) \equiv L(\theta) + \rho(\theta)$, $n\mathcal{L}^*(\theta) = \log u(\theta) + L(\theta) + \rho(\theta)$, $\tilde{\theta}^*$ maximizes $\mathcal{L}^*(\theta)$, θ is the posterior mode and therefore maximizes $\mathcal{L}(\theta)$, and

$$\sigma^{-2} = -\left(\frac{\partial^2 \mathcal{L}(\theta)}{\partial \theta^2}\right)\bigg|_{\theta=\tilde{\theta}}, \quad \sigma^{*-2} = -\left(\frac{\partial^2 \mathcal{L}^*(\theta)}{\partial \theta^2}\right)\bigg|_{\theta=\tilde{\theta}^*}.$$

Proof
See Tierney and Kadane (1986).

REMARK 1: The result in Equation (7.8) involves only second derivatives of $L(\theta)$, whereas the analogous result in Equation (7.5) also involves third derivatives of $L(\theta)$.

REMARK 2: The terms omitted in the approximation in Equation (7.8) are $O(1/n^2)$, as in the result in Equation (7.3).

REMARK 3: While Theorem 7.2 holds for arbitrary functions $u(\theta)$, Theorem 7.3 holds for $u(\theta) > 0$ (or $u(\theta) < 0$). It has also been shown that the theorem can be modified so that it will hold for general $u(\theta)$. (See Tierney et al. (1989).)

REMARK 4: It may be necessary to use numerical techniques to evaluate $\tilde{\theta}$ and $\tilde{\theta}^*$ in order to apply the theorem.

Example 7.3

As an illustration of Theorem 7.3, we consider the following example based upon the Poisson sampling distribution. We can consider the same context as in Example 7.1, where the likelihood function is

$$f(x_1, \ldots, x_n \mid \theta) \propto e^{-n\theta}\theta^{\sum_1^n x_j}.$$

While we adopted a log-normal prior density in Example 7.1, here we adopt (along with Tierney and Kadane, (1986)) a gamma density prior for θ, so that the posterior density is given by

$$h(\theta \mid x_1, \ldots, x_n) \propto \theta^{\sum x_i + \alpha - 1} e^{-(\beta + n)\theta}$$

where the prior density is

$$g(\theta) \propto \theta^{\alpha-1}e^{-\beta\theta}, \qquad \alpha > 0, \beta > 0.$$

Thus, if $\tilde{\alpha} \equiv \sum_1^n x_i + \alpha$ and $\tilde{\beta} \equiv \beta + n$, the posterior density becomes

$$h(\theta \mid x_1, \ldots, x_n) \propto \theta^{\tilde{\alpha}-1}e^{-\tilde{\beta}\theta}.$$

The exact posterior mean is given by

$$E(\theta \mid x_1, \ldots, x_n) = \frac{\tilde{\alpha}}{\tilde{\beta}}.$$

Suppose we wanted to approximate the posterior mean of θ by using Theorem 7.3. First we note from the structure of Equation (7.1) that we only need to use the kernel of the likelihood function, as well as the kernel of the prior, rather than the entire normalized densities, since the proportionality constants of both cancel out in the ratio. Accordingly, set the log of the kernel of the likelihood equal to

$$L(\theta) = -n\theta + \sum x_i \log\theta,$$

and set the log of the kernel of the prior equal to

$$\rho(\theta) = (\alpha - 1)\log\theta - \beta\theta.$$

Thus,

$$\begin{aligned} n\mathcal{L}(\theta) &\equiv L(\theta) + \rho(\theta) \\ &= (\textstyle\sum x_i + \alpha - 1)\log\theta - \theta(n + \beta) \\ &\equiv (\tilde{\alpha} - 1)\log\theta - \theta\tilde{\beta}. \end{aligned}$$

Similarly, taking $u(\theta) \equiv \theta$ in equation 7.1, take

$$\begin{aligned} n\mathcal{L}^*(\theta) &\equiv \log\theta + L(\theta) + \rho(\theta) \\ &= \log\theta + n\mathcal{L}(\theta) \\ &= \log\theta + (\tilde{\alpha} - 1)\log\theta - \theta\tilde{\beta} \\ &= \tilde{\alpha}\log\theta - \theta\tilde{\beta}. \end{aligned}$$

Next note the $n\mathcal{L}(\theta)$ is maximized at $\theta = \hat{\theta}^* = (\tilde{\alpha} - 1)/\tilde{\beta}$ and that $n\mathcal{L}^*(\theta)$ is maximized at $\theta = \hat{\theta}^* = \tilde{\alpha}/\tilde{\beta}$. Moreover, $\sigma^2 = (\tilde{\alpha} - 1)/\tilde{\beta}^2$, and $\sigma^{*2} = \tilde{\alpha}/\tilde{\beta}^2$. The posterior mean approximation is then found by substituting into Equation (7.1). We find the approximation

$$\begin{aligned} \hat{E}(\theta \mid x_1, \ldots, x_n) &\equiv I(x_1, \ldots, x_n) \\ &= \frac{(\tilde{\alpha}^{0.5}/\tilde{\beta})(\tilde{\alpha}/\tilde{\beta})^{\tilde{\alpha}} \exp\{-\tilde{\alpha}\}}{[(\tilde{\alpha} - 1)^{0.5}/\tilde{\beta}][(\tilde{\alpha} - 1)/\tilde{\beta}]^{\tilde{\alpha}-1} \exp\{-(\tilde{\alpha} - 1)\}} \\ &= \left(\frac{\tilde{\alpha}}{\tilde{\beta}}\right)[\tilde{\alpha}/(\tilde{\alpha} - 1)]^{\tilde{\alpha}-0.5} e^{-1} \\ &= [E(\theta \mid x_1, \ldots, x_n)][\tilde{\alpha}/(\tilde{\alpha} - 1)]^{\tilde{\alpha}-0.5} e^{-1}. \end{aligned}$$

Note that we must have $\tilde{\alpha} > 1$ for the approximation to be applicable (otherwise we are estimating a positive quantity by one which is negative). Note that the relative error of the approximation depends only upon $\tilde{\alpha} = \sum x_i + \alpha$, and not upon sample size n. Table 7.1 shows relative errors in the approximation and was presented by Tierney and Kadane (1986). Thus, as $\tilde{\alpha}$ increases from 2 to 10, the relative error (the approximate posterior mean, $\hat{E}$, compared with the true posterior mean, E) decreases from 4.05 to 0.097 percent.

REMARK: For Theorem 7.3 to be applicable we must have a unimodal $\mathcal{L}(\theta)$, as we had in this example. Unfortunately, however, some complicated posterior distributions are multimodal in all but extremely large samples, rendering the theorem inapplicable in such cases. Moreover, even when $\mathcal{L}(\theta)$ and $\mathcal{L}^*(\theta)$ are both unimodal, it may not be easy to find the modal values.

Table 7.1 Relative Errors in the Approximation

$\tilde{\alpha}$	2	3	4	6	8	10
$\hat{E}(\theta \mid x_1, \ldots, x_n)/E(\theta \mid x_1, \ldots, x_n)$ Relative error	1.0405	1.0138	1.0069	1.0028	1.0015	1.00097

7.3.3 Naylor–Smith Approximation

Analysis of Bayesian estimators in large samples, and comparison of them with maximum likelihood estimators, shows that the asymptotic normality assumption can be quite misleading. In many cases the asymptotic posterior distribution becomes symmetric only very slowly, so that a seemingly "large" sample size does not generate normality of the posterior. Naylor (1982) explored Gaussian quadrature methods of numerical integration to evaluate Bayesian integrals in several dimensions efficiently. Some of this work is described in Naylor and Smith (1982) (see also Naylor and Smith, 1983; Smith et al., 1985). Classical quadrature techniques are presented, for example, in Davis and Rabinowitz (1967) and in Cohen et al. (1973). We summarize the Naylor–Smith approach below.

A classical result on numerical integration is that if $p_{2n-1}(x)$ denotes any polynomial of degree $(2n-1)$ in x, which is well defined in $[a, b]$, and $\{f_n(x)\}$ denotes a sequence of orthogonal polynomials in $[a, b]$ relative to a positive weighting function $w(x)$, such that $f_n(x)$ is a polynomial of degree n, then

$$\int_a^b w(x) p_{2n-1}(x)\, dx = \sum_{k=1}^{n} \alpha_k p_{2n-1}(x_k), \tag{7.9}$$

where the x_ks are the roots of $f_n(x) = 0$, and where the α_k coefficients are given by

$$\alpha_k = \int_a^b \frac{w(x) f_n(x)}{(x - x_k) f_n'(x_k)}\, dx.$$

Proof
See Cohen et al. (1973, pp. 87 and 88).

Various systems of orthogonal polynomials have been used. We select the Hermite class of orthogonal polynomials that are orthogonal relative to the weighting function

$$w(t) = \exp(-x^2),$$

on the interval $[a, b] = [-\infty, \infty]$. In this case, Equation (7.9) becomes the approximation

$$\int_{-\infty}^{+\infty} e^{-x^2} f(x) \simeq \sum_{k=1}^{n} \alpha_k f(x_k),$$

where $x_1, \ldots, x_n$ are the roots of the Hermite polynomial equation $H_n(x) = 0$, and

$$\alpha_k = \frac{2^{n-1} n! \sqrt{\pi}}{n^2 [H_{n-1}(x_k)]^2}.$$

Table 7.2 Zeros of Hermite Polynomials and Coefficients for Gauss–Hermite Quadrature

n	x_k	α_k
1	0	1.7724539
2	± 0.7071068	0.8862269
3	0	1.1816359
	± 1.2247449	0.2954100
4	± 0.5246476	0.8049141
	± 1.6506801	0.0813128
5	0	0.9453087
	± 0.9585725	0.3936193
	± 2.0201829	0.0199532

Integration methods of this (Hermite) form, based upon Equation (7.9), are called *Gaussian formulae.* If $f(t)$ is a polynomial of degree at most $(2n - 1)$, the approximation is exact. The first five zeros of $H_n(x)$ and the coefficient values of α_k are given in Table 7.2.

More generally, if $h(x)$ is a suitably regular function, and for some (μ, σ^2),

$$g(x) = \frac{h(x)}{(2\pi\sigma^2)^{0.5}} \exp\left\{-0.5\left(\frac{x-\mu}{\sigma}\right)^2\right\}.$$

Naylor and Smith (1982) give the approximate relation

$$\int_{-\infty}^{\infty} g(x)\,dx \simeq \sum_{k=1}^{n} m_k g(z_k), \tag{7.10}$$

where

$$m_k = \alpha_k \exp\left\{\frac{x_k^2}{\sigma\sqrt{2}}\right\}, \qquad z_k = (\mu + \sigma x_k \sqrt{x}).$$

Tables of x_k, α_k [and $\alpha_k \exp(x_k^2)$] that are more extensive than Table 7.2 are available for $n = 1(1)20$ in Salzer et al. (1952). The error will be small if $h(z)$ is approximately a polynomial. The precision of the approximation in Equation (7.10) depends upon the choices of μ and σ^2. A simple choice is to take (μ, σ^2) as the MLEs, but any prior guess could be used. Equation (7.10) could be used to evaluate posterior means or variances by taking $s(x)$ to be x^k times the posterior density, for $k = 1, 2$, and the approximation will be good as long as the posterior density is well approximated by the product of a normal density and a polynomial of degree at most $(2n - 3)$.

Readers interested in the extension of these results to problems with more than one parameter in the sampling density (several dimensions of numerical integration) should consult Naylor and Smith (1982).

7.4 IMPORTANCE SAMPLING

Suppose that we wish to evaluate the integral I approximately, where

$$I(\mathcal{D}) \equiv I(x_1, \ldots, x_n) \equiv \int u(\theta)h(\theta \mid x_1, \ldots, x_n)\, d\theta. \tag{7.11}$$

$u(\theta)$ denotes some function to be defined, and $h(\theta \mid x_1, \ldots, x_n)$ denotes a posterior density.

Note that the integral is also given by the population mean: $I = E\{u(\theta)\}$. Suppose we could sample from the posterior distribution to obtain values $\theta_1, \ldots, \theta_M$. Then, by the law of large numbers, an approximate value for I would be the sample mean $\hat{I} = 1/M \sum_{i=1}^{M} u(\theta_i)$.

Let $g^*(\theta)$ denote a "generating density" called the *importance function*. This k-dimensional density will be used to generate M points whose ordinates will be averaged to approximate $I(\mathcal{D})$. Let $\theta_1, \ldots, \theta_M$ be M points generated independently from $g^*(\theta)$. This is known as *importance sampling*. $g^*(\theta)$ is generally chosen to approximate the posterior density, but it is also chosen so that the θ_ms can be easily generated. Equation (7.11) can be rewritten

$$\begin{aligned} I &= \frac{\int u(\theta) f(\mathcal{D} \mid \theta) p(\theta)\, d\theta}{\int f(\mathcal{D} \mid \theta) p(\theta)\, d\theta} \\ &= \int u(\theta) w(\theta)\, d\theta, \end{aligned} \tag{7.12}$$

where

$$w(\theta) = \frac{f(\mathcal{D} \mid \theta) p(\theta)}{\int f(\mathcal{D} \mid \theta) p(\theta)\, d\theta}.$$

As an approximation to $I(\cdot)$ in Equation (7.12) we take the weighted average of the $u(\theta_m)$s, namely

$$\hat{I}(\mathcal{D}) = \sum_{m=1}^{M} \hat{w}(\theta_m) u(\theta_m), \tag{7.13}$$

where the weights $\hat{w}(\theta_m)$ are

$$\hat{w}(\theta_m) = \frac{f(\mathcal{D} \mid \theta_m)p(\theta_m)/g^*(\theta_m)}{\sum_{m=1}^{M} [f(\mathcal{D} \mid \theta_m)p(\theta_m)/g^*(\theta_m)]}.$$

Note that the weights $\hat{w}(\theta_m)$ sum to unity. We also note that because $f(\mathcal{D} \mid \theta_m)$ and $p(\theta_m)$ occur in ratio form, we will require only their kernels rather than their complete densities (since the normalizing constants of both cancel out in the ratio). Stewart (1983) points out that under easily attainable conditions, $\hat{I}(\mathcal{D})$ will converge almost surely to I, as M approaches infinity. The precision of the approximation will be heavily dependent upon the choice of $g^*(\theta)$, however. For additional discussion on how to choose the importance function, see Kloek and Van Dijk (1978, pp. 316 and 317).

SUMMARY

In this chapter we have focused upon the importance of approximations, numerical methods, and computer-program-assisted solutions to problems of Bayesian inference. We presented the large-sample normal approximation to posterior distributions (the effect of the prior disappears in very large samples). Then we showed that the large-sample distribution could be improved with one that also took the prior into account. We presented approximations to Bayesian integrals depending upon (1) numerical integration (in a few dimensions) and (2) Monte Carlo importance sampling (for many-dimensional integrals). We showed how to study posterior distributions by means of simulations.

EXERCISES

7.1* Let $X_1, \ldots, X_N$ denote i.i.d. observations from the distribution

$$f(x \mid \beta) = \beta e^{-\beta x}, \qquad x > 0, \beta > 0.$$

Adopt a natural conjugate prior distribution for β and find:
(a) the posterior density kernel for β given $x_1, \ldots, x_N$;
(b) the large-sample posterior normal distribution for β, given $x_1, \ldots, x_N$.

7.2* Use the Lindley approximation of Theorem 7.2 to approximate the posterior mean in Exercise 7.1, part (a).

7.3* Give the Tierney–Kadane approximation for the posterior variance in Exercise 7.1, part (a).

* Solutions for asterisked exercises may be found in Appendix 7.

7.4* Explain the method of Gauss–Hermite quadrature, and explain how you would use it to evaluate

$$I = \int_0^1 \frac{e^{-x}}{(1+x)^{10}}\, dx,$$

using a three-point quadrature grid. (*Hint*: take $n = 3$ in Table 7.2.)

7.5* Explain what is meant by "importance sampling." When would you use it?

7.6* Explain how you would use simulation of the posterior distribution to evaluate the posterior distribution of σ^3, where

$$y_i \mid x_i = a + bx_i + \varepsilon_i, \quad i = 1, \ldots, n,$$

$\varepsilon_i \sim N(0, \sigma^2)$, the ε_i are uncorrelated, (a, b) are unknown coefficients, and the prior distribution is

$$\begin{aligned} g(a, b, \sigma^2) &= g_1(a, b) g_2(\sigma^2), \\ g_1(a, b) &\propto \text{constant}, \\ g_2(\sigma^2) &\propto \frac{1}{\sigma^2}. \end{aligned}$$

7.7 If you did not use a program you wrote yourself, which commercial computer program would you use if you intended to carry out a Bayesian analysis of the coefficients in a univariate multiple regression? Why?

FURTHER READING

Cohen, A. M., Cutts, J. F., Fielder, R., Jones, D. E., Ribbans, J. and Stuart, E. (1973). *Numerical Analysis*, New York, Halsted Press (John Wiley and Sons, Inc.).

Davis, P. J. and Rabinowitz, P. (1967). *Numerical Integration*, Waltham, MA, Blaisdell.

Dempster, A. P. (1980). "Bayesian Inference in Applied Statistics", in J. M. *Bayesian Statistics*, Bernardo, M. H. DeGroot, D. V. Lindley and A. F. M. Smith (Eds.), Valencia, Spain, University Press, 266–291.

Erdelyi, A. (1956). *Asymptotic Expansions*, New York, Dover Publications.

Hyde, C. C. and Johnstone, I. M. (1979). "On Asymptotic Posterior Normality for Stochastic Processes" *J. Royal Statist. Soc.* (*B*), **41**, 184–189.

Kloek, T. and Van Dijk, H. K. (1978). "Bayesian Estimates of Equation System Parameters: An Application of Integration by Monte Carlo," *Econometrica*, **46**, 1–19.

Le Cam, L. (1956). "On the Asymptotic Theory of Estimation and Testing Hypotheses," *Proc. Third Berkeley Symp. on Math. Statist. and Prob.*, **1**, Berkeley, University of California Press, 129–156.

Lindley, D. V. (1965). *Introduction to Probability and Statistics, Part 2: Inference*, Cambridge, Cambridge University Press.

Lindley, D. V. (1980). "Approximate Bayesian Methods", in J. M. Bernardo, M. H. DeGroot, D. V. Lindley and A. F. M. Smith (Eds.), Bayesian Statistics, Valencia, Spain, University Press, 223–245.

Naylor, J. C. (1982). "Some Numerical Aspects of Bayesian Inference," Unpublished Ph.D. Thesis, University of Nottingham.

Naylor, J. C. and Smith, A. F. M. (1982). "Applications of a Method for the Efficient Computation of Posterior Distributions," *Appl Statist.*, **31**(3), 214–225.

Naylor, J. C. and Smith, A. F. M. (1983). "A Contamination Model in Clinical Chemistry," in A. P. Dawid and Smith, Eds., Practical Bayesian Statistics, Harlow, England, Longman.

Salzer, H. E., Zucker, R. and Capuano, R. (1952). "Tables of the Zeros and Weight Factors of the First Twenty Hermite Polynomials," *J. Res. Nat. Bur. of Standards*, **48**, 111–116.

Smith, A. F. M., Skene, A. M., Shaw, J. E. H., Naylor, J. C. and Dransfield, M. (1985). "The Implementation of the Bayesian Paradigm," *Comm. in Statist: Theory and Methods*, **14**(5), 1079–1102.

Stewart, L. T. (1983), "Bayesian Analysis Using Monte Carlo Integration—A Powerful Methodology For Handling Some Difficult Problems," in *Practical Bayesian Statistics*, Edited by David, A. P., and Smith, A. F. M., Harlow, England, Longman.

Tierney, L. and Kadane, J. B. (1986). "Accurate Approximations for Posterior Moments and Marginals," *J. Am. Statist. Assoc.*, **81**, 82–86.

Tierney, L., Kass, R. E. and Kadane, J. B. (1989). "Fully Exponential Laplace Approximations of Expectations and Variances of Non-Positive Functions," *J. Am. Statist. Assoc.*, **84**, 710–716.

PART III

Bayesian Statistical Inference and Decision Making

CHAPTER 8

Bayesian Estimation

8.1 INTRODUCTION

In this chapter we will explain the procedures for estimating unknown quantities associated with probability distributions from a Bayesian point of view. We will examine various distributions from both univariate and multivariate points of view, and we will treat both point and interval estimation, including both credibility intervals and highest posterior density intervals. We will treat a variety of discrete and continuous data distributions.

The most frequently occurring type of Bayesian point estimation required is point estimation in one dimension. While point estimators are most frequently used, it is generally the case that point estimators should be accompanied by interval estimators. That is, point estimators alone will usually not suffice; usually we *also* want to have a measure of the interval of uncertainty. In Section 8.2 we focus on point estimation in univariate data distributions; we treat multivariate data distributions in Section 8.3; and we treat interval estimation in Section 8.4. Section 8.5 provides a brief summary of *empirical Bayes estimation* (see also, Section 14.5.1).

8.2 UNIVARIATE (POINT) BAYESIAN ESTIMATION

In the absence of any guidance from decision theory, to channel our estimation procedures along specific lines so that the estimators will be associated with certain types of costs or gains in risky situations (see Chapter 11), Bayesian point estimators are most frequently taken to be the means of one-dimensional marginal posterior distributions. Much less often, they are taken to be the modes of one-dimensional marginal posterior distributions, or even less often, their medians are used. Other Bayesian estimators are sometimes also used, and are described in Chapter 11. In this chapter we will focus on the *means* of posterior distributions.

When we have developed a joint posterior distribution for several unknown parameters of a probability distribution, and theoretically, we could take as an estimate of a parameter θ, the mean of θ with respect to the joint posterior distribution, we usually prefer to take the mean of the marginal posterior distribution of θ. In some situations, of course, these means will be identical (when the posterior distributions are symmetric), but that will not always be the case, and there might be advantages of using one type of posterior mean versus the other (avoidance of nuisance parameters, focus on specific types of properties of the one versus the other, ease of computation, and so on). We will examine various well-known distributions and we will give the posterior mean Bayes' estimators in each case, under both vague (objective) and natural conjugate (subjective) prior distributions. First we will treat some discrete data distributions, and then some continuous data distributions.

8.2.1 Binomial Distribution

Vague Prior

Let X denote the number of successes in n independent Bernoulli trials where the outcome on each trial is a *success* or a *failure*, and the probability of success on a single trial is denoted by θ, $0 < \theta < 1$. The probability mass function for X is given by

$$f(x \mid \theta, n) = \binom{n}{x} \theta^x (1-\theta)^{n-x} \quad 0 < \theta < 1, \tag{8.1}$$

for $x = 0, 1, 2, \ldots, n$, and f is zero otherwise. A *vague prior* probability density function for θ is given by:

$$g(\theta) = 1, \qquad \text{for } 0 < \theta < 1 \tag{8.2}$$

and

$$g(\theta) = 0, \qquad \text{otherwise.}$$

The posterior probability density function (pdf) for θ is therefore given (Bayes' theorem) by

$$h(\theta \mid x, n) \propto \theta^x (1-\theta)^{n-x}, \tag{8.3}$$

a beta-distribution kernel, so that the full density is given by

$$h(\theta \mid x, n) = \frac{1}{B(x+1, n-x+1)} \theta^{(x+1)-1} (1-\theta)^{(n-x+1)-1}. \tag{8.4}$$

A point estimator for θ is therefore given by

$$\hat{\theta} = E(\theta \mid x, n) = \frac{x+1}{(x+1)+(n-x+1)} = \frac{x+1}{n+2}. \tag{8.5}$$

Natural Conjugate Prior

A natural conjugate prior density for θ may be given in the beta-density kernel form:

$$g(\theta) \propto \theta^{\alpha-1}(1-\theta)^{\beta-1}, \qquad 0 < \alpha,\ 0 < \beta,\ 0 < \theta < 1. \tag{8.6}$$

α and β are hyperparameters that must be assessed. So they are assumed to be known.

Multiplying Equation (8.6) by Equation (8.1), suppressing the binomial coefficient, which is just a proportionality constant, gives the posterior density kernel

$$h(\theta \mid x, n, \alpha, \beta) \propto \theta^{(x+\alpha)-1}(1-\theta)^{(n-x+\beta)-1}. \tag{8.7}$$

As in the case of the vague prior, we recognize this kernel as that of a beta distribution, but this time, the distribution has parameters $(x+\alpha, n-x+\beta)$, so that the complete posterior beta distribution has density given by

$$h(\theta \mid x, n, \alpha, \beta) = \frac{1}{B(x+\alpha, n-x+\beta)}\theta^{(x+\alpha)-1}(1-\theta)^{(n-x+\beta)-1}. \tag{8.8}$$

In this case, the Bayes' estimator of θ is given by

$$\hat{\theta} = E(\theta \mid x, n, \alpha, \beta) = \frac{x+\alpha}{(x+\alpha)+(n-x+\beta)} = \frac{x+\alpha}{\alpha+\beta+n}. \tag{8.9}$$

We note that if $\alpha = \beta = 1$, the point Bayes' estimators in the vague and natural conjugate cases (Equations (8.5) and (8.9)) become identical.

8.2.2 Poisson Distribution

Suppose the number of events in a given time interval, X, follows a Poisson distribution with parameter θ. Then, the probability mass function for X is given by

$$f(x \mid \theta) = \frac{e^{-\theta}\theta^x}{x!}, \qquad x = 0, 1, 2, \ldots, \quad 0 < \theta. \tag{8.10}$$

Vague Prior

A vague prior density for θ is given by the (improper) density:

$$g(\theta) \propto \frac{1}{\theta}. \tag{8.11}$$

The kernel of the pdf for θ is given by the product of Equations (8.10) and (8.11) as

$$h(\theta \mid x) \propto \theta^{x-1}e^{-\theta}. \tag{8.12}$$

Thus, we recognize the kernel of a gamma posterior distribution with probability density

$$h(\theta \mid x) = \frac{e^{-\theta}\theta^{x-1}}{\Gamma(x)}, \qquad 0 < \theta < \infty, \tag{8.13}$$

and zero, otherwise. The Bayes' estimator of θ is given by the posterior mean:

$$\hat{\theta} = E(\theta \mid x) = x. \tag{8.14}$$

Note that the sample mean is given by the reciprocal results $E(X \mid \theta) = \theta$.

Natural Conjugate Prior

A natural conjugate prior density for θ is the gamma density given by

$$g(\theta) \propto \theta^{\alpha-1}e^{-\beta\theta}, \qquad 0 < \alpha, \;\; 0 < \beta, \;\; 0 < \theta < \infty, \tag{8.15}$$

and zero, otherwise. Multiplying Equations (8.10) and (8.15) gives the kernel of the posterior density:

$$h(\theta \mid x, \alpha, \beta) \propto \theta^{x+\alpha-1}e^{-(\beta+1)\theta}. \tag{8.16}$$

The complete posterior density is given by

$$h(\theta \mid x, \alpha, \beta) = \frac{(\beta+1)^{x+\alpha}\theta^{x+\alpha-1}e^{-(\beta+1)\theta}}{\Gamma(x+\alpha)}, \qquad 0 < \theta < \infty, \tag{8.17}$$

and zero, otherwise. The Bayes' estimator of θ is given by the posterior mean:

$$\hat{\theta} = E(\theta \mid x, \alpha, \beta) = \frac{x+\alpha}{1+\beta}. \tag{8.18}$$

8.2.3 Negative Binomial (Pascal) Distribution

The probability mass function (pmf) for the distribution of the number of failures, X, observed before r successes in independent Bernoulli trials of an experiment in which θ denotes the probability of success on a single trial, is given by:

$$f(x \mid \theta, r) = \binom{x+r-1}{r-1}\theta^r(1-\theta)^x, \qquad x = 0, 1, \ldots, \quad 0 < \theta < 1. \tag{8.19}$$

Vague Prior

A vague prior for θ is given by the uniform prior density:

$$g(\theta) = 1, \qquad 0 < \theta < 1. \tag{8.20}$$

The kernel of the posterior distribution for θ is given by

$$h(\theta \mid x, r) = \theta^{(r+1)-1}(1-\theta)^{(x+1)-1}, \tag{8.21}$$

and the full beta posterior distribution is given by

$$h(\theta \mid x, r) = \frac{1}{B(r+1, x+1)}\theta^{(r+1)-1}(1-\theta)^{(x+1)-1}, \qquad 0 < \theta < 1. \tag{8.22}$$

The Bayes' estimator of θ is given by

$$\hat{\theta} = E(\theta \mid x, r) = \frac{r+1}{r+x+2}. \tag{8.23}$$

Natural Conjugate Prior

A natural conjugate prior for θ is given by the beta prior density kernel:

$$g(\theta) \propto \theta^{\alpha-1}(1-\theta)^{\beta-1}, \qquad 0 < \alpha, \quad 0 < \beta, \quad 0 < \theta < 1. \tag{8.24}$$

The posterior density kernel is found as

$$h(\theta \mid x, r, \alpha, \beta) \propto \theta^{(r+\alpha+1)-1}(1-\theta)^{(x+\beta+1)-1}, \tag{8.25}$$

and the full posterior distribution is given by the beta density:

$$h(\theta \mid x, r, \alpha, \beta) = \frac{1}{B(r+\alpha+1, x+\beta+1)}\theta^{(r+\alpha+1)-1}(1-\theta)^{(x+\beta+1)-1}, \qquad 0 < \theta < 1. \tag{8.26}$$

The Bayes' estimator for θ is given by

$$\hat{\theta} = E(\theta \mid x, r, \alpha, \beta) = \frac{r+\alpha+1}{r+x+\alpha+\beta+2}. \tag{8.27}$$

8.2.4 Univariate Normal Distribution (Unknown Mean, but Known Variance)

Let $X_1, \ldots, X_N$ denote independent and identically distributed observations from a $N(\theta, \sigma^2)$ distribution. Assume that the variance $\sigma^2 = \sigma_0^2$, a known quantity (see also, Section 4.6.1).

Vague (Flat) Prior

Suppose that you have very little information about the unknown mean θ, so that for you, over a rather large interval, all values of θ are equally likely. Accordingly, you decide to adopt a uniform prior density for θ over a large interval. In fact, since you really do not know how to define the endpoints of this large interval, as an approximation, you decide to replace this proper, uniform prior by an improper prior, one that is constant over the entire real line. Such a prior density cannot integrate to one, so it is called improper. But it is generally a good approximation to the uniform prior density over a large range, and it simplifies the mathematics of the computation. Such a prior density that is constant over the entire real line is often called a *flat*, *vague*, *non-informative*, *diffuse* or *default* prior density (see Chapter 5 for additional discussion of these priors). We express the flat prior density in the form:

$$g(\theta) \propto \text{constant}, \tag{8.28}$$

for all θ, $-\infty < \theta < +\infty$. Of course it can be thought of approximately as:

$$g(\theta) = \lim_{a \to \infty} \left(\frac{1}{2a} \right), \qquad -a < \theta < a.$$

Now express the kernel of the posterior density as the product of the likelihood and the prior density:

$$h(\theta \mid x_1, \ldots, x_n) \propto l(x_1, \ldots, x_n \mid \theta) g(\theta), \tag{8.29}$$

and substituting in Equation (8.28) gives

$$h(\theta \mid x_1, \ldots, x_n) \propto l(x_1, \ldots, x_n \mid \theta). \tag{8.30}$$

Now write the likelihood function in terms of the sample mean, which is sufficient here. Since $(\bar{X} \mid \theta) \sim N(\theta, \sigma_0^2/n)$,

$$l(x_1, \ldots, x_n \mid \theta) = l(\bar{x} \mid \theta) = \exp\left\{ -\left(\frac{n}{2\sigma_0^2} \right) (\bar{x} - \theta)^2 \right\}. \tag{8.31}$$

Substituting Equation (8.31) into Equation (8.30) gives

$$h(\theta \mid \bar{x}) \propto \exp\left\{ \left(-\frac{n}{2\sigma_0^2} \right) (\theta - \bar{x})^2 \right\}. \tag{8.32}$$

Inspecting Equation (8.32) shows that the posterior density of θ follows the distribution:

$$(\theta \mid \bar{x}) \sim N(\bar{x}, \sigma_0^2/n), \tag{8.33}$$

and of course, the Bayes' estimator of θ is just $\bar{x}$.

Note that although the prior distribution was improper, the posterior distribution is proper, and is also normal. Moreover, since you brought very little prior information to bear on this problem, not surprisingly, the posterior distribution is centered at the sample mean, which is also the maximum likelihood estimator, and is the best information that you have. In this instance we see the advantage to be gained by adopting the form of Bayes' theorem that uses proportionality; it was never necessary to carry out a formal integration since we could recognize the form of the posterior distribution family from the *kernel* of the posterior distribution (the form of the distribution without its normalizing constant that makes it integrate to one).

Normal Distribution Prior

Suppose that historical information about θ is available. To find your posterior distribution in this case, you decide to adopt this historical information for your prior distribution. If your prior distribution for θ is $N(m, \tau^2)$, and (m, τ^2) is known from your historical information, the prior density is given by

$$g(\theta) \propto \exp\left\{-0.5\left(\frac{\theta - m}{\tau}\right)^2\right\}. \tag{8.34}$$

The likelihood function is given in Equation (8.31), so that multiplying by Equation (8.34) gives the posterior density kernel:

$$h(\theta \mid \bar{x}, m, \tau) \propto \exp\left\{-0.5\left[\left(\frac{\theta - m}{\tau}\right)^2 + n\left(\frac{\theta - \bar{x}}{\sigma_0}\right)^2\right]\right\}. \tag{8.35}$$

Completing the square in θ in the exponent of Equation (8.35) gives

$$h(\theta \mid \bar{x}, m, \tau) \propto \exp\left\{-0.5\left(\frac{\theta - \tilde{\theta}}{\omega}\right)^2\right\} \tag{8.36}$$

where

$$\frac{1}{\omega^2} = \frac{1}{\tau^2} + \frac{1}{(\sigma_0^2/n)}, \tag{8.37}$$

and

$$\tilde{\theta} = \left(\frac{(1/\tau^2)}{(1/\tau^2) + (n/\sigma_0^2)}\right)m + \left(\frac{(n/\sigma_0^2)}{(1/\tau^2) + (n/\sigma_0^2)}\right)\bar{x}. \tag{8.38}$$

That is, $(\theta \mid \bar{x}, m, \tau) \sim N(\tilde{\theta}, \omega^2)$. We may now see that when the sample mean $\bar{x}$ follows a normal distribution with mean θ and known variance, and when the prior for θ also follows a normal distribution, the posterior distribution of θ is also normal with a mean that is a linear function of the prior mean and the sample mean. Moreover, if we define the *precision* of a distribution as the reciprocal of its variance, the posterior precision, $1/\omega^2$, is equal to the sum of the precision of the prior distribution, $1/\tau^2$, and the precision of the sampling distribution n/σ_0^2. The Bayes' estimator of θ is the posterior mean, $\tilde{\theta}$, which is a convex combination or weighted average of the prior mean, m, and the sample mean, $\bar{x}$, and the weights are proportions of precision corresponding to the prior distribution and the sampling distribution, respectively (the *total precision* is the quantity in the denominators of the weights). A more simple expression is obtained in the following way.

Let h_{prior} denote the precision of the prior distribution, so that $h_{\text{prior}} = 1/\tau^2$, and let h_{data} denote the precision of the data distribution, so that $h_{\text{data}} = n/\sigma_0^2$. If h_{total} denotes the total precision, so that $h_{\text{total}} = h_{\text{prior}} + h_{\text{data}}$, the Bayes' estimator of θ is given by the suggestive form:

$$\hat{\theta}_{\text{Bayes}} = \tilde{\theta} = \left(\frac{h_{\text{prior}}}{h_{\text{total}}}\right)m + \left(\frac{h_{\text{data}}}{h_{\text{total}}}\right)\bar{x}. \tag{8.39}$$

REMARK: Note that if we define $\alpha(h_{\text{prior}}/h_{\text{total}})$, Equation (8.39) may be written in the suggestive convex combination form:

$$\tilde{\theta} = (\alpha)m + (1-\alpha)\bar{x},$$

where $0 \le \alpha \le 1$.

8.2.5 Univariate Normal Distribution (Unknown Mean and Unknown Variance)

Suppose the data in an experiment follow a normal distribution with general unknown mean and unknown variance. Suppose you have independent observations $X_1, \ldots, X_n$ that are identically distributed as $N(\theta, h^{-1})$, where h denotes the precision. The likelihood function for the data is given by:

$$f(x_1, \ldots, x_n \mid \theta, h) \propto h^{0.5n} \exp\left\{(-0.5h)\sum_{i=1}^{n}(x_i - \theta)^2\right\}. \tag{8.40}$$

Note that we can write:

$$\begin{aligned}\sum_{i=1}^{n}(x_i - \theta)^2 &= \sum_{i=1}^{n}[(x_i - \bar{x}) + (\bar{x} - \theta)]^2 \\ &= \sum_{i=1}^{n}(x_i - \bar{x})^2 + n(\theta - \bar{x})^2.\end{aligned}$$

Then, setting $(n-1)s^2 = \sum_{i-1}^{n}(x_i - \bar{x})^2$, the likelihood function may be written

$$f(x_1, \ldots, x_n \mid \theta, h) \equiv l(\bar{x}, s^2 \mid \theta, h) = h^{0.5n} \exp\{-0.5h[(n-1)s^2 + n(\theta - \bar{x})^2]\}. \tag{8.41}$$

Bayes' theorem gives for the posterior density

$$q(\theta, h \mid \text{data}) \propto l(\bar{x}, s^2 \mid \theta, h)g(\theta, h) \tag{8.42}$$

where the prior density is denoted by $g(\theta, h)$.

Vague Prior Distribution

Adopt the vague prior densities for (θ, h):

$$g(\theta, h) = g_1(\theta)g_2(h), \tag{8.43}$$

where $g_1(\theta) \propto$ constant, and $g_2(h) \propto 1/h$. That is, θ and h are assumed to be *a priori* independent, we adopt a flat prior for θ, and we adopt a flat prior for [ln h], which implies that $g_2(h) \propto 1/h$ (see Section 5.3.1 for discussion of this prior). Therefore, a vague prior for (θ, h) corresponds to the (improper) density

$$g(\theta, h) \propto \frac{1}{h}. \tag{8.44}$$

Substituting Equations (8.41) and (8.44) into Equation (8.42) gives the joint posterior density

$$q(\theta, h \mid \text{data}) \propto \left[\frac{1}{h}\right] \left[h^{0.5n} \exp\left\{-0.5h \sum_{i=1}^{n}(x_i - \theta)^2\right\}\right].$$

The joint posterior density may be written

$$q(\theta, h \mid \text{data}) \propto \left[\frac{1}{h}\right] [h^{0.5n} \exp\{-0.5h[(n-1)s^2 + n(\theta - \bar{x})^2]\} \tag{8.45}$$

Rewriting this equation in a more recognizable form gives:

$$\begin{aligned} q(\theta, h \mid \text{data}) &= q_1(\theta \mid h, \text{data})q_2(h \mid \text{data}) \\ &\propto [h^{0.5} \exp\{-0.5nh(\theta - \bar{x})^2] \, [h^{0.5(n-3)} \exp\{-0.5h(n-1)s^2\}]. \end{aligned} \tag{8.46}$$

Thus, by inspection, the conditional posterior distribution of $(\theta \mid h, \bar{x}, s^2) \sim N(\bar{x}, 1/hn)$; and if $h \equiv 1/h\sigma^2$,

$$(\theta \mid \sigma^2, \bar{x}, s^2) \sim N\left(\bar{x}, \frac{\sigma^2}{n}\right); \tag{8.47}$$

sand the marginal posterior distribution of h is gamma $[(n-1)/2, 2/\{(n-1)s^2\}]$. That is, its posterior density is given by

$$r(h \mid \bar{x}, s^2) \propto h^{0.5(n-1)-1} \exp\{-0.5h(n-1)s^2\}. \tag{8.48}$$

Recall that if Z follows a gamma (a, b) distribution, its density is given by

$$r(z) = \frac{1}{b^a \Gamma(a)} z^{a-1} \exp\{-z/b\}, \qquad 0 < z < \infty. \tag{8.49}$$

Moreover, $E(Z) = ab$, and $var(Z) = ab^2$. Next, find the marginal posterior density of θ by integrating h out of Equation (8.46). Thus,

$$p(\theta \mid \text{data}) = \int_0^\infty q(\theta, h)dh \propto \int s_0^\infty h^{0.5(n-2)} \exp\{-hA\}dh, \tag{8.50}$$

where $A = 0.5[(n-1)s^2 + n(\theta - \bar{x})^2$, and A does not depend on h. Changing variables in Equation (8.50) by letting $t = hA$ gives

$$p(\theta \mid \text{data}) \propto \frac{1}{A^{0.5n}} \int_0^\infty t^{0.5(n-2)} \exp(-t)dt. \tag{8.51}$$

Since the remaining integral is just a constant (it is a gamma function), we can incorporate it into the proportionality constant to get

$$p(\theta \mid \text{data}) \propto \frac{1}{\{(n-1)s^2 + n(\theta - \bar{x})^2\}^{0.5n}}. \tag{8.52}$$

We recognize this kernel as that of a Student t-distribution. That is, marginally, the posterior distribution of θ is Student t, centered at the sample mean, with $(n-1)$ degrees of freedom. It is now clear that as in the case of known variance of the data, when the data are normally distributed, and the prior distribution on the parameters is vague, the Bayes' estimator of the population mean, θ, is given by the sample mean, $\bar{x}$.

REMARK: The traditional Student t-distribution has density

$$f_0(t) = \frac{c}{\{n + t^2\}^{0.5(n+1)}}, \qquad -\infty < t < \infty,$$

where

$$c = \frac{n^{0.5n}}{B(0.5, 0.5n)},$$

and B denotes the beta function. A generalization of this distribution to one with arbitrary location and scale is the one with density

$$f(t) = \frac{c\sigma^{-1}}{\left\{n + \left(\frac{t-\theta}{\sigma}\right)^2\right\}^{0.5(n+1)}}.$$

Its first two moments are $E(t) = \theta$, and $\mathrm{var}(t) = \left(\frac{n}{n-2}\right)\sigma^2$.

Natural Conjugate Prior Distribution

We adopt the natural conjugate prior distribution for $(\theta, h) : (\theta \mid h) \sim N(\theta_0, Kh^{-1})$, and $h \sim$ gamma (a, b), where (θ_0, K, a, b) are hyperparameters (parameters of the prior distribution that must be assessed from prior information). In the gamma distribution, as $a \to 0$, and $b \to \infty$, the density approaches the vague prior density adopted in the case of vague prior distribution. In terms of densities, the joint natural conjugate prior density is now given by:

$$\begin{aligned} g(\theta, h) &= g_1(\theta \mid \theta_0, K, h) g_2(h \mid a, b) \\ &\propto \left[h^{0.5} \exp\left\{\left(-\frac{h}{2K}\right)(\theta - \theta_0)^2\right\}\right] \left[h^{a-1} \exp\left\{-\left(\frac{h}{2b}\right)\right\}\right] \end{aligned}$$

or, after combining terms, the joint natural conjugate prior density is given by

$$g(\theta, h) \propto h^{0.5(2a-1)} \exp\left\{(-h)\left(\frac{1}{2K}\right)(\theta - \theta_0)^2 + \frac{1}{2b}\right\}. \tag{8.53}$$

Note that *a priori*, θ and h are not taken to be independent. Substituting the likelihood function and the natural conjugate prior into Bayes' theorem gives the joint posterior density

$$\begin{aligned} h(\theta, h \mid \bar{x}, s^2, n, \theta_0, a, b, K) &\propto [h^{0.5n} \exp\{-0.5h[n(\bar{x} - \theta)^2 + (n-1)s^2]\}] \\ &\quad \cdot \left[h^{0.5(2a-1)} \exp\left\{(-h)\left(\frac{1}{2K}\right)(\theta - \theta_0)^2 + \frac{1}{2b}\right\}\right]. \end{aligned}$$

Combining terms gives

$$h(\theta, h \mid \bar{x}, s^2, n, \theta_0, a, b, K) \propto h^{0.5(2a+n-1)}$$
$$\cdot \exp -0.5h\left\{\left[n(\bar{x}-\theta)^2 + (n-1)s^2 + \frac{1}{K}(\theta-\theta_0)^2 + \frac{1}{b}\right]\right\}. \quad (8.55)$$

Next, complete the square in θ in the exponent of the last equation and factor the expression, to obtain:

$$h(\theta, h \mid \bar{x}, s^2, n.a, b, K) \propto \left[h^{0.5} \exp\left\{-0.5h\left(n + \frac{1}{K}\right)(\theta - \tilde{\theta})^2\right\}\right]$$
$$\cdot [h^{0.5(2a+n-2)} \exp\{-0.5h\lambda\}], \quad (8.56)$$

where

$$\tilde{\theta} \equiv \frac{n\bar{x} + (\theta_0/K)}{n + (1/K)}, \quad (8.57)$$

and

$$\lambda = \frac{(n/K)}{n + (1/K)}(\bar{x} - \theta_0)^2 + (n-1)s^2 + (1/b). \quad (8.58)$$

That is, if $\sigma^2 \equiv (1/h)$, the conditional posterior distribution of $\theta \mid h$ is

$$(\theta \mid \sigma^2, \bar{x}, n, K) \sim N\left(\tilde{\theta}, \frac{\sigma^2}{n + (1/K)}\right) \quad (8.59)$$

and the posterior distribution of h is gamma $[0.5(2a+n), (2/\lambda)]$. Note that the joint prior and the joint posterior densities are in the same family, namely the normal-gamma family. (Such will always be the case when natural conjugate prior distributions are used.) Note also (from Equations (8.57) and (8.59)) that as in the case of known variance of the data, the (conditional) posterior mean is again a convex combination of the prior mean and the sample mean (compare with Equations (8.38) or (8.39)). Now find the marginal posterior distribution for θ by integrating q out of Equation (8.56). We find that if

$$\tilde{A} = 0.5\left[n(\bar{x}-\theta)^2 + (n-1)s^2 + \frac{1}{K}(\theta-\theta_0)^2 + \frac{1}{b}\right] \quad (8.60)$$

the marginal posterior density of θ becomes

$$\tilde{p}(\theta \mid \bar{x}, s^2, n, \theta_0, a, b, K) = \int_0^\infty h^{0.5(2a+n-1)} \exp\{-h\tilde{A}\} dh. \quad (8.61)$$

Making a change of variable and integrating gives

$$\tilde{p}(\theta \mid \bar{x}, s^2, n, \theta_0, a, b, K) \propto \frac{1}{(\tilde{A})^{0.5(2a+n+1)}}$$
$$= \frac{1}{\left\{n(\bar{x}-\theta)^2 + (n-1)s^2 + \frac{1}{K}(\theta-\theta_0)^2 + \frac{1}{b}\right\}^{0.5(2a+n+1)}}. \tag{8.62}$$

Completing the square in θ in Equation (8.62) gives

$$\tilde{p}(\theta \mid \bar{x}, s^2, n, \theta_0, a, b, K) \propto \frac{1}{\{\alpha^2 + (\theta - \bar{\theta})^2\}^{0.5(2a+n+1)}} \tag{8.63}$$

where the Bayesian estimator of θ is the marginal posterior mean $\bar{\theta}$, where

$$\bar{\theta} \equiv \left(\frac{n}{n+(1/K)}\right)\bar{x} + \left(\frac{(1/K)}{n+(1/K)}\right)\theta_0, \qquad \alpha^2 \equiv \frac{(1/b) + (\theta_0^2/K) + \sum_{i=1}^{n} x_i^2}{n+(1/K)}. \tag{8.64}$$

That is, the marginal posterior distribution of θ is Student t, centered at $\bar{\theta}$, a convex combination of the sample mean and the prior mean, with $(2a+n)$ degrees of freedom. The Student t-distribution was discussed in the Remark following Equation (8.52).

8.3 MULTIVARIATE (POINT) BAYESIAN ESTIMATION

8.3.1 Multinomial Distribution

Suppose we are concerned with an experiment that has p possible outcomes. The experiment is repeated independently n times. Let X_j denote the number of times the experiment resulted in category $j, j = 1, \ldots, p$. We must have $X_1 + \cdots + X_p = n$. Let θ_j denote the probability that the outcome of any of the repeated experiments gives a result in category $j, j = 1, \ldots, p$. We must have $\theta_1 + \cdots + \theta_p = 1$; $0 < \theta_j < 1$, $j = 1, \ldots, p$. In such a situation, the (multinomial) probability mass function for the data is given by

$$f(x_1, \ldots, x_n \mid \theta_1, \ldots, \theta_{p-1}, n)$$
$$= \frac{n!}{x_1!x_2!\cdots x_p!}\theta_1^{x_1}\theta_2^{x_2}\cdots\theta_{p-1}^{x_{p-1}}(1-\theta_1-\theta_2-\cdots-\theta_{p-1})^{x_p}. \tag{8.65}$$

Vague Prior

A vague prior density for the parameters is given by

$$g(\theta_1, \ldots, \theta_{p-1}) = \prod_{j=1}^{p-1} g_j(\theta_j) = 1,$$
$$0 < \theta_j < 1, \quad j = 1, \ldots, p-1, \qquad \sum_{j=1}^{p-1} \theta_j < 1 \quad (8.66)$$

where $g_j(\theta_j) = 1, j = 1, \ldots, p-1$. Multiplying Equations (8.65) and (8.66) gives the posterior density kernel

$$h(\theta_1, \ldots, \theta_{p-1} \mid x_1, \ldots, x_p, n) \propto \theta_1^{x_1}\theta_2^{x_2} \cdots \theta_{p-1}^{x_{p-1}}(1 - \theta_1 - \theta_2 - \cdots - \theta_{p-1})^{x_p}. \quad (8.67)$$

We might now recognize the density kernel in Equation (8.67) as that of a Dirichlet distribution (see for example, Press, 1982, p. 143) whose complete distribution is given by:

$$h(\theta_1, \ldots, \theta_{p-1} \mid x_1, \ldots, x_p, n) = \frac{\Gamma[(x_1+1) + \cdots + (x_p+1)]}{\Gamma(x_1+1)\Gamma(x_2+1)\cdots\Gamma(x_p+1)} \cdot \theta_1^{(x_1+1)-1}\theta_2^{(x_2+1)-1} \cdots \theta_{p-1}^{(x_{p-1}+1)-1} \cdot (1 - \theta_1 - \theta_2 - \cdots - \theta_{p-1})^{(x_p+1)-1}. \quad (8.68)$$

The Bayes' estimators of the unknown parameters are given by the marginal posterior means

$$\hat{\theta}_j = E(\theta_j \mid x_1, \cdots x_p, n) = \frac{(x_j+1)}{\sum_{k=1}^{p}(x_k+1)} = \frac{x_j+1}{n+p}, \qquad j = 1, \ldots, p. \quad (8.69)$$

Natural Conjugate Prior

A natural conjugate prior for the parameters is given by the Dirichlet distribution density kernel:

$$g(\theta_1, \ldots, \theta_{p-1}) \propto \theta_1^{\alpha_1-1}\theta_2^{\alpha_2-1} \cdots \theta_{p-1}^{\alpha_{p-1}-1}(1 - \theta_1 - \theta_2 - \cdots - \theta_{p-1})^{\alpha_p-1}, \qquad 0 < \alpha_j, \quad (8.70)$$

for $j = 1, \ldots, p$. Multiplying the kernel in Equation (8.65) by that in Equation (8.70) gives the Dirichlet posterior density kernel representation:

$$h(\theta_1, \ldots, \theta_{p-1} \mid x_1, \ldots, x_p, n, \alpha_1, \ldots, \alpha_p) \propto \theta_1^{(x_1+\alpha_1)-1}\theta_2^{(x_2+\alpha_2)-1} \cdots \theta_{p-1}^{(x_{p-1}+\alpha_{p-1})-1} \cdot (1 - \theta_1 - \theta_2 - \cdots - \theta_{p-1})^{(x_p+\alpha_p)-1}. \quad (8.71)$$

Thus, it is clear that the Bayes' estimators of the unknown parameters in the data distribution are given by the marginal posterior means:

$$\hat{\theta}_j = E(\theta_j \mid x_1, \ldots, x_p, n, \alpha_1, \ldots, \alpha_p) = \frac{x_j + \alpha_j}{(x_1 + \alpha_1) + \cdots + (x_p + \alpha_p)}$$

$$= \frac{x_j + \alpha_j}{n + \sum_{k=1}^{p} \alpha_k}, \tag{8.72}$$

for $j = 1, \ldots, p$.

8.3.2 Multivariate Normal Distribution with Unknown Mean Vector and Unknown Covariance Matrix

Vague Prior Distribution

We would like to evaluate the Bayes' estimator for the case of multivariate normal data under a vague prior distribution. We will follow the approach developed in Example 5.6. Suppose that $(\mathbf{X}_1, \ldots, \mathbf{X}_n$ are an i.i.d. sample of p-dimensional vectors that follow the p-dimensional normal distribution: $N(\boldsymbol{\theta}, \Sigma)$. The data density function is

$$f(\mathbf{x}_1, \ldots, \mathbf{x}_n \mid \boldsymbol{\theta}, \Sigma) = \frac{1}{(2\pi)^{0.5np} \mid \Sigma \mid^{0.5n}} \exp\left\{-0.5 \sum_{i=1}^{n} (\mathbf{x}_i - \boldsymbol{\theta})' \Sigma^{-1} (\mathbf{x}_i - \boldsymbol{\theta})\right\}. \tag{8.73}$$

The log density is therefore

$$L \equiv \log f(\mathbf{x}_1, \ldots, \mathbf{x}_n \mid \theta, \Sigma)$$

$$= \log(2\pi)^{-0.5np} - 0.5n \log \mid \Sigma \mid - 0.5 \sum_{i=1}^{n} (\mathbf{x}_i - \boldsymbol{\theta})' \Sigma^{-1} (\mathbf{x}_i - \boldsymbol{\theta})\}. \tag{8.74}$$

We will develop the Jeffreys' invariant prior density by applying the procedure to one matrix parameter at a time.

First assume that Σ is a known constant matrix. Then, using vector differentiation and noting that the derivative of a scalar L with respect to a vector θ is the vector $\partial L / \partial \theta$ of derivatives of L with respect to each element of the vector $\theta = (\theta_i), i = 1, \ldots, p$:

$$\frac{\partial L}{\partial \boldsymbol{\theta}} \equiv \left(\frac{\partial L}{\partial \theta_i}\right) = -n\Sigma^{-1}(\boldsymbol{\theta} - \bar{\mathbf{x}}). \tag{8.75}$$

Similarly,

$$-E\left\{\frac{\partial^2 L}{\partial \boldsymbol{\theta} \partial \boldsymbol{\theta}'}\right\} = n\Sigma^{-1} = \text{constant}. \tag{8.76}$$

So the information matrix is constant, and the Jeffreys' invariant prior for θ is given by:

$$g(\boldsymbol{\theta}) \propto \text{constant}. \tag{8.77}$$

Now assume $\boldsymbol{\theta}$ is a known constant vector. Write L in the more convenient form:

$$\begin{aligned} L &\equiv \log f(\mathbf{x}_1, \ldots, \mathbf{x}_n \mid \boldsymbol{\theta}, \Sigma) \\ &= \log(2\pi)^{-0.5np} + 0.5n \log \mid \Lambda \mid -0.5 \operatorname{tr}\Lambda \left\{ \sum_{i=1}^{n} (\mathbf{x}_i - \boldsymbol{\theta})(x_i - \boldsymbol{\theta})' \right\}, \end{aligned} \tag{8.78}$$

where "tr" denotes the trace of a square matrix (the sum of its diagonal elements), and we have reparameterized in terms of the precision matrix $\Lambda \equiv \Sigma^{-1}$. Forming the first two derivatives of L with respect to $\Lambda \equiv (\lambda_{ij})$ gives (see, for example, Press, 1982, p. 79 for details):

$$\begin{aligned} \frac{\partial L}{\partial \Lambda} &= n\Lambda^{-1} - 0.5 \operatorname{diag}(\Lambda^{-1}) - 0.5 \sum_{i=1}^{n} (\mathbf{x}_i - \boldsymbol{\theta})(\mathbf{x}_i - \boldsymbol{\theta})' \}, \\ \frac{\partial^2 L}{\partial \lambda_{ij} \partial \lambda_{kl}} &\propto \frac{1}{\mid \Lambda \mid^{(p+1)}}. \end{aligned} \tag{8.79}$$

Taking expectations in the second derivatives with respect to the data vectors leaves the result unchanged. So the Jeffreys' invariant prior density for Λ is given by:

$$g(\Lambda) \propto \frac{1}{\mid \Lambda \mid^{0.5(p+1)}}. \tag{8.80}$$

If $\boldsymbol{\theta}$ and Λ are *a priori* independent, the invariant prior density implied by the Jeffreys' argument is

$$g(\boldsymbol{\theta}, \boldsymbol{\Lambda}) \propto \frac{1}{\mid \Lambda \mid^{0.5(p+1)}}. \tag{8.81}$$

Equation (8.81) gives the multivariate generalization of the vague (and identical Jeffreys) prior density developed for the univariate case. It is the objective prior density most frequently used in the multivariate case to express knowing little, or vagueness. There are no hyperparameters to assess.

The joint posterior density kernel is found by multiplying Equations (8.74) and (8.81) to get

$$h(\boldsymbol{\theta}, \Lambda \mid \mathbf{x}_1, \ldots, \mathbf{x}_n) \propto |\Lambda|^{0.5(n-p-1)} \exp\left\{-0.5 \sum_{j=1}^{n} (\mathbf{x}_j - \boldsymbol{\theta})'\Lambda(\mathbf{x}_j - \boldsymbol{\theta})\right\}. \tag{8.82}$$

Now note that we can write the sum in the exponent in Equation (8.82) as:

$$\begin{aligned} \sum_{j=1}^{n} (\mathbf{x}_j - \boldsymbol{\theta})'\Lambda(\mathbf{x}_j - \boldsymbol{\theta}) &= \sum_{j=1}^{n} \operatorname{tr}(\mathbf{x}_j - \boldsymbol{\theta})(\mathbf{x}_j - \boldsymbol{\theta})'\Lambda] \\ &= \operatorname{tr}\left\{\Lambda\left[\sum_{j=1}^{n}(\mathbf{x}_j - \boldsymbol{\theta})(\mathbf{x}_j - \boldsymbol{\theta})'\right]\right\}. \end{aligned} \tag{8.83}$$

Also note that

$$\begin{aligned} \sum_{j=1}^{n} (\mathbf{x}_j - \boldsymbol{\theta})(\mathbf{x}_j - \boldsymbol{\theta})' &= \sum_{j=1}^{n} [(\mathbf{x}_j - \bar{\mathbf{x}}) + (\bar{\mathbf{x}} - \boldsymbol{\theta})][(\mathbf{x}_j - \bar{\mathbf{x}}) + (\bar{\mathbf{x}} - \boldsymbol{\theta})] \\ &= \mathbf{V} + n(\bar{\mathbf{x}} - \boldsymbol{\theta})(\bar{\mathbf{x}} - \boldsymbol{\theta})' \equiv \mathbf{A}, \end{aligned} \tag{8.84}$$

where $\mathbf{V} = (\mathbf{x}_j - \mathbf{x})(\mathbf{x}_j - \mathbf{x})'$. Substituting Equations (8.83) and (8.84) into Equation (8.85), below, gives the joint posterior density kernel:

$$h(\boldsymbol{\theta}, \Lambda \mid \mathbf{x}_1, \ldots, \mathbf{x}_n) \propto |\Lambda|^{0.5(n-p-1)} \exp\{-0.5\operatorname{tr}(\Lambda)(\mathbf{A})\}. \tag{8.85}$$

Note that $\mathbf{A}$ does not depend upon Λ, but only upon $\boldsymbol{\theta}$. The marginal posterior density kernel of $\boldsymbol{\theta}$ is found by integrating Equation (8.85) with respect to Λ. It is given by:

$$h(\boldsymbol{\theta} \mid \mathbf{x}_1, \ldots, \mathbf{x}_n) \propto \int |\Lambda|^{0.5(n-p-1)} \exp\{-0.5 \operatorname{tr} (\Lambda)(\mathbf{A})\} d\Lambda. \tag{8.86}$$

The integration is readily carried out by noting that the integrand is the kernel of a Wishart distribution density. The result is (see, for example, Press, 1982, p. 108, Equation (5.1.1)):

$$h(\boldsymbol{\theta} \mid \mathbf{x}_1, \ldots, \mathbf{x}_n) \propto \frac{1}{|\mathbf{A}|^{0.5n}} = \frac{1}{|\mathbf{V} + n(\boldsymbol{\theta} - \bar{\mathbf{x}})(\boldsymbol{\theta} - \bar{\mathbf{x}})'|^{0.5n}}. \tag{8.87}$$

Then,

$$h(\boldsymbol{\theta} \mid \mathbf{x}_1, \ldots, \mathbf{x}_n) \propto \frac{1}{|\mathbf{I} + n\mathbf{V}^{-1}(\boldsymbol{\theta} - \bar{\mathbf{x}})(\boldsymbol{\theta} - \bar{\mathbf{x}})'|^{0.5n}}. \tag{8.88}$$

Recall that for any two matrices $\mathbf{R}$: $(k \times m)$ and $\mathbf{W}$: $(m \times k)$, $| \mathbf{I}_k + \mathbf{RW} | = | \mathbf{I}_m + \mathbf{WR} |$. Applying this property in Equation (8.88) gives, for the marginal posterior density kernel,

$$h(\boldsymbol{\theta} \mid \mathbf{x}_1, \ldots, \mathbf{x}_n) \propto \frac{1}{\{1 + n(\boldsymbol{\theta} - \bar{\mathbf{x}})'\mathbf{V}^{-1}(\boldsymbol{\theta} - \bar{\mathbf{x}})\}^{0.5n}}. \tag{8.89}$$

The kernel in Equation (8.89) may be recognized as the kernel of a multivariate Student t-distribution with mean at $\bar{\mathbf{x}}$. That is, the Bayesian estimator of $\boldsymbol{\theta}$ is the marginal posterior mean $\hat{\boldsymbol{\theta}} = E(\boldsymbol{\theta} \mid \text{data})\bar{\mathbf{x}}$.

Natural Conjugate Prior Distribution

It would be interesting to examine the case of natural conjugate families of prior distributions for $(\boldsymbol{\theta}, \Sigma)$ in the multivariate situation. Such discussion is somewhat beyond the scope of this volume. Results may be found in, for example, Press, 1982, Section 7.1.6.

8.4 INTERVAL ESTIMATION

8.4.1 Credibility Intervals

Bayesian interval estimators for a scalar parameter θ are obtained from the posterior distribution for θ (assumed to be continuous). Suppose that the posterior cdf for θ is given by $F(\theta \mid x_1, \ldots, x_n) \equiv F(\theta \mid \text{data})$. Then, for some preassigned α, we can find an interval (a, b) such that:

$$1 - \alpha = P\{a < \theta < b \mid \text{data}\} = F(b) - F(a). \tag{8.90}$$

The interval (a, b) is called a *credibility interval* for θ, at credibility level α (the term "credibility interval" appears to have been first used by Edwards, Lindman et al. 1963). Suppose, for example, $\alpha = 0.05$. Then, we refer to such an interval as a 95 percent credibility interval. Some people even refer to (a, b) as a 95 percent Bayesian confidence interval, attempting to distinguish it from a frequentist confidence interval.

For example, suppose $(x_1, \ldots, x_n \mid \theta)$ denotes a sample from $N(\theta, 1)$. $\bar{x}$ is sufficient for θ. If we adopt a vague prior distribution for θ, the posterior density is given, as in Section 8.2.4, Equation (8.33), by

$$(\theta \mid \bar{x}) \sim N\left(\bar{x}, \frac{\sigma_0^2}{n}\right) = N\left(\bar{x}, \frac{1}{n}\right), \tag{8.91}$$

where we have taken $\sigma_0^2 = 1$. A 95 percent symmetric credibility interval for θ is given by:

$$0.95 = P\left\{\bar{x} - \frac{2}{\sqrt{n}} \le \theta \le \bar{x} + \frac{2}{\sqrt{n}} \middle| \bar{x}\right\}. \tag{8.92}$$

8.4.2 Credibility Versus Confidence Intervals

The interpretation of a credibility interval is straightforward. Such is not the case for a confidence interval. The interpretation of a confidence interval involves *indirect* statements about the true value. Since θ has a probability distribution from a Bayesian point of view, we can make probability statements *directly* about θ in the usual way. Thus, the interpretation of Equation (8.90) is that given the observed data, the probability of the event $\{a < \theta < b\}$ is $(1 - \alpha)$. For the example of the normal distribution, Equation (8.92) asserts that conditional on the data (summarized by $\bar{x}$), θ lies in the interval, $(\bar{x} - 2/\sqrt{n}, \bar{x} + 2/\sqrt{n})$ with 95 percent probability. This is exactly the kind of statement a person working on an application of statistics would like to make. The probability statement is based upon the current set of data, $\bar{x}$, and not upon data *not* observed. Moreover, it is a probability statement on the actual quantity of interest. Note that we need not be dealing with large numbers of repeated trials even conceptually.

By contrast, a *confidence interval* for θ at 95 percent level of confidence is given by

$$0.95 = P\left\{\bar{x} - \frac{2}{\sqrt{n}} \le \theta \le \bar{x} + \frac{2}{\sqrt{n}} \middle| \theta\right\}. \tag{8.93}$$

While the intervals involved in Equation (8.92) and (8.93) are identical there is an enormous difference in interpretation. Because we have conditioned on θ in Equation (8.93), a probability statement for the equation cannot be made directly about θ, but can only be made about the observed random variable, $\bar{x}$; $\bar{x}$ is the only random variable involved from a frequentist point of view. How should we interpret the (*confidence interval*) probability statement in Equation (8.93)?

To interpret the probability statement in Equation (8.93) we adopt the *long run* interpretation of probability. Suppose an experiment is carried out in which, for a preassigned value of θ, a random variable X is observed n times, and its average value, $\bar{x}$, is calculated. Let A denote the event that $\{\bar{x} - 2/\sqrt{n} \le \theta \le \bar{x} + 2/\sqrt{n}\}$ occurs (either A occurs or it does not). If A occurs we will call this experiment a "success"; otherwise, a failure. Imagine that this same experiment were carried out K times, and that each time we calculated a new $\bar{x}$, and each time we calculated a new interval $(\bar{x} - 2/\sqrt{n}, \bar{x} + 2/\sqrt{n})$, and determined whether θ belonged to this interval. Let r denote the number of times in K trials that the success A occurs. Imagine that we take a larger and larger number of independent trials of this experiment, so that

$K \to \infty$. Then we know from the Bernoulli theorem that:

$$P\{A\} = \lim_{K\to\infty}\left(\frac{r}{K}\right) = P\left\{\bar{x} - \frac{2}{\sqrt{n}} \leq \theta \leq \bar{x} + \frac{2}{\sqrt{n}} \,\middle|\, \theta\right\}.$$

So, to interpret Equation (8.93) we must make a statement such as: If we were to carry out an experiment analogous to the current one many times, and each time we took a sample of size n, and each time we calculated a new sample mean, $\bar{x}$, in about 95 percent of the experiments, θ would be included within the 95 percent confidence interval: $(\bar{x} - 2/\sqrt{n}, \bar{x} + 2/\sqrt{n})$. Note that for such a frequentist interpretation to make sense we need to be contemplating a large number of independent, repeatable, experiments; moreover, the interval depends on $\bar{x}$s that we have never observed.

However, we have only carried out one experiment, and we really want to make inferences about θ that depend only upon the $\bar{x}$ actually observed. In the sequel we will be concerned only with credibility intervals.

8.4.3 Highest Posterior Density Intervals and Regions

A highest posterior density (HPD) interval is a specific credibility interval that has certain important properties. We note from Equation (8.90) that the credibility interval at credibility level $(1 - \alpha)$ is given by the interval (a, b) such that $(1 - \alpha) = F(b) - F(a)$, where F denotes the posterior cdf. But when we preassign α, for known functional form F, this does not specify the interval (a, b). Which interval (a, b) should be chosen? It is not uniquely defined. So we choose the specific $(1 - \alpha)$ interval that contains most of the posterior probability. To do so, we choose the smallest interval (a, b) to satisfy two properties (to be specific in this illustration, we work with 95 percent credibility, but the formalism is the same for more general levels of credibility):

Property (1): $F(b) - F(a) = 0.95$;

Property (2): If $h(\theta \mid x_1, \ldots, x_n)$ denotes the posterior density, for $a \leq \theta \leq b$, $h(\theta \mid x_1, \ldots, x_n)$ has a greater value than that for any other interval for which Property (1) holds.

The posterior density for every point inside the HPD interval is greater than that for every point outside the interval. Also, for a given credibility level, the HPD interval is as small as possible.

In higher dimensions, we would define the credibility content $(1 - \alpha)$ of a multi-dimensional region R (rather than the credibility level of an interval). Then, the conditions required to define the HPD region become:

1. $P\{\theta \epsilon R \mid \text{data}\} = 1 - \alpha$;
2. For all $\theta_1 \in R$, and all $\theta_2 \notin R$, $h(\theta 1 \mid \text{data}) \geq h(\theta_2 \mid \text{data})$.

The HPD region has smallest possible volume among all regions of content $(1 - \alpha)$.

Formal Statement for HPD Intervals

The HPD interval (region) always exists, and it is unique, as long as for all intervals (regions) of content $(1 - \alpha)$, the posterior density is never uniform in any interval (region) of the space of θ (see Box and Tiao, 1973, Section 2.8).

Proof (Zellner, 1971, p. 27)

We want to find the smallest interval (a, b) with the constraining property that $\int_a^b h(\theta \mid \text{data})\, d\theta = 1 - \alpha$, for posterior density $h(\theta \mid \text{data})$, for preassigned α. Assume that the posterior density is unimodal. Form the Lagrangian (with Lagrange multiplier λ)

$$\mathcal{L} = (b - a) + \lambda \left\{ \int_a^b h(\theta \mid \text{data})\, d\theta - (1 - \alpha) \right\},$$

differentiate with respect to a and b separately, setting the results equal to zero, and solve using the constraining property. Accordingly,

$$\frac{\partial \mathcal{L}}{\partial a} = -1 - \lambda h(a \mid \text{data}) = 0, \text{ which implies that } h(a \mid \text{data}) = -\frac{1}{\lambda}; \text{ analogously,}$$

$$\frac{\partial \mathcal{L}}{\partial b} = 1 + \lambda h(b \mid \text{data}) = 0, \text{ which implies that } h(b \mid \text{data}) = -\frac{1}{\lambda}.$$

Note that since h is a probability density, we must have $h > 0$, so that we must have $\lambda < 0$ The second order conditions give:

$$\frac{\partial^2 \mathcal{L}}{(\partial a)^2} = -\lambda \frac{\partial h(a \mid \text{data})}{\partial a}; \qquad \frac{\partial^2 \mathcal{L}}{(\partial b)^2} = \lambda \frac{\partial h(b \mid \text{data})}{\partial b}; \qquad \frac{\partial^2 \mathcal{L}}{(\partial a)(\partial b)} = \frac{\partial^2 \mathcal{L}}{(\partial b)(\partial a)} = 0.$$

Since the posterior density is unimodal, we must have $\partial h(a \mid \text{data})/\partial a > 0$, so that $(\partial^2 \mathcal{L}/(\partial a)^2 > 0$. Analogously, because of unimodality, we must have $\partial h(b \mid \text{data})/\partial b < 0$, so that $\partial^2 \mathcal{L}/(\partial b)^2 > 0$. So the Hessian matrix of second derivatives is positive definite, which implies that we have achieved a minimum for the interval (a, b).

Now assume the posterior density is multimodal. We can always take the highest mode and adopt the above argument. If the posterior density has no modes at all, for the HPD interval we merely take the interval where the posterior density is highest. If there is an interval for which the posterior density is constant, clearly there will be ambiguity about where the HPD interval is, so there will be many possibilities. Analogous arguments apply to HPD regions.

If the posterior density is unimodal the matter of finding HPD intervals or regions is fairly simple (choose an interval an equal distance from either side of the mode). If the posterior distribution is symmetric the HPD interval or region is also symmetric. If, however, the posterior distribution is multimodal, finding the HPD interval or region can be a complicated numerical problem. In such situations, find the highest modes first, and work down to lower modes.

8.5 EMPIRICAL BAYES' ESTIMATION

Another type of Bayesian estimation involves estimating a parameter θ of a data distribution without knowing or assessing the parameters of the prior (subjective) distribution for θ. Sometimes the parameters of the prior distribution for θ are assigned their own higher stage prior distribution; the collection of stages constitutes a hierarchical model (such models are considered in detail in Chapter 14). Other times the parameters of the (subjective) prior distribution for θ are estimated from the data. In the latter case, this method of estimation is called *empirical Bayes estimation* (see also, Robbins, 1955; Maritz, 1970; Efron and Morris, 1973; 1977; Berger, 1980; Lwin et al., 1989; Carlin and Louis, 2000). The term "empirical Bayes" was coined by Robbins (1955), who used the term in the context in which the same type of estimator or decision was required repeatedly in a large sequence of problems.

In the most frequently adopted version of empirical Bayes' estimation the Bayesian estimation structure is used with a preassigned family of prior distributions to obtain the Bayes' estimator. But the parameters of the prior distribution (the hyperparameters) are not assessed subjectively; rather, they are estimated from the current data set by using the marginal distribution of the data, given the hyperparameters. Often, the hyperparameters are estimated by maximum likelihood or by sample moments. The data-based estimators of the hyperparameters are then substituted for the hyperparameters in the prior distribution, as well as in the Bayesian estimator, to obtain an *approximate Bayesian estimator*, an *empirical Bayes estimator*. This approach often has the virtue of providing estimators that may be shown to be *admissible* with respect to quadratic loss functions (sum of squares of estimation errors), or other reasonable loss functions (admissibility and loss functions are discussed in more detail in Chapter 11). Admissibility is a property of estimators (or more generally, *decision rules*) that involves their expected loss being smaller than that of all other estimators for at least some value of θ, and at least as small for all values of θ. Loss functions for estimators are a measure of the *value* of the estimators to the analyst or decision maker. How much is lost by using this estimator compared with how little might have been lost had the best possible estimator been used? The empirical Bayes approach yields estimators that bound the risk when using quadratic (or other unbounded) loss functions.

Unfortunately, the empirical Bayes' procedure usually violates Bayes' theorem (it violates Bayes' theorem whenever the current data set is used to estimate the hyperparameters) and therefore, the laws of probability, which require that the prior distribution not depend upon the current data set. That is, the assumptions of conditional probability, upon which Bayes' theorem is founded, require that the prior distribution depend only upon its (preassigned) parameters and not upon the data; otherwise the theorem doesn't hold. Thus, even though it is incoherent to use these estimators, empirical Bayes' estimators can be viewed as approximate Bayes' estimators and sometimes, such approximate Bayes' estimators are as close to Bayesian estimation as the researcher can reasonably get, given the complexity of the problem. That is, in some applications (for example, in problems of image segmentation and reconstruction using contextual classification), the Bayesian estimators are so complicated to compute that empirical Bayesian estimators are generally preferable. One problem with these estimators is that because there is no natural standard error, it is difficult to find credibility intervals or test hypotheses.

Another problem with empirical Bayes estimation is that by evaluating estimators on the basis of the criterion of admissibility, it requires that we average over all possible values of the data, including values that we have never observed. Such a property makes the procedure less desirable because as Bayesians we would like our estimators to depend only upon data that we have actually observed. But sometimes, when all else fails, empirical Bayes estimators provide a useful fall-back approach.

As an illustration, suppose $\mathbf{x} \mid \boldsymbol{\theta} \sim N(\boldsymbol{\theta}, \mathbf{I}_k)$, and we have a sample of size one. As a prior distribution, we take $(\boldsymbol{\theta} \mid \boldsymbol{\mu}, \tau^2) \sim N(\boldsymbol{\mu}, \tau^2\mathbf{I}_k)$. The posterior distribution becomes $\boldsymbol{\theta} \mid \mathbf{x} \sim N(\boldsymbol{\theta}_B, \phi^2\mathbf{I}_k)$, where the Bayes' estimator (the posterior mean vector of the k means) is given by the weighted average of the sample mean vector and the prior mean vector:

$$\boldsymbol{\theta}_B = \left[\frac{1}{1+(1/\tau^2)}\right]\mathbf{x} + \frac{(1/\tau^2)}{1+(1/\tau^2)}\Bigg]\boldsymbol{\mu}, \quad \phi^2 = \frac{1}{1+(1/\tau^2)}, \tag{8.94}$$

and the weights are precision proportions. To evaluate this Bayes' estimator it is necessary to assign values to the hyperparameters (μ, τ^2). The orthodox Bayesian will assess (μ, τ^2). However, sometimes it is extremely difficult or costly to assess these hyperparameters. In such cases it is prudent to evaluate the empirical Bayes' estimator from the data.

Recall there is only one vector data point, $\mathbf{x}$. Usually, $\boldsymbol{\mu} \equiv (\mu_i)$ is estimated by taking μ_i to be the common grand mean of all k data point components $\mu_i = (1/k)\sum_{j=1}^{k} x_j$. That is, if we define the column k-dimensional vector $\mathbf{e} \equiv (1, \ldots, 1)'$, we take:

$$\begin{aligned} \boldsymbol{\mu} &= \mu\mathbf{e}, \\ \mu &= \mu_i = \frac{1}{k}\sum_{j=1}^{k} x_j. \end{aligned} \tag{8.95}$$

To estimate τ^2, note that unconditionally, $\mathbf{x} \sim \mathbf{N}(\boldsymbol{\mu}, (1+\tau^2)\mathbf{I}_k)$. So $\mathbf{x} - \boldsymbol{\mu} \sim N[0, (1+\tau^2)\boldsymbol{I}_k]$. Define the sum of squares of the components: $v = \sum_{j=1}^{k}(x_j - \mu_j)^2$. Note that $v \sim (1+\tau^2)\chi_k^2$, and therefore, if $z = v^{-1}$, z follows an inverted gamma distribution with pdf:

$$p(z) \propto \frac{\exp\left\{-\dfrac{1}{z(1+\tau^2)}\right\}}{(1+\tau^2)^{k/2} z^{(k+2)/2}},$$

with mean $E(z) = 1/(k-2)(1+\tau^2)$. (See, for example, Press, 1982, p 199.) Therefore,

$$E\left[\frac{k-2}{v}\right] = \frac{1}{1+\tau^2}.$$

Taking the sample mean equal to the population mean as an approximation gives:

$$\frac{k-2}{\nu} = \frac{1}{1+\tau^2}.$$

Substituting this approximation into Equation (8.94) gives the empirical Bayes' estimator:

$$\hat{\boldsymbol{\theta}}_{eBe} = \mu + \left(1 - \frac{k-2}{\nu}\right)(\mathbf{x} - \boldsymbol{\mu}). \tag{8.96}$$

Often, the expected loss of the empirical Bayes' estimator is less than that of the maximum likelihood estimator, but it is never less than that of the Bayes' estimator. Some caviats about the use of the empirical Bayes' estimator are that:

1. The components of the $\mathbf{x}$ vector should be expressed in the same units;
2. For a good estimator of $\boldsymbol{\theta} \equiv (\theta_j)$, the components θ_j should be close to one another.

8.6 BAYESIAN ROBUSTNESS IN ESTIMATION

Bayesian robustness refers to the need for Bayesian scientists to account for uncertainty about either the likelihood or the prior distribution in a Bayesian analysis. In a decision problem there will usually be uncertainty about the loss (utility) function as well. The prior may be known only to lie in some pre-assigned class, for example. The likelihood function may be known to be based upon very subjectively-determined data. We have elected to treat this subject here only in sufficient detail to provide the flavor of the problem. For more extensive discussions see, for example, Insua and Ruggeri, 2000.

The simplest way to study robustness of the prior distribution is to consider several priors and then to compute their corresponding posteriors. The posteriors are then compared to study the different implications. A mapping can be developed to show how the different priors correspond to different posterior implications (see also Section 5.2.1, and Dickey, 1973).

Alternatively, sometimes it is useful to consider the class of ε-containment priors. This is the class of prior probability densities the form:

$$\tilde{g}(\theta) = (1 - \varepsilon)g(\theta) + \varepsilon\Pi(\theta),$$

where ε denotes the contamination parameter, $g(\theta)$ denotes the candidate prior density being proposed for analysis, and $\Pi(\theta)$ denotes a class of contamination prior densities. Berger and Berliner (1986) proposed this prior family (with $\varepsilon = 0.2$), and it was used to study robustness in Bayesian factor analysis (see Chapter 15, and Lee and Press (1998)). For example, $g(\theta)$ might be a normal distribution with mean θ, and $\Pi(\theta)$ might be a family of Cauchy distribution densities with median θ.

Often families of priors are studied that correspond to the same elicited prior information. For example, perhaps we should consider families all of which have the same first two moments. Sometimes size of the class of priors may be reduced by trying to make the prior distribution less sensitive to outliers; in other situations, the class of priors is narrowed down by considering the extremes corresponding to the maximum and minimum obtainable under all possible priors being considered.

Sensitivity analysis with respect to estimation in Bayesian regression and the analysis of variance are often carried out. In that context error distributions alternative to the normal are used to see what predictive distributions correspond; to see how the residuals change; also, to see which observations become influential in the regression. These sensitivity analyses are another form of Bayesian robustness analysis.

SUMMARY

This chapter has demonstrated how to estimate parameters of probability distributions. Both point and interval estimation procedures were explicated for both univariate and multivariate distributions, under both vague and natural conjugate prior distributions. For univariate data distributions we treated estimation in the binomial, Poisson, negative binomial (Pascal), and normal with both known and unknown variances. For multivariate data distributions, we treated both the multinomial and the normal with unknown mean vector and unknown covariance matrix. We discussed credibility intervals and regions, compared credibility and confidence intervals, and treated highest posterior density intervals and regions. Finally, we discussed empirical Bayes estimation, and robust Bayesian estimation.

EXERCISES

8.1 Describe how you would go about finding a Bayesian estimator for a data distribution not discussed in this chapter, such as a gamma-data distribution with known scale factor.

8.2 Find a Bayes' estimator for a univariate normal distribution with mean zero but unknown variance, under a vague prior.

8.3* Explain the use of a "highest posterior density" (HPD) credibility interval.

8.4 Explain how to find an HPD credibility interval for a bimodal posterior distribution.

8.5* What is an empirical Bayes' estimator?

*Solutions for asterisked exercises may be found in Appendix 7.

8.6* What are the differences between a confidence interval and a credibility interval.

8.7 Why is *admissibility* a questionable property of an estimator for a Bayesian?

8.8 How does the expected loss of an empirical Bayes' estimator compare with that of a Bayesian estimator? With that of an MLE?

8.9 Give a Bayesian estimator for the parameter of an exponential distribution under a natural conjugate prior.

8.10 Suppose you have a sample of size n from $N(\theta, \mathbf{I}_p)$, and $\boldsymbol{\theta} = 5(\phi, \ldots, \phi)'$. Adopt a vague prior distribution for ϕ, and give a Bayes' estimator for $\boldsymbol{\theta}$.

8.11 Explain the form of the Bayesian estimator for the sample mean in the case of a univariate normal data distribution with known variance, under a natural conjugate prior distribution for the sample mean.

8.12* Suppose $X \sim N(\theta, 1)$, and we take three independent observations, $X_1 = 2, X_2 = 3, X_3 = 4$. If we adopt a vague prior for θ, then:

(a) Find the posterior distribution for θ given the data;

(b) Find a two-tailed 95 percent credibility interval for θ.

FURTHER READING

Berger, J. O. (1980). *Statistical Decision Theory and Bayesian Analysis*, 2nd Edition, New York, Springer-Verlag.

Box, G. E. P. and Tiao, G. C. (1973). *Bayesian Inference in Statistical Analysis*, Reading, MA., Addison-Wesley Publishing Co.

Carlin, B. P. and Louis, T. A. (2000). *Bayes and Empirical Bayes Methods for Data Analysis*, 2nd Edition, Boca Raton, Florida, Chapman and Hall.

Edwards, W., Lindman, H. and Savage, L. J. (1963). "Bayesian Statistical Inference for Psychological Research," *Psychol. Revi.*, **70**, 193.

Efron, B. and Morris, C. (1973). "Stein's Estimation Rule and its Competitors—An Empirical Bayes Approach", *J. Am. Statist. Assoc.*, **68**, 117–130.

Efron, B. and Morris, C. (1977). "Stein's Paradox in Statistics," *Scientific American*, **236**, 119–127.

Lwin, T. and Maritz, J. S. (1989). *Empirical Bayes Methods*, CRC Press (Ringbound).

Maritz, J. S. (1970). *Empirical Bayes Methods*, London, Methuen.

Press, S. J. (1982). *Applied Multivariate Analysis: Including Bayesian and Frequentist Methods of Inference*, Malabar, Florida, Krieger Publishing Co.

Robbins, H. (1955). "An Empirical Bayes Approach to Statistics," In *Proceedings of the Third Berkeley Symposium on Mathematical Statistics and Probability*, **1**, Berkeley, University of California Press, 157–164.

Zellner, A. (1971). *An Introduction to Bayesian Inference in Econometrics*, New York, John Wiley and Sons, Inc.

CHAPTER 9

Bayesian Hypothesis Testing

9.1 INTRODUCTION

In this chapter we discuss Bayesian hypothesis testing. We begin with some historical background regarding how hypothesis testing has been treated in science in the past, and show how the Bayesian approach to the subject has really provided the statistical basis for its development. We then discuss some of the problems that have plagued frequentist methods of hypothesis testing during the twentieth century. We will treat two Bayesian approaches to the subject:

1. The vague prior approach of Lindley (which is somewhat limited but easy to implement); and
2. The very general approach of Jeffreys, which is the current, commonly accepted Bayesian method of hypothesis testing, although it is somewhat more complicated to carry out.

9.2 A BRIEF HISTORY OF SCIENTIFIC HYPOTHESIS TESTING

There is considerable evidence of ad hoc tests of hypotheses that were developed to serve particular applications (especially in astronomy), as science has developed. But there had been no underlying theory that could serve as the basis for generating appropriate tests in general until Bayes' theorem was expounded. Moreover, researchers had difficulty applying the theorem even when they wanted to use it.

Karl Pearson (1892), initiated the development of a formal theory of hypothesis testing with his development of chi-squared testing for multinomial proportions. He liked the idea of applying Bayes' theorem to test hypotheses, but he could not quite figure out how to generate prior distributions to support the Bayesian approach. Moreover, he did not recognize that consideration of one or more alternative hypotheses might be relevant for testing a basic scientific hypothesis.

"Student" (William Sealy Gosset, 1908) in developing his t-test for the mean of a normal distribution, and for his work with the sample correlation coefficient, claimed that he would have preferred to use Bayes' theorem (he referred to it as "inverse probability"), but he did not know how to set his prior distribution.

Fisher (1925) developed a formal theory of hypothesis testing that would serve for a variety of scientific situations (although Fisher, like Karl Pearson, also did not consider alternative hypotheses; that modification would wait for Neyman and Pearson, 1933). Fisher attempted to develop an approach that would be "objective" in some sense, and would compare the actual observed data with how data might look if they were generated randomly. Fisher's approach to scientific hypothesis testing was totally non-Bayesian; it was based upon the thinking of his time that was dominated by the influential twentieth-century philosopher, Karl Popper, (1935, 1959). Popper advocated a theory of "falsification" or "refutability" to test scientific theories. As Popper saw it, a scientific theory should be tested by examining evidence that could, in principle, refute, disconfirm, or falsify the theory. Popper's idea for testing a scientific theory was for the scientist to set up a *strawman hypothesis* (a hypothesis opposite to what the scientist truly believes, but under consideration to see if it can be destroyed), and then show that the strawman hypothesis is indeed false. Then the theory based upon the strawman hypothesis could be discarded. Otherwise, one had to wait for additional evidence before proceeding to accept the strawman hypothesis. Fisher adopted this falsification/strawman position.

For example, Popper suggests that a scientist might believe that he/she has a strong theory about why some phenomenon takes place. The scientist might set up a hypothesis that implies that the phenomenon takes place, say, at random. The hypothesis of randomness is then tested (that's the strawman hypothesis). The scientist may then find that empirical data do not support the randomness hypothesis. So it must be rejected. But the scientist's real hypothesis that he/she believes in cannot yet be accepted as correct, if that is the alternative hypothesis. It will require more testing before it can be accepted.

This process was formalized by Fisher suggesting beginning with a null hypothesis, a hypothesis that the researcher believes *a priori* to be false, and then carrying out an experiment that will generate data that will show the null hypothesis to be false. Fisher proposed a "test of significance" in which if an appropriate test statistic exceeds some special calculated value, based upon a pre-assigned significance level, the null hypothesis is rejected. But if it turns out that the null hypothesis cannot be rejected, no conclusion is drawn. For Fisher, there was no criterion for accepting a hypothesis. In science, we never know when a theory is true in some sense; we can only show that a theory may be false because we can find contradictions or inconsistencies in its implications. Fisher also suggested that one could alternatively compute a "p-value," that is, the probability of observing the actually observed value of the test statistic, or anything more extreme, assuming the null hypothesis is true. Some frequentists also think of the p-value as a "sample significance level." (We will discuss below how p-values relate to posterior probabilities.)

To the extent that scientists should reject a hypothesis (theory) if experimental data suggest that the alternative hypothesis (theory) is more probable, this is very sensible; it is the way science proceeds. So Fisher appropriately suggested that we not accept the alternative hypothesis when we reject the null hypothesis. We should just postpone a decision until we have better information. We will see that in Bayesian hypothesis testing, we compute the weight of the experimental evidence as measured by how probable it makes the main hypothesis relative to the alternative hypothesis.

The Popper/Fisher approach to hypothesis testing was not probability based. That is, probabilities were not placed on the hypothesis being tested. For Fisher, one could not place a probability on an hypothesis because an hypothesis is not a random variable in the frequentist sense. For a Bayesian, however, there is no problem placing a probability on an hypothesis. The truthfulness of the hypothesis is unknown, so a researcher can put his/her subjective probability on it to express his/her degree of uncertainty about its truthfulness. We will see how this is done in the Jeffreys approach to Bayesian hypothesis testing.

Jerzy Neyman and Egon Pearson, in a series of papers starting in 1928 developed a theory of hypothesis testing that modified and extended Fisher's ideas in various ways (see, for example, Neyman and Pearson, 1966, where these papers are collected). Neyman and Pearson introduced the idea of alternative hypotheses. In addition, they introduced the notions of Type One and Type Two errors that could be made in testing statistical hypotheses. They defined the concept of "power of a test," and proposed that the ratio of the likelihood of the null hypothesis to the likelihood of the alternative hypothesis be used to compare a simple null hypothesis against a simple alternative hypothesis (Neyman–Pearson Lemma). But the theory was all still embedded in the frequentist falsification notions of Popper and Fisher.

Wald (1939) proposed a theory of decision making that incorporated statistical inference problems of hypothesis testing. We will be discussing decision theory in Chapter 11. This theory suggested, as did the ideas of Neyman and Pearson, that hypotheses should be tested on the basis of their consequences. Bayesian methodology shaped the development of this approach to hypothesis testing in that it was found that Bayesian-derived decision making procedures are the ones that generate the very best decision rules (see Chapter 11). When it was found to be difficult to assign prior distributions to develop the optimal procedures, given the limited development of prior distribution theory at that time, alternative *minimax* (and other) procedures were developed to generate useful decision rules. But the Bayes procedures, that is, procedures found by minimizing the expected loss, or maximizing the expected utility, of a decision rule, were seen to be the best decision rules that could be found (in contexts in which the decision maker had to make decisions in the face of uncertain outcomes of experiments that were decided by nature (this was shown in Savage, 1954). In contexts (games) in which decisions were to be made with respect to other human decision makers, minimax, or some other non-expected-loss criterion might be appropriate). There is additional discussion in Chapter 11.

Lehmann (1959) summarized the Fisher/Neyman–Pearson/Wald frequentist methods of hypothesis testing procedures in his book on hypothesis testing.

Throughout, he used the long-run frequency interpretation of probability rather than the Bayesian or subjective probability notion.

In keeping with Bayesian thinking, Good (1950, 1965) proposed and further codified the testing process, by suggesting that to compare scientific theories, scientists should examine *the weight of the evidence* favoring each of them, and he has shown that this concept is well defined in terms of a conditioning on prior information. Good (1983) developed these ideas further.

This was the state of statistical and scientific hypothesis testing until Jeffreys (1961) and Lindley (1965) proposed their Bayesian approaches that are outlined in Sections 9.4 and 9.5.

9.3 PROBLEMS WITH FREQUENTIST METHODS OF HYPOTHESIS TESTING

There are a variety of problems, difficulties, and inconsistencies associated with frequentist methods of testing hypotheses that are overcome by using Bayesian methods. Some of these problems are enumerated below.

1. *Bayesians have infrequent need to test.* To begin, hypothesis testing per se is something Bayesian scientists do only infrequently. The reason is that once the posterior distribution is found, it contains all the information usually required about an unknown quantity. The posterior distribution can and should be used to learn, and to modify or update earlier held beliefs and judgments. We do not normally need to go beyond the posterior distribution. In some situations, however, such as where the researcher is attempting to decide whether some empirical data conform to a particular theory, or when the researcher is trying to distinguish among two or more theories that might reasonably explain some empirical data, the Bayesian researcher does need to test among several hypotheses or theories. We also need to go beyond the posterior distribution in experimental design situations (see, for example, Raiffa and Schlaifer, 1961, who discuss *the value of sample information*, and *preposterior analysis*).
2. *Problems with probabilities on hypotheses.* The frequentist approach to hypothesis testing does not permit researchers to place probabilities of being correct on the competing hypotheses. This is because of the limitations on mathematical probabilities used by frequentists. For the frequentist, probabilities can only be defined for random variables, and hypotheses are not random variables (they are not observable). But (Bayesian) subjective probability is defined for all unknowns, and the truthfulness of the hypotheses is unknown. This limitation for frequentists is a real drawback because the applied researcher would really like to be able to place a degree of belief on the hypothesis. He or she would like to see how the weight of evidence modifies his/her degree of belief (probability) of the hypothesis being true. It is subjective probabilities of the competing hypotheses being true that are

compared in the *subjective* Bayesian approach. Objective Bayesians, as well as frequentists, have problems in hypothesis testing with odds ratios (as with Bayes' factors) because odds ratios are not defined when improper prior probabilities are used in both the numerator and denominator in the odds ratio.

3. *Problems with preassigned significance levels.* Frequentist methods of hypothesis testing require that a level of significance of the test (such as 5 percent) be preassigned. But that level of significance is quite arbitrary. It could just as easily be less than 1, 2, 4, 6 percent, etc. Where should the line be drawn and still have the result be significant statistically for a frequentist? The concept is not well defined. In the Bayesian approach we completely obviate the necessity of assigning such arbitrary levels of significance. If the weight of the evidence favors one hypothesis over another, that is all we need to know to decide in favor of that hypothesis.

4. *Inadequacy of frequentist testing of a sharp null hypothesis.* Suppose we wish to test $H_0 : \theta = \theta_0$ vs. $H_1 : \theta \neq \theta_0$, for some known θ_0, and we decide to base our test on a statistic $T \equiv T(X_1, \ldots, X_n)$. It is usually the case that in addition to θ being unknown, we usually cannot be certain that $\theta = \theta_0$ precisely, even if it may be close to θ_0. (In fact, in the usual frequentist approach, we often start out believing $\theta \neq \theta_0$, which corresponds to some intervention having had an effect, but we test a null hypothesis H_0 that $\theta = \theta_0$; that is, we start out by disbelieving $\theta = \theta_0$, and then we test it.) Suppose θ is actually ε away from θ_0, for some $\varepsilon > 0$, and ε is very small. Then, by *consistency* of the testing procedure, for sufficiently large n we will reject H_0 with probability equal to one. So, depending upon whether we want to reject H_0, or whether we want to find that we cannot reject H_0, we can choose n accordingly. This is a very unsatisfactory situation. (The same argument applies to all significance testing.)

5. *Frequentist use of possible values never observed.* Jeffreys (1961, p. 385) points out that frequentist hypothesis testers have to rely upon values of observables never observed. He says:

   > What the use of the (p-value) implies, therefore, is that a hypothesis that may be true may be rejected because it has not predicted observable results that have not occurred.

   Jeffreys is reacting to the fact that the frequentist divides the sample space into a critical region for which the null hypothesis will be rejected if the test statistic fall into it, and a complementary region for which the null hypothesis will not be rejected if the test statistic falls into it. But these regions contain values of possible test statistics never actually observed. So the test depends upon values that are observable, but have never actually been observed. Tests that depend upon values that have never actually been observed violate the likelihood principle (see Chapter 3).

6. *Problems with p-values.* The p-value is generally the value reported by researchers for the statistical significance of an experiment they carried out. If p is very low ($p \leq 0.05$), the result found in the experiment is considered

statistically significant by frequentist standards. But the p-value depends upon the sample size used in the experiment. By taking a sufficiently large sample size we can generally achieve a small p-value.

Berger and Selke (1987) and Casella and Berger (1987) compared p-values with the posterior distribution for the same problem. They found for that problem that the evidence against the null hypothesis based upon the posterior distribution is generally weaker than that reflected by the p-value. That is, the p-value suggests rejecting the null hypothesis more often than the Bayesian approach would suggest. In this sense, the Bayesian hypothesis test is more conservative than is the frequentist hypothesis test. The example used by Berger and Selke (1987) is presented in the following.

Suppose the probability density function for an observable X is given by $f(x \mid \theta)$. We are interested in testing the sharp null hypothesis $H_0 : \theta = \theta_0$ versus the alternative hypothesis that $H_1 : \theta \neq \theta_0$. Let $T(X)$ denote an appropriate test statistic. Denote the p-value:

$$p \equiv P\{T(X) \geq t \mid \theta = \theta_0\}. \tag{9.1}$$

While the result will be general, for concreteness, we will make the example very specific. Suppose we have a sample of independent and identically distributed data $X = (X_1, \ldots, X_n)$, and $(X_i \mid \theta) \sim N(\theta, \sigma_0^2)$, for known $\sigma_0^2, i = 1, \ldots, n$. The usual test statistic (sufficient) in this problem is:

$$G(X) = \frac{\sqrt{n \mid \bar{X} - \theta_0 \mid}}{\sigma_0}. \tag{9.2}$$

Define

$$g \equiv G(x) = \frac{\sqrt{n \mid \bar{x} - \theta_0 \mid}}{\sigma_0}. \tag{9.3}$$

The p-value becomes:

$$\begin{aligned} p &= P\{\mid G(X) \mid > g\} \\ &= P\{[G(X) < -g] \bigcup [G(X) > g]\} \\ &= \Phi(-g) + [1 - \Phi(g)] = 2[1 - \Phi(g)]. \end{aligned} \tag{9.4}$$

$\Phi(g)$ denotes the cdf of the standard normal distribution. Suppose, in the interest of fairness, a Bayesian scientist assigns 50 percent prior probability to H_0 and 50 percent prior probability to H_1, but he/she spreads the mass on H_1 out according to

$N(\theta_0, \sigma_0^2)$. The posterior probability on the null hypothesis is shown below to be given by:

$$P\{H_0 \mid x\} = \frac{1}{1 + \frac{1}{\sqrt{n+1}} \exp\left\{\frac{g^2}{2(1+1/n)}\right\}}. \tag{9.5}$$

Proof

By Bayes' theorem,

$$P\{H_0 \mid x\} = \frac{f(x \mid \theta_0)\pi_0}{m(x)},$$

where

$$m(x) = f(x \mid \theta_0)\pi_0 + (1 - \pi_0)m_g(x),$$

and

$$m_g(x) = \int f(x \mid \theta)g(\theta)\, d\theta, \quad g(\theta) \sim N(\theta_0, \sigma_0^2).$$

π_0 denotes the prior probability of H_0. Equivalently,

$$P\{H_0 \mid x\} = \frac{1}{1 + \left(\frac{1-\pi_0}{\pi_0}\right)\frac{m_g(x)}{f(x \mid \theta_0)}}.$$

Note that since

$$(\bar{X} \mid \theta) \sim N(\theta, \sigma_0^2/n), \quad \theta \sim N(\theta_0, \sigma_0^2),$$

marginally,

$$\bar{X} \sim N\left[\theta_0, \left(1 + \frac{1}{n}\right)\sigma_0^2\right].$$

Simplifying gives the result in equation (9.5).

We provide the values of the posterior probability $P\{H_0 \mid x\}$ in Table 9.1, from which may be seen that, for example, for $n = 50$, the frequentist researcher could reject H_0 at $p = 0.050$ (5 percent), since $g = 1.96$, whereas the posterior probability $P\{H_0 \mid x\} = 0.52$, so actually, H_0 is favored over the alternative. At $n = 50$, the frequentist approach to hypothesis testing suggests that the null hypothesis be rejected at the 5 percent level of significance, whereas from a Bayesian point of view, for $n = 50$ or more, the posterior probability says that we should not reject the

Table 9.1 Values of the Posterior Probability $P\{H_0 \mid x\}$

p-value	g	$n = 1$	$n = 5$	$n = 10$	$n = 20$	$n = 50$	$n = 100$	$n = 1000$
0.100	1.645	0.42	0.44	0.47	0.56	0.65	0.72	0.89
0.050	1.960	0.35	0.33	0.37	0.42	0.52	0.60	0.82
0.010	2.576	0.21	0.13	0.14	0.16	0.22	0.27	0.53
0.001	3.291	0.086	0.026	0.024	0.026	0.034	0.045	0.124

null hypothesis. So for the Bayesian hypothesis tester the evidence against the null hypothesis is weaker.

9.4 LINDLEY'S VAGUE PRIOR PROCEDURE FOR BAYESIAN HYPOTHESIS TESTING

A procedure for testing a hypothesis H_0 against an alternative hypothesis H_1 from a Bayesian point of view was suggested by Lindley, 1965, Vol. 2, p. 65. The test procedure is readily understood through a simple example.

Suppose we have the independent and identically distributed data $X_1, \ldots, X_n$, and $(X_i \mid \theta) \sim N(\theta, 1), i = 1, \ldots, n$. We wish to test $H_0 : \theta = \theta_0$, versus $H_1 : \theta \neq \theta_0$. We first recognize that $\bar{X}$ is sufficient for θ, and that $(\bar{X} \mid \theta) \sim N(\theta, 1/n)$. Adopt the vague prior density for θ, $g(\theta) \propto$ constant. The result is that the posterior density for θ is given by $(\theta \mid \bar{x}) \sim N(\bar{x}, 1/n)$. Next, develop a credibility interval for θ at level of credibility α, where say $\alpha = 5$ percent. This result is:

$$P\left\{\bar{x} - \frac{1.96}{\sqrt{n}} \leq \theta \leq \bar{x} + \frac{1.96}{\sqrt{n}} \mid \bar{x}\right\} = 95 \text{ percent.} \tag{9.6}$$

Now examine whether the interval includes the null hypothesis value $\theta = \theta_0$. If θ_0 is not included within this 95 percent credibility interval, this is considered evidence against the null hypothesis, and H_0 is rejected. Alternatively, if θ_0 is found to lie within the 95 percent credibility interval, we cannot reject the null hypothesis.

Lindley (1965) actually proposed this frequentist-like Bayesian testing procedure by using a frequentist confidence interval approach instead of the credibility interval approach outlined here. But under a vague prior density for θ the approaches are equivalent.

We note here that these ideas result from attempts to avoid totally abandoning the frequentist notions of Popper/Fisher/Neyman–Pearson. It is still required that: We preassign a level of significance for the test; we still adopt a strawman null hypothesis; and we still do not place probabilities on the competing hypotheses. The procedure also requires that we adopt a vague prior density for the unknown, θ (and for the credibility interval to be sensible, the prior density must be smooth in the vicinity of θ_0). However, suppose we have meaningful prior information about θ, and we would like to bring that information to bear on the problem. The procedure does not afford us any way to bring the information into the problem. Even worse, if the prior information is mixed continuous and discrete, the testing approach will not be applicable. Regardless, for simple situations where vague prior densities are sensible, the Lindley hypothesis testing procedure provides a rapid, and easily applied, testing method. A more general hypothesis testing procedure is found in the Jeffreys (1961) approach described in Section 9.5.

9.4.1 The Lindley Paradox

The Bayesian approach to hypothesis testing when little prior information is available received substantial interest when Lindley (1957) called attention to the paradoxical result that a frequentist scientist could strongly reject a sharp (null) hypothesis H_0, while a Bayesian scientist could put a lump of prior probability on H_0 and then spread the remaining prior probability out over all other values in a "vague" way (uniformly) and find there are high posterior odds in favor of H_0. This paradox is equivalent to the discussion surrounding Table 9.1.

9.5 JEFFREYS' PROCEDURE FOR BAYESIAN HYPOTHESIS TESTING

Jeffreys (1961, Chapters 5 and 6) suggested a totally Bayesian approach to hypothesis testing that circumvents the inadequacies of the frequentist test procedures. This procedure is outlined below.

9.5.1 Testing a Simple Null Hypothesis Against a Simple Alternative Hypothesis

First we consider the case of testing a simple null hypothesis $H_0 : \theta = \theta_0$ against a simple alternative hypothesis $H_1 : \theta = \theta_1$, where, θ_0 and θ_1 are preassigned constants (recall that a *simple* hypothesis is one for which there is only one possible value for the unknown). We assume that H_0 and H_1 are mutually exclusive and exhaustive hypotheses. Let $T \equiv T(X_1, \ldots, X_n)$ denote an appropriate test statistic based upon a sample of n observations. Then, by Bayes' theorem, the posterior probability of H_0, given the observed data T, is

$$P\{H_0 \mid T\} = \frac{P\{T \mid H_0\}P\{H_0\}}{P\{T \mid H_0\}P\{H_0\} + P\{T \mid H_1\}P\{H_1\}}, \tag{9.7}$$

where $P\{H_0\}$ and $P\{H_1\}$ denote the researcher's prior probabilities of H_0 and H_1. Similarly, for hypothesis H_1, we have:

$$P\{H_1 \mid T\} = \frac{P\{T \mid H_1\}P\{H_1\}}{P\{T \mid H_0\}P\{H_0\} + P\{T \mid H_1\}P\{H_1\}} \tag{9.8}$$

Note that $P\{H_0 \mid T\} + P\{H_1 \mid T\} = 1$. Equations (9.7) and (9.8) can be combined to form the ratio:

$$\frac{P\{H_0 \mid T\}}{P\{H_1 \mid T\}} = \left[\frac{P\{H_0\}}{P\{H_1\}}\right]\left[\frac{P\{T \mid H_0\}}{P\{T \mid H_1\}}\right]. \tag{9.9}$$

Recall that if two probabilities sum to one, their ratio is called the *odds* in favor of the event whose probability is in the numerator of the ratio. Therefore, Equation (9.9) may be interpreted to state that the posterior odds ratio in favor of H_0 is equal to the product of the prior odds ratio in favor of H_0 and the likelihood ratio.

Jeffreys' Hypothesis Testing Criterion

The Jeffreys' criterion for hypothesis testing becomes, in a natural way:

> If the posterior odds ratio exceeds unity, we accept H_0; otherwise, we reject H_0 in favor of H_1.

It is not necessary to specify any particular level of significance. We merely accept or reject the null hypothesis on the basis of which posterior probability is greater; equivalently, we accept or reject the null hypothesis on the basis of whether the posterior odds ratio is greater or less than one. (If the posterior odds ratio is precisely equal to one, no decision can be made without additional data or additional prior information.) Note that if there were several possible hypotheses, this approach would extend in a natural way; we would find the hypothesis with the largest posterior probability.

REMARK: When we accept the null hypothesis because the weight of the evidence shows the null hypothesis to be favored by the data over the alternative hypothesis, we should recognize that we are merely doing so on the basis that the null hypothesis is the one to be entertained until we have better information or a modified theory. We are not assuming that the null hypothesis is true, merely that with the present state of knowledge, the null hypothesis is more probable than the alternative hypothesis.

Bayes' Factors

We note from equation (9.9) that the ratio of the posterior odds ratio to the prior odds ratio, called the *Bayes' factor*, is a factor that depends only upon the sample data. The Bayes' factor reflects the extent to which the data themselves (without prior information) favor one model over another. In the case of testing a simple null hypothesis against a simple alternative hypothesis the Bayes' factor is just the likelihood ratio, which is also the frequentist test statistic for comparing two simple hypotheses (the result of the Neyman—Pearson Lemma). It becomes a bit more complicated (instead of just the simple likelihood ratio) in the case of a simple null hypothesis versus a composite alternative hypothesis. Because the prior odds ratio is often taken to be one by the objective Bayesian, the Bayes' factor acts as an objectivist Bayesian's answer to how to compare models. The subjectivist Bayesian scientist needs only the posterior odds ratio to compare models (which may differ from the Bayes' factor depending upon the value of the prior odds ratio).

As an example, suppose $(X \mid \theta) \sim N(\theta, 1)$, and we are interested in testing whether $H_0 : \theta = 0$, versus $H_1 : \theta = 1$, and these are the only two possibilities. We take a random sample $X_1, \ldots, X_N$ and form the sufficient statistic $T = \bar{X} = (1/N)\sum_{j=1}^{N} X_j$. We note that $(T \mid H_0) \sim N(0, 1/N)$, and $(T \mid H_1) \sim N(1, 1/N)$. Assume that, *a priori*, $P\{H_0\} = P\{H_1\} = 0.5$. Then, the posterior odds ratio is given by

$$\frac{P\{H_0 \mid T\}}{P\{H_1 \mid T\}} = \left(\frac{0.5}{0.5}\right)\left[\frac{\left(\dfrac{N}{2\pi}\right)^{0.5} \exp\{(-0.5N)\bar{x}^2\}}{\left(\dfrac{N}{2\pi}\right)^{0.5} \exp\{-0.5N(\bar{x}-1)^2\}}\right] \tag{9.10}$$

$$= \exp\{-0.5N[\bar{x}^2 - (\bar{x}-1)^2]\}$$
$$= \exp\{-0.5N(2\bar{x}-1)\}. \tag{9.11}$$

Suppose our sample is of size $N = 10$; and we find $\bar{x} = 2$. Then, the posterior odds ratio becomes:

$$\frac{P\{H_0 \mid T\}}{P\{H_1 \mid T\}} = e^{-0.5N(2\bar{x}-1)} = 3.1 \times 10^{-7}. \tag{9.12}$$

Since the posterior odds ratio is so small, we must clearly reject H_0 in favor of $H_1 : \theta = 1$. Because the prior odds ratio is unity in this case, the posterior odds ratio is equal to the Bayes' factor.

Note that comparing the posterior odds ratio with unity is equivalent to choosing the larger of the two posterior probabilities of the hypotheses. If we could assign losses to the two possible incorrect decisions, we would choose the hypothesis with the smaller expected loss. (See Chapter 11 for the role of loss functions in decision making)

9.5.2 Testing a Simple Null Hypothesis Against a Composite Alternative Hypothesis

Next we consider the more common case of testing a simple hypothesis H_0 against a composite hypothesis H_1. Suppose there is a parameter θ (possibly vector valued) indexing the distribution of the test statistic $T = T(X_1, \ldots, X_N)$. Then, the ratio of the posterior density of H_0 compared with that of H_1 is:

$$\frac{P\{H_0 \mid T\}}{P\{H_1 \mid T\}} = \frac{P\{T \mid H_0\}P\{H_0\}}{P\{T \mid H_1\}P\{H_1\}} = \left[\frac{P\{H_0\}}{P\{H_1\}}\right]\frac{P\{T \mid H_0, \theta\}}{\int P\{T \mid H_1, \theta\}g(\theta)\, d\theta}, \tag{9.13}$$

where: $g(\theta)$ denotes the prior density for θ under H_1. Thus, the posterior odds ratio, in the case of a composite alternative hypothesis, is the product of the prior odds ratio times the ratio of the averaged or marginal likelihoods under H_0 and H_1. (Note that under H_0, because it is a simple hypothesis, the likelihood has only one value, so its average is that value. If the null hypothesis were also composite, we would need to use an integral average in that case as well.) We assume, of course, that these integrals converge. (In the event $g(\theta)$ is an improper density, the integrals will not always exist.) Note also that in this case the Bayes' factor is the ratio of the likelihood under H_0 to the averaged likelihood under H_1.

We have assumed there are no additional parameters in the problem. If there are, we deal with them by integrating them out with respect to an appropriate prior distribution.

For example, suppose $(X \mid \theta, \sigma^2) \sim N(\theta, \sigma^2)$, and we are interested in testing the hypothesis $H_0 : \{\theta = 0, \sigma^2 > 0\}$, versus the alternative hypothesis $H_1 : \{\theta \neq 0, \sigma^2 > 0\}$. If $X_1, \ldots, X_n$ are i.i.d., $(\bar{X}, s^2)$ is sufficient for (θ, σ^2), where s^2 is the sample variance. Then, the posterior odds ratio for testing H_0 versus H_1 is:

$$\frac{P\{H_0 \mid \bar{x}, s^2\}}{P\{H_1 \mid \bar{x}, s^2\}} = \left[\frac{P\{H_0\}}{P\{H_1\}}\right] \times \left[\frac{\int f_1(\bar{x} \mid \theta = 0) f_2(s^2 \mid \sigma^2) g_2(\sigma^2)\, d(\sigma^2)}{\iint f_1(\bar{x} \mid \theta) f_2(s^2 \mid \sigma^2) g_1(\theta) g_2(\sigma^2)\, d(\sigma^2)\, d\theta}\right], \tag{9.14}$$

for appropriate prior densities, $g_1(\theta)$ and $g_2(\sigma^2)$.

As an example of a simple versus composite hypothesis testing problem in which there are no additional parameters, suppose $(X \mid \theta) \sim N(\theta, 1)$, and we are interested in testing $H_0 : \theta = 0$, versus $H_1 : \theta \neq 0$. We take a random sample $X_1, \ldots, X_{10}$, of size $N = 10$, and form the sufficient statistic $T = \bar{X}$, and assume $\bar{X} = 2$. Assume $P\{H_0\} = P\{H_1 = 0.5$. We note that $(\bar{X} \mid \theta) \sim N(\theta, 1/N)$, and so

$$P\{T \mid H_0, \theta\} = \left(\frac{N}{2\pi}\right)^{0.5} e^{-(0.5N)\bar{x}^2}, \tag{9.15}$$

and

$$P\{T \mid H_1, \theta\} = \left(\frac{N}{2\pi}\right)^{0.5} e^{-(0.5N)(\bar{x}-\theta)^2}. \tag{9.16}$$

As a prior distribution for θ under H_1 we take $\theta \sim N(1, 1)$. Then,

$$g(\theta) = \frac{1}{\sqrt{2\pi}} e^{-0.5(\theta-1)^2}.$$

The posterior odds ratio becomes:

$$\frac{P\{H_0 \mid T\}}{P\{H_1 \mid T\}} = \frac{\left(\frac{N}{2\pi}\right)^{0.5} e^{-0.5N\bar{x}^2}}{\int \left(\frac{N}{2\pi}\right)^{0.5} e^{-0.5N(\bar{x}-\theta)^2} \times \frac{1}{\sqrt{2\pi}} e^{-0.5(\theta-1)^2} d\theta}$$

$$= \frac{\sqrt{2\pi} e^{-(0.5N)\bar{x}^2}}{\int e^{-0.5[(\theta-1)^2 + N(\theta-\bar{x})^2]} d\theta}$$

$$= \sqrt{(N+1)} \exp\left\{(-0.5)\left[\frac{(N\bar{x}+1)^2}{(N+1)} - 1\right]\right\}.$$

Since $N = 10$, and $\bar{x} = 2$, we have

$$\frac{P\{H_0 \mid T\}}{P\{H_1 \mid T\}} = 1.1 \times 10^{-8}. \tag{9.18}$$

Thus, we reject $H_0 : \theta = 0$ in favor of $H_1 : \theta \neq 0$. That is, the evidence strongly favors the alternative hypothesis H_1.

9.5.3 Problems With Bayesian Hypothesis Testing with Vague Prior Information

Comparing models or testing hypotheses when the prior information about the unknowns is weak or vague presents some difficulties. Note from Equation (9.14) that as long as $g_1(\theta)$ and $g_2(\sigma^2)$ are proper prior densities, the integrals and the Bayes' factor are well defined. The subjectivist Bayesian scientist who uses subjective information to assess his/her prior distributions will have no difficulty adopting Jeffreys' method of testing hypotheses or comparing models and theories. But the objectivist Bayesian has considerable problems, as will be seen from the following discussion.

Suppose, for example, that $g_2(\sigma^2)$ is a vague prior density so that:

$$g_2(\sigma^2) \propto \frac{1}{\sigma^2}.$$

The proportionality constant is arbitrary. So, in this situation, in the ratio of integrals in Equation (9.14), there results an arbitrary ratio of constants, rendering the criterion for decision arbitrary.

A solution to this problem was proposed by Lempers (1971, Section 5.3). He suggested that in such situations, the data could be divided into two parts, the first of which is used as a *training sample*, and the remaining part for hypothesis testing. A two-step procedure results. For the first step, the training sample is used with a vague prior density for the unknowns and a posterior density is developed in the usual way. This posterior distribution is not used for comparing models. In the second step, the posterior density developed in the first step is used as the (proper) prior density for model comparison with the remaining part of the data. Now there are no hypothesis testing problems. The resulting Bayes' factor is now based upon only part of the data (the remaining part of the data after the training data portion is extracted) and accordingly is called a *partial Bayes' factor.*

A remaining problem is how to subdivide the data into two parts. Berger and Pericchi (1996) suggested that the training data portion be determined as the smallest possible data set that would generate a proper posterior distribution for the unknowns (a proper posterior distribution is what is required to make the Lempers proposal operational). There are, of course, many ways to generate such a minimal training data set. Each such training data set would generate a feasible Bayes' factor. Berger and Pericchi call the average of these feasible Bayes' factors the *intrinsic Bayes' factor.* If the partial Bayes' factor is robust with respect to which training data set is used (so that resulting posterior probabilities do not vary much) using the intrinsic Bayes' factor is very reasonable. If the partial Bayes' factor is not robust in this sense, there can still be problems.

O'Hagen (1993) considers the case in which there are very large data sets so that asymptotic behavior can be used. For such situations he defines a *fractional Bayes' factor* that depends upon the fraction "b" of the total sample of data that has not been used for model comparison. It is not clear that the use of fractional Bayes' factors will improve the situation in small or moderate size samples.

The Bayesian (Jeffreys) approach is now the preferred method of comparing scientific theories. For example, in the book by Mathews and Walker (1965, pp. 361–370), in which in the Preface the authors explain that the book was an outgrowth of lectures by Richard Feynman at Cornell University, Feynman suggests that to compare contending theories (in physics) one should use the Bayesian approach. (This fact was called to my attention by Dr. Carlo Brumat.)

SUMMARY

This chapter has presented the Bayesian approach to hypothesis testing and model comparison. We traced the development of scientific hypothesis testing from the approach of Karl Pearson to that of Harold Jeffreys. We showed how the Bayesian approach differs from the frequentist approach, and why there are problems with the frequentist methodology. We introduced both the Lindley vague prior approach to hypothesis testing as well as the Jeffreys general prior approach to testing and model comparison. We examined the testing of simple null hypotheses against simple alternative hypotheses as well as the testing of simple versus composite hypotheses. We discussed Bayes' factors, partial Bayes' factors, intrinsic Bayes' factors, and fractional Bayes' factors.

EXERCISES

9.1 Suppose that $X_1, \ldots, X_{10}$ are independent and identically distributed as $N(\theta, 4)$. Test the hypothesis that $H_0 : \theta = 3$ versus the alternative hypothesis $H_0 : \theta \neq 3$. Assume that you have observed $\bar{X} = 5$, and that your prior probabilities are $P\{H_0\} = 0.6$, and $P\{H_1\} = 0.4$. Assume that your prior probability for θ follows the law $N(1,1)$. Use the Jeffreys' testing procedure.

9.2 Give the Bayes' factor for Exercise 9.1.

9.3 Suppose the probability mass function for an observable variable X is given by: $f(x \mid \lambda) = (e^{-\lambda}\lambda^x)/x!, x = 0, 1, \ldots, \lambda > 0$. Your prior density for λ is given by: $g(\lambda) = 2e^{-2\lambda}$. You observe $X = 3$. Test the hypothesis $H_0 : \lambda = 1$, versus the alternative hypothesis $H_1 : \lambda \neq 1$. Assume that your prior probabilities on the hypotheses are: $P\{H_1\} = P\{H_0\} = 1/2$.

9.4 Explain the use of the *intrinsic Bayes' factor.*

9.5 Explain the difference between the Lindley and Jeffreys methods of Bayesian hypothesis testing.

9.6* What is meant by "Lindley's paradox"?

9.7 Explain some of the problems associated with the use of p-values and significance testing.

9.8 Explain how frequentist hypothesis testing violates the likelihood principle.

9.9 Suppose $X_1, \ldots, X_n, n = 50$, are i.i.d. observations from $N(\theta, 7)$. You observe $\bar{X} = 2$. Suppose your prior distribution for θ is vague. Use the Lindley hypothesis testing procedure to test $H_0 : \theta = 5$, versus $H_1 : \theta \neq 5$.

9.10 Suppose that $X_1, \ldots, X_n$ are i.i.d. following the law $N(\theta, \sigma^2)$. We assume that σ^2 is unknown. Form the sample mean and variance: $\bar{X} = 1/n \sum_1^n X_i$ and $s^2 = 1/n \sum_1^n (X_i - \bar{X})^2$. You observe $\bar{x} = 5$, $s^2 = 37$, for $n = 50$. You adopt the prior distributions: $\theta \sim N(1, 1)$ and $g(1/\sigma^2) \propto (\sigma^2)^4 e^{-2\sigma^2}$, with θ and σ^2 *a priori* independent. Assume that $P\{H_0\} = 0.75, P\{H_1\} = 0.25$. Test the hypothesis $H_0 : \theta = 3$, versus $H_1 : \theta \neq 3$.

9.11 Find the Bayes' factor for the hypothesis testing problem in Exercise 9.10.

9.12* Suppose r denotes the number of successes in n trials, and r follows a binomial distribution with parameter p. Carry out a Bayesian test of the hypothesis $H : p = 0.2$, versus the alternative $A : p = 0.8$, where these are the only two possibilities. Assume that $r = 3$, and $n = 10$, and that the prior probabilities of H and A are equal.

FURTHER READING

Berger, J. O. and Pericchi, L. R. (1996). "The Intrinsic Bayes Factor for Model Selection and Prediction," *J. Am. Statist. Assoc.*, **91**, 109–122.

*Solutions for asterisked exercises may be found in Appendix 7.

Berger, J. O. and Selke, T. (1987). "Testing A Point Null Hypothesis: The Irreconcilability of p-Values and Evidence", *Jr. Am. Statist. Assoc.*, **82**(397), 112–122.

Casella, G. and Berger, R. L. (1987). "Reconciling Bayesian and Frequentist Evidence in the One-Sided Testing Problem," *Jr. Am. Statist. Assoc.*, **82**(397), 106–111.

Fisher, R. A. (1925) 1970. *Statistical Methods for Research Workers*, 14th Edition, New York; Hafner; Edinburgh, Oliver and Boyd.

Good, I. J. (1950). *Probability and the Weighting of Evidence*, London, Charles Griffin and Co., Ltd.

Good, I. J. (1965). *The Estimation of Probabilities: An Essay on Modern Bayesian Methods*, Research Monograph #30, Cambridge, MA, The MIT Press.

Good, I. J. (1983). *Good Thinking: The Foundations of Probability and its Applications*, Minneapolis, University of Minnesota Press.

Jeffreys, H. (1939), (1948), (1961). *Theory of Probability*, 3rd Edition, Oxford, The Clarendon Press.

Lehmann, E. L. (1959). *Testing Statistical Hypotheses*, New York, John Wiley and Sons, Inc.

Lempers, F. B. (1971). *Posterior Probabilities of Alternative Linear Models*, Rotterdam, University Press.

Lindley, D. V. (1957). "A Statistical Paradox," *Biometrika*, **44**, 187–192.

Lindley, D. V. (1965). *Introduction to Probability and Statistics (Part 1—Probability, and Part 2—Inference)*, Cambridge, Cambridge University Press.

Mathews, J. and Walker, R. L. (1965). *Mathematical Methods of Physics*, New York, W. A. Benjamin, Inc.

Neyman, J. and Pearson, E. S. (1933). "On the Testing of Statistical Hypotheses in Relation to Probability A Priori," *Proc. Of the Cambridge Phil. Soc.*, **29**, 492–510.

Neyman, J. and Pearson, E. S. (1966). *Joint Statistical Papers of J. Neyman and E. S. Pearson*, Berkeley, CA, University of California Press (10 papers).

O'Hagen, A. (1993). "Fractional Bayes Factors for Model Comparison," Statistical Research Report 93-6, University of Nottingham.

Pearson, K. (1892). *The Grammar of Science*, London, Adam and Charles Black.

Popper, K. (1935), (1959). *The Logic of Scientific Discovery*, New York, Basic Books; London, Hutchinson.

Raiffa, H. and Schlaifer, R. (1961). *Applied Statistical Decision Theory*, Boston, Graduate School of Business Administration, Harvard University.

Savage, L. J. (1954). *The Foundation of Statsistics*, New York; John Wiley & Sons Inc.

"Student" (William Sealy Gosset) (1908). *Biometrika*, **6**, 1–25. (Paper on the Student t-distribution.)

Wald, A. (1939). "Contributions to the Theory of Statistical Estimation and Testing Hypotheses," *Ann. Math. Statist.*, **10**, 299–326.

CHAPTER 10

Predictivism

10.1 INTRODUCTION

This chapter introduces the idea of *predictivism*, from a Bayesian point of view. We discuss the philosophy behind predictivism, the formalism of Bayesian predictive distributions, the role of exchangeability in prediction, de Finetti's theorem, and the de Finetti Transform. Finally, we will relate predictive distributions to classification and spatial imaging.

10.2 PHILOSOPHY OF PREDICTIVISM

Predictivism involves the use of a belief system about observables and unobservables in science, and a philosophy of scientific methodology that implements that belief system. In the predictivism philosophy, science is about prediction. It is about developing mathematical and statistical models and hypotheses, based upon observed data, that can be used to predict new results under modified conditions. There are some who would say, "If you can't predict with it, the model (theory) isn't any good." There are also some who would say, "Never put a (prior) distribution on an unobservable." The only things that matter to such scientists are quantities that can be measured (observables). But as we shall see later in this chapter (Section 10.5), de Finetti's theorem demonstrates the intimate connection between unobservables (parameters, and their prior distributions) and observables (quantities that generate the likelihood function, and quantities that generate the predictive distribution). We adopt an eclectic, nondoctrinaire position, congruent with de Finetti's theorem, in which both parameters and their prior distributions, as well as predictive distributions are useful for many purposes.

Predicting new observational outcomes of scientific experiments has been the principal objective of experimental science for centuries. When the variable "time" is involved, the prediction is called *forecasting*. Bayesian modeling for linear forecasting problems was treated, for example, in Broemeling(1985). When the model itself is evolving in time, a new collection of problems develops, leading to a study of sets of models changing in time, called *dynamic linear modeling* (West and Harrison, 1997). Predictivism is the focus of the book by Geisser (1993). For us, the (Bayesian) predictivism approach will be implemented with methods that involve predictive distributions.

10.3 PREDICTIVE DISTRIBUTIONS/COMPARING THEORIES

Frequently, scientists test a theory by using a mathematical formulation of the theory, called "the model," and then predict the value of future observations from the model. Of course it is rare that the observed values of the future observations are exactly the same as the values predicted by the mathematical model of the theory. Failure of exact prediction arises for at least two reasons. First, nature is generally much more complex than any mathematical formulation, however sophisticated, is able to capture. Second, all future observations must be measured, and measurements always have measurement error associated with them, attributable to inaccuracies of various kinds in every device used for measurement. The discrepancy between predicted and observed data values is frequently referred to as "prediction error." With a good mathematical formulation of a theory, most nuances of the theory will be included within the mathematical model. In a so-called "good experiment," measurement error is very small. So under good conditions, prediction error can be held to a very small value.

The quality of any scientific theory is measured by how well the theory predicts future observations. If the theory predicts poorly, it does not matter how well it performs in terms of fitting all previous data; it is not an acceptable theory. Sometimes the theory predicts well, but performs poorly in fitting previous data well. Such performance is preferable to poor prediction, although ideally, the theory would perform well at both prediction and fitting previous data.

Once the current experiment has been performed the scientist must evaluate the prediction error. Is the prediction error too large for the theory to be acceptable? How large is too large? These questions involve inference. Mere calculation of the frequently used root mean-squared-error of the prediction (which only looks at how far a particular point prediction is from what is believed to be *truth*) is often inadequate for meaningful evaluation of the quality of the prediction (as compared with a complete distribution for the predicted quantity).

A measure of how well one theory compares with another is sometimes determined by comparing the divergence between their predictive distributions using, for example, the *Kullback–Leibler divergence*, $d(f_1, f_2)$, where f_i, $i = 1, 2$ denotes the predictive probability density for a continuous observation x being predicted under theory i, $i = 1, 2$; (Kullback, 1959). This divergence measure is defined as:

$$d(f_1, f_2) \equiv \int f_1(x) \ln [f_1(x)/f_2(x)]\, dx.$$

If x were discrete with possible values $(x_1, \ldots, x_q)$, we would use the corresponding divergence measure:

$$d(f_1, f_2) \equiv \sum_{j=1}^{q} f_1(x_j) \ln[f_1(x_j)/f_2(x_j)].$$

The Kullback–Leibler divergence will be discussed later in this chapter in terms of information and entropy. We next provide a brief summary of how to evaluate the predictive distribution of the experimental data in the case of a discrete random variable.

10.3.1 Predictive Distribution for Discrete Random Variables

Suppose that a scientist is planning to carry out an experiment associated with a particular physical phenomenon. In connection with this experiment, the scientist wants to predict the value he/she expects to observe in the experiment. Call this as-yet-unobserved value, Y. Suppose further that there are just two competing theories for generating Y; call them *theory A* and *theory B*. (In a more general case, there might be more than two theories to compare.) Finally, suppose that last year the scientist (or some other scientist) carried out a similar experiment, and observed the outcome of that similar experiment to be X. Now, based upon all the information the scientist has about this phenomenon, he wants to predict Y. Moreover, the scientist wants more than just the prediction of a single value; he wants to find the probabilities that the future observation is likely to fall into particular ranges.

The *predictive probability* of the future observation Y, given the earlier observed observation X, is a weighted average of the predicted values of Y assuming that *theory A* holds, and the predicted values of Y given that *theory B* holds. We have assumed for simplicity of explication that only *theory A* or *theory B* can hold. (This is the most simple case; more interesting cases occur when more than two theories have been proposed.) The weighted average for the predictive probability is given by

$$P\{Y \mid X\} = P\{Y \mid \textit{theory } A\} \times P\{\textit{theory } A \mid X\} + P\{Y \mid \textit{theory } B\} \times P\{\textit{theory } B \mid X\}.$$

Note that the weights here, $P\{\textit{theory } A \mid X\}$ and $P\{\textit{theory } B \mid X\}$, are the posterior probabilities (updated degrees of belief of the experimenter) obtained from Bayes' formula (following the previous experiment) of *theories A* and *B*, given that the result of that experiment was the datum X.

Discrete Data Example: Comparing Theories Using the Binomial Distribution

Suppose it has already been found that the posterior probabilities of *theory A* and *theory B*, given the earlier observation X, are given by:

$$P\{\textit{theory } A \mid X\} = 0.2, \quad P\{\textit{theory } B \mid X\} = 0.8.$$

These are the weights that will be used in the weighted average. Then,

$$P\{Y \mid X\} = P\{Y \mid \textit{theory } A\} \cdot (0.2) + P\{Y \mid \textit{theory } B\} \cdot (0.8).$$

Now suppose that a particular experiment can result in only one of two ways: Either a specific effect is found (we call this outcome a success) or the specific effect is not found (we call this outcome a failure). The experiment might be repeated independently many times. Let Y denote the number of successes in (for simplicity) two independent trials of the experiment. Assume that the probability p of success under *theory A* is the same in both trials ($p = 0.1$ if *theory A* is true), and that the probability of success under *theory B* is the same in both trials ($p = 0.5$ if *theory B* is true). The binomial pmf gives:

$$P\{Y \mid \textit{theory } A\} = P\{Y \mid p = 0.1\} = \binom{2}{Y}(0.1)^Y(0.9)^{2-Y},$$

and

$$P\{Y \mid \textit{theory } B\} = P\{Y \mid p = 0.5\} = \binom{2}{Y}(0.5)^2.$$

Combining terms gives for the predictive probability for the unknown Y, conditional on X:

$$P\{Y \mid X\} = (0.2)\binom{2}{Y}(0.1)^Y(0.9)^{2-Y} + (0.8)\binom{2}{Y}(0.5)^2.$$

Since there are assumed to be just two trials, Y, the number of successes observed, can be 0, 1, or 2 (0 successes, 1 success, or 2 successes). We can evaluate the predictive probabilities for these three cases by substituting $Y = 0$, $Y = 1$, and $Y = 2$ in the last formula. They are given by

$$P\{Y = 0 \mid X\} = 0.362; \quad P\{Y = 1 \mid X\} = 0.436; \quad P\{Y = 2 \mid X\} = 0.202.$$

Note that these predictive probabilities sum to 1, as indeed they must if they are to represent a (discrete) probability distribution.

It is now clear that although the experiment has not yet been carried out, if it were to be carried out, the most likely value of Y is $Y = 1$ (i.e., the most likely number of successes is 1), since its predictive probability, 0.436, is greater than the predictive probabilities for the other possible values, 0.362, or 0.202. Moreover,

$$\mathrm{P}\{(Y \leq 1) \mid X\} = \mathrm{P}\{(Y = 0 \mid X\} + \mathrm{P}\{Y = 1) \mid X\} = 0.798 \cong 80 \text{ percent.}$$

Also,

$$P\{(Y \geq 1) \mid X\} = P\{(Y = 1) \mid X\} + P\{(Y = 2) \mid X\} = 0.638 \cong 64 \text{ percent.}$$

So there is an 80 percent chance that there will be zero successes or one success, while there is only a 64 percent chance that there will be one or two successes. But most likely, there will be just one success.

10.3.2 Predictive Distribution for a Continuous Random Variable

Suppose now that $X_1, \ldots, X_n$ denotes an i.i.d. sample from a distribution indexed by θ, with probability density function $f(x \mid \theta)$. For prior density $g(\theta)$, the posterior probability density function is given by

$$h(\theta \mid X_1, \ldots, X_n) \propto \prod_{i=1}^{n} f(x_j \mid \theta) g(\theta).$$

Suppose we wish to predict a new observation y. The predictive density for y is given by

$$p(y \mid x_1, \ldots, x_n) = \int f(y \mid \theta) h(\theta \mid x_1, \ldots, x_n)\, d\theta.$$

Continuous Data Example: Exponential Data

Suppose $(X \mid \theta)$ follows an exponential distribution with probability density function:

$$f(x \mid \theta) = \theta \exp\{-\theta x\} \quad 0 < \theta \quad 0 < x < \infty.$$

The likelihood for the data is

$$L(\theta; X_1, \ldots, X_n) = \theta^n \exp\{n\bar{x}\theta\}.$$

If we adopt a vague prior density for θ, $[g(\theta) \propto \theta^{-1}]$, the posterior density for θ is given by:

$$h(\theta \mid \bar{x}) \propto [\theta^n \exp\{n\bar{x}\theta\}][\theta^{-1}] = \theta^{n-1} \exp\{n\bar{x}\theta\}.$$

The predictive density for a new observation Y, becomes

$$p(y \mid x_1, \ldots, x_n) \propto \int [\theta^{n-1} \exp\{n\bar{x}\theta\}][\theta \exp\{-\theta y\}]\, d\theta = \int \theta^n \exp\{-\theta(n\bar{x} + y)\}\, d\theta.$$

The last integral is reducible to that of a gamma density. The result is that the predictive density for Y, given the earlier X data, is given by:

$$p(y \mid x_1, \ldots, x_n) = \frac{n(n\bar{x})^n}{(y + n\bar{x})^{n+1}}, \quad 0 < y < \infty.$$

This result provides an entire distribution for Y, not just a point estimate. So interval estimates can be made, and the tightness of the distribution and other characteristics can be studied.

10.3.3 Assessing Hyperparameters From Predictive Distributions

Because it is difficult to know how to think about parameters of prior probability distributions directly (they are called *hyperparameters*, and are unobservable), it can be difficult to assess these hyperparameters, especially in the multivariate case. For this reason, methods have been developed for assessing such hyperparameters by using an associated predictive distribution involving only observables (Kadane et al., 1980). A normal, linear, multiple regression context is assumed, and questions are asked about the quantiles of the distribution of the dependent variable, conditional on the independent variables. To be specific, adopt the usual (Gauss–Markov) regression model,

$$\mathbf{y} = \mathbf{X}\boldsymbol{\beta} + \mathbf{u}, \quad \mathbf{u} \sim N(\mathbf{0}, \sigma^2 \mathbf{I}_n),$$

where $\mathbf{y}$ denotes a vector of n independent observations of a dependent variable, $\mathbf{X}$ denotes a regressor matrix of explanatory variables, and $\mathbf{u}$ denotes an n-vector of disturbances. $(\boldsymbol{\beta}, \sigma^2)$ are unknown parameters. Adopt a natural conjugate family of prior distributions (see Chapter 5) for $(\boldsymbol{\beta}, \sigma^2)$, so that $(\boldsymbol{\beta} \mid \sigma^2)$ follows a normal prior distribution, and σ^2 follows an inverted gamma prior distribution. The parameters of these prior distributions are the hyperparameters in question. They are usually assessed by asking the individual who possesses substantive information to specify these hyperparameters. Alternatively, Kadane et al. (1980) developed a computer routine to help elicit answers to questions to that individual about his/her predictive distribution for a new observation y^*, conditional on specified values of the explanatory variables. Asking for medians repeatedly, and averaging answers generates assessments for the hyperparameters. As long as the natural conjugate prior family and other model assumptions are satisfied, this approach yields useful assessments of the hyperparameters.

10.4 EXCHANGEABILITY

An infinite, or potentially infinite, sequence of random variables, $\theta_1, \ldots, \theta_q, \ldots$ (such as a sequence of coin flips) is said to be *exchangeable*, if the joint probability distribution of any finite subset of the θ_is is invariant under permutations of the subscripts.

The concept has also been extended to strictly finite sequences, but finite exchangeability does not possess the nice properties of infinite exchangeability (de Finetti's theorem holds only in the infinite exchangeability case, for example). A special case occurs when the θ_is are i.i.d. Then the variables are certainly exchangeable (but exchangeable random variables are not necessarily independent). That is, independence implies exchangeability, but not conversely. So exchangeability is less demanding than complete independence.

10.5 DE FINETTI'S THEOREM

10.5.1 Summary

Bayesian analysis of a problem often requires that subjective information be introduced into the model. Sometimes it is more useful or convenient to bring that information to bear through the prior distribution; other times it is preferred to bring it to bear through the predictive distribution. This section develops procedures for assessing exchangeable, proper, predictive distributions that, when desired, reflect "knowing little" about the model, i.e., exchangeable, maximum entropy distributions. But what do such distributions imply about the model and the prior distribution? We invoke de Finetti's theorem to define the "de Finetti transform," which permits us, under many frequently occurring conditions (such as natural conjugacy), to find the unique associated prior distribution for those parameters.

10.5.2 Introduction and Review

Bruno de Finetti expounded his representation theorem on exchangeability in 1931 (de Finetti, 1931, 1937, 1974, Chapter 9; Gandolfo, 1987). He stated the theorem for the case of Bernoulli random variables in his 1937 paper.

Many of Bruno de Finetti's papers are reproduced in both Italian and English in the book *Statistica*, 1993. We give below a brief tour through some of the earlier research on de Finetti's theorem.

Since de Finetti's original work a number of authors have extended the theorem in various ways, to finite sequences of exchangeable random variables (rather than the infinite number required by de Finetti's theorem), to partial exchangeability, and to more abstract spaces (e.g., Diaconis, 1977; Diaconis and Freedman, 1981; Freedman, 1962; Hewitt and Savage, 1955; Jaynes, 1983; Olshen, 1973). It was not until the Hewitt and Savage paper, in 1955, that the *uniqueness* of the representation given in the theorem was explicitly recognized and proven. A relatively simple proof based upon urn models and conditioning on the order statistic was given by Heath and Sudderth (1976) for binary random variables. Diaconis (1987) showed that de Finetti's theorem for binary variables is equivalent to the Hausdorff (1921) moment problem.

The uniqueness of the representation theorem of de Finetti is closely related to the notion of identifiability of mixtures (Rao, 1992; Teicher, 1961).

General discussions of de Finetti's theorem may be found, for example, in Berger (1985), Bernardo and Smith (1994), Kingman (1978), Florens, et al. (1990), and Press (1989). Some earlier research on characterizations relating to mixtures in de Finetti's theorem was carried out by Smith (1981, involving rotationally symmetric data), Diaconis and Ylvisaker (1985, involving finite mixtures of conjugate priors), and Morris, (1982, 1983, involving natural exponential families with quadratic variance functions).

Kingman (1978) gave some uses of the theorem in genetics, and Aldous (1981) gave some applications in probability. Lindley and Phillips (1976) discussed the theorem for Bernoulli processes, Jaynes (1986) presented some applications in physics.

Bayes' theorem in terms of probability densities asserts that

$$p(x \mid \omega) = \frac{f(x \mid \omega)g(\omega)}{h(x)}$$

where $f(x \mid \omega)$ denotes the model or likelihood function, $g(\omega)$ denotes the prior density for ω, and $h(x)$ denotes the unconditional or predictive density for x. To make inferences about ω given the observed data x we need to specify f, g, and h. But what is the best way to do this?

We always must have $\int p(\omega \mid x)\,d\omega = 1$, so $h(x)$ is just the normalizing constant

$$h(x) = \int f(x \mid \omega)g(\omega)\,d\omega.$$

But suppose we also know that $h(x)$ is the density of exchangeable random variables. Do we gain anything? De Finetti's theorem will show us what we gain.

10.5.3 Formal Statement of De Finetti's Theorem

Let $X_{(1)}, \ldots, X_{(n)}, \ldots$ denote an exchangeable sequence of real valued random variables in $\mathbb{R}^p$ (or more generally, in a complete, separable, metric space) with probability measure P. Then, there exists a probability measure Q over the space of all distribution functions $\mathfrak{I}$ such that the joint distribution function of any finite sequence of size n, say, $X_1, \ldots, X_n$, has the form

$$P(X_1, \ldots, X_n) = \int_{\mathfrak{I}} \prod_{i=1}^{n} F(x_i)\,dQ(F)$$

where $Q(F)$ denotes the mixing distribution,

$$Q(F) = \lim_{n \to \infty} P(F_n),$$

as n approaches infinity, and F_n denotes the empirical cdf of $X_1, \ldots, X_n$, and this result holds almost surely.

This form of the presentation of the theorem goes back to de Finetti (1937). It was given in Chow and Teicher (1978), and explicitly in this form in Bernardo and Smith (1994, Chap. 4). In this form the theorem applies in very general spaces. We plan, however, to concentrate on the special case where there are densities with respect to Lebesgue or counting measures.

10.5.4 Density Form of De Finetti's Theorem

Suppose the X random variables are absolutely continuous with respect to Lebesgue measure, with density $f(x \mid \omega)$, for some finite dimensional vector parameter, ω, rather than the infinite dimensional label F. Let $X_{(1)}, \ldots, X_{(n)}, \ldots$ denote an exchangeable sequence of random variables. Let $h(x_1, \ldots, x_n)$ denote the predictive (unconditional) joint density with respect to Lebesgue measure of the subset of the n exchangeable random variables $X_1, \ldots, X_n$ in the infinite sequence.

Olshen (1973) pointed out that this "exchangeable sequence is distributed as if a probability on the range of the data were chosen at random, and then independent, identically distributed data subsequently were generated according to the chosen probability." Accordingly, let the parameter ω (possibly vector valued) index the sampling distribution of the X_i.s

We do not attempt to state the most general conditions for the existence of a density type of de Finetti theorem. Diaconis (1994) pointed out that for preassigned predictive density with respect to Lebesgue measure, it is an open question as to the conditions under which there is a corresponding mixing distribution that has a density with respect to Lebesgue measure. Alternatively, we will assign mixing densities with respect to Lebesgue measure (or counting measures), and then find the corresponding predictive densities.

Once the predictive density for h is specified, there exists a unique conditional sampling density $f(x \mid \omega)$ and a unique cdf $G(\omega)$, a prior (mixing) distribution, such that h may be expressed as the average of the joint density of n i.i.d. random variables, conditional on a parameter ω, weighted by a prior (mixing) distribution $G(\omega)$. Symbolically, we have, almost surely,

$$h(x_1, \ldots, x_n) = \int f(x_1 \mid \omega) \cdots f(x_n \mid \omega)\, dG(\omega),$$

with

$$G(\omega) = \lim G_n(\omega),$$

as n approaches infinity, and $G_n(\omega)$ denotes the empirical cdf of $(X_1, \ldots, X_n)$. The representation must hold for all n. (This assertion follows from the fact, subsumed within the definition of exchangeability, that there is an infinite sequence of such random variables.)

Conversely, the theorem asserts that if the conditional sampling density f and the prior G are specified, then h is determined uniquely.

If the prior parameter ω is also absolutely continuous with respect to Lebesgue measure, with density $g(\omega)$, we replace $dG(\omega)$ by $g(\omega)\, d\omega$. From here on, except for counting measure cases, we assume generally that we have this special case of the $g(\omega)$ form, so that almost surely,

$$h(x_1, \ldots, x_n) = \int f(x_1 \mid \omega) \cdots f(x_n \mid \omega) g(\omega)\, d\omega,$$

and

$$g(\omega) = \frac{d}{d\omega}\left[\lim_{n\to\infty} G_n(\omega)\right].$$

10.5.5 Finite Exchangeability and de Finetti's Theorem

Following work of de Finetti, and Feller (1971), Diaconis (1977), has shown that there are exceptions to the representation theorem if the sequence is permitted to be just finite (finite exchangeability). In addition, Diaconis and Freedman (1981) showed that even if we have only finite exchangeability, de Finetti's theorem holds approximately, in the sense that the distance between the distribution function corresponding to the unconditional density $h(x_1, \ldots, x_n)$ for an infinite sequence of exchangeable random variables, and that for the finite sequence, can be made arbitrarily small (the absolute value of the difference can be made less than a constant times $1/n$).

10.6 THE DE FINETTI TRANSFORM

We think of $h(x)$ as the transform of $g(\omega)$, for preassigned $f(x \mid \omega)$ and correspondingly, $g(\omega)$ as the inverse transform of $h(x)$ (See Press, 1996).

We can relate $g(\omega)$ and $h(x)$ in various ways, depending upon the objective in a given problem. For example, if there is substantive information about $h(x)$ we can assess a predictive distribution for it (Kadane et al., 1980), and then attempt to find the inverse transform $g(\omega)$ that corresponds (for preassigned $f(x \mid \omega)$).

In some situations, we might know very little about $h(x)$, and might therefore wish to invoke maxent (maximum entropy) considerations to specify $h(x)$ as the density of a maxent distribution.

In other situations, $g(\omega)$ or $h(x)$ might be assessed in other ways. There might, for example, be moment-type constraints, tail-type constraints, probability content constraints, algebraic or differential equation-type constraints, or whatever. In any case, we can introduce them through the de Finetti transform.

Here we give two examples of de Finetti transforms for illustrative purposes. We will see that the exchangeability constraint implies very special predictive densities $h(x) = h(x_1, \ldots, x_n)$ that have rotational symmetries. We will then provide a summary table including a variety of transform pairs.

Example 10.1 Binomial Sampling Distribution with Uniform Prior

The case of the binomial sampling distribution is the one that de Finetti originally considered. It was also the case that Thomas Bayes studied when he developed his eponymous theorem (Bayes, 1763, Proposition 9). Bayes was really using a version of what was later to be called de Finetti's theorem when he specified his prior distribution.

Suppose we are dealing with the binomial distribution with r successes in n independent Bernoulli trials, where n is potentially infinite, and the probability of success p is the same from trial to trial. Thus, the trials are exchangeable. De Finetti's theorem applied to this case asserts that there is a unique h corresponding to the binomial probability mass function for binary random variables so that:

$$h(x_1, \dots, x_n) = \int_0^1 \prod_1^n p^{x_i}(1-p)^{1-x_i} g(p)\, dp.$$

Equivalently, for $r = \sum_1^n x_i$,

$$h(r) = \binom{n}{r} \int_0^1 p^r (1-p)^{n-r} g(p)\, dp.$$

Also,

$$g(p) = \frac{d}{dp} \lim_{n \to \infty} P\left\{\frac{r}{n} \le p\right\}.$$

The theorem now asserts that g uniquely determines h, and conversely. In this situation, Bayes either selected the uniform prior for $g(p)$, or equivalently, he selected the discrete uniform prior for $h(r)$. See also Stigler (1982) for a discussion of whether Bayes was actually selecting g or h. In either case, one determines the other, and of course, the unconditional density being generated is readily found to be

$$h(r) = \frac{1}{n+1}, \qquad \text{for } r = 0, 1, \dots, n.$$

This relation must hold for all n, in order to satisfy the requirement of exchangeability.

Example 10.2 Normal Distribution with Both Unknown Mean and Unknown Variance

Now we suppose the $X_{(i)}$s to be exchangeable, and that any finite subset $X_{(1)}, \dots, X_{(n)}$ are i.i.d. with common conditional sampling density, $N(\theta, \sigma^2)$, with both θ and σ^2 unknown. Adopt the joint natural conjugate prior density

$$g(\theta, \sigma^2) = g_1(\theta \mid \sigma^2) g_2(\sigma^2)$$

with $(\theta \mid \sigma^2) \sim N(\phi, \sigma^2)$, and ϕ preassigned, and

$$g_2(\sigma^2) \propto (\sigma^2)^{-0.5v} \exp\left\{-\frac{\tau^2}{2\sigma^2}\right\},$$

with τ^2 and G preassigned (σ^2 follows an inverted gamma distribution). It is straightforward to carry out the required integrations in de Finetti's theorem. The (first) integration with respect to θ merely involves integrating a normal density. The integration with respect to σ^2 involves integrating an inverted gamma density. The result is that unconditionally, the X_is have the joint multivariate Student t-density given by

$$h(x) \propto [\tau^2 + (x - \alpha)'\Sigma^{-1}(x - \alpha)]^{-0.5n+v-2},$$

where: $\alpha = \phi(n + 1)^{-1}\Sigma e_n$, and Σ denotes the intraclass covariance matrix $\Sigma = I_n + e_n e'_n$, e_n denotes an n-vector of ones. The distribution has mean α, and v degrees of freedom, and of course, must hold for all n. It is straightforward to check that the density h is invariant with respect to permutation of the subscripts of the x_is Note that $h(x)$ is rotationally symmetric (Freedman, 1962; Smith, 1981; Bernardo and Smith, 1994, Section 4.4).

Note: In this example, θ and σ_2 were not taken to be independent. If we were to take them to be independent *a priori*, there would be no finite-sample closed-form representation for $h(x)$ with a natural conjugate prior, only large sample approximations. So it follows that there cannot be a closed form unique prior of natural conjugate form.

In Table 10.1, we present a brief summary of some readily calculable de Finetti transform pairs for problems in which the model for the data, $f(x \mid \omega)$, might be binomial, normal, Poisson, exponential, gamma, or Pareto. Prior distributions are often taken to be natural conjugate. All predictive and prior densities given in Table 10.1 are proper.

10.6.1 Maxent Distributions and Information

Shannon information

As mentioned above, there are some scientists who would prefer not to put distributions on parameters, but would rather put distributions only on observables. Such people may prefer to select a predictive density, $h(x)$, that expresses the idea of *knowing little*, that is, a maximum entropy (maxent) or minimum information prior distribution. Assuming that exchangeability is meaningful in a given situation, they could accordingly examine a table of de Finetti transforms, and select an $h(x)$, and therefore a prior $g(\theta)$, and a sampling density $f(x \mid \theta)$, that correspond closely to the situation at hand.

The basic notion of "information" used here was introduced by Shannon (1948), and was explicated, expanded, and applied by many others, The Shannon "entropy" in a distribution, with pmf $h(x)$, is defined for discrete random variables (Shannon originally discussed entropy and information only for discrete random variables) as:

$$H[h(x)] = -\sum_{i-1}^{n} h(x_i) \log h(x_i) = -E[\log h(x)].$$

For variables that are absolutely continuous with respect to Lebesgue measure, Jaynes (1963) extended that definition using the notion of "relative entropy," that

Table 10.1 Brief Table of De Firetto Transforms

No.	MODEL Sampling Density $f(x \mid \text{parameter})$	TRANSFORM Predictive Density $h(x)$ Kernel	INVERSE TRANSFORM Prior Density Kernel g (parameter)
1	binomial $\binom{n}{r}p^r(1-p)^{n-r}$ $r = 0, 1, \ldots, n$ (unknown p)	discrete uniform $h(r) = \frac{1}{n+1}$, $r = \sum_1^n x_i$, $r = 0, 1, \ldots, n$	uniform $g(p) = 1, \quad 0 < p < 1$
2	binomial $\binom{n}{r}p^r(1-p)^{n-r}$ $r = 0, 1, \ldots, n$ (unknown p	beta binomial $\frac{\binom{n}{r}\Gamma(\alpha+\beta)\Gamma(\alpha+r)\Gamma(n+\beta-r)}{\Gamma(\alpha+\beta+n)\Gamma(\alpha)\Gamma(\beta)}$ $r = \sum_1^n x_i, \ r = 0, 1, \ldots, n$	beta $p^{\alpha-1}(1-p)^{\beta-1}$, $0 < p < 1$ $0 < \alpha, 0 < \beta$
3	normal $N(0, \sigma^2)$ (unknown σ^2)	multivariate Student t $\frac{1}{\{v - 2 + x' \sum^{-1} x\}^{(v-2+n)/2}}$ $\sum = (r/v - 2)I_n$	inverted gamma $\frac{1}{(\sigma^2)^{v/2}} e^{-r^2/2\sigma^2}$
4	normal $N(\theta, r^2)$ (r^2 known; θ unknown)	normal $N(\phi e_n, \Sigma)$ $\Sigma = \sigma^2[(1-p)I_n + \rho e_n e_n']$, $\sigma^2 = r^2 + \omega^2$, $\rho = \frac{\omega^2}{r^2 + \omega^2}$ $e_n = (1, \ldots, 1)', e_n : (n \times 1)$	normal $N(\phi, \omega^2)$
5	normal $N(\theta, \sigma^2)$ (θ, σ^2 both unknown)	multivariate Student t $\frac{1}{\{r^2 + (x-\alpha)'\Sigma^{-1}(x-\alpha)\}^{\frac{n+v-2}{2}}}$ $\Sigma = I_n + e_n e_n'$ $e_n = (1, \ldots, 1)' : (n \times 1)$ $\alpha = \frac{\phi}{n+1}\Sigma e_n$	normal-inverted gamma $(\theta \mid \sigma^2) \sim N(\phi, \sigma^2)$ $\sigma^2 \sim (\sigma^2)^{-\frac{v}{2}} e^{-\tau^2/2\sigma^2}$
6	Poisson $\frac{e^{-\lambda}\lambda^x}{x!}$ (unknown λ)	$\frac{\Gamma[\alpha - 1 + \Sigma_1^n x_i]}{(n+\beta)^{\alpha+\Sigma_1^n x_i - 1}\Pi_1^n\Gamma(x_i+1)}$	gamma $\lambda \sim \lambda^{\alpha-1}e^{-\beta\lambda}, \quad 0 < \lambda$ $0 < \alpha, \quad 0 < \beta$
7	exponental $\lambda e^{-\lambda x}$ (unknown λ)	$\frac{\beta^\alpha\Gamma(\alpha+n)}{\Gamma(\alpha)(\beta + \Sigma_1^n x_i)^\alpha}$	gamma $\frac{\beta^\alpha}{\Gamma(\alpha)}\lambda^{\alpha-1}e^{-\lambda\beta}$

Table 10.1 ***(continued)***

No.	MODEL Sampling Density $f(x \mid \text{parameter})$	TRANSFORM Predictive Density $h(x)$ Kernel	INVERSE TRANSFORM Prior Density Kernel g (parameter)
8	gamma $\frac{\beta^{\alpha} x^{\alpha-1} e^{-\beta x}}{\Gamma(\alpha)}$ (known α; and unknown β)	$\frac{b^{\alpha}(\Pi_1^n x_i)^{\alpha-1}\Gamma(n\alpha+a)}{\Gamma^n(\alpha)\Gamma(a)(b+\Sigma_1^n x_i)^{n\alpha+a}}$	gamma $\frac{b^a \beta^{a-1} e^{-b\beta}}{\Gamma(a)}$
9	Pareto $\frac{\alpha x_0^a}{x^{a+1}}, x > x_0$ (unknown shape a) (known threshhold x_0)	$\frac{\beta^a \Gamma(n+\alpha)}{\Gamma(\alpha)(\Pi_1^n x_i)\left[\beta + \log\left(\frac{\Pi_1^n x_i}{x_0^n}\right)\right]^{n+\alpha}}$ for $\min_i(x_i) > x_0$	gamma $\frac{\beta^{\alpha}}{\Gamma(\alpha)} a^{\alpha-1} e^{-\beta a}$

is, entropy of distribution relative to another distribution (also called the "Kullback–Leibler divergence" (Kullback, 1959)

$$H[h(x)] = -\int h(x) \log[h(x)/m(x)]\, dx,$$

where $m(x)$ denotes a benchmark distribution against which $h(x)$ is measured. For additional discussion of this idea in Bayesian context, see Kapur, 1989, and Soofi, 1992. $m(x)$ is sometime thought of as a prior distribution, and entropy so defined is sometimes referred to as Bayesian entropy. We will adopt the uniform distribution for the benchmark measure, $m(x)$. So take $m(x) \propto$ constant.

We note that $H[h(x)]$, defined for relative entropy, is both location and scale invariant, for both discrete and continuous random variables, whereas the apparent extension of Shannon's original definition for discrete random variables to the case of variables absolutely continuous with respect to Lebesgue measure is only location invariant.

Kapur (1989) and Athreya (1991) give constraint conditions for various distributions, such the normal and Poisson, to be maxent for various $m(x)$ prior distributions.

The information in the distribution is defined as $-H[h(x)]$.

Akaike (1978) suggested that we might generalize Shannon's definition somewhat by considering the entropy of $h_i(x)$ with respect to $h_2(x)$ as $E\{[h_1(x)/h_2(x)] \log[h_1(x)/h_2(x)]\}$. Where expectation is taken with respect to $h_2(x)$. He also suggested other definitions of information (Akaike, 1973, 1977, 1978). A good discussion and a rationale for the use of entropy is given in Jaynes, 1994, Chapter 11. For additional relevant discussions, see for example, Cover and Thomas (1991) for an engineering approach to information), Gokhale and Kullback (1978, for an information approach to inference), Harris (1982, for an encyclopedia discussion), Jaynes (1957), Levine and Tribus (1979) and Zellner (1977, 1988, 1993, for a development of the concept of maximal data information prior distributions). An appealing geometric interpretation is given in Soofi (1993). We will use the Shannon information definition throughout.

Suppose we wish to choose the unconditonal data density, $h(x) = h(x_1, \ldots, x_n)$, so that for every n, the distribution with density h contains "as little information as possible." One formulation would be to select $h(x)$ so that the distribution has maximum entropy (maxent) in the Shannon sense, subject to some prespecified constraints. We might wish to impose some constraints about moments of h, or about moments of various functions of x.

In the example involving binomial sampling, the data are discrete. Using maxent considerations, we know that r follows the discrete uniform distribution with density $h(r) = (1+n)^{-1}, r = 0, 1, \ldots, n$. Moreover, from de Finetti's theorem, we know that $g(\theta)$ is uniquely specified. Also, we know (from Section 10.6, Example 10.1) that $g(\theta)$ is the density of the uniform distribution. Note that while $g(p)$ is the minimum information prior (in the Shannon information sense, extended to absolutely continuous variables), $h(r)$ is the minimum information predictive density for r.

More generally, once $h(x)$ has been fixed by maxent considerations, for all n, $g(\theta)$ for fixed $f(x/\theta)$ (and also $f(x \mid \theta)$ for fixed $g(\theta)$) is uniquely determined by de Finetti's theorem. But finding the corresponding $g(\theta)$ for a given $h(x)$ may not always be an easy task (very generally, it is a problem in the theory of "integral equations").

Jaynes (1986) proposed a solution to this problem for the case of an "N-point distribution"; that is, he examined the case of finite exchangeability (see also, Heath and Sudderth, 1976). He proposed expanding the prior density in terms of Legendre polynomials, and found that for this case one could uniquely determine the expansion coefficients that define the relationship. But to accomplish this, he required that the mixing distribution (which he called the generating function) be permitted to be negative, an impossible condition to prevail if the mixing distribution is to be a bona fide probability distribution.

10.6.2 Characterizing $h(x)$ as a Maximum Entropy Distribution

It remains to discuss how to specify that a preassigned $h(x)$ is the density of a maxent distribution. This is the opposite problem to the usual one of finding the maxent distribution subject to some prespecified constraints. Our problem here is, given a preassigned density function that we would like to be the density of a maxent distribution, are there constraints that would produce such a result, and if so, what are they? Problems of this type are commonly approached using the Calculus of Variations (see, for example Bliss, 1946; or Courant and Hilbert, 1953, Chapter IV).

First of all, there is no guarantee that constraints will always exist to make any preassigned distribution a maxent distribution. However, it turns out that we can do it for many distributions (including the ones with common names that are usually of interest as natural conjugate distributions; see Table 10.2). We will adopt moment-type constraints in all cases.

Lisman and van Zuylen (1972) considered this problem (in both the usual and the inverse directions). They provided a table of appropriate (moment-type) constraints

Table 10.2 Exchangeable Maximum Entropy Distributions

Case No.	Maxent Distribution pdf/pmf	Secondary Constraints	Domains
1	Uniform $h(x) = \text{constant}$		$x = (x_i)$, $\alpha_i < x_i < b_i$ $i = 1, \ldots, n$
2	Exponenial $h(x) = \phi^n e^{-\phi \sum_1^n x_i}$	$EX_i = \phi, i = 1, \ldots, n$	$0 < x_i < \infty$
3	Normal $h(x) = \dfrac{1}{(2\pi)^{0.5n} \mid \Sigma \mid^{0.5}}$ $\exp\{-0.5(x-\theta)'\Sigma^{-1}(x-\theta)\}$ for $\theta \equiv \phi e_n, \quad e_n = (1, \ldots, 1)'$, $e_n : (n \times 1)$, $\Sigma \equiv \sigma^2[(1-\rho)I_n + \rho e_n e_n']$ $-\dfrac{1}{n-1} < \rho < 1$	$EX = \theta$ $\text{var}(X) = \Sigma$	$x = (x_i)$, $-\infty < x_i < \infty$, $i = 1, \ldots, n$ $\theta = (\theta_i)$ $-\infty < \theta_i < \infty$ $i = 1, \ldots, n$, $\Sigma > 0$.
4	Student t $h(x) = \dfrac{C_n \mid \Sigma \mid^{-0.5}}{[\nu + (x-\alpha e_n)'\Sigma^{-1}(x-\alpha e_n)]^{\frac{\nu+n}{2}}}$ $e_n : (n \times 1), e_n \equiv (1, \ldots, 1)'$, $\Sigma = \sigma^2[(1-\rho)I_n + \rho e_n e_n']$, $-\dfrac{1}{n-1} < \rho < 1$	$E\log[\nu + (x-\alpha e_n)'$ $\Sigma^{-1}(x-\alpha e_n)]$ $= \text{constant}$	$x = (x_i)$, $-\infty < x_i, < \infty$ $i = 1, \ldots, n$ $\Sigma > 0$
5	Dirichlet $h(x \mid \phi_0, \phi_1) \propto$ $[\prod_{j=1}^n x_j^{\phi_0 - 1}](1 - \sum_1^n x_j)^{\phi_1 - 1}$	$E[\log X_j] = g_0$, $E[\log(1 - \sum_1^n X_j) = g$, for constants g_0, g	$x = (x_j)$, $j = 1, \ldots, n$ $\sum_1^n x_j < 1$ $0 < \phi_0, 0 < \phi_1$
6	Weibull $h(x) = a^n \prod_1^n x_1^{\alpha-1} e^{-\sum_1^n x^\alpha}$	$E[X_i^\alpha] = 1$ $E[\log X_i] = g$	$0 < x_i < \infty$ $x = (x_i)$ $0 < a$
7	Cauchy $h(x) =$ $\dfrac{C_n \mid \Sigma \mid^{-0.5}}{[1 + (x-\alpha e_n)'\Sigma - 1(x-\alpha e_n)]^{\frac{n+1}{2}}}$ $e_n : (n \times 1)$, $e_n = (1, \ldots, 1)'$, $\Sigma = \sigma^2[(1-\rho)I_n + \rho e_n e_n']$ $-\dfrac{1}{n-1} < \rho < 1, C_n = \dfrac{\Gamma\left(\dfrac{n+1}{2}\right)}{\pi^{(n+1)/2}}$	$E[\log\{1 + \alpha e_n)'$ $\Sigma^{-1}x - \alpha e_n)\}] =$ constant	$x = (x_i)$ $-\infty < x_i < \infty$ $i = 1, \ldots, n$, $\Sigma > 0$

(continued)

Table 10.2 (*continued*)

Case No.	Maxent Distribution pdf/pmf	Secondary Constraints	Domains
8	Wishart x_i are i.i.d. $N(0, \Sigma)$ $V = \sum_1^v x_i x_i'$, $x_i : (n \times 1)$ $h(V) = h(x_1, \ldots, x_v)$ $= \frac{C \mid V \mid^{(v-n-1)/2} e^{-0.5tr \sum^{-1} v}}{\mid \Sigma \mid v/2}$ $\Sigma = \sigma^2[(1-\rho)I_n + \rho e_n e_n']$, $-\frac{1}{n-1} < \rho < 1$	$E\left[\frac{V}{v}\right] = \text{constant}$ $E[\log \lvert V \rvert] = \text{constant}$	$V_i(n \times n) > 0$ $v \geq n$ $\Sigma : (n \times n) > 0$
9	Gamma $h(x) = C_n(\prod_1^n x_i^{\alpha-1})e^{-\beta \sum_1^n x_i}$,	$EX_i = \text{constant}$ $E[\log X_i] = \text{constant}$ $i = 1 \ldots, n$	$0 < x_i < \infty$ $x = (x_i)$ $0 < \alpha < 0 < \beta$
10	Laplace $H(x) = \frac{a^n}{2^n} e^{-\alpha \sum_1^n \mid X_i \mid}$	$E\{\mid X_i \mid\} = \text{constant}$ $i = 1, \ldots, n$	$x = (x_i)$ $-\infty < x_i < \alpha$ $a > 0$
11	Poission $h(x) = e^{-\lambda} \lambda^{\sum_{i=1}^n x_i} / \prod_{i=1}^n (x_i!)$	$E(X_i) = \lambda_i = \text{constant}$ $E \log(X_i!)\lambda_2 = \text{constant}$	$X_i = 0, 1, 2, \ldots$ $\lambda > 0$

that would generate the maxent distribution when the prespecified distribution is the uniform (both the discrete and continuous cases), the exponential, the geometric, the normal, the Cauchy, gamma and chi-squared, beta, Weibull, the Pearson, and several others without names. They have given the formulation that if $h(x)$ is the preassigned distribution that is to have maximum entropy, and if we adopt the moment constraints

$$\int f_j(x)h(x)\,dx = C_j, \, (j = 1, \ldots, n)$$

where the C_js are constants, and the $f_j(x)$s are constraint functions, the calculus of variation provides the result that for maximum entropy of $h(x)$ we must have $h(x)$ expressed in the form

$$\log[h(x)] = (\lambda_1 - 1) + \sum_{j=2}^{n} [\lambda_j f_j(x)]$$

where the λ_js are Lagrange multipliers (constants) to be determined from the constraint conditions. To solve the inverse problem, we merely set $h(x)$ to some desired density function, and fix the $f_j(x)$s, and the corresponding λ_js (and n) accordingly.

Kagan et al. (1973) used the same approach as Lisman and van Zuylen to characterize distributions on the basis of maxent, subject to (moment-type) constraints. They also provided a table, as in Lisman and van Zuylen, but their table was abbreviated; they added the Laplace distribution to the table and added the multivariate normal distribution. Soofi (1992) extended this line of research to the Student distribution, and he too provided a summary table.

Gokhale (1975) considered the reverse problem in a more general form. He showed the following. Suppose that $f(x)$ is a pdf with x possibly vector valued, and f satisfies the moment-type constraints:

$$E[T_1(X)] = \eta_1, \quad E[T_j(X; \eta_1, \ldots, \eta_{j-1})] = \eta_j, \qquad j = 2, \ldots, m. \tag{10.1}$$

Let $\eta = (\eta, \ldots, \eta_m)$. Assume $p(x, \theta)$ denotes a family of distributions that depends upon a vector of parameters $\Theta = (\theta_1, \ldots, \theta_m)$, and that p may be expressed in the form:

$$p(x, \theta) = c(\theta) \exp[u_1(\theta)T_1(X) + \sum_{j=2}^{m} u_j(\theta)T_j(x; \eta_1(\theta), \ldots, \eta_{j-1}(\theta))] \tag{10.2}$$

where:

$$E[T_1(X)] = \eta_1(\theta), \quad E[T_j(X; \eta(\theta), \ldots, \eta_{j-1}(\theta))] = \eta_j(\theta), \quad j = 2, \ldots, m.$$

Let $\eta(\theta) = [\eta_1(\theta), \ldots, \eta_m(\theta)]$. Then, for some $\xi \in \Theta$, $p(x, \xi)$ maximizes the entropy of $f(x)$ among all densities f, subject to Equation (10.1), and having the same support as $p(x, \xi)$ almost everywhere. Conversely, if f maximizes the entropy subject to Equation (10.1), and there exists a $\xi \in \Theta$ such that $p(x, \xi)$ and f have the same support, $f(x) = p(x, \xi)$, almost everywhere.

The exponential family is a special case. In that situation,

$$p(x, \theta) = c(\theta) \exp\left[\sum_{j=1}^{m} u_j(\theta)T_j(x) \right],$$

where $c(\theta)$ and $u_j(\theta)$ are $1:1$ functions of $E_\theta[T_j(X)]$. This work added the Dirichlet and the Wishart to the family of "known" maxent distributions.

Athreya (1991) showed that every probability density is the unique maximizer of relative entropy in an appropriate class, and gave the constraints for the Poisson distribution (see Table 10.2, Case 11).

For convenience we have collected in Table 10.2 some of the distributions in the maxent family for n exchangeable variables, along with the moment-type constraints that must he imposed to yield them. In general, we must have the maxent density integrate to unity as the primary constraint (or the pmf must sum to unity in the discrete case). In addition, however, we usually have one or more secondary constraints that must be imposed. Finally, we must have the distribution invariant with respect to permutations of the subscripts.

The direct maxent problem is solved in the following way. Let $h(x)$, $x : (n \times 1)$ denote the unknown maxent density (with respect to Lebesgue measure, for example). Then, we want to find $h(x)$ so that it minimizes

$$\int F[x, h(x)]\, dx$$

where $F(x) \equiv -h(x) \log h(x)$, subject to the K moment constraints

$$\int G_j[x, h(x)]\, dx = a_j, \quad j = 1, \ldots, K,$$

and where $G_j(x) = f_j(x)h(x)$, with $f_1(x) = 1$, and $a_1 = 1$. It is well known that a necessary condition for a stationary point for the integral over F is that the Euler equation be satisfied, namely

$$\frac{\partial}{\partial h}\left[F + \sum_{j=1}^{k} \lambda_j G_j\right] = 0,$$

where the λ_js are unknown Lagrange multipliers to be determined from the boundary and constraint conditions. This approach was used, in part, to create Table 10.2.

In summary, for many situations, we are now able to specify moment-type constraints on a preassigned unconditional distribution with density $h(x)$, such that h corresponds to "knowing little," that is, a maxent distribution. Then, by taking the inverse de Finetti transform we can determine the proper prior distribution and the sampling distribution that correspond. Conversely, for a given sampling distribution, and a given proper prior distribution (perhaps a natural conjugate distribution) we can often find the maxent distribution that corresponds.

Note that in Table 10.2 all distributions in the maxent family must satisfy the primary constraint that they integrate, or sum, to unity; and $\Gamma_n(a)$ denotes the multivariate gamma function, where $\Gamma_n(a) = \pi^{n(n-1)/4} \Pi_{j=1}^{n} \Gamma(a - (j-1)/2)$.

Arbitrary Priors

Most of the prior density families considered so far have been related to natural conjugate distributions. Suppose we have a general arbitrary prior density $g(\omega)$. Then we know (Diaconis and Ylvisaker, 1985) that any prior (with or without a density) can be well approximated by a finite mixture of conjugate priors. Accordingly, assume there is density $g(\omega)$, and let

$$g(\omega) = \sum_{i=1}^{N} a_i g_i(\omega),$$

where $g_i(\omega)$ could be, for example, beta or normal densities, $i = 1, \ldots, N$. Then, from de Finetti's Theorem for densities,

$$h(x) = \int f(x|\omega) \sum_{i=1}^{N} a_i g_i(\omega)\, d\omega$$

That is, $h(x)$ can be represented as:

$$\sum_{i=1}^{N} a_i h_i(x) = \sum_{i=1}^{N} a_i \int f(x|\omega) g_i(\omega)\, d\omega$$

where

$$h_i(\underline{x}) \equiv \int f(\underline{x} \mid \underline{\omega}) g_i(\underline{\omega})\, d\underline{\omega}$$

$i = 1, \ldots, N$. Suppose we went the other way by taking the $h_i(\underline{x})$ to be Dirichlet, or multivariate normal, and required them to be maxent. We could find appropriate constraints by using the separate constraints for which each $h_i(\underline{x})$ is maxent. Then, by de Finetti's theorem, the corresponding $h(\underline{x})$ gives $g(\underline{\omega})$ and $f(\underline{x} \mid \underline{\omega})$.

10.6.3 Applying De Finetti Transforms

Suppose $(X, \ldots, X_n)$ denotes a finite subset of observable variables from an (infinitely) exchangeable set; this is our data. Suppose further that $f(\underline{x} \mid \underline{\omega})$ is an appropriate model for the data. The next question is, if we are to carry out Bayesian inference, how should we chose $g(\underline{\omega})$ and $h(\underline{x})$?

In such a situation we turn to Table 10.1 to find the model corresponding to $f(\underline{x} \mid \underline{\omega})$. Suppose, for example, the model we decide upon is $N(\theta, \sigma^2)$, for both θ and σ^2 unknown. Referring to Table 10.1 we see that for the normal model with both parameters unknown (Case 5), the de Finetti transform density $h(\underline{x})$ is the special multivariate Student t-density shown.

Next, go to Table 10.2 (Case 4) and note that the multivariate Student t-distribution is maxent for $E\{\log[\nu + (\underline{x} - \alpha\underline{e}_n)'\Sigma^{-1}(\underline{x} - \alpha\underline{e}_n)]\} = \text{constant}$. In many situations, so little is known about $\underline{X}$ that there is no reason not to believe that such a constraint is appropriate.

If we believe subjectively that this moment constraint is reasonable in this problem, we can adopt the multivariate Student t-distribution as the maxent distribution for $h(\underline{x})$. The note from Table 10.1 that the inverse de Finetti transform is

$$g(\underline{\omega}) = g(\theta, \sigma^2) = g_1(\theta \mid \sigma^2) g_2(\sigma^2),$$

where

$$\theta \mid \sigma^2 \sim N(\phi, \sigma^2), \qquad g_2(\sigma^2) \propto \frac{1}{(\sigma^2)^{0.5\nu}} e^{-\tau^2/2\sigma^2}$$

Now we must simply assess the hyperparameters (ϕ, ν, τ^2). This might be accomplished using either $g(\theta, \sigma^2)$ or $h(\underline{x})$, perhaps with the assistance of the method of Kadane et al. (1980).

10.6.4 Some Remaining Questions

Much more work relating to de Finetti transforms needs to be carried out. For example, it is important to expand Tables 10.1 and 10.2, to establish closer linkages among natural conjugate priors and maxent distributions, to investigate what happens when improper prior distributions are introduced, to explore the relationships among prior distributions derived by de Finetti transform considerations and those derived by other procedures, and to examine methods for handling data distributions where infinite exchangeability cannot be assumed. In the last instance, if at least finite exchangeability can be assumed, then perhaps we can use approximate priors, perhaps mixtures of normals as approximations for distributions with rotational symmetry. Such results might be derived from the work of Diaconis (1977), Diaconis and Freedman (1980), or Diaconis et al. (1992).

10.7 PREDICTIVE DISTRIBUTIONS IN CLASSIFICATION AND SPATIAL AND TEMPORAL ANALYSIS

Predictive distributions are traditionally used to compare scientific theories and to predict the next observation in a general linear model. But they are also used in spatial and temporal image analysis. For example, some artificial Earth satellites and high flying aircraft carry sensors that can receive signals reflected in many frequencies of scenes on the ground; scenes such as lakes, forests, oceans, deserts, urban populations, underground weapon silos, and so on. Satellites that repeatedly cover approximately the same ground on successive passes over the Earth can take advantage of both spatial and temporal correlations. These multidimensional signals (often assumed to be normally distributed) can be analysed in a Bayesian way to classify the nature of the ground scene. The ground scene is subdivided into small elements called picture elements (pixels), and clumps of nearby pixels are all grouped together to take advantage of their mutual correlations. This is called contextual classification (see also Section 16.9.3). The predictive distributions of new observation vectors are used to classify new ground scenes using data from ground scenes with known structures (*ground truth*). For reviews of some of this literature see, for example, Chapter 16, and Klein and Press (1996), and McLachlan (1992).

10.8 BAYESIAN NEURAL NETS

Another application of predictive distributions involves Bayesian modeling using artificial neural networks. Artificial neural networks (an artificial network is a graphical model) of the brain appear to have been first discussed by McCulloch and Pitts (1943, and reproduced in Anderson and Rosenfeld, 1988). In this section we will suppress the word "artificial" since it is clearly understood. More modern accounts may be found, for example, in Hastie et al. (2001), Jensen (1996), Jensen (2001), Lauritzen (1996), and Neal 1996. A commonly used neural network is called the *single hidden layer back-propagation network,* also sometimes referred to as the *single layer perceptron*. More generally, there are multilayer perceptron networks (models for the neuron interactions in the brain). In a neural network there are observable inputs, hidden intermediate variables, outputs, and targets. Denote the p inputs by: $X_1, \ldots, X_p$. From linear combinations of the inputs there are r derived hidden (latent) variables in the first hidden layer: $Z_1, \ldots, Z_r$, where

$$Z_j(X) = g\left(a_j + \sum_i b_{ij} X_i\right),$$

$j = 1, \ldots, r$, and $g(\cdot)$ denotes some appropriate function. The a_j and b_j are unknown parameters on which we will perform Bayesian inference. These derived variables are called hidden because they are not observed. A multilayer network will contain additional layers of derived variables. The number of such hidden layers to use depends on how many it takes to fit the data well. (See also Chapter 14.)

For example, in a two layer network, the first layer may be denoted by the Z_j variables, as developed above; then, each derived variable in the second layer is some function $h(\cdot)$ of a linear combination of the Z_js. We might have many layers developed in this way. We will illustrate this with a single layer network. A network is represented by a graph (see Fig. 10.1). The *activation* functions $g(\cdot)$, $h(\cdot)$, ... most commonly used to represent the hidden variables are usually taken to be the logistic cdf. Thus,

$$g(t) = h(t) = \cdots = \frac{1}{1 + \exp(-t)}.$$

In the special case in which the activation function is taken to be the identity function, so that $g(t) = t$, the nonlinear regression reduces to linear regression. The k (unknown) outputs are the functions $f_1(X), f_2(X), \ldots, f_k(X)$. The target values, $Y_1, \ldots, Y_m$ will depend upon these outputs. The graph of the network in Figure 10.1 is *acyclic*, in that it has no loops or cycles. Moreover, it is a *directed graph* in that the arcs (lines) connecting the *nodes* making up the graph are ordered (the lines end with arrows). There is a node for every random variable in the model, and the nodes in the directed acyclic graph (DAG) in Figure 10.1 are represented by the circles. In Figure 10.1, the inputs are the *parents* of the hidden variables (the children), and the hidden variables are the *parents* of the outputs. This terminology

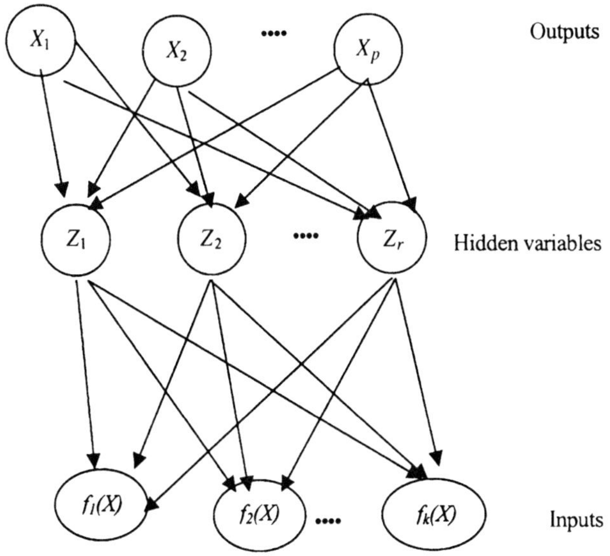

Figure 10.1 Single hidden layer neural network

derives from the fields of artificial intelligence and machine learning. The DAG models (see also Complement A to Chapter 6 where we use the related idea of *doodle bugs*) are used, for example, for nonlinear regression and classification (Bayesian neural networks), Bayesian model selection and uncertainty (see Chapter 13), and studies of causality (causal networks); see Geiger and Heckerman (1999), Thiesson et al. (1997), Heckerman et al. (1997), Heckerman and Chickering (1996), Geiger et al. (1996), and Heckerman et al. (1994).

It is useful to think of Bayesian regression (Chapter 12) and Bayesian classification (Chapter 16) as neural net applications. For Bayesian regression problems, the neural network extends the approach to nonlinear models. For univariate regression, the number of targets $k = 1$; for multivariate regression, $k > 1$. For classification applications, the targets are the k classes or populations into which unknown observations must be classified.

For example, in a (nonlinear) regression context, we might take $(Y_k \mid X) \sim N[f_k(X), \sigma_k^2]$, with $X \equiv (X_1, \ldots, X_p)$, and the Y_ks are conditionally independent given X. The σ_k^2s are taken to be hyperparameters. Then, the joint pdf of the targets $Y \equiv (Y_1, \ldots, Y_m)$ is

$$q(y) = \prod_{k=1}^{m} \left\{ \frac{1}{\sigma_k \sqrt{2\pi}} \exp\left\{ \left(-\frac{1}{2\sigma_k^2} \right) [y_k - f_k(X)]^2 \right\} \right\}.$$

In a classification context we might take

$$P\{Y = k \mid X = x\} = \frac{\exp\{f_k(x)\}}{\sum_{j=1}^{m} f_j(x)}.$$

The unknown parameters in neural networks must be estimated. In non-Bayesian neural network analyses, the unknown parameters are estimated by minimizing an objective function that is typically the sum of squared differences between the outputs and the targets. In Bayesian neural networks, prior information is used effectively in the estimation process.

It is assumed that there are *training data* available, that is, a set of pairs of known values: $(x_1, y_1), \ldots, (x_n, y_n)$. In regression contexts, these are the data upon which the model is built; in the classification context, these are data whose correct classification is known with probability one (with remotely sensed data from satellites, these data are called *ground truth*).

The Bayesian approach to learning in neural networks uses the training data and prior distributions on the parameters to form posterior distributions for the unknown parameters in the usual way, as described earlier in this volume. Then we form the predictive distribution for a new target value, Y_{n+1}, corresponding to a new input X_{n+1}. Point estimates for the new target value is taken to be the mean of the predictive distribution.

As mentioned above, a neural network may involve *hidden Markov models* (HMMs). In HMMs, the variables are connected temporally or spatially, and the Markov property holds. For example, referring to the neural network depicted in Figure 10.1, suppose $r = p = T$, t denotes time, and we have T timepoints. The Zs are assumed to be discrete. Suppose further that the hidden Zs follow a first order Markov process, so that:

$$p(X_1, \ldots, X_T, Z_1, \ldots, Z_T) = p(Z_1)p(X_1 \mid Z_1) \prod_{t=2}^{T} P(X_t \mid Z_t)p(Z_t \mid Z_{t-1}).$$

Of course Markov models higher than first order might be appropriate as well. HMMs add substantial flexibility and richness to a neural network. From a Bayesian point of view, when HMMs are present in the network, there are many fewer parameters present, and, therefore, fewer prior distributions to assess. How many HMMs to include in a model is an open question.

If conditional probabilities for the sets of variables representing the nodes, the parents and the children, are provided at each of the nodes, richer neural networks can be generated. Moreover, the graphical representations can be made richer by using, for example, squares to denote discrete random variables, and circles to denote continuous random variables, and shading the squares or circles to denote observed random variables, and using an unshaded or clear circle or square to represent a hidden or latent random variable. Finally, it is suggested that neural networks may be used to establish causality among some of the variables (rather than assuming relationships reflect correlation or association) (Pearl, 2001).

Many references to available computer software to implement Bayesian neural network models may be found at the web site:
http://www.cs.berkeley.edu/~murphyk/Bayes/economist.html.

SUMMARY

This chapter presented the notion of a Bayesian predictive distribution. The predictive distribution was discussed for both discrete and continuous random variables. We discussed the philosophy of predictivism and the fact that it implies that analysts should depend only upon observable variables. We discussed exchangeability and de Finetti's theorem, and that de Finetti's theorem connects unobservables with observables. We also introduced the notion of the de Finetti transform, the operation that connects the prior distribution and the predictive distribution in a one-to-one relationship. We explained that the predictive distribution can be used to assess hyperparameters, and finally, we pointed out that predictive distributions are regularly used in spatial and temporal classification problems. We discussed Bayesian neural networks, hidden variables, directed acyclic graphs, and HMMs, and explained that predictive distributions are used to predict new values of the target variables in Bayesian neural networks.

EXERCISES

10.1 Suppose that X denotes the number of successes in n repetitions of an experiment carried each time independently of previous trials. Let θ denote the probability of success of the experiment in a single trial. Suppose that your prior beliefs about θ are well represented by a beta distribution with pdf:

$$g(\theta) \propto \theta^4(1-\theta)^2.$$

Find the pdf for the unconditional distribution of the number of successes, assuming that there were six successes.

10.2* Suppose $X \mid \theta \sim N(\theta, 1)$, and the prior density for θ is $N(0, 1)$. Give the predictive density for a new observation.

10.3 What is de Finetti's theorem?

10.4 Suppose that for given $\theta, X_1, \ldots, X_n$ are i.i.d. observations from $N(\theta, 7)$, and that your prior density for θ is $N(5, 3)$. Find your predictive distribution for a new data point, y.

10.5* Provide conditions under which a set of exchangeable random variables might be treated approximately as if they were independent. (*Hint*: think about finite versus infinite exchangeability.)

10.6 Explain what a de Finetti transform is.

10.7 Suppose the sampling distribution for some experimental data is given by $N(\theta, \sigma^2)$, where both parameters are unknown. If the prior distribution for

*Solutions for asterisked exercises may be found in Appendix 7.

these parameters is in the natural conjugate prior family (normal-inverted gamma), find the de Finetti transform.

10.8 Suppose you are carrying out an experiment in which the response data follow a binomial distribution with probability of success on a single trial $= p$. You search the literature and find that another researcher has carried out an analogous experiment in which there were m replications of her experiment, with X successes. To predict the results you might expect, you decide to adopt a vague prior density for p to find the posterior probability for p given X. If Y denotes the number of successes you are likely to have in your own repeated experiments that you decide to run n times, calculate your predictive distribution for Y given X.

10.9* Go to the original sources and explain the notion of "partial exchangeability." (*Hint*: see Diaconis and Freedman, 1981).

10.10 Among all distributions that have a finite first moment, what is the maxent distribution?

10.11 Among all distributions that have a finite second moment, what is the maxent distribution?

10.12* What can be said about de Finetti's theorem for finitely exchangeable events? (*Hint*: see Diaconis, 1977.)

10.13 Suppose $f_1(x)$ and $f_2(x)$ denote two pdfs for two continuous random variables. Give the *Kullback–Leibler divergence* for the first density function relative to the second. Why might this divergence function be a useful construct?

10.14 How are the *entropy* in a distribution and the *information* in a distribution related?

10.15 Find the entropy in the normal distribution, $N(\theta, 1)$, relative to a uniform distribution.

10.16* Following the suggestion at the end of Section 12.4.9, assume that μ and α_i are independent *a priori*, assume that the α_is are exchangeable, and that $\alpha_i \sim N(\xi^*, \Phi^*)$, for all i.

(1) Find the prior density of θ_i.

(2) Find the prior density for B.

(3) Find the joint posterior density of (B, Σ).

10.17 What is the role of exchangeability in de Finetti's theorem?

10.18 A simple regression has been carried out on a set of response variables, $(y_1, \ldots, y_n)$, and a corresponding set of explanatory variables, $(x_1, \ldots, x_n)$. There is just one explanatory variable. The Gaussian linear regression model has been adopted. Adopt a vague prior density for the regression coefficients and the error variance.

(a) Find the joint posterior distribution for the unknown parameters.

(b) Find the predictive distribution and an associated 95 percent credibility interval for $E(y^* \mid (x_1, \ldots, x_n))$, where y^* denotes a new response variable observation predicted from a new value of the explanatory variable, x^*.

(c) Find the predictive distribution and an associated 95 percent credibility interval for y^*.

(d) Why are the lengths of the credibility intervals in parts (b) and (c) different?

10.19 Assume that $X_1, \ldots, X_n$ denotes an i.i.d. sample from $N(\theta, 1)$, so that $\bar{x} \mid \theta \sim N(\theta, 1/n)$. Let Y denote a new observation from the same distribution as the Xs. Adopt a vague prior density for θ, and show that the resulting predictive density for $(y \mid \bar{x})$ is $N(\bar{x}, 1 + 1/n)$.

FURTHER READING

Akaike, H. (1973). "Information Theory and an Extension of the Maximum Likelihood Principle," *2nd Int. Symp. on Information Theory*, B. N. Petrov and F. Csaki (Eds.), Budapest, Akademiai Kiado, 267–281.

Akaike, H. (1977). "On Entropy Maximization Principle," in *Proceedings of the Symposium on Applications of Statistics*, P. R. Krishnaiah (Ed.), Amsterdam, North Holland, 27–47.

Akaike, H. (1978). "A New Look at the Bayes Procedure," *Biometrika*, **65**(1) 53–59.

Aldous, D. (1981). "Representations for Partially Exchangeable Arrays," *J. Multivariate Analysis*, 581–598.

Anderson, J. and Rosenfeld, E. (Ed.) (1988). *Neurocomputing: Foundations of Research*, Cambridge, MA, MIT Press.

Athreya, K. B. (1991). "Entropy maximization," manuscript, Iowa State University, Department of Mathematics, Math. Subject Classification 60 E.

Bayes, T. (1763). "An Essay Towards Solving a Problem in the Doctrine of Chances," *Philos. Trans. Royal Soc. London*, V61.53, 370–418, Reprinted in Appendix 4.

Berger, J. O. (1985). *Statistical Decision Theory and Bayesian Analysis*, Second Edition, New York, Springer-Verlag.

Bernardo, J. M. and Smith A. F. M. (1994). *Bayesian Theory*, New York, John Wiley and Sons, Inc.

Bliss, G. A. (1946). *Lectures on the Calculus of Variations*, Chicago, University of Chicago Press.

Broemeling, L. D. (1985). *Bayesian Analysis of Linear Models*, New York, Marcel Dekker.

Chow, Y. S. and Teicher, H. (1978). *Probability Theory*, Berlin, Springer-Verlag.

Cover, T. M. and Thomas, J. A. (1991). *Elements of Information Theory*, New York, John Wiley and Sons, Inc.

Courant, R. and. Hilbert, D, (1953). *Methods of Mathematical Physics*, Vol. 1, in English, New York, Interscience Publishers, Inc.

Dawid, A. P. (1978). "Extendability of Spherical Matrix Distributions," *J. Multivariate Analysis*, **8**, 559–566.

de Finetti, B. (1931). "Funzioni Caxatteristica di un Fenomeno Aleatorio," *Atti della R. Academia Nazionale dei Lincii*, Ser. 6, Memorie, classe di Scienze, Fisiche, Matematiche e Naturali, Vol. 4, 251–299.

De Finetti, B. (1937). "La Prévision: Ses Lois Logique, Ses Sources Subjectives," *Ann. de l'institut Henri Poincaré*, Vol. 7, 1–68, Translated in *Studies in Subjective Probability*, H. Kyberg and H. Smokler, Eds., New York, John Wiley and Sons, 1964.

De Finetti, B. (1974). *Probability, Induction and Statistics*, New York, John Wiley and Sons, Inc.

Diaconis, P. (1977). Finite Forms of de Finetti's Theorem on Exchangeability," *Synthese*, **36**, 271–281.

Diaconis, P. (1987). "Application of the Method of Moments in Probability and Statistics," *Proc. Symposia in Applied Math.*, **37**, in *Moments in Mathematics*, by H. J. Landau, Ed., 125–142.

Diaconis, P. (1994), Personal Communication.

Diaconis, P. and Freedman, D. (1978). "de Finetti's Generalizations of Exchangeability," R. Jeffrey, Ed. *Studies in Inductive Logic and Probability.*

Diaconis, P. and Freedman, D. (1980). "Finite exchangeable sequences," *Annals of Probability*, **8**(4) 745–764.

Diaconis, P. and Freedman, D. (1981). "Partial exchangeability and sufficiency," in *Proc. Indian Statistical Institute Golden Jubilee International Conference on Statistics and New Directions*, Calcutta, Indian Statistical Institute, 205–236.

Diaconis, P., Eaton, M. L., and Lauritzen S. L. (1992). "Finite de Finetti Theorems in Linear Models and Multivariate Analysis," *Scand. J. Statist.*, **19**, 289–315.

Diaconis, P. and Ylvisaker, D. (1985). "Quantifying Prior Opinion," in J. M. Bernardo, M. H. De Groot, D. B. Lindley, A. F. M. Smith, Eds., Bayesian statistics 2, Amsterdam, Elsevier Science Publishers B. V. (North Holland).

Feller, W. (1971). An Introduction to Probability Theory and its Applications, Vol II, 2nd Edition, New York, John Wiley and Sons, Inc.

Florens, J. P., Mouchart, M. and Rolin, J. M. (1990). *Elements of Bayesian Statistics*, New York, Marcel Dekker, Inc.

Freedman, D. (1962). "Invariants Under Mixing Which Generalize de Finetti's Theorem," *Ann. Math. Statist.*, **33**, 916–923.

Gandolfo, G. (1987). "de Finetti, Bruno," in J. Eatwell, M. Milgate, and P. Newman, Eds., *The New Palgrave: A Dictionary of Economics*, **1**, 765–766.

Geiger, D. and Heckerman, D. (January 1999). "Parameter Priors for Directed Acyclic Graphical Models and the Characterization of Several Probability Distributions," Technical Report. MSR-TR-98-67, Microsoft Research.

Geiger, D., Heckerman, D. and Meek, C. (May, 1996). "Asymptotic Model Selection for Directed Networks with Hidden Variables," Technical Report. MSR-TR-96-07, Microsoft Research.

Geisser, S. (1993). *Predictive Inference: An Introduction*, New York, Chapman and Hall.

Gokhale, D. B. (1975), "Maximum Entropy Characterizations of Some Distributions," in G. Patil et al. (Eds.), *Statistical Distributions in Scientific Work*, **3**, 299–304, Dordrecht, Netherlands, Reidel Publishing Co.

Gokhale, D. B. and Kullback S. (1978). *The Information in Contingency Tables*, New York, Marcel Dekker, Inc.

Harris, B. (1982), "Entropy," in S. Kotz, N. L. Johnson and C. B. Read, Eds. *Encyclopedia of Statistical Sciences*, V61.2, 512–516.

Hastie, T., Tibshirani, R. and Friedman, J. (2001). *The Elements of Statistical Learning: Data Mining, Inference, and Prediction*, New York, Springer-Verlag.

Hausdorff, F. (1921). "Summationsmethoden und Momentfolgen," *Math. Zeit.*, **9**, 74–109, 281–299.

Heath, D. and Sudderth, W. (1976). "de Finetti's Theorem on Exchangeable Variables," *Am. Statistist.*, 30, 188–189.

Heckerman, D., Geiger, D. and Chickering, D. (December 1994). "Learning Bayesian Networks: The Combination of Knowledge and Statistical Data," Technical Report. MSR-TR-94-09, Microsoft Research.

Heckerman, D. and Chickering, D. (November 1996). "A Comparison of Scientific and Engineering Criteria for Bayesian Model Selection," Techcnical Report. MSR-TR-96-07, Microsoft Research.

Heckerman, D., Meek, C. and Cooper, G. (February, 1997). "A Bayesian Approach to Causal Discovery," Technical Report. MSR-TR-97-05, Microsoft Research.

Hewitt, E. and Savage, I. J. (1955). "Symmetric Measures on Cartesian Products," *Trans. Am, Math. Soc.*, **80**, 470–501.

Jaynes, E. (1957). "Information Theory and Statistical Mechanics I," *Phys. Rev.*, **106**, 620–630.

Jaynes, E. (1963). "Brandeis Lectures." Reprinted in E. T. Jaynes: Papers on Probability, Statistics, and Statistical Physics, 1983, by R. D. Rosenkranz (Ed.), Dordrecht, Holland, D. Reidel Publishing Co., 39–76. Also, "Information Theory and Statistical Mechanics," in Statistical Physics, by K.W. Ford (Ed.), W. A. Benjamin, Inc., New York, 181–218.

Jaynes, E. T. (1983). *Papers on Probability*, Statistics and Statistical Physics of E. T. Jaynes, by R. D. Rosenkranz (Ed.), Dordrecht, Holland, D. Reidel Publishing Co.

Jaynes, E. (1986). "Some Applications and Extensions of the de Finetti Representation Theorem," in P. Goel and A. Zellner, Eds., *Bayesian Inference and Decision Techniques with Applications: Essays in Honor of Bruno de Finetti*, Amsterdam, North Holland Publishing Co.

Jaynes, E. (1994). *Probability Theory: The Logic of Science*, book manuscript.

Jensen, F. V. (1996). *Introduction to Bayesian Networks*, New York, Springer-Verlag.

Jensen, F. V., and Jensen, F. B. (2001). *Bayesian Networks and Decision Graphs*, New York, Springer-Verlag.

Johnson, N. L., and Kotz, S. (1969). *Distributions in Statistics: Discrete Distributions*, New York, Houghton Mifflin Co.

Kadane, J. B., Dickey, J. M., Winkler, R. L., Smith, W. S. and Peters, S. C. (1980). "Interactive Elicitation of Opinion of a Normal Linear Model," *J. Am. Statist. Assoc.*, **75**, 845–851.

Kagan, A. M., Linnik, Y. V. and Rao, C. R. (1973). *Characterization Problems in Mathematical Statistics*, New York, John Wiley and Sons, Inc.

Kapur, J. N. (1989). *Maximum Entropy Models in Science and Engineering*, New York, John Wiley and Sons, Inc.

Kingman, J. F. C. (1978). "Uses of Exchangeability," *Ann. Prob.*, **6**, 183–197.

Kullback, S. (1959). *Information Theory and Statistics*, New York, John Wiley and Sons, Inc.

Lauritzen, S. L. (1982). *Statistical Models as Extremal Families*, Aalborg, Aalborg University Press.

Lauritzen, S. L. (1996). *Graphical Models*, Oxford: Oxford University Press.

Levine, R. D. and Tribus, M., Eds. (1979). *The Maximum Entropy Formalism*, Cambridge, MA, The MIT Press.

Lindley, D. V. and Phillips, L. D. (1976). "Inference for a Bernoulli Process: A Bayesian View," *Am. Statist.*, **30**(3), 112–119.

Lisman, J. H. C. and van Zuylen, M. C. A. (1972). "Note on the Generation of Most Probable Frequency Distributions," *Statistica. Neerlandica*, Bol. **26**, 19–23.

McCulloch, W. and Pitts, W. (1943). "A Logical Calculus of the Ideas Immanent in Nervous Activity," *Bull. Math. Biophys.*, **5**, 115–133, 96–104, reprinted in Anderson and Rosenfeld, 1988.

Morris, C. N. (1982). "Natural Exponential Families with Quadratic Variance Functions," *Ann. Statist.*, **10**, 65–80.

Morris, C. N. (1983). "Exponential Families with Quadratic Variance Functions: Statistical Theory," *Ann. Statist.*, **11**(2), 515–529.

Neal, R. M. (1996). *Bayesian Learning for Neural Networks*, New York, Springer-Verlag.

Olshen, R. (1973). "A Note on Exchangeable Sequences," *Z. Wahrscheinlichkeitstheorie und Verw. Gebeite*, **28**, 317–321.

Pearl, J. (2001). *Causality, Reasoning, and Inference*, Cambridge, Cambridge University Press.

Press, S. J. (1989). *Bayesian Statistics: Principles, Models and Applications* (1st Edition), New York, John Wiley and Sons, Inc.

Press, S. James (1996) "The de Fixetti Transform," in *Maximum Entropy and Bayesian Methods*, Kittanson, and Silver, R. (Editors), Dortrect: Kluer Pub. Co., pp. 101–108.

Rao, B. L. S. P. (1992). *Identifiability in Stochastic Models*, San Diego, CA, Academic Press, Inc.

Shannon, C. E. (1948). "The Mathematical Theory of Communication," *Bell System Technical Journal*, July–October 1948, reprinted in C. E. Shannon and W. Weaver (1949), *The Mathematical Theory of Communication*, University of Illinois Press, 3–91.

Smith, A. F. M. (1981). "On Random Sequences with Centered Spherical Symmetry," *J. Royal Statist. Soc. B*, **43**, 208–209.

Soofi, E. S. (1992). "Information Theory and Bayesian Statistics," Manuscript, School of Business Administration, University of Wisconsin, Milwaukee, WI.

Soofi, E. S. (1993). "Capturing the Intangible Concept of Information," Manuscript, School of Business Administration, University of Wisconsin, Milwaukee, WI.

Statistica (1993). Anno LII, Supplemento al n.3, 1992, *Probabilita E Inducione: Bruno de Finetti*, Biblioteca di Statistica, Cooperativa Libraria Universitaria Editrice Bologna, 40126, Bologna–Via Marsala 24 (in English and Italian).

Stigler, S. M. (1982). "Thomas Bayes and Bayesian Inference," *J. Royal Statist. Soc., A*, **145** (2), 250–258.

Teicher, H. (1961). "Identifiability of mixtures," *Ann. Math. Statist.*, **32**, 244–248.

Thiesson, B., Meek, C., Chickering, D. and Heckerman, D. (December 1997). "Learning Mixtures of DAG Models," Tech. Report. MSR-TR-97-30, Microsoft Research.

West, M. and Harrison J. (1997). *Bayesian Forecasting and Dynamic Models*, 2nd Edition, New York, Springer-Verlag.

Zellner, A. (1977). "Maximal Data Information Prior Distributions," in A. Aykac and C. Brumat, Eds., *New Developments in the Applications of Bayesian Methods*, Amsterdam, North Holland Publishing Co., 211–232.

Zellner, A. (1988). "Optimal Information Processing and Bayes' Theorem," *The Am. Statist.*, **42**(4), 278–284.

Zellner, A. (1993). "Models, Prior Information, and Bayesian Analysis," Manuscript, Graduate School of Business, University of Chicago, April 1993.

CHAPTER 11

Bayesian Decision-Making

11.1 INTRODUCTION

The statistical (and economic) theory of decision making was propounded in the books of von Neumann and Morgenstern (1944), and Wald (1950), and in their associated papers. Their work emerged out of *game theory* considerations; that is, viewing the risk of making a particular decision in an uncertain situation as a game in which the results of the game would be economic gains or losses. The remainder of the twentieth century then saw a large literature of books and papers that further developed the theory.

Savage (1954) put forward the Bayesian approach to the subject. The Bayesian approach to decision making was later developed further in a *business context* by Schlaifer (1959, 1961, 1969), Pratt (1965), Raiffa and Schlaifer (1961), and Raiffa (1968). Since then, the principles of Bayesian statistical decision making have found their way into applications in economics and business (management science, marketing, finance, and portfolio analysis), medicine (diagnosis, treatment, and disease management), law (voir dire selection and exclusion of potential jurors), engineering (deciding where and how to build, and whether a particular engineering design justifies the potential costs and benefits), military tracking systems, and many other areas of application. It is also being applied in sample survey contexts as respondents decide how to answer questions (Press and Tanur, 2002).

11.1.1 Utility

The fundamental idea behind Bayesian decision making is that there exists a construct called a *utility function*, $U(X)$, that expresses numerically (usually on a scale from zero to one) how an individual, the decision maker, *values* the possible consequences of any action he/she may take in an uncertain situation. We use upper case X to denote that the consequence may be a random quantity. We usually take "zero" to be the value of the decision maker's utility when the worst possible consequence occurs (when the least gain will be made), and we take "one" to be the value of his utility function when he obtains the best possible consequence (when the decision will result in the greatest possible gain). So for a particular situation in which a decision must be made, $0 \leq U(X) \leq 1$, for all possible random consequences, X, and utilities, $U(X)$. In general, X will be vector-valued, while $U(X)$ will always be a number on the unit interval. The shape of a utility function for all intermediate values of $U(X)$ on the unit interval, for a particular individual, will depend upon how that individual views his risk of gain or loss relative to his current wealth or utility position. Decision makers with very little wealth, such as those at the bottom of the economic ladder, will value a gain or loss of, say \$1000, far more than will chief executive officers of multinational corporations; that is, decision makers who possess substantial wealth.

11.1.2 Concave Utility

Decision makers generally have concave utility functions. The concavity property is discussed below. A real valued function $U(x)$ on an interval (a, b) of the real line is *concave* if for any two points x_1 and x_2 in the interval (a, b), and any number $p, 0 \leq p \leq 1$,

$$U[px_1 + (1-p)x_2] \geq pU(x_1) + (1-p)U(x_2).$$

Equivalently, for any two points x_1 and x_2 on the graph of the function, the line segment connecting the two points lies on or below the graph (Fig. 11.1). A utility function with this shape is called a *concave utility function*.

Now consider the utility function, $U(x)$, in the situation in which x denotes monetary value, and let p denote a probability (Fig. 11.1). The quantity $[px_1 + (1-p)x_2]$ denotes some point in the interval (x_1, x_2), say x_0. We can think of the situation as an experiment with an uncertain outcome, characterized as a gamble, for which the outcome random variable $X = x_1$ has probability p, and $X = x_2$ has probability $(1-p)$. So the gambler (decision maker) receives x_1 with probability p, and receives x_2 with probability $(1-p)$. *Expected monetary value* for

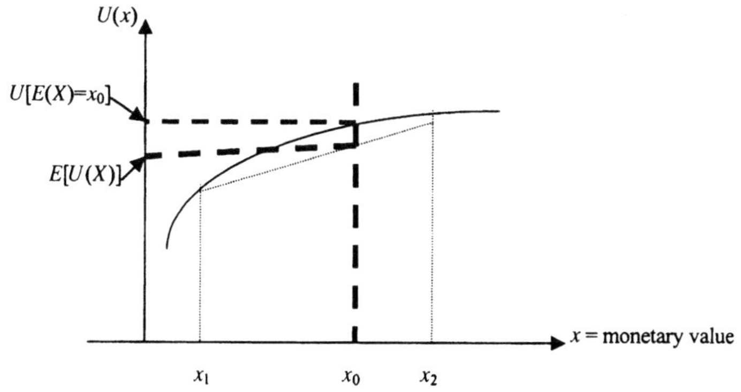

Figure 11.1 Concave utility function (risk-averse decision maker)

this gamble is given by $E(X) = [px_1 + (1 - p)x_2]$, a value that also lies in the interval (x_1, x_2). If the decision maker decides to take this gamble, his utility increase will be the utility of his expected gain, $U[E(X)]$. A gamble in which the expected winnings is zero is called a *fair bet*.

11.1.3 Jensen's Inequality

A result attributed to Jensen (1906) is that for real-valued concave functions, $U(X)$, of a random vector X for which $E(X)$ is finite,

$$U[E(X)] \geq E[U(X)],$$

for X belonging to (a, b) with probability one. This property must hold, in particular, for concave utility functions.

This means that a decision maker with a concave utility function will prefer to select a certain (for sure) increase in monetary value $X = x_0$ (because he perceives that accepting money with certainty will result in greater utility for him), compared with taking a gamble that will yield a random gain $E(X) = x_0$ (Fig. 11.1). Such a conservative decision maker, like most of us, is also called *risk-averse*. His utility function is generally concave-shaped in the region of a gamble.

11.1.4 Convex Utility

Decision makers who are prone to taking big risks for a large possible gain generally have *convex utility* functions (the line segment connecting any two points c and d on the utility function graph lies on or above the graph; (Fig. 11.2).

11.1.5 Linear Utility

Linear utility functions are both concave and convex.

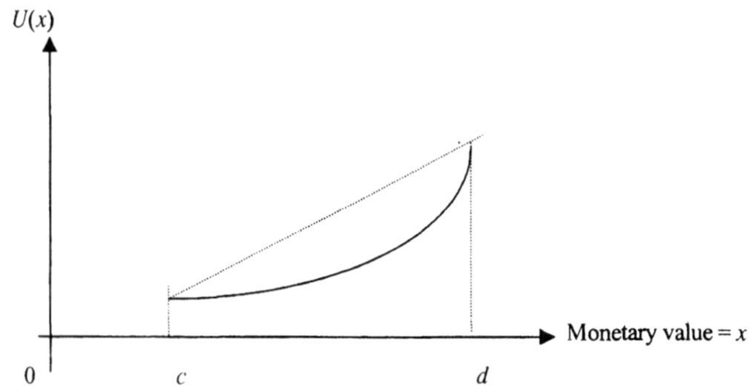

Figure 11.2 Convex utility function (risk-prone decision maker)

11.1.6 Optimizing Decisions

Savage, 1954 developed a normative set of reasonable axioms (discussed in Chapter 2) that decision makers would be expected to follow when they are involved in making decisions under uncertainty. He then showed that as long as decision makers obey these axioms in decision-making situations, they should act to maximize the expected value of their utility functions (with respect to their posterior distributions; see below, in this section).

The uncertainty in a decision making situation is often associated with a parameter θ (generally unknown) of the decision-making problem, and in some problems, θ may take on a finite number of possible values, and in other problems, it may assume any value in a continuum. θ is sometimes called the state of nature, or state of the world. In the decision problem it may be possible to take a finite or an infinite number of possible actions. For each combination of state of nature and action there will be a consequence. The consequence might be some large monetary value, such as \$100,000, or it might be an improved position in the market-place for a particular product sold by some corporation, or it might be that a careful treatment of some cancer has become possible because of a correct, early diagnosis. So utility is not always expressible in terms of monetary value. In any case, the setting for a decision-making problem is one in which there is a decision maker who must make a decision under uncertainty about the possible state of the world when the decision will have to be made, and the decision maker's total utility will drop to some lower value for some decisions, and will increase to some greater value for other decisions.

Bayesian decision making involves selecting decisions by using prior information and observational data, when they are available, to maximize the expected utility of the decision maker. When observational data are not available, the decision maker should use his prior information about which state of nature θ is likely to prevail at the time that the decision must be made (his prior knowledge is the best information he has), in order to decide which action will maximize his expected utility. When observable data are available, the decision maker should combine the data with his prior distribution for θ, via Bayes' theorem, to obtain his posterior distribution for θ; then he should make that decision that maximizes his expected utility with respect to the posterior distribution of θ. In this chapter we will demonstrate how to make decisions as a Bayesian scientist under a variety of decision-making conditions.

11.2 LOSS FUNCTIONS

A *loss function* for a decision maker may be defined in terms of his utility function. Suppose $U^*(x)$ denotes the best possible consequence that can occur in a potential gamble, and let $U(x)$ denote the decision maker's utility for the consequence he receives from the actual decision he makes. Then the decision maker suffers a loss by taking any decision different from the one that would yield the best possible consequence. Let

$$L(x) = \textit{regret} = \textit{opportunity loss} = U^*(x) - U(x) \geq 0.$$

The loss function is the objective function generally used in Bayesian statistical analysis. It must be nonnegative, by definition. The objective in decision problems is to choose that action that minimizes the expected value of the loss function with respect to the posterior distribution, if data are available. If not, then the expected loss should be minimized with respect to the prior distribution. The loss function is generally expressed in terms of the state of nature, θ, and the action to be taken, a. So we might write $L(\theta, a)$. Concave utility functions imply convex loss functions, and, conversely, convex utility functions imply concave loss functions.

Estimation is the special case of decision making in which the decision is to choose that estimator that minimizes the expected loss. In estimation problems, the action to be taken is to choose an estimator, $\hat{\theta}$, so the action $a = \hat{\theta}$. In estimation problems, we write the loss function as: $L(\theta, \hat{\theta})$.

11.2.1 Quadratic Loss Functions

In estimation problems, the loss function is perhaps most often taken to be of quadratic form, namely,

$$L(\theta, \hat{\theta}) = c(\theta - \hat{\theta})^2,$$

where c denotes a constant. (If θ and $\hat{\theta}$ are multidimensional, the equivalent loss function may be written as a positive semidefinite quadratic form: $L(\theta, \hat{\theta}) = (\theta - \hat{\theta})' A(\theta - \hat{\theta})$, for some positive semidefinite matrix **A**, and where the prime denotes transpose.) Note that this form also represents the loss function as a symmetric function, in which case, underestimates of θ are equally consequential with overestimates. It is not always appropriate to use symmetric loss functions. We will discuss asymmetric loss functions below. The constant c must be taken to be positive for conservative decision makers, so that the loss function is convex (the utility function will be concave).

Why Use Quadratic Loss?

1. One of the reasons the quadratic loss function is used so frequently is that for a positive constant multiplier of the quadratic loss function, the loss function is convex, so that the loss function is meaningful for conservative risk takers.
2. A second reason is that the most commonly used estimate of error in many branches of science is the root mean squared error (rms error), but the quadratic loss is just the squared value of the rms error.

3. Finally, another reason for the popularity of the quadratic loss function may be seen from the following argument.

Expanding the loss function in Taylor series about $\hat{\theta} = \theta$ gives

$$L(\theta, \hat{\theta}) = L(\hat{\theta}, \hat{\theta}) + \frac{(\theta - \hat{\theta})}{1!} \frac{\partial L}{\partial \theta}\bigg|_{\theta=\hat{\theta}} + \frac{(\theta - \hat{\theta})^2}{2!} \frac{\partial^2 L}{\partial \theta^2}\bigg|_{\theta=\hat{\theta}} + \cdots$$

In an estimation problem, $L(\hat{\theta}, \hat{\theta}) = 0$. Moreover, since in that case,

$$E[L(\theta, \hat{\theta})] = E[0] + E(\theta, \hat{\theta}) \frac{\partial L}{\partial \theta}\bigg|_{\theta=\hat{\theta}} + \frac{E(\theta - \hat{\theta})^2}{2!} \frac{\partial^2 L}{\partial \theta^2}\bigg|_{\theta=\hat{\theta}} + \cdots$$

and since $E(\theta - \hat{\theta}) = 0$, if $c \equiv 0.5\partial^2 L/\partial\theta^2 \mid_{\theta=\hat{\theta}}$,

$$E[L(\theta, \hat{\theta}] = cE(\theta - \hat{\theta})^2 + \cdots$$

Finally, since by consistency of $\hat{\theta}$, we must have in probability, $\hat{\theta} \to \theta$, as $n \to \infty$, where n denotes the size of the sample, the higher order terms must disappear for large sample sizes. So in estimation problems, as a first approximation, we often take $L(\theta, \hat{\theta}) = c(\theta - \hat{\theta})^2$, and we try to choose $\hat{\theta}$ so as to minimize $cE(\theta - \hat{\theta})^2$.

Suppose there is a data set $\mathbf{X} = (x_1, \ldots, x_n)$ and we wish to specify a Bayesian estimator $\hat{\theta}(\mathbf{X}) \equiv \hat{\theta}$, depending upon $\mathbf{X}$. The Savage axioms (Chapter 2) imply that a Bayesian decision maker should minimize the expected loss with respect to the decision maker's posterior distribution. In the case of the quadratic loss function, the Bayesian decision maker should choose $\hat{\theta}$ according to $\min_{\hat{\theta}} E[c(\theta - \hat{\theta})^2]$. The solution is obtained by differentiating with respect to $\hat{\theta}$, setting the result equal to zero, and solving for $\hat{\theta}$. Accordingly,

$$\begin{aligned}
\frac{d}{d\hat{\theta}}\{E[c(\theta - \hat{\theta})^2] \mid \mathbf{X}\} &= c\frac{d}{d\hat{\theta}}\{E[\theta^2 + \hat{\theta}^2 - 2\theta\hat{\theta}] \mid \mathbf{X}\} \\
&= c\frac{d}{d\hat{\theta}}\{[E(\theta^2) + \hat{\theta}^2 - 2\hat{\theta}E(\theta)] \mid \mathbf{X}\} \\
&= c\{[0 + 2\hat{\theta} - 2E(\theta)] \mid \mathbf{X}\} = 0.
\end{aligned}$$

Solving gives the Bayes' estimator, $\hat{\theta} = \theta_{\text{Bayes}} = E\{\theta \mid \mathbf{X}\}$. That is:

> The Bayes' estimator with respect to a quadratic loss function is the mean of the posterior distribution. If no data have been observed, the posterior distribution reduces to the prior distribution, and the Bayes' estimator becomes the mean of the prior distribution.

11.2.2 Linear Loss Functions

Suppose there are just two possible actions in a decision problem, a_1, a_2, and θ denotes the state of nature (assumed to be continuous). The loss function is assumed to be given, for $b_2 > b_1$, by

$$L(a_1, \theta) = a_1 + b_1\theta,$$
$$L(a_2, \theta) = a_2 + b_2\theta.$$

The risks (expected losses) for taking the two actions are:

$$\rho(a_1) = a_1 + b_1 E(\theta),$$
$$\rho(a_2) = a_2 + b_2 E(\theta).$$

If there are relevant data available, the expectations are taken over the appropriate posterior distribution; otherwise, they are taken over the prior distribution. The Bayesian decision is to take action a_1 if $\rho(a_1) > \rho(a_2)$; if the reverse is true, take action a_2. Otherwise (in the case of equal risks), flip a fair coin. The break-even point is

$$\theta_0 = \theta_{\text{break even}} = \frac{a_1 - a_2}{b_2 - b_1},$$

so that the decision rule becomes: choose action a_1 *if* $E(\theta) < \theta_0$, and choose action a_2 *if* $E(\theta) > \theta_0$. Otherwise, flip a fair coin.

11.2.3 Piecewise Linear Loss Functions

In an estimation context, suppose the loss function is piecewise linear, and is given, for $k_1 > 0$, and $k_2 > 0$, by

$$L(\theta, \hat{\theta}) = \begin{cases} k_1(\hat{\theta} - \theta), & \theta \le \hat{\theta} \\ k_2(\theta - \hat{\theta}), & \theta \ge \hat{\theta}. \end{cases}$$

A graph of this function is shown in Figure 11.3. Note that this is an *asymmetric loss function* (for $k_1 \neq k_2$). We will also encounter the asymmetric Linex loss function, discussed in Section 11.2.4, below. Let $h(\theta \mid \mathbf{X})$ denote the posterior density for the

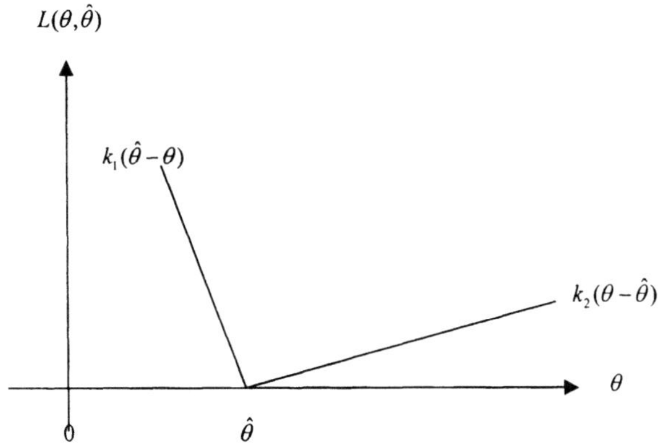

Figure 11.3 Piecewise linear loss function

decision maker's belief about θ, based upon a sample $\mathbf{X} \equiv (x_1, \ldots, x_n)$, and let $H(\theta \mid \mathbf{X})$ denote the corresponding cdf. Define the *risk function*:

$$\rho(\hat{\theta}) = E[L(\theta, \hat{\theta})] = \int_{-\infty}^{\infty} L(\theta, \hat{\theta}) h(\theta \mid \mathbf{X})\, d\theta$$
$$= \int_{-\infty}^{\hat{\theta}} k_1(\hat{\theta} - \theta) h(\theta \mid \mathbf{X})\, d\theta + \int_{\hat{\theta}}^{\infty} K_2(\theta - \hat{\theta}) h(\theta \mid \mathbf{X})\, d\theta.$$

Thus, the risk function is given by

$$\rho(\hat{\theta}) = \hat{\theta}[(k_1 + k_2)H(\hat{\theta} \mid \mathbf{X}) - k_2]$$
$$- \left\{ k_1 \int_{-\infty}^{\hat{\theta}} \theta h(\theta \mid \mathbf{X})\, d\theta + k_2 \int_{\infty}^{\hat{\theta}} \theta h(\theta \mid \mathbf{X})\, d\theta \right\}.$$

Now minimize the risk with respect to $\hat{\theta}$. The result is the Bayes' estimator.

$$\frac{d\rho(\hat{\theta})}{d\hat{\theta}} = [(k_1 + k_2)H(\hat{\theta} \mid \mathbf{X}) - k_2] + \hat{\theta}[(k_1 + k_2)h(\hat{\theta} \mid \mathbf{X})]$$
$$- \{k_1 \hat{\theta} h(\hat{\theta} \mid \mathbf{X}) + k_2 \hat{\theta} h(\hat{\theta} \mid \mathbf{X})\}.$$

Setting this result equal to zero and solving, gives

$$[(k_1 + k_2)H(\hat{\theta} \mid \mathbf{X})] - k_2 = 0.$$

Therefore, the Bayes' estimator is the value of $\hat{\theta}$ that satisfies the equation:

$$H(\hat{\theta} \mid \mathbf{X}) = \frac{k_2}{k_1 + k_2}.$$

In other words:

> When the loss function is piecewise linear, the Bayes' estimator is the $[k_2/(k_1 + k_2)]$ fractile of the posterior distribution.

REMARK: In the special case in which $k_1 = k_2 = k$, the Bayes' estimator becomes the *median* of the posterior distribution. In this case, the loss function is the *absolute error loss*,

$$L(\theta, \hat{\theta}) = k \mid \theta - \hat{\theta} \mid .$$

In summary:

> The Bayes's estimator with respect to an absolute error loss function is the posterior median.

11.2.4 Zero/One Loss Functions

As discussed in Chapter 9, Bayesian hypothesis testing does not usually involve testing against a *sharp* null hypothesis (testing $H: \theta = \theta_0 =$ some preassigned value). More typically, the Bayesian scientist is concerned with whether some unknown parameter is, within reasonable limits, in some preassigned interval (within measurement error and other unaccounted factors). Accordingly, define, for $c > 0$, and $d > 0$, the zero/one loss function:

$$L(\theta, \hat{\theta}) = \begin{cases} 0, & \text{if } (\hat{\theta} - c) \leq \theta \leq (\hat{\theta} + d) \\ 1, & \text{otherwise.} \end{cases}$$

This loss function might be as depicted in Figure 11.4.

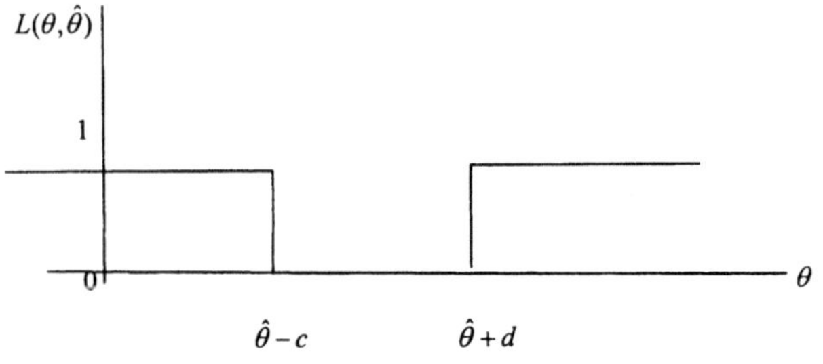

Figure 11.4 Zero/one loss function

The risk function is

$$\rho(\hat{\theta}) = E\{L(\theta, \hat{\theta})\} = \int L(\theta, \hat{\theta}) h(\theta \mid \mathbf{X})\, d\theta$$

$$= \int_{-\infty}^{\hat{\theta}-c} h(\theta \mid \mathbf{X})\, d\theta + \int_{\hat{\theta}+d}^{\infty} h(\theta \mid \mathbf{X})\, d\theta.$$

Differentiating the risk function gives

$$\frac{d\rho(\hat{\theta})}{d\hat{\theta}} = h[(\hat{\theta} - c) \mid \mathbf{X}] - h[(\hat{\theta} + d) \mid \mathbf{X}]. \tag{11.2}$$

So far we have said nothing about the shape of the posterior density $h(\theta \mid \mathbf{X})$. Suppose first that h has no modal value, θ_0; that is, no value of $\theta = \theta_0$ such that $h(\theta)$ is less than $h(\theta_0)$ to the left of θ_0, and $h(\theta)$ is less than $h(\theta_0)$ to the right of θ_0; suppose for example, the posterior distribution is exponential. Then, there will be no mode, and no value of $\hat{\theta}$ for which Equation (11.2) can be set equal to zero since there will be no stationary points of the posterior density. $\hat{\theta}$ should then be selected so as to make the risk in Equation (11.1) as small as possible. Now suppose that the posterior density has exactly one mode. Then there will be exactly one stationary point for which Equation (11.2) can be set equal to zero, so that

$$h[(\hat{\theta} - c) \mid \mathbf{X}] = h[(\hat{\theta} + d) \mid \mathbf{X}]. \tag{11.3}$$

For preassigned c and d, there are many values of $\hat{\theta}$ for which Equation (11.3) would be satisfied. Note that it is possible that the stationary point corresponds to a minimum value of the posterior density, as with beta distribution posteriors for which both parameters are less than one. If the stationary point corresponds to a maximum of the posterior density, this makes the interval $I_{\hat{\theta}} = \{\theta : \hat{\theta} - c \le \theta \le \hat{\theta} + d\}$ a *modal interval* (the mode will lie within that interval). That is, for all $\theta \in I_{\hat{\theta}}$, the posterior density, $h(\theta \mid \mathbf{X})$, is greater than the posterior density outside that interval. It is also an HPD interval (see Chapter 8, Section 8.4). Moreover, as $c \to 0, d \to 0, \hat{\theta}$ becomes the posterior mode.

For unimodal posterior distributions, the Bayes' estimator with respect to a zero/one loss function is the mode of the posterior distribution.

If the posterior distribution is multimodal, there will not be a unique solution for the mode (there will be various local modes and perhaps a unique global mode as well, but not necessarily). There may be many values of $\hat{\theta}$ that satisfy Equation (11.3).

REMARK: *Finding Posterior Modes.* It is not always convenient to define the posterior mode explicitly. Alternatively, it is sometimes useful to determine the mode of the posterior distribution numerically by using the *EM algorithm* (Dempster et al. 1977). The approach involves finding the posterior mode using a two-step numerical procedure. There is the E-step (Expectation), followed by the M-step (Maximization). The procedure is begun with an initial guess for the mode, say θ_1. The data are augmented (see Chapter 6). Then, we find the expectation of the log of the posterior density evaluated at $\theta = \theta_1$. Next, we find the value of θ in the augmented data that maximizes the expectation found in the first step. The procedure is then iterated. The result converges to the posterior mode.

11.2.5 Linex Asymmetric Loss Functions

The asymmetric Linex loss function was put forth by Varian (1975). The loss function is defined, for $b > 0$ and $a \neq 0$, by:

$$L(\theta, \hat{\theta}) = L(\hat{\theta} - \theta) = b\{e^{a(\hat{\theta}-\theta)} - a(\hat{\theta} - \theta) - 1\}.$$

This loss function is depicted in Figures 11.5a and b. For $\hat{\theta} - \theta = 0$, the loss is zero. For $a > 0$, the loss declines almost exponentially for $(\hat{\theta} - \theta) > 0$, and rises approximately linearly when $(\hat{\theta} - \theta) < 0$. For $a < 0$, the reverse is true. It is straightforward to find that the Bayes' estimator is given by

$$\hat{\theta} = \theta_{\text{Bayes}} = -\frac{1}{a} \log E[e^{-a\theta}]. \tag{11.4}$$

The expectation is taken with respect to the posterior distribution.

To evaluate Equation (11.4) explicitly, assume, for example, that $(\theta \mid \mathbf{X}) \sim N(\bar{\theta}, \tau^2)$. Then, the Bayes' estimator becomes

$$\theta_{\text{Bayes}} = \bar{\theta} - \frac{a\tau^2}{2}.$$

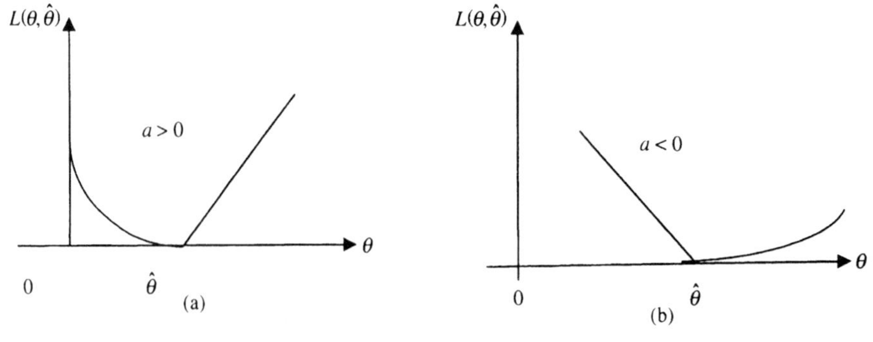

Figure 11.5 (a) Linex loss function for $a > 0$; (b) Linex loss function for $a < 0$

Thus, $\theta_{\text{Bayes}} < \bar{\theta},\ a > 0$, and $\theta_{\text{Bayes}} > \bar{\theta},\ a < 0$. That is, the Bayes' estimator is less than the posterior mean for positive a, and greater than the posterior mean for negative a.

11.3 ADMISSIBILITY

One of the important concepts of (frequentist) decision theory is that of *admissibility*. It is defined in a somewhat negative way. An estimator, $\hat{\theta}_A$, is *admissible* if there is no other estimator, $\hat{\theta}$, such that:

1. $\int L(\theta, \hat{\theta}) f(\mathbf{X} \mid \theta)\, d\mathbf{X} \leq \int L(\theta, \hat{\theta}_A) f(\mathbf{X} \mid \theta)\, d\mathbf{X}$, for all θ, and
2. $\int L(\theta, \hat{\theta}) f(\mathbf{X} \mid \theta)\, d\mathbf{X} < \int L(\theta_0, \hat{\theta}_A) f(\mathbf{X} \mid \theta)\, d\mathbf{X}$, for some $\theta = \theta_0$.

So $\hat{\theta}_A$ is admissible if there does not exist another estimator $\hat{\theta}$ that has at least as small a frequentist risk (loss averaged over the entire sample space) for all θ, and a strictly smaller frequentist risk for some particular $\theta = \theta_0$. Note that the definition of admissibility requires that we average over all possible values of the data, $\mathbf{X}$, not just the data values we happened to observe in a given experiment. This construct therefore violates the likelihood principle (see Chapter 3). So we would not normally advocate the use of the admissibility construct. It is interesting, however, to call attention to an important result of decision theory as it relates to Bayesian analysis. Directly from the definition of admissibility, it is straightforward to show the following result.

Theorem 11.1 (Bayes Rules and Admissibility). As long as the prior distribution is proper (it must integrate to one), and the corresponding Bayes' procedure is unique, all resulting Bayesian estimators are admissible.

The implication of this theorem is that optimal estimators, in terms of admissibility, are Bayes' estimators. Moreover, constructively, optimal (from a frequentist standpoint) estimators can be found by assigning proper prior distributions and evaluating the corresponding Bayes' estimators.

REMARK 1: The converse result is that any admissible procedure (decision rule) must be Bayes with respect to some prior distribution (Wald, 1950). The prior distribution might not be proper, however.

REMARK 2: Improper prior distributions sometimes lead to inadmissible estimators. For example, the sample mean vector for a p-dimensional normal distribution is an inadmissible estimator of the population mean vector for $p > 2$, assuming a quadratic loss function (James and Stein, 1961).

REMARK 3: Bayes' procedures need not be unique. For example, consider the following decision problem with a finite number of possible actions. The losses are arrayed in the simple matrix below.

Loss Matrix

Prior Probability	States of Nature	Action 1	Action 2	Action 3
$\frac{1}{2}$	θ_1	4	1	3
$\frac{1}{2}$	θ_2	1	4	3

Thus, there are three possible actions, two possible states of nature, and the prior probabilities for the states of nature are $(\frac{1}{2}, \frac{1}{2})$.

For example, the loss under Action 2 and state of nature θ_2 is equal to 4. The risk for taking Action $j, j = 1, 2, 3$ is:

$$\begin{aligned}\rho(j) &= L(1, \theta_1)P\{\theta_1\} + L(1, \theta_2)P\{\theta_2\} \\ &= 0.5\{L(1, \theta_1) + L(1, \theta_2)\}.\end{aligned}$$

Therefore,

$$\rho(1) = 5/2; \quad \rho(2) = 5/2; \quad \rho(3) = 3.$$

To minimize the risk, choose either Action 1 or Action 2, since either one has the same minimum risk of 5/2. So the Bayes risk is not unique in this case. We might as well flip a fair coin to decide between Action 1 or Action 2 (but either one is better than Action 3).

SUMMARY

In this chapter we have discussed how to make decisions under uncertainty from a Bayesian point of view. We introduced the notions of utility and loss functions, and studied the decision-making behavior of conservative decision makers; that is, those who have convex loss functions (concave utility functions). We examined quadratic loss functions (for which the Bayes' estimator is the posterior mean), piecewise linear and absolute error loss functions (for which the Bayes' estimator is a fractile of the posterior distribution), zero/one loss functions (for which the Bayes' estimator is the mode of the posterior distribution), Linex loss functions, and finite action with finite number of states of nature decision-making problems. We also examined the notion of admissibility, and its relationship with Bayes' decision rules, and pointed out that admissible (optimal) estimators must be Bayes' estimators.

EXERCISES

11.1 Let θ denote the proportion of defectives in a large population of cell phone parts. A sample of size 15 is taken from the population and three defectives are found. Suppose the prior distribution for θ is beta, with density:

$$g(\theta) \propto \theta^{4-1}(1-\theta)^{9-1}, \quad 0 \le \theta \le 1.$$

Adopt the loss function:

$$L(\theta, \hat{\theta}) = \begin{cases} 8(\hat{\theta} - \theta), & \text{if } \theta \le \hat{\theta} \\ \theta - \hat{\theta}, & \text{if } \theta \ge \hat{\theta}. \end{cases}$$

Give a Bayes' estimate for θ.

11.2 Suppose you are considering the purchase of a certain stock on the New York Stock Exchange, tomorrow. You will take either action a_1: buy the stock tomorrow, or action a_2: not to buy the stock tomorrow. For these two actions, you have the following loss function:

$$L(a_1, \theta) = 500 + 0.2\theta, \quad L(a_2, \theta) = 300 + 0.4\theta.$$

Assume there are no data available, but your prior distribution for θ is $N(1100, 2500)$. As a Bayesian, should you buy the stock?

11.3 Suppose you wish to estimate the mean of the distribution $N(\theta, 10)$, and you have a loss function $L(\theta, \hat{\theta}) = 5(\theta - \hat{\theta})^2$. You take a sample of data and find $\bar{x} = 17$. Your prior belief is that $\theta \sim N(6, 11)$. Find a Bayesian estimate of θ.

11.4* Explain the importance of Bayesian estimators with respect to proper prior distributions in decision theory.

11.5 You have data from a multivariate normal distribution $N(\boldsymbol{\theta}, \boldsymbol{\Sigma})$, with $\Sigma = 5\mathbf{I}$. From your data you find that $\bar{\mathbf{x}} = (1, 3, 4, 6)'$. Adopt the loss function $L(\theta, \hat{\theta}) = (\theta - \hat{\theta})'A(\theta - \hat{\theta})$, with $A = 4\mathbf{I}$. Your prior distribution for θ is $\theta \sim N(\phi, 2\mathbf{I})$, with $\phi = (2, 2, 3, 7)'$. Find a Bayesian estimator for θ.

11.6 Your data $(x_1, \ldots, x_n)$ arise from a distribution with probability density $f(x \mid \theta) = \theta e^{-\theta x}, x > 0$. Adopt a natural conjugate prior density for θ, and the loss function, $L(\theta, \hat{\theta}) = 4 \mid \theta - \hat{\theta} \mid$, and find a Bayesian estimator for θ.

11.7* Why is admissibility not usually an important criterion for a Bayesian scientist for choosing estimators which have good decision–theoretic properties?

*Solutions for asterisked exercises may be found in Appendix 7.

11.8 Suppose you have data $(\mathbf{x}_1, \ldots, \mathbf{x}_n)$, all p-dimensional, from a multivariate normal distribution $N(\boldsymbol{\theta}, \boldsymbol{\Sigma})$, with Σ unknown. Adopt the loss function:

$$L(\theta, \hat{\theta}) = (\theta - \hat{\theta})' \mathbf{S}^{-1} (\theta - \hat{\theta}),$$

where **S** denotes the sample covariance matrix. Adopt a natural conjugate prior distribution for $(\boldsymbol{\theta}, \boldsymbol{\Sigma})$ and develop the appropriate Bayes' estimator for θ.

11.9 Adopt a zero/one loss function and estimate the mean in the context in which we have a sample $(x_1, \ldots, x_n)$ from $N(\theta, 5)$, with $\bar{x} = 8$, and a normal prior density $\theta \sim N(4, 9)$.

11.10 For the problem in Exercise 11.9, suppose there were no time to collect data that would bear on the problem (ignore the observation that $\bar{x} = 8$), and a decision had to be made about θ. Find a Bayes' estimate of θ.

11.11* For the model in Section 12.2.1, give:

(a) The modal Bayesian estimator for β_2;

(b) The minimum risk Bayesian estimator with respect to a quadratic loss function for β_2;

(c) The minimum risk Bayesian estimator with respect to an absolute value loss function for β_2.

11.12 Suppose there are 4 possible actions, $a_i, i = 1, 2, 3, 4$, and three possible states of nature, $\theta_i, i = 1, 2, 3$, in a decision problem, and we have the loss matrix:

Loss

	a_1	a_2	a_3	a_4
θ_1	5	2	6	1
θ_2	4	3	1	2
θ_3	0	8	2	6

Suppose the prior probabilities for θ are as shown in the following table:

Prior Probabilities

	θ_1	θ_2	θ_3
Prior Probability	0.3	0.2	0.5

First note that action a_4 *dominates* action a_2 since the losses are smaller, regardless of the state of nature. So a_2 need not be considered further in any decision problem. Assuming no data are available, find a suitable Bayes' decision rule.

11.13 [Research Exercise] It may be noted from Section 11.2.1 that quadratic loss functions are unbounded in the parameter θ. This may lead to gambling gains in utility that can be infinite. This is known as the St. Petersburg Paradox. Explain. (*Hint*: see, for example, Shafer, 1988.)

11.14 [Research Exercise] What if there were more than one rational decision maker involved in the decision-making process. How might the decision be made? (*Hint*: see, for example, Kadane et al., 1999.)

FURTHER READING

Some books on Bayesian decision theory that provide additional discussion about the topics in this chapter are referenced below (note suggested levels of difficulty): Berger (1985), advanced; Blackwell and Gershick (1954), advanced; Chernoff and Moses (1959), elementary; DeGroot (1970), intermediate; LaValle (1978), advanced; Lindley (1985), intermediate; Martin (1967), advanced; Morgan (1968), elementary; Robert (2001), advanced; and Winkler (1972), elementary.

Berger, J. O. (1985). *Statistical Decision Theory and Bayesian Analysis*, New York, Springer-Verlag.

Blackwell, D. and Gershick, M. A. (1954). *Theory of Games and Statistical Decisions*, New York, John Wiley and Sons, Inc.

Chernoff, H. and Moses, L. E. (1959). *Elementary Decision Theory*, New York, John Wiley and Sons, Inc.

De Groot, M. H. (1970). *Optimal Statistical Decisions*, New York, McGraw-Hill.

Dempster, A. P., Laird, N. M. and Rubin, D. B. (1977). "Maximum Likelihood from Incomplete Data via the EM Algorithm," (with discussion) *Jr. Royal Statist. Soc., B*, **39**, 1–38.

James, W. and Stein, C. (1961). "Estimation With Quadratic Loss," in J. Neyman and E.L. Scott, Eds., *Proceedings of the Fourth Berkeley Symposium*, **1**, Berkeley, University of California Press, 361–380.

Jensen, J. L. W. V. (1906). "Sur les Functions Convexes et les Inegalites Entre les Valeurs Moyennes," *Acta Math.*, **30**, 175–193.

Kadane, J. B., Schervish, M. J. and Seidenfeld, T. (1999). *Rethinking the Foundations of Statistics*, Cambridge, Cambridge University Press.

LaValle, I. H. (1978). *Fundamentals of Decision Analysis*, New York, Holt, Rinehart and Winston.

Lindley, D. V. (1985). *Making Decisions*, London, John Wiley and Sons, Inc.

Martin, J. J. (1967). *Bayesian Decision Problems and Markov Chains*, New York, John Wiley and Sons, Inc.

Morgan, B. W. (1968). *An Introduction to Bayesian Statistical Decision Processes*, Englewood Cliffs, New Jersey, Prentice-Hall, Inc.

Pratt, J. W., Raiffa, H. and Schlaifer, R. (1965). *Introduction to Statistical Decision Theory*, New York, McGraw-Hill.

Press, S. J. and Tanur, J. M. (2002). "Decision-Making of Survey Respondents," *Proceedings of the Annual Meetings of the American Statistical Association in New York City*, August 2002.

Raiffa, H. (1968). *Decision Analysis: Introductory Lectures on Choices under Uncertainty*, Reading, MA, Addison-Wesley.

Raiffa, H. and Schlaifer, R. (1961). *Applied Statistical Decision Theory*, Graduate School of Business Administration, Harvard University, Boston, MA.

Robert, C. (2001). *The Bayesian Choice: From Decision—Theoretic Foundations to Computational Implementation*, Second Edition, New York, Springer-Verlag.

Savage, L. J. (1954). *The Foundations of Statistics*, New York, John Wiley and Sons, Inc.

Schlaifer, R. (1959). *Probability and Statistics for Business Decisions: An Introduction to Managerial Economics Under Uncertainty*, New York, McGraw-Hill.

Schlaifer, R. (1961). *Introduction to Statistics for Business Decisions*, New York, McGraw-Hill.

Schlaifer, R. (1969). *Analysis of Decisions Under Uncertainty*, New York, McGraw Hill.

Shafer, G. (1988). "The St. Petersburg Paradox," in S. Kotz and N. Johnson, Eds., *Encyclopedia of Statistical Sciences*, **8**, 865–870.

Varian, H. R. (1975). "A Bayesian Approach to Real Estate Assessment," in S.E. Fienberg and A. Zellner, Eds., *Studies in Bayesian Econometrics and Statistics in Honor of Leonard J. Savage*, Amsterdam, North Holland Publishing Co., 195–208.

von Neumann, J. and Morgenstern, O. (1st Edition, 1944; 2nd Edition 1947; 3rd Edition, 1953; renewed copyright 1972), *Theory of Games and Economic Behavior*, Princeton, New Jersey, Princeton University Press.

Wald, A. (1950). *Statistical Decision Functions*, New York, John Wiley and Sons, Inc.

Winkler, R. (1972). *An Introduction to Bayesian Inference and Decision*, New York, Holt, Rinehart and Winston, Inc.

PART IV

Models and Applications

CHAPTER 12

Bayesian Inference in the General Linear Model

12.1 INTRODUCTION

This chapter presents some basic applications of Bayesian inference to problems of interest in regression, and in the analysis of variance and covariance. Both univariate and multivariate cases are treated. There will be some discussion of fixed effects models, random effects models, and mixed models. The focus will be on the behavior of posterior and predictive distributions of coefficients and effects rather than on hypothesis testing.

12.2 SIMPLE LINEAR REGRESSION

12.2.1 Model

The simple linear regression model relates a single dependent variable to a single independent variable, linearly in the coefficients. The model is called "simple" if there is only one independent variable, and "multiple" if there is more than one independent variable. The model is called "univariate" if there is only one dependent variable, and "multivariate" if there is more than one dependent variable. The univariate simple linear regression model is:

$$y_i \mid x_i = \beta_1 + \beta_2 x_i + u_i, \qquad i = 1, \ldots, n,$$

where y_i denotes the ith observation on the dependent variable, x_i denotes the ith observation on the independent variable, u_i denotes the ith disturbance or error, and (β_1, β_2) are unknown parameters.

Note: Assume $u_1, \ldots, u_n$ are independent $N(0, \sigma^2)$, so that the unknown parameters of the model are $(\beta_1, \beta_2, \sigma^2)$.

12.2.2 Likelihood Function

Let $\boldsymbol{y} = (y_1, \ldots, y_n)'$ and let $\boldsymbol{x} = (x_1, \ldots, x_n)'$. Ignoring a proportionality constant, the likelihood function becomes

$$L(\boldsymbol{y} \mid \boldsymbol{x}, \beta_1, \beta_2, \sigma) = \frac{1}{\sigma^n} \exp\left\{ -\frac{1}{2\sigma^2} \sum_1^n (y_i - \beta_1 - \beta_2 x_i)^2 \right\}.$$

12.2.3 Prior

Suppose we adopt the vague prior density

$$\begin{aligned} g(\beta_1, \beta_2, \sigma) &= g_1(\beta_1) g_2(\beta_2) g_3(\sigma), \\ g_1(\beta_1) &\propto \text{constant}, \\ g_2(\beta_2) &\propto \text{constant}, \\ g_3(\sigma) &\propto \frac{1}{\sigma}. \end{aligned}$$

12.2.4 Posterior Inferences About Slope Coefficients

By Bayes' theorem, the joint posterior density is

$$h(\beta_1, \beta_2, \sigma \mid x, y) \propto \frac{1}{\sigma} \cdot \frac{1}{\sigma^n} \exp\left\{ -\frac{1}{2\sigma^2} \sum_1^n (y_i - \beta_1 - \beta_2 x_i)^2 \right\},$$
$$-\infty < \beta_1, \beta_2 < +\infty, \quad 0 < \sigma < \infty.$$

Equivalently,

$$h(\beta_1, \beta_2, \sigma \mid \boldsymbol{x}, \boldsymbol{y}) \propto \frac{1}{\sigma^{n+1}} \exp\left\{ -\frac{1}{2\sigma^2} \sum_1^n (y_i - \beta_1 - \beta_2 x_i)^2 \right\}.$$

The MLEs are

$$\hat{\beta}_1 = \bar{y} - \hat{\beta}_2 \bar{x}, \qquad \hat{\beta}_2 = \frac{\sum (x_i - \bar{x})(y_i - \bar{y})}{\sum (x_i - \bar{x})^2},$$

where $\bar{x} = n^{-1} \sum x_i, \bar{y} = n^{-1} \sum y_i$, and an unbiased estimator of σ^2 is

$$\hat{\sigma}^2 = s^2 = \frac{1}{n-2} \sum (y_i - \hat{\beta}_1 - \hat{\beta}_2 x_i)^2.$$

Writing

$$\sum_1^n (y_i - \beta_1 - \beta_2 x_i)^2 = \sum_1^n [(y_i - \hat{\beta}_1 - \hat{\beta}_2 x_i) - (\beta_1 - \hat{\beta}_1) - (\beta_2 - \hat{\beta}_2)x_i]^2,$$

combined with integrating with respect to σ, gives the two-dimensional posterior density

$$h_1(\beta_1, \beta_2 \mid y, x) \propto \{(n-2)s^2 + n(\beta_1 - \hat{\beta}_1)^2 + (\beta_2 - \hat{\beta}_2)^2 \sum x_i^2 + 2(\beta_1 - \hat{\beta}_1)(\beta_2 - \hat{\beta}_2)\sum x_i\}^{-0.5n}.$$

That is, β_1 and β_2 given $\boldsymbol{x}$ and $\boldsymbol{y}$, jointly follow a bivariate Student's t-distribution, so that marginally they each follow univariate Student's t-posterior distributions (see, for example, Press, 1982, Section 6.2).

Marginally, we have

$$\left[\frac{\sum(x_i - \bar{x})^2}{(s^2/n)\sum x_i^2}\right]^{0.5} \cdot (\beta_1 - \hat{\beta}_1) \mid \boldsymbol{y}, \boldsymbol{x} \sim t_{n-2}$$

and

$$\left.\frac{\beta_2 - \hat{\beta}_2}{\left(s/[\sum(x_i - \bar{x})^2]^{0.5}\right)}\right| \boldsymbol{y}, \boldsymbol{x} \sim t_{n-2}.$$

12.2.5 Credibility Intervals

For example, we can make credibility statement about the slope coefficient β_2 (Fig. 12.1):

$$P\{\hat{\beta}_2 - z_\alpha \delta \leq \beta_2 \leq \hat{\beta}_2 + z_\alpha \delta \mid \boldsymbol{y}, \boldsymbol{x}\} = 1 - 2\alpha,$$

where z_α is the α-fractile point of the Student's t-distribution, $f(t)$ is its density, and

$$\delta \equiv \frac{s}{[\sum(x_i - \bar{x})^2]^{0.5}}.$$

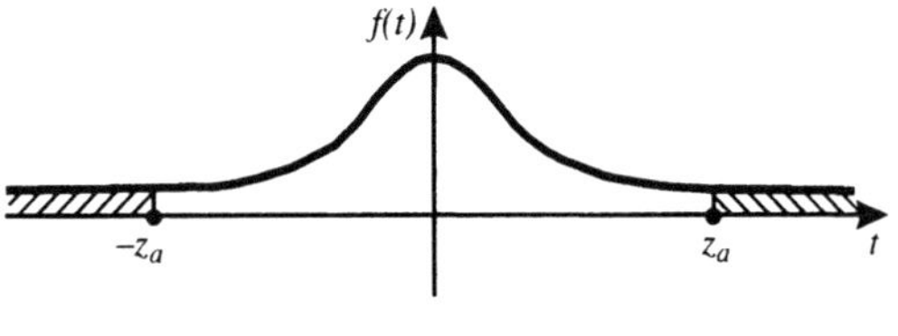

Figure 12.1 Posterior density of β_2

12.2.6 Example

Suppose y_i denotes average microprocessor speed of personal computers (in megahertz), as perceived by a company in the computer industry in year i, and x_i denotes average megabytes of RAM memory of personal computers in year i, and we have 21 years of data starting in year 1980. The data are given in Table 12.1.

The MLEs and other standard estimators are given by

$$\hat{\beta}_1 = -2.650, \qquad \hat{\beta}_2 = 2.375, \qquad s = 46.469,$$

$$\bar{x} = 107.3, \qquad s_x = \left\{ \frac{1}{n} \sum_{i=1}^{21} (x_i - \bar{x})^2 \right\}^{0.5.}$$

Then, the (marginal) posterior densities are given in Figures 12.2 and 12.3. The posterior means are equal to the MLEs and are given by

$$E(\beta_1 \mid x, y) = -2.650, \qquad E(\beta_2 \mid x, y) = 2.375.$$

We use these types of results to make predictions of future microprocessor speeds.

Table 12.1 Data for Example

y	x	Year
25	0.10	1980
25	0.15	1981
25	0.20	1982
50	10.00	1983
50	20.00	1984
50	32.00	1985
50	40.00	1986
100	50.00	1987
100	60.00	1988
100	64.00	1989
100	64.00	1990
120	75.00	1991
200	75.00	1992
300	100.00	1993
350	128.00	1994
350	128.00	1995
400	150.00	1996
500	256.00	1997
700	300.00	1998
700	300.00	1999
1000	400.00	2000

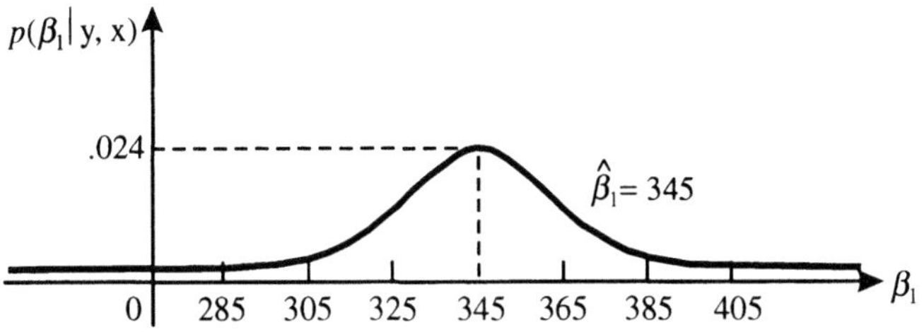

Figure 12.2 Posterior density of β_1

12.2.7 Predictive Distribution

Suppose we wish to find predictive intervals for a new output y^*, based upon a new observation x^* and past experience. We know

$$y^* \mid x^* = \beta_1 + \beta_2 x^* + u, \qquad u \sim N(0, \sigma^2).$$

The predictive density of y^* is given by

$$p(y^* \mid x^*) = \iiint L(y^* \mid x^*, \beta_1, \beta_2, \sigma) h(\beta_1, \beta_2, \sigma \mid \boldsymbol{x}, \boldsymbol{y}) d\beta_1, d\beta_2, d\sigma.$$

It is straightforward to show [see, for example, Press, 1982, p. 258] that

$$\left. \frac{y^* - \hat{\beta}_1 - \hat{\beta}_2 x^*}{\hat{\sigma}\left[1 + \dfrac{1}{n} + \dfrac{(x^* - \bar{x})^2}{\sum_1^n (x_i - \bar{x})^2}\right]^{0.5}} \right| \text{sample} \sim t_{n-2}$$

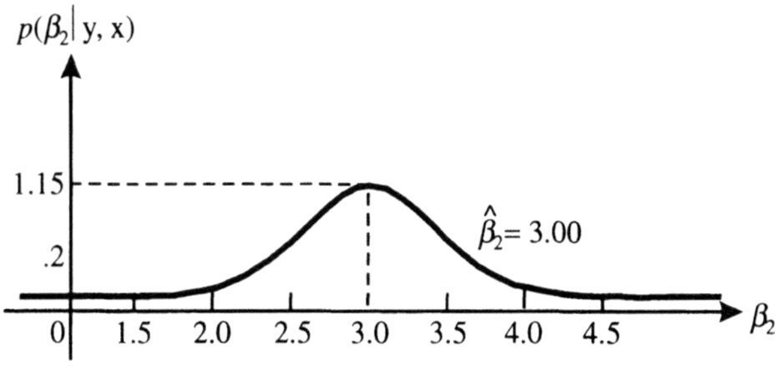

Figure 12.3 Posterior density of β_2.

Assuming $x^* = 500$, the data from the example in Section 12.2.6 gives

$$\left.\frac{y^* - 1184.85}{59.08}\right| \text{sample} \sim t_{19}.$$

At a credibility level of 95 percent this gives (Fig. 12.4):

$$P\{1061.2 \leq y^* \leq 1308.5 \mid \text{sample}\} = 95 \text{ percent}.$$

We can, of course, find smaller intervals by lowering the credibility level to less than 95 percent.

12.2.8 Posterior Inferences About the Standard Deviation

It is straightforward to make posterior inferences about σ from the marginal posterior density. It is given by (using h generically)

$$h(\sigma \mid \boldsymbol{x}, \boldsymbol{y}) = \iint h(\beta_1, \beta_2, \sigma \mid \boldsymbol{x}, \boldsymbol{y}) d\beta_1 d\beta_2,$$

or, using the functional form given in Section 12.2.4, we find for the marginal posterior density of σ,

$$h(\sigma \mid \boldsymbol{x}, \boldsymbol{y}) \propto \iint \frac{1}{\sigma^{n+1}} e^{-(0.5\sigma^2)q(\beta_1, \beta_2)} d\beta_1 d\beta_2,$$

where

$$q(\beta_1, \beta_2) \equiv (n-2)s^2 + n(\beta_1 - \hat{\beta}_2)^2 + (\beta_2 - \hat{\beta}_2) \sum x_i^2 + 2(\beta_1 - \hat{\beta}_1)(\beta_2 - \hat{\beta}_2) \sum x_i.$$

Carrying out the integration gives (Fig. 12.5)

$$h(\sigma \mid \boldsymbol{x}, \boldsymbol{y}) \propto \frac{1}{\sigma^{n-1}} e^{-(n-2)s^2/2\sigma^2}, \qquad 0 < \sigma < \infty, \quad 2 < n,$$

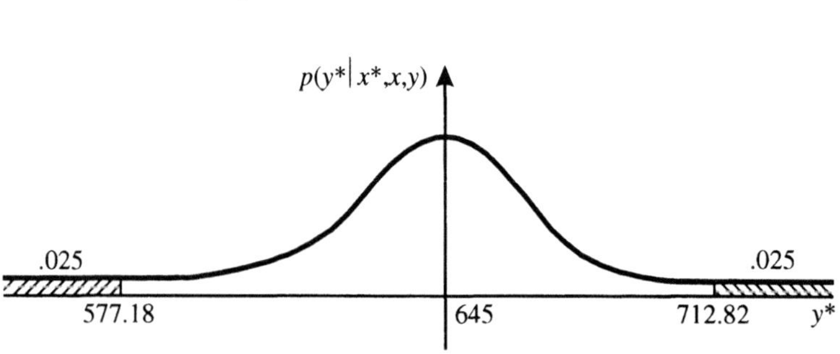

Figure 12.4 Predictive density

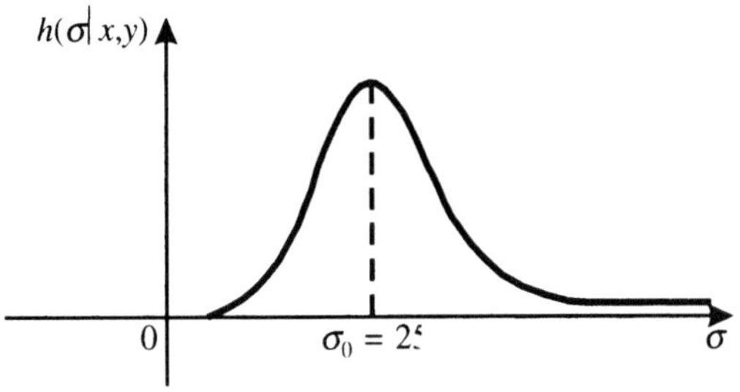

Figure 12.5 Posterior density of σ

an inverted gamma distribution density (i.e., $1/\sigma$ follows a gamma distribution). Thus,

$$E(\sigma \mid \boldsymbol{x}, \boldsymbol{y}) = \frac{\Gamma[0.5(n-3)]}{\Gamma[0.5(n-2)]}[0.5(n-2)]^{0.5}s,$$

$$\operatorname{var}(\sigma \mid \boldsymbol{x}, \boldsymbol{y}) = \frac{(n-2)s^2}{(n-4)} - [E(\sigma \mid \boldsymbol{x}, \boldsymbol{y})]^2$$

Furthermore,

$$\sigma_0 = \text{modal value of } \sigma = s\sqrt{\frac{n-2}{n-1}}$$

for $s = 46.469$, $n = 21$, $\sigma_0 = 45.29$.

12.3 MULTIVARIATE REGRESSION MODEL

12.3.1 The Wishart Distribution

Let V denote a $p \times p$, symmetric, positive definite matrix of random variables, and we write $V > 0$. This means that $V = V'$ and that all of the latent roots of V are positive. V follows a Wishart distribution if the elements are jointly continuous with multivariate pdf:

$$p(V \mid \boldsymbol{\Sigma}, n) \propto \frac{|V|^{0.5(n-p-1)}}{|\boldsymbol{\Sigma}|^{0.5n}} \exp\{-0.5 \operatorname{tr} \boldsymbol{\Sigma}^{-1} V\}, \qquad p \le n, \quad \boldsymbol{\Sigma} > 0,$$

and we write

$$V \sim W(\boldsymbol{\Sigma}, p, n).$$

Note that $\boldsymbol{\Sigma}$ is a scale matrix, n is a degrees-of-freedom parameter, and p is the dimension. The proportionality constant depends upon (p, n) only, and is given by:

$$c^{-1} = 2^{(np)/2}\pi^{p(p-1)/4}\prod_{j=1}^{p}\Gamma\left(\frac{n+1-j}{2}\right).$$

12.3.2 Multivariate Vague Priors

Suppose we want to express "knowing little" about the values of the elements of $\boldsymbol{\Sigma}$ with a vague prior. We would use the improper prior density (see Chapter 5, Section 5.4):

$$p(\boldsymbol{\Sigma}) \propto \frac{1}{|\boldsymbol{\Sigma}|^{0.5(p+1)}}.$$

12.3.3 Multivariate Regression

The multivariate regression model relates p correlated dependent variables to q independent variables using linear relationships. The model is

$$\begin{array}{ccccccc} \underset{(N\times 1)}{\boldsymbol{y}_1} & |\boldsymbol{X} = & \underset{(N\times q)}{\boldsymbol{X}} & \cdot & \underset{(q\times 1)}{\boldsymbol{\beta}_1} & + & \underset{(N\times 1)}{\boldsymbol{u}_1} \\ \vdots & & & \vdots & & & \vdots \\ \underset{(N\times 1)}{\boldsymbol{y}_p} & |\boldsymbol{X} = & \underset{(N\times q)}{\boldsymbol{X}} & \cdot & \underset{(q\times 1)}{\boldsymbol{\beta}_p} & + & \underset{(N\times 1)}{\boldsymbol{u}_p} \end{array},$$

where

$$\boldsymbol{u}_j \sim N(\boldsymbol{0}, \sigma_{jj}\boldsymbol{I}_N), \qquad j = 1, \ldots, p,$$

and the $\boldsymbol{u}_j$s are correlated. Evidently, if $\boldsymbol{Y} \equiv (\boldsymbol{y}_1, \ldots, \boldsymbol{y}_p)$, $\boldsymbol{U} \equiv (\mathbf{u}_1, \ldots, \mathbf{u}_p)$, $\boldsymbol{B} = (\boldsymbol{\beta}_1, \ldots, \boldsymbol{\beta}_p)$, the model becomes

$$\underset{(N\times p)}{\boldsymbol{Y}} = \underset{(N\times q)}{\boldsymbol{X}}\ \underset{(q\times p)}{\boldsymbol{B}} + \underset{(N\times p)}{\boldsymbol{U}},$$

$$\boldsymbol{U}' \equiv (\boldsymbol{v}_1, \ldots, \boldsymbol{v}_N),$$

$$\boldsymbol{v}_j \sim N(\boldsymbol{0}, \boldsymbol{\Sigma}), \qquad \text{i.i.d.}, \boldsymbol{\Sigma} > 0.$$

Thus, a multivariate regression model involves the study of several univariate regressions taken simultaneously, because the disturbance terms in the several regressions are mutually correlated. So information in one regression can be used to estimate coefficients in other regressions (and the same is true for predictions of future response variable values). Bayesian inference in the multivariate regression model involves use of the multivariate normal distribution, as well as the Wishart distribution.

12.3.4 Likelihood Function

The likelihood function of U' (same as that of U) is the joint density of the $\mathbf{v}_j$s and, by independence, is given by

$$p(U \mid \Sigma) = p(v_1, \ldots, v_N \mid \Sigma) = \prod_1^N p(v_j \mid \Sigma),$$

or

$$p(U \mid \Sigma) \propto \frac{1}{\mid \Sigma \mid^{0.5N}} \exp\{-0.5 \mathrm{tr}\, \Sigma^{-1} U'U\},$$

since $U'U = \sum_1^N v_j v_j'$. Since $(Y \mid X, B) = XB + U$, and since the Jacobian of the transformation[1] is unity, transforming variables from U to Y gives

$$p(Y \mid X, B, \Sigma) \propto \frac{1}{\mid \Sigma \mid^{0.5N}} \exp\{-0.5 \mathrm{tr} \Sigma^{-1} (Y - XB)'(Y - XB)\}.$$

Orthogonality Property of Least-Squares Estimators

Let $\hat{B}$ denote the least-squares estimator. By adding and subtracting XB, we can find

$$\begin{aligned} A &\equiv (Y - XB)'(Y - XB) \\ &= [(Y - X\hat{B}) + X(\hat{B} - B)]'[(Y - X\hat{B}) + X(\hat{B} - B)]. \end{aligned}$$

But it is easy to check that the least-squares estimator is

$$\hat{B} = (X'X)^{-1} X'Y.$$

It may be found that $X'\hat{U} \equiv X'(Y - X\hat{B}) = 0$, the orthogonality property of least-squares estimators, so that the cross-product terms in (A) vanish, and we obtain

$$A = V + (B - \hat{B})'(X'X)(B - \hat{B}),$$

where

$$V \equiv (Y - X\hat{B})'(Y - X\hat{B}).$$

[1]The Jacobian of the one-to-one transformation $Y = f(X)$, where the distinct elements of Y are $(y_1, \ldots, y_n)$, and where those of X are $(x_1, \ldots, x_n)$, is $\mid \det(A) \mid^{-1}$, where $A \equiv (a_{ij})$ and $a_{ij} = \partial y_i / \partial x_j$; $\det(A)$ denotes determinant of the matrix A. In this case, Y and U are related by an additive constant, so the Jacobian is unity.

Then, the likelihood function becomes

$$p(\boldsymbol{Y} \mid \boldsymbol{X}, \boldsymbol{B}, \boldsymbol{\Sigma}) \propto \frac{1}{\mid \boldsymbol{\Sigma} \mid^{0.5N}} e^{-\text{tr}\,\boldsymbol{\Sigma}^{-1} 0.5[\boldsymbol{V}+(\boldsymbol{B}-\hat{\boldsymbol{B}})'(\boldsymbol{X}'\boldsymbol{X})(\boldsymbol{B}-\hat{\boldsymbol{B}})]}.$$

12.3.5 Vague Priors

Adopting a vague prior distribution on $(\boldsymbol{B}, \boldsymbol{\Sigma})$ gives

$$p(\boldsymbol{B}, \boldsymbol{\Sigma}) = p(\boldsymbol{B})p(\boldsymbol{\Sigma}),$$

$$P(\boldsymbol{B}) \propto \text{constant},$$

$$p(\boldsymbol{\Sigma}) \propto \frac{1}{\mid \boldsymbol{\Sigma} \mid^{0.5(p+1)}},$$

$$p(\boldsymbol{B}, \boldsymbol{\Sigma}) \propto \frac{1}{\mid \boldsymbol{\Sigma} \mid^{0.5(p+1)}}.$$

12.3.6 Posterior Analysis for the Slope Coefficients

By Bayes' theorem, the joint posterior density is

$$p(\boldsymbol{B}, \boldsymbol{\Sigma} \mid \boldsymbol{X}, \boldsymbol{Y}) \propto \frac{1}{\mid \boldsymbol{\Sigma} \mid^{0.5(n+p+1)}} e^{-0.5\text{tr}\,\boldsymbol{\Sigma}^{-1}\boldsymbol{A}}.$$

Posterior inferences about $\boldsymbol{B}$ are made most easily from the marginal posterior density, given by

$$p(\boldsymbol{B} \mid \boldsymbol{X}, \boldsymbol{Y}) \propto \int_{\boldsymbol{\Sigma}>0} \frac{1}{\mid \boldsymbol{\Sigma} \mid 0.5^{(N+p+1)}} e^{-0.5\text{tr}\,\boldsymbol{\Sigma}^{-1}\boldsymbol{A}} d\boldsymbol{\Sigma}.$$

This integration may be carried out readily by recognizing that if $\boldsymbol{\Sigma}^{-1}$ followed a Wishart distribution, its density would be proportional to that in the integrand. The result is

$$p(\boldsymbol{B} \mid \boldsymbol{X}, \boldsymbol{Y}) \propto \frac{1}{\mid \boldsymbol{V} + (\boldsymbol{B} - \hat{\boldsymbol{B}})'(\boldsymbol{X}'\boldsymbol{X})(\boldsymbol{B} - \hat{\boldsymbol{B}}) \mid^{0.5N}}.$$

That is, $\boldsymbol{B}$ follows a matrix $\boldsymbol{T}$-distribution. Therefore its columns (and rows) marginally follow multivariate Student's t-distributions, so that for $j = 2, \ldots, p$ we have for the columns of $\boldsymbol{B}$,

$$p(\boldsymbol{\beta}_j \mid \boldsymbol{X}, \boldsymbol{Y}) \propto \frac{1}{\{v_{jj} + (\boldsymbol{\beta}_j - \hat{\boldsymbol{\beta}}_j)'(\boldsymbol{X}'\boldsymbol{X})(\boldsymbol{\beta}_j - \hat{\boldsymbol{\beta}}_j)\}^{0.5(N-p+1)}},$$

where

$$V \equiv (v_{ij}).$$

Joint posterior inferences about all of the coefficients in a given regression equation may be made from the result:

$$[(\boldsymbol{\beta}_j - \hat{\boldsymbol{\beta}}_j)' \boldsymbol{G}_j (\boldsymbol{\beta}_j - \hat{\boldsymbol{\beta}}_j) \mid \boldsymbol{X}, \boldsymbol{Y}] \sim F_{q+1, N-p-1},$$

where

$$\boldsymbol{G}_j \equiv \frac{\boldsymbol{X}'\boldsymbol{X}(N-p-q)}{(N-p-q+1)(q+1)s_j^2}, \qquad s_j^2 = \frac{v_{jj}}{N-p-q+1}.$$

Posterior inferences about specific regression coefficients are made from the marginal univariate Student's t-distributions:

$$\left[\frac{\beta_{ij} - \hat{\beta}_{ij}}{s_j \sqrt{k_{ii}}} \middle| \hat{\beta}_{ij}, s_j, \boldsymbol{X} \right] \sim t_{N-p-q+1},$$

where β_{ij} denotes the ith element of $\boldsymbol{\beta}_j$, and

$$\boldsymbol{K} \equiv (\boldsymbol{X}'\boldsymbol{X})^{-1} = (k_{ij}).$$

12.3.7 Posterior Inferences About the Covariance Matrix

It is straightforward to check, by integrating the joint posterior density of $(\boldsymbol{B}, \boldsymbol{\Sigma} \mid \boldsymbol{X}, \boldsymbol{Y})$ with respect to $\boldsymbol{B}$ and by transforming $\boldsymbol{\Sigma} \to \boldsymbol{\Sigma}^{-1}$ (the Jacobian is $\mid \boldsymbol{\Sigma} \mid^{-(p+1)}$; see, for example, Press, 1982, p. 47), that the marginal density of $\boldsymbol{\Sigma}^{-1}$ is

$$\boldsymbol{\Sigma}^{-1} \mid \boldsymbol{X}, \boldsymbol{Y}k \sim W(\boldsymbol{V}^{-1}, p, N-q).$$

Thus, inferences about variances (diagonal elements of $\boldsymbol{\Sigma}$ can be made from inverted gamma distributions (the marginal distributions of the diagonal elements $\boldsymbol{\Sigma}$).

12.3.8 Predictive Density

In the multivariate regression model, suppose that we have a new independent variable observation $\boldsymbol{w}$: $(q \times 1)$ and that we wish to predict the corresponding dependent variable observation, say, $\boldsymbol{z}$: $(p \times 1)$. Extending the univariate regression approach to the multivariate case, it is not hard to show (see, for example, Press, 1982, p. 420) that the density of the predictive distribution is given by

$$p\left(\underset{(p\times 1)}{z} \mid X, Y\right) \propto \frac{1}{\{\nu + (z - \hat{B}\omega)' H (z - \hat{B}\omega)\}^{0.5(\nu+p)}},$$

where

$$\nu = N - q - p + 1, \qquad H \equiv \frac{\nu V^{-1}}{1 + \omega' D^{-1} \omega}, \qquad D \equiv \sum_1^N x_j x_j',$$

where x_j denotes the jth independent variable observation.

12.4 MULTIVARIATE ANALYSIS OF VARIANCE MODEL

12.4.1 One-Way Layout

Adopt the p-dimensional analysis of variance model:

$$\underset{(p\times 1)}{y_{ij}} = \underset{(p\times 1)}{\mu} + \underset{(p\times 1)}{\alpha_i} + \underset{(p\times 1)}{u_{ij}}, \qquad i = 1, \ldots, q, \quad j = 1, \ldots, n_i,$$

where y_{ij} denotes a p-vector of observations representing the jth replication in the ith population, μ denotes the grand mean, α_i denotes the main effect due to population i, and u_{ij} denotes a disturbance term. The univariate analysis of variance model is obtained by setting $p = 1$.

12.4.2 Reduction to Regression Format

Let $N = \sum_{i=1}^{n} n_i$, and define

$$\begin{aligned}
\underset{(p\times N)}{Y'} &= [y_{11}, \ldots, y_{1n_1}; \ldots; y_{q1}, \ldots, y_{qn_q}] \\
\underset{(p\times N)}{U'} &= [u_{11}, \ldots, u_{1n_1}; \ldots; u_{q1}, \ldots, u_{qn_q}] \\
\underset{(p\times N)}{\theta_i} &= \mu + \alpha_i, \\
\underset{(p\times q)}{B'} &= (\theta_i, \ldots, \theta_q),
\end{aligned}$$

$$\underset{(N\times q)}{X} = \begin{pmatrix} e_{n_1} & & 0 \\ & \ddots & \\ 0 & & e_{n_q} \end{pmatrix}, \qquad \underset{(n\times 1)}{e_n} = \begin{pmatrix} 1 \\ \vdots \\ 1 \end{pmatrix}.$$

The model becomes

$$\underset{(N\times p)}{Y} = \underset{(N\times q)}{X} \cdot \underset{(q\times p)}{B} + \underset{(N\times p)}{U}.$$

12.4.3 Likelihood

Assume the $\boldsymbol{u}_{ij}$s are mutually independent and

$$\boldsymbol{u}_{ij} \sim N(\mathbf{0}, \Sigma), \qquad \Sigma > 0.$$

The likelihood function is then given by

$$p(Y \mid X, B, \Sigma) = \frac{1}{|\Sigma|^{0.5N}} e^{-0.5 \operatorname{tr}[V+(B-\hat{B})'S(B-\hat{B})]\Sigma^{-1}},$$

where

$$\underset{(q\times p)}{\hat{B}} = (X'X)^{-1}X'Y, \qquad \underset{(q\times q)}{S} = X'X, \qquad \underset{(p\times q)}{V} = (Y - X\hat{B})'(Y - X\hat{B}).$$

12.4.4 Priors

1. Assume that B and Σ are independent, that is,

$$p(B, \Sigma) = p(B)p(\Sigma).$$

2. Assume that Σ follows an inverted Wishart distribution (and therefore, Σ^{-1} follows a Wishart distribution), so that for some hyperparameters (υ, H) we have

$$p(\Sigma) \propto \frac{1}{|\Sigma|^{0.5\upsilon}} e^{-0.5 \operatorname{tr} \Sigma^{-1} H}.$$

3. For $B' \equiv (\theta, \ldots, \theta_q)$, take the θ_i to be i.i.d. (and therefore exchangeable; see Chapter 10) and normally distributed. Note that if the correlations between pairs of θ_i vectors are small, exchangeable θ_is will be approximately independent (when they are $N(\xi, \Phi)$). Note that we are not assuming vague prior distributions. Thus,

$$p(B) = \prod_{i=1}^{q} p(\theta_i),$$

$$(\theta_i \mid \xi, \Phi) \sim N(\xi, \Phi), \qquad \Phi > 0,$$

so that

$$p(B \mid \xi, \Phi) \propto \frac{1}{|\Phi|^{0.5q}} \exp\left\{-0.5 \sum_{1}^{q} (\theta_i - \xi)'\Phi^{-1}(\theta_i - \xi)\right\}.$$

12.4.5 Practical Implications of the Exchangeability Assumption in the MANOVA Problem

Assume that the prior distribution for the mean vectors of the populations in Section 12.4.4 is invariant under reordering of the populations (the populations are exchangeable; see Chapter 10). Under nonexchangeability, the matrix of coefficients

$$\underset{(p\times q)}{B'} \equiv (\theta_1, \ldots, \theta_q)$$

has (pq) elements. If they are assumed to be jointly normally distributed, there would be a total (mean + variances and covariances) of $[pq + 0.5pq(pq + 1)]$ hyperparameters to assess. For example, if $q = 3$ and $p = 5$ without exchangeability, there would be 135 parameters to assess.

Under exchangeability we must assess only ξ: $(p \times 1)$ and Φ: $(p \times p)$ for a total of $[p + 0.5p(p + 1)]$ hyperparameters. For $p = 5$, this implies 20 assessments instead of 135.

Other Implications

Without exchangeability, as the number of populations, q, increases, the number of hyperparameters that must be assessed increases beyond the 135 in this example; with exchangeability we never have to assess more than 20 (for $p = 5$), regardless of the size of q.

Without exchangeability, the elements of B' are taken to be correlated; under exchangeability the columns of B' are assumed to be independent. This takes advantage of the basic population structure in the problem (by assuming that the means of the populations are independent, *a priori*). Define

$$\underset{(pq\times 1)}{\theta} = \begin{pmatrix} \theta_1 \\ \vdots \\ \theta_q \end{pmatrix}, \qquad \underset{(pq\times 1)}{\theta^*} = e_q \otimes \xi = \begin{pmatrix} \xi \\ \vdots \\ \xi \end{pmatrix}$$

and note that

$$\sum_1^q (\theta_i - \xi)'\Phi^{-1}(\theta_i - \xi) = (\theta - \theta^*)'(I_q \otimes \Phi^{-1})(\theta - \theta^*).$$

We are using the notation $\otimes$ for the direct product of two matrices. It is defined by the following. For $\underset{(p\times q)}{A} \equiv (a_{ij})$, and $\underset{(r\times s)}{B} \equiv (b_{ij})$, $A \otimes B = C{:}\,(pr \times qs)$, where:

$$C = \begin{pmatrix} a_{11}B & a_{12}B & \cdots & a_{1q}B \\ \vdots & \vdots & & \vdots \\ a_{p1}B & a_{p2}B & \cdots & a_{pq}B \end{pmatrix}.$$

12.4.6 Posterior

Joint Posterior

The joint posterior density of $(\boldsymbol{B}, \boldsymbol{\Sigma})$ is given by:

$$p(\boldsymbol{B}, \boldsymbol{\Sigma} \mid \xi, \boldsymbol{\Phi}, \boldsymbol{X}, \boldsymbol{Y}) \propto \frac{1}{|\boldsymbol{\Phi}|^{0.5q}} e^{-0.5\,\mathrm{tr}(\boldsymbol{B}-\boldsymbol{B}^*)(\boldsymbol{B}-\boldsymbol{B}^*)'\boldsymbol{\Phi}^{-1}} \times \frac{1}{|\boldsymbol{\Sigma}|^{0.5(N+\upsilon)}} e^{-0.5\,\mathrm{tr}[\boldsymbol{V}+\boldsymbol{H})+(\boldsymbol{B}-\hat{\boldsymbol{B}})'\boldsymbol{S}(\boldsymbol{B}-\hat{\boldsymbol{B}})]\boldsymbol{\Sigma}^{-1}},$$

with $\boldsymbol{S} = \boldsymbol{X}'\boldsymbol{X}$, or, after completing the square in $\boldsymbol{B}$,

$$p(\boldsymbol{B}, \boldsymbol{\Sigma} \mid \xi, \boldsymbol{\Phi}, \boldsymbol{X}, \boldsymbol{Y}) \propto \frac{1}{|\boldsymbol{\Sigma}|^{0.5(N+\upsilon)}} e^{-0.5\{\mathrm{tr}(\boldsymbol{H}+\boldsymbol{V})(\boldsymbol{\Sigma}^{-1})+C(\boldsymbol{\Sigma})\}} \times e^{-0.5(\boldsymbol{\theta}-\bar{\boldsymbol{\theta}})'[(\boldsymbol{I}_q\otimes\boldsymbol{\Phi}^{-1})+(\boldsymbol{S}\otimes\boldsymbol{\Sigma}^{-1})](\boldsymbol{\theta}-\bar{\boldsymbol{\theta}})}$$

where

$$C(\boldsymbol{\Sigma}) \equiv (\hat{\boldsymbol{\theta}} - \boldsymbol{\theta}^*)'\left\{(\boldsymbol{I}_q \otimes \boldsymbol{\Phi}^{-1})[(\boldsymbol{I}_q \otimes \boldsymbol{\Phi}^{-1}) + (\boldsymbol{S} \otimes \boldsymbol{\Sigma}^{-1})]^{-1}(\boldsymbol{S} \otimes \boldsymbol{\Sigma}^{-1})\right\}(\hat{\boldsymbol{\theta}} - \boldsymbol{\theta}^*)$$

and

$$\bar{\boldsymbol{\theta}} = [(\boldsymbol{I}_q \otimes \boldsymbol{\Phi}^{-1}) + (\boldsymbol{S} \otimes \boldsymbol{\Sigma}^{-1})]^{-1}[(\boldsymbol{I}_q \otimes \boldsymbol{\Phi}^{-1})\boldsymbol{\theta}^* + (\boldsymbol{S} \otimes \boldsymbol{\Sigma}^{-1})\hat{\boldsymbol{\theta}}].$$

Conditional Posterior

Thus, conditionally,

$$(\boldsymbol{\theta} \mid \boldsymbol{\Sigma}, \xi, \boldsymbol{\Phi}, \boldsymbol{X}, \boldsymbol{Y}) \sim N\left\{\bar{\boldsymbol{\theta}}, [(\boldsymbol{I}_q \otimes \boldsymbol{\Phi}^{-1}) + (\boldsymbol{S} \otimes \boldsymbol{\Sigma}^{-1})]^{-1}\right\}.$$

Note: If

$$\boldsymbol{Q} \equiv [(\boldsymbol{I} \otimes \boldsymbol{\Phi}^{-1}) + (\boldsymbol{S} \otimes \boldsymbol{\Sigma}^{-1})]^{-1}(\boldsymbol{I} \otimes \boldsymbol{\Phi}^{-1})$$

so

$$(\boldsymbol{I} - \boldsymbol{Q}) \equiv [(\boldsymbol{I} \otimes \boldsymbol{\Phi}^{-1}) + (\boldsymbol{S} \otimes \boldsymbol{\Sigma}^{-1})]^{-1}(\boldsymbol{S} \otimes \boldsymbol{\Sigma}^{-1}),$$

then

$$\bar{\boldsymbol{\theta}} = \boldsymbol{Q}\boldsymbol{\theta}^* + (\boldsymbol{I} - \boldsymbol{Q})\hat{\boldsymbol{\theta}},$$

which is a matrix weighted average, Stein-type estimator (see Berger, 1980; Efron and Morris, 1973; and James and Stein, 1961).

Marginal Posterior

Integrating $\boldsymbol{\Sigma}$ out of the joint posterior for $(\boldsymbol{B}, \boldsymbol{\Sigma})$ (by noting that $\boldsymbol{\Sigma}^{-1}$ in the integral has a Wishart density) gives

$$p(\boldsymbol{B} \mid \boldsymbol{X}, \boldsymbol{Y}) \propto \frac{e^{-0.5\sum_{i=1}^{q}(\theta_i - \xi)'\boldsymbol{\Phi}^{-1}(\theta_i - \xi)}}{|\,\boldsymbol{\Phi}\,|^{0.5q}|\,(\boldsymbol{V} + \boldsymbol{H}) + (\boldsymbol{B} - \hat{\boldsymbol{B}})'\boldsymbol{S}(\boldsymbol{B} - \hat{\boldsymbol{B}})|^{0.5\delta}}$$

where $\delta = N + \upsilon - p - 1$.

Equivalently, since $\boldsymbol{\Phi}$ is assumed to be a known hyperparameter, we have

$$p(\boldsymbol{B} \mid \boldsymbol{X}, \boldsymbol{Y}) \propto \frac{e^{-0.5(\theta - \theta^*)\boldsymbol{I}_q \otimes \boldsymbol{\Phi}^{-1}(\theta - \theta^*)}}{|\,(\boldsymbol{V} + \boldsymbol{H}) + (\boldsymbol{B} - \hat{\boldsymbol{B}})'\boldsymbol{S}(\boldsymbol{B} - \hat{\boldsymbol{B}})\,|^{0.5\delta}}.$$

We have shown (Press, 1982, p. 255), where $\boldsymbol{B}$ plays the role of $\boldsymbol{B}'$ here, that in large samples, it is approximately true that

$$(\boldsymbol{\theta} \mid \boldsymbol{X}, \boldsymbol{Y}) \sim N(\bar{\boldsymbol{\theta}}, \boldsymbol{M}),$$

where

$$\bar{\boldsymbol{\theta}} = \boldsymbol{K}\boldsymbol{\theta}^* + (\boldsymbol{I} - \boldsymbol{K})\hat{\boldsymbol{\theta}},$$
$$\boldsymbol{K} = \boldsymbol{M}(\boldsymbol{I}_q \otimes \boldsymbol{\Phi}^{-1}),$$
$$\boldsymbol{M} = \left\{(\boldsymbol{I}_q \otimes \boldsymbol{\Phi}^{-1}) + (\boldsymbol{S} \otimes \hat{\boldsymbol{\Sigma}}^{-1})\right\}^{-1},$$
$$\hat{\boldsymbol{\Sigma}} = \frac{\boldsymbol{V} + \boldsymbol{H}}{N}.$$

Thus, in large samples, the posterior mean, $\bar{\boldsymbol{\theta}}$, is a matrix weighted average of the prior mean and the MLE (that is, a Stein-type estimator).

12.4.7 Balanced Design

Take $n_1 = n_2 = \cdots = n_q = n$. Then,

$$\boldsymbol{S} = \boldsymbol{X}'\boldsymbol{X} = n\boldsymbol{I}_q$$

and

$$\bar{\boldsymbol{\theta}} = \boldsymbol{K}\boldsymbol{\theta}^* + (\boldsymbol{I} - \boldsymbol{K})\bar{\boldsymbol{\theta}},$$

where

$$\boldsymbol{K} = \boldsymbol{I}_q \otimes \left\{\boldsymbol{I}_p + n\boldsymbol{\Phi}\hat{\boldsymbol{\Sigma}}^{-1}\right\}^{-1}.$$

So

$$\bar{\theta}_i = (\boldsymbol{I}_p + n\,\boldsymbol{\Phi}\hat{\boldsymbol{\Sigma}}^{-1})^{-1}\left\{\xi + n\,\boldsymbol{\Phi}\hat{\theta}^{-1}\hat{\theta}_i\right\}, \qquad i = 1, \ldots, q.$$

That is, the posterior mean of the ith group is a matrix weighted average of the prior mean ξ and the MLE in large samples. Also, for large n and $\boldsymbol{\Omega} \equiv [\boldsymbol{\Phi}^{-1} + n\hat{\boldsymbol{\Sigma}}^{-1}]^{-1}$, we have

$$(\boldsymbol{\theta}_i \mid \text{data}) \sim (\bar{\boldsymbol{\theta}}_i, \boldsymbol{\Omega}).$$

Case of $p = 1$

Take $\boldsymbol{\Phi} = \phi^2, \boldsymbol{\Sigma} = \sigma^2$. Then, in terms of precisions, $(\phi^2)^{-1}$ and $(\hat{\sigma}^2/n)^{-1}$, we have

$$\bar{\boldsymbol{\theta}}_i = \left[\frac{(\phi^2)^{-1}}{(\phi^2)^{-1} + (\hat{\sigma}^2/n)^{-1}}\right]\xi + \left[\frac{(\hat{\sigma}^2/n)^{-1}}{(\phi^2)^{-1} + (\hat{\sigma}^2/n)^{-1}}\right]\hat{\theta}_i.$$

Interval Estimation

Since in large samples $\boldsymbol{\theta}_i = \boldsymbol{\mu} + \boldsymbol{\alpha}_i, i = 1, \ldots, q$, is normally distributed, interval estimates of ξ or $\boldsymbol{\Phi}$ are known only imperfectly, we use matrix $\boldsymbol{T}$-distributions for interval estimation (that is, we use the marginals of matrix $\boldsymbol{T}$-distributions, which are multivariate t-distributions).

12.4.8 Example: Test Scores

Suppose $\boldsymbol{y}_{ij}$ denotes the bivariate test scores (verbal and quantitative) of the jth student in a sample of students in state i taking a national test (say $i = 1, 2, 3; j = 1, \ldots, 100; p = 2; q = 3$). We want to compare mean scores of students across three states.

Model

$$\underset{(2\times 1)}{\boldsymbol{y}_{ij}} = \boldsymbol{\mu} + \boldsymbol{\alpha}_i + \boldsymbol{\theta}_i + \boldsymbol{\mu}_{ij}, \qquad \text{with } \boldsymbol{\theta}_i \equiv \boldsymbol{\mu} + \boldsymbol{\alpha}_i,$$

and

$$\boldsymbol{u}_{ij} \sim N(\boldsymbol{0}, \boldsymbol{\Sigma}), \qquad \text{i.i.d.}, \boldsymbol{\Sigma} > 0.$$

We propose to carry out a one-way-classification fixed-effects MANOVA, from a Bayesian point of view. Define

$$\underset{(2\times 3)}{\boldsymbol{B}'} = (\boldsymbol{\theta}_1, \boldsymbol{\theta}_2, \boldsymbol{\theta}_3)$$

and assume $p(\boldsymbol{B}, \boldsymbol{\Sigma}) = p(\boldsymbol{B})p(\boldsymbol{\Sigma})$,

$$\boldsymbol{\Sigma} \sim \boldsymbol{W}^{-1}(\boldsymbol{H}, 2, v), \qquad \boldsymbol{H} = (h_{ij}) > 0.$$

Suppose we assess

$$\xi = \begin{pmatrix} 425 \\ 425 \end{pmatrix}, \qquad \Phi = \begin{pmatrix} 2500 & 1250 \\ 1250 & 2500 \end{pmatrix} = \begin{pmatrix} \sigma^2 & \sigma^2\rho \\ \sigma^2\rho & \sigma^2 \end{pmatrix}$$

($\sigma = 50, \rho = 0.5$); we also assess

$$E(\boldsymbol{\Sigma}) = \frac{\boldsymbol{H}}{v - 2p - 2} = \frac{\boldsymbol{H}}{v - 6} = \begin{pmatrix} 10{,}000 & 5000 \\ 5000 & 10{,}000 \end{pmatrix}.$$

Based upon informed guesses about the variances of the elements of $\boldsymbol{\Sigma}$, we take $v = 100$. Then,

$$\boldsymbol{H} = 10^4 \begin{pmatrix} 94 & 47 \\ 47 & 94 \end{pmatrix}.$$

We have assessed nine parameters.

Since the samples are large we have, approximately, the posterior distribution

$$(\boldsymbol{\theta} \mid \text{data}) \sim N(\bar{\boldsymbol{\theta}}, \boldsymbol{M}).$$

Suppose that the MLEs are given by

$$\hat{\boldsymbol{\theta}}_1 = \begin{pmatrix} 460 \\ 450 \end{pmatrix}, \qquad \hat{\boldsymbol{\theta}}_2 = \begin{pmatrix} 420 \\ 430 \end{pmatrix}, \qquad \hat{\boldsymbol{\theta}}_3 = \begin{pmatrix} 390 \\ 460 \end{pmatrix}$$

and that the sum-of-squares-of-residuals matrix is

$$\boldsymbol{V} = 10^5 \begin{pmatrix} 14.60 & 7.30 \\ 7.30 & 11.60 \end{pmatrix}.$$

Then,

$$\hat{\boldsymbol{\Sigma}} = \frac{\boldsymbol{V} + \boldsymbol{H}}{N} = 10^3 \begin{pmatrix} 8 & 4 \\ 4 & 7 \end{pmatrix},$$

where

$$N = \sum_1^3 n_i, \qquad n_i = 100, \qquad v = 100, \qquad q = 3.$$

Assume the design is balanced. Then, from the result in Section 12.4.7, we have

$$(\boldsymbol{\theta}_i \mid \text{data}) \sim N(\bar{\boldsymbol{\theta}}_i, \boldsymbol{\Omega}),$$

where

$$\bar{\theta}_1 = \begin{pmatrix} 459 \\ 449 \end{pmatrix}, \quad \bar{\theta}_2 = \begin{pmatrix} 420 \\ 430 \end{pmatrix}, \quad \bar{\theta}_3 = \begin{pmatrix} 391 \\ 459 \end{pmatrix}, \quad \mathbf{\Omega} = \begin{pmatrix} 78 & 39 \\ 39 & 68 \end{pmatrix}.$$

Note that each element of $\bar{\theta}_i$ lies between the corresponding elements of ξ and $\hat{\theta}_i$, $i = 1, 2, 3$ (we have deleted the final decimals).

Contrasts

$$(\theta_1 - \theta_2) \mid \text{data} \sim N(\bar{\theta}_1 - \bar{\theta}_2, 2\mathbf{\Omega}).$$

For example,

$$(\theta_1 - \theta_2) \mid \text{data} \sim N\left[\begin{pmatrix} 39 \\ 19 \end{pmatrix}, \begin{pmatrix} 156 & 78 \\ 78 & 136 \end{pmatrix}\right].$$

Thus,

$$P\left\{\begin{matrix}\text{Mean verbal score in state 1} \\ \text{exceeds mean verbal score in state 2}\end{matrix} \middle| \text{data}\right\} = 99.91 \text{ percent},$$

and we can compute any other credibility intervals as easily.

12.4.9 Posterior Distributions of Effects

Since

$$\theta_i = \mu + \alpha_i, \qquad i = 1, \ldots, q,$$

the μ and α_i effects are confounded, as is well known in MANOVA.

If we adopt the usual identifying constraint,

$$\frac{1}{q} \sum_1^q \alpha_i = 0,$$

we see that:

$$(1) \qquad \mu = \frac{1}{q} \sum_1^q \theta_i,$$

and

(2) $$\boldsymbol{\alpha}_i = \boldsymbol{\theta}_i - \frac{1}{q}\sum_1^q \boldsymbol{\theta}_i.$$

In the case of a balanced design (for example, in large samples), since we know that

$$(\boldsymbol{\theta}_i \mid \text{data}) \sim N(\bar{\boldsymbol{\theta}}_i, \boldsymbol{\Omega}),$$

the posterior distributions of the effects are therefore:

(1) $$(\boldsymbol{\mu} \mid \text{data}) \sim N\left(\frac{1}{q}\sum_1^q \bar{\boldsymbol{\theta}}_i, \frac{\boldsymbol{\Omega}}{q}\right)$$

and

(2) $$(\boldsymbol{\alpha}_i \mid \text{data}) \sim N\left(\bar{\boldsymbol{\theta}}_i, \boldsymbol{\Omega}\left(1 - \frac{1}{q}\right)\right).$$

Alternatively, we could adopt the assumption of exchangeability of the $\boldsymbol{\alpha}_i$s instead of using the constraints. That is, instead of adopting the normal prior on $\boldsymbol{\theta}_i$, as in Section 12.4.4, we could assume that $\boldsymbol{\mu}$ and $\boldsymbol{\alpha}_i$ in Section 12.4.1 are independent, *a priori*, and that each is normally distributed. Their distributions are then combined, and analysis proceeds as before. (See Exercise 12.15.)

12.5 BAYESIAN INFERENCE IN THE MULTIVARIATE MIXED MODEL[2]

In this section we consider the multivariate mixed model analysis of variance from the Bayesian inference point of view. The multivariate mixed model is a generalization of the usual fixed effects multivariate linear model.

12.5.1 Introduction

Mixed effects models are increasingly being adopted (see, for example, Hulting and Harville, 1991; Rajagopalan and Broemeling, 1983; Robinson, 1991). The mixed model MANOVA combines fixed and random effects MANOVA models. The random effects model was well explicated by Scheffé (1959) and Rao and Kleffe (1980). The model is useful for analysing experiments that arise, for example, in agriculture (see our numerical example later in this section). Robinson (1991) showed that linear relationships among the fixed and random effects are sometimes also of interest (see also Gelfand et al., 1990).

[2] This section is adapted from Jelenkowska and Press, 1997, with tthe permission of American Sciences Press, Inc.

The random effects model includes unobservable effects that are assumed to be random in the sampling theory approach. From a Bayesian point of view these unobservable effects are unknown and are accordingly assigned a (prior) distribution. It is traditional to assign a distribution with mean zero to the random effects. The error covariance matrices are of cardinal interest (as to whether they are zero). That is, are the putative random effects really fixed, or do they vary with the experiment performed?

From the Bayesian point of view, since all unknown quantities are given a prior distribution, the distribution of the so-called "random effects" is really part of the specification of the prior distributions for the unknowns. Moreover, we assign prior distributions to the error covariance matrices so that we can study the marginal posterior distributions of the error covariance matrices (we will also obtain their posterior means), as well as the posterior distributions of both the fixed and the random effects.

We consider cases of both exchangeable and nonexchangeable prior distributions for various model parameters. We will develop posterior distributions for the unknown quantities, and show that the marginal posterior distribution of the random effects is multiple matrix T. We will then find a suitable normal approximation to the multiple matrix T-distribution. The approximation generalizes to the matrix case the approximation adopted by Rajagopalan and Broemeling (1983) for the multiple multivariate t-distribution, wherein multivariate Student t kernels were approximated by normal kernels. Rajagopalan and Broemeling found (1983, p. 712) that, "generally, the approximations are fairly close to the true distributions." They also found that even in cases where the approximation was not strong, the posterior means for the variance components were well approximated. We will also provide approximate posterior means for the error covariance matrices. The mixed model will be illustrated with an agricultural example.

12.5.2 Model

The multivariate mixed linear model with $(c + 1)$ unknown covariance matrices is given by

$$\boldsymbol{Y} = \boldsymbol{X\theta} + \sum_{i=1}^{c} \boldsymbol{U}_i \boldsymbol{B}_i + \boldsymbol{E} = \boldsymbol{X\theta} + \boldsymbol{UB} + \boldsymbol{E}, \tag{12.1}$$

where

- $\boldsymbol{Y}$ is an $(n \times p)$ matrix of n observations on the p dependent variables,
- $\boldsymbol{X}$ is an $(n \times q)$ known design matrix of rank q,
- $\boldsymbol{\theta}$ is a $(q \times p)$ matrix of unobservable fixed effects (possibly including covariates),
- $\boldsymbol{U}_i$ is an $(n \times m_i)$ known design matrix $i, i = 1, \ldots, c$,
- $\boldsymbol{B}_i$ is an $(m_i \times p)$ matrix of unobservable random effects, $i = 1, \ldots, c, c$ is the number of multivariate random effects,

$\boldsymbol{E}$ is an $(n \times p)$ matrix of error vectors.

These matrices may be written in the partitioned forms:

$$\underset{n\times p}{\boldsymbol{Y}} = (\boldsymbol{y}_1, \ldots, \boldsymbol{y}_n)', \qquad \underset{n\times p}{\boldsymbol{E}} = (\boldsymbol{e}_1, \ldots, \boldsymbol{e}_n)'$$

$$\underset{m\times p}{\boldsymbol{B}} = (\boldsymbol{B}_1', \ldots, \boldsymbol{B}_c'), \qquad \underset{n\times m}{\boldsymbol{U}} = (\boldsymbol{U}_1, \ldots, \boldsymbol{U}_c),$$

where $\boldsymbol{y}_j'$ denotes the jth row of $\boldsymbol{Y}, j = 1, \ldots, n$, and prime denotes transpose. Also, for the matrices $\boldsymbol{E}$, $\boldsymbol{\theta}$, and $\boldsymbol{B}$, the vectors $\boldsymbol{e}_j'$, $\boldsymbol{\theta}_k'$ denote the jth, and kth row of matrix $\boldsymbol{E}$ and $\boldsymbol{\theta}$, respectively, $j = 1, \ldots, n$, $k = 1, \ldots, q$, and $\boldsymbol{B}_i'$ denotes the ith block matrix of $\boldsymbol{B}$. Also, $m \equiv \sum_1^c m_i$, where m_i denotes the number of treatment levels for random effect i.

We assume that:

1. $E(\boldsymbol{e}_j) = 0$, $\mathrm{var}(\boldsymbol{e}_j) = \mathrm{cov}(\boldsymbol{e}_j, \boldsymbol{e}_j') = \Sigma$ for all j. Σ is assumed to be an arbitrary, positive definite, symmetric matrix of order p. It is one of the variance components.
2. The $\boldsymbol{e}_j$ vectors are mutually independent. Then $\mathrm{cov}(\boldsymbol{e}_j, \boldsymbol{e}_{j'}) = 0$ for $j \neq j'$; that is, the vectors $\boldsymbol{e}_j$ are mutually uncorrelated.
3. $e_j \sim N_p(0, \Sigma)$ for all $j = 1, \ldots, n$; that is the vector $\boldsymbol{e}_j$ is normal with mean zero and covariance matrix Σ.
4. The rows of each $\boldsymbol{B}_i$ are assumed to be independent and $N_p(0, \Sigma_i)$. The Σ_is are the remaining (multivariate) variance components. We rewrite the assumption in matrix form as $\mathrm{vec}(\boldsymbol{B}_i) \sim N_p(0, \boldsymbol{I}_{m_i} \otimes \boldsymbol{\Sigma}_i)$, $i = 1, \ldots, c$, so that each matrix $\boldsymbol{B}_i$: $(m_i \times p)$ of $\boldsymbol{B}$: $(m \times p)$ has a p-dimensional normal distribution with mean matrix zero and covariance matrix $\boldsymbol{I}_{m_i} \otimes \boldsymbol{\Sigma}_i$; each Σ_i is a positive definite symmetric matrix of order p. While this assumption really involves prior information, we include it here because, traditionally, it is part of the mixed model. It should be noted that the $\boldsymbol{e}_j$ are independent normal random vectors, and that the probability density of the error matrix is

$$p(\boldsymbol{E} \mid \boldsymbol{\Sigma}) = p(\boldsymbol{e}_1, \ldots, \boldsymbol{e}_n \mid \Sigma) = \prod_{j=1}^{n} p(\boldsymbol{e}_j \mid \Sigma).$$

Since

$$p(\boldsymbol{e}_j \mid \Sigma) \propto \mid \Sigma \mid^{0.5} \exp\{-0.5(\boldsymbol{e}_j'\Sigma^{-1}\boldsymbol{e}_j)\},$$

$$p(E \mid \Sigma) \propto \ \Sigma \mid^{-0.5n} \exp\left\{-0.5 \sum_{j=1}^{n}(\boldsymbol{e}_j'\Sigma^{-1}\boldsymbol{e}_j)\right\},$$

where $\propto$ denotes proportionality. Since $\boldsymbol{E}'\boldsymbol{E} = \sum_{j=1}^{n} \boldsymbol{e}_j\boldsymbol{e}_j'$, the likelihood becomes

$$p(\boldsymbol{E} \mid \boldsymbol{\Sigma}) \propto \mid \boldsymbol{\Sigma} \mid^{-0.5n} \exp\{-0.5 \,\mathrm{tr}\, \boldsymbol{E}'\boldsymbol{E}\,\boldsymbol{E}^{-1}\},$$

where tr(·) denotes the trace operation. Now change variables to find the density of Y. Since $\boldsymbol{E} = \boldsymbol{Y} - \boldsymbol{X\theta} - \boldsymbol{UB}$, the Jacobian of the transformation is unity. The likelihood function becomes

$$p(\boldsymbol{Y} \mid \boldsymbol{\theta}, \boldsymbol{B}, \boldsymbol{\Sigma}) \propto |\, \boldsymbol{\Sigma}\, |^{-0.5n} \exp\left\{-0.5\,\mathrm{tr}(\boldsymbol{Y} - \boldsymbol{X\theta} - \boldsymbol{U}\,\boldsymbol{B})'(\boldsymbol{Y} - \boldsymbol{X\theta} - \boldsymbol{UB})\boldsymbol{\Sigma}^{-1}\right\}.$$

It can be found that the maximum likelihood estimators (MLEs) of matrix $\boldsymbol{B}$ and matrix $\boldsymbol{\theta}$ have the form $\hat{\boldsymbol{\beta}} = (\boldsymbol{U}'\boldsymbol{RU})^{-}\boldsymbol{U}'\boldsymbol{RY}$, and $\hat{\boldsymbol{\theta}} = (\boldsymbol{X}'\boldsymbol{X})^{-1}\boldsymbol{X}'(\boldsymbol{Y} - \boldsymbol{U}\hat{\boldsymbol{B}})$, where $\boldsymbol{R} - \boldsymbol{I}_n - \boldsymbol{X}(\boldsymbol{X}'\boldsymbol{X})^{-1}\boldsymbol{X}'$, and where $(\boldsymbol{U}'\boldsymbol{RU})^{-}$ is the Moore–Penrose generalized inverse of $\boldsymbol{U}'\boldsymbol{RU}$. (If the rank of any one of the $\boldsymbol{U}_i$s is less than full, $(\boldsymbol{U}'\boldsymbol{RU})$ will be singular.) Thus, we can write

$$\begin{aligned}(\boldsymbol{Y} - \boldsymbol{X\theta} - \boldsymbol{UB})'(\boldsymbol{Y} - \boldsymbol{X\theta} - \boldsymbol{UB}) &= \boldsymbol{Y}'\boldsymbol{RY} - \hat{\boldsymbol{B}}\boldsymbol{U}'\boldsymbol{RU}\hat{\boldsymbol{R}} + (\boldsymbol{B} - \hat{\boldsymbol{B}})'\boldsymbol{U}'\boldsymbol{RU}(\boldsymbol{B} - \hat{\boldsymbol{B}}) \\ &\quad + (\boldsymbol{\theta} - \hat{\boldsymbol{\theta}})'\boldsymbol{X}'\boldsymbol{X}(\boldsymbol{\theta} - \hat{\boldsymbol{\theta}}). \qquad (12.2)\end{aligned}$$

The likelihood function may be written in terms of the MLEs as

$$\begin{aligned}p(\boldsymbol{Y} \mid \boldsymbol{\theta}, \boldsymbol{B}, \boldsymbol{\Sigma}) \propto |\, \boldsymbol{\Sigma}\, |^{-0.5n} \exp\Big\{&-0.5\,\mathrm{tr}[\boldsymbol{Y}'\boldsymbol{RY} - \hat{\boldsymbol{B}}'\boldsymbol{U}'\boldsymbol{RU}\hat{\boldsymbol{B}} \\ &+ (\boldsymbol{B} - \hat{\boldsymbol{B}}'\boldsymbol{U}'\boldsymbol{RU}(\boldsymbol{B} - \hat{\boldsymbol{B}}) + (\boldsymbol{\theta} - \hat{\boldsymbol{\theta}})'\boldsymbol{X}'\boldsymbol{X}(\boldsymbol{\theta} - \hat{\boldsymbol{\theta}})]\boldsymbol{\Sigma}^{-1}\Big\}. \qquad (12.3)\end{aligned}$$

Unknown parameters of the model are $\boldsymbol{\theta}, \boldsymbol{B}, \boldsymbol{\Sigma}, \boldsymbol{\Sigma}_1, \ldots, \boldsymbol{\Sigma}_c$. Thus, we have two levels of parameters in the mixed linear model. The first of them contains the matrix of fixed $\boldsymbol{\theta}$, the matrix of random effects $\boldsymbol{B}$, and the covariance matrix of errors vectors, $\boldsymbol{\Sigma}$. The second level of parameters of interest includes the covariance matrices associated with the random effects, $\boldsymbol{\Sigma}_1, \ldots, \boldsymbol{\Sigma}_c$. The $\boldsymbol{\Sigma}_j$s are the main parameters of interest in the mixed linear model but their estimation in closed form is very complicated.

12.5.3 Prior Information

We will introduce two-stage prior information. First, we assume that $\boldsymbol{B}_i$, $\boldsymbol{\theta}$, and $\boldsymbol{\Sigma}$ are independent *a priori* and the condition of $\boldsymbol{B}_i$ given $\boldsymbol{\Sigma}_i$ is normal, so that

$$p(\boldsymbol{B}_i \mid \boldsymbol{\Sigma}_i) \propto |\, \boldsymbol{\Sigma}_i\, |^{-0.5m_i} \exp\left\{-0.5\,\mathrm{tr}(\boldsymbol{B}_i'\boldsymbol{B}_i\boldsymbol{\Sigma}_i^{-1})\right\}, \qquad i = 1, \ldots, c. \qquad (12.4)$$

A vague Jeffreys' prior density function will be used for the fixed effects; that is, $p(\boldsymbol{\theta}) \propto$ constant. The Jeffreys' prior is chosen so that the data may "speak for themselves," as described by Box and Tiao (1973). Further, we assume that the precision matrix of the errors vectors, $\boldsymbol{\Sigma}^{-1}$, has a p-dimensional Wishart distribution

$$p(\boldsymbol{\Sigma}^{-1} \mid \boldsymbol{V}, d) \propto |\boldsymbol{\Sigma}^{-1}|^{0.5(d-p-1)} \exp\left\{-0.5\,\mathrm{tr}(\boldsymbol{V}\boldsymbol{\Sigma}^{-1})\right\}, \qquad (12.5)$$

where $\boldsymbol{V} > 0, d > p$.

We next include prior information for the scale matrices associated with the random effects.

A. Nonexchangeable Case

Assume the $\sum_i$ s are independent, with

$$p(\boldsymbol{\Sigma}_i^{-1} \mid \boldsymbol{V}_i, \delta_i) \propto |\boldsymbol{\Sigma}_i^{-1}|^{0.5(\delta_i-p-1)} \exp\left\{-0.5\,\mathrm{tr}(\boldsymbol{V}_i\boldsymbol{\Sigma}_i^{-1})\right\}, \qquad i = 1, \ldots, c. \qquad (12.6)$$

Note that $(\boldsymbol{V}, \boldsymbol{V}_i, d, \delta_i)$ are hyperparameters of the prior distribution that must be assessed for the decision maker on the basis of his specific prior information. The assessment problem is not simple and must be carried out carefully. There are c pairs $(\boldsymbol{V}_i, \delta_i)$ that must be assessed, as well as $(\boldsymbol{V}, d)$. We now go forward in the analysis assuming $(\boldsymbol{V}, \boldsymbol{V}_i, d, \delta_i)$ are known for a given decision maker; $i = 1, \ldots, c$.

Thus, the joint prior density function of the parameters $\boldsymbol{\theta}, \boldsymbol{B}, \boldsymbol{\Sigma}^{-1}$ is

$$p(\boldsymbol{\theta}, \boldsymbol{B}, \boldsymbol{\Sigma}^{-1} \mid \boldsymbol{\Sigma}_1^{-1}, \ldots, \boldsymbol{\Sigma}_c^{-1}) \propto |\boldsymbol{\Sigma}^{-1}|^{0.5(d-p-1)} \prod_{i=1}^{c} |\boldsymbol{\Sigma}_i|^{-0.5m_i} \cdot \exp\left\{-0.5\,\mathrm{tr}\left(\sum_{i=1}^{c} \boldsymbol{B}_i'\boldsymbol{B}_i\boldsymbol{\Sigma}_i^{-1} + \boldsymbol{V}\boldsymbol{\Sigma}^{-1}\right)\right\} \qquad (12.7)$$

and the prior density function of the second level parameters is

$$p(\boldsymbol{\Sigma}_1^{-1}, \ldots, \boldsymbol{\Sigma}_c^{-1}) \propto \prod_{i=1}^{c} |\boldsymbol{\Sigma}_i^{-1}|^{0.5(\delta_i-p-1)} \exp\left\{-0.5\,\mathrm{tr}\left(\sum_{i=1}^{c} \boldsymbol{V}_i\boldsymbol{\Sigma}_i^{-1}\right)\right\}. \qquad (12.8)$$

B. Exchangeable Cases

In some problems it is reasonable to assume that certain parameters have (finitely) exchangeable distributions. There are several case combinations we might consider in this context

1. Case 1—the columns of $\boldsymbol{\theta}$ are exchangeable;
2. Case 2—the $\boldsymbol{B}_i$ matrices have an exchangeable distribution;
3. Case 3—the $\boldsymbol{\Sigma}_i^{-1}$s are exchangeable;
4. Case 1 and Case 2 hold;
5. Case 1 and Case 3 hold;
6. Case 2 and Case 3 hold;
7. Cases 1, 2, and 3 all hold.

When any of these cases is a reasonable expression of minimal information about the underlying parameters, helpful reductions in the number of parameters that must be assessed in the problem result.

For example, suppose Case 3 is applicable. Then, we might assume that for all $i = 1, \ldots, c$,

$$p(\boldsymbol{\Sigma}_i^{-1} \mid \boldsymbol{\Omega}, \delta) \propto |\, \boldsymbol{\Sigma}_i^{-1} \,|^{0.5(\delta-p-1)} \exp\left\{-0.5\, \mathrm{tr}(\boldsymbol{\Omega}\boldsymbol{\Sigma}_i^{-1})\right\}, \tag{12.9}$$

for $\boldsymbol{\Omega} > 0$, $\delta > p$. Here, we have assumed that instead of the δ_is, there is a common δ to be assessed, similarly, a common $\boldsymbol{\Omega}$ instead of the $\boldsymbol{V}_i$s. Now, instead of having to assess the parameters $(\boldsymbol{V}_i, \delta_i)$, $i = 1, \ldots, c$ in Equation (12.6), we need only assess $(\boldsymbol{\Omega}, \delta)$. Since there are $c\{0.5[(p)(p+1)] + 1\}$ in Equation (12.6), and only $\{0.5[p(p+1) + 1]$ in Equation (12.9), there is a respectable reduction in effort required.

For example, suppose $p = 4$, and $c = 3$. Then there are 33 hyperparameters to assess in Equation (12.6), and only 11 in Equation (12.9).

If all three types of exchangeability are applicable (Case 7), additional substantial reductions in assessment effort will result.

We now go forward assuming no exchangeability at all; the exchangeable cases really reduce to special cases of no exchangeability.

12.5.4 Posterior Distributions

Using Equations (12.3), (12.7) and (12.8) and Bayes' theorem the joint posterior density of $\boldsymbol{\theta}, \boldsymbol{B}, \boldsymbol{\Sigma}^{-1}, \boldsymbol{\Sigma}_1^{-1}, \ldots, \boldsymbol{\Sigma}_c^{-1}$ is obtained as

$$\begin{aligned}
p(\boldsymbol{\theta}, \boldsymbol{B}, \boldsymbol{\Sigma}_1^{-1}, \boldsymbol{\Sigma}_2^{-1}, \ldots, \boldsymbol{\Sigma}_c^{-1}, \boldsymbol{\Sigma}^{-1} \mid \boldsymbol{y}) \\
&\propto |\, \boldsymbol{\Sigma}^{-1} \,|^{0.5(n+d-p-1)} \\
&\cdot \{-0.5\, \mathrm{tr}[\boldsymbol{A} + (\boldsymbol{B} - \hat{\boldsymbol{B}})'\boldsymbol{U}'\boldsymbol{R}\boldsymbol{U}'(\boldsymbol{B} - \hat{\boldsymbol{B}}) \\
&+ (\boldsymbol{\theta} - \hat{\boldsymbol{\theta}})'\boldsymbol{X}'\boldsymbol{X}(\boldsymbol{\theta} - \hat{\boldsymbol{\theta}})]\boldsymbol{\Sigma}^{-1}\} \\
&\cdot \prod_{i=1}^{c} |\, \boldsymbol{\Sigma}_i^{-1} \,|^{0.5(m_i+\delta_i-p-1)} \exp\left\{-0.5 \sum_{i=1}^{c} [\mathrm{tr}(\boldsymbol{V}_i + \boldsymbol{B}_i'\boldsymbol{B}_i)\boldsymbol{\Sigma}_i^{-1}]\right\},
\end{aligned} \tag{12.10}$$

where $\boldsymbol{A} = \boldsymbol{V} + \boldsymbol{Y}'\boldsymbol{R}\boldsymbol{Y} - \hat{\boldsymbol{B}}'\boldsymbol{U}'\boldsymbol{R}\boldsymbol{U}\hat{\boldsymbol{B}}$.

Integrating Equation (12.10) with respect to $\boldsymbol{\Sigma}^{-1}, \boldsymbol{\Sigma}_1^{-1}, \ldots, \boldsymbol{\Sigma}_c^{-1}$ the joint density of $(\boldsymbol{\theta}, \boldsymbol{B} \mid \boldsymbol{Y})$ is given by

$$\begin{aligned}
p(\boldsymbol{\theta}, \boldsymbol{B} \mid \boldsymbol{Y}) \propto |\, \boldsymbol{A} + (\boldsymbol{B} - \hat{\boldsymbol{B}})'\boldsymbol{U}'\boldsymbol{R}\boldsymbol{U}(\boldsymbol{B} - \hat{\boldsymbol{B}}) + (\boldsymbol{\theta} - \hat{\boldsymbol{\theta}})'\boldsymbol{X}'\boldsymbol{X}(\boldsymbol{\theta} - \hat{\boldsymbol{\theta}}) \,|^{-0.5(n+d)} \\
\cdot \prod_{i=1}^{c} |\, \boldsymbol{V}_i + \boldsymbol{B}_i'\boldsymbol{B}_i \,|^{-0.5(m_i+\delta_i)}.
\end{aligned} \tag{12.11}$$

This proves the following.

Theorem 12.1

(a) The posterior distribution of $\boldsymbol{\theta}$, conditional on $\boldsymbol{B}$, is a matrix $\boldsymbol{T}$-distribution with degrees of freedom $v = (n + d - p)$, matrix scale parameters $\hat{\boldsymbol{\theta}}$, $(\boldsymbol{X}'\boldsymbol{X})^{-1}$ and location matrix $\boldsymbol{T}^*$ (for notation, see for example, Press, 1982, p. 138), where:

$$\boldsymbol{T}^* = [\boldsymbol{A} + (\boldsymbol{B} - \hat{\boldsymbol{B}})'\boldsymbol{U}'\boldsymbol{R}\boldsymbol{U}(\boldsymbol{B} - \hat{\boldsymbol{B}});$$

(b) The joint posterior density of $(\boldsymbol{\Sigma}^{-1}, \boldsymbol{\Sigma}_1^{-1}, \ldots, \boldsymbol{\Sigma}_c^{-1}, \boldsymbol{B} \mid \boldsymbol{Y})$ is given by:

$$p(\boldsymbol{\Sigma}^{-1}, \boldsymbol{\Sigma}_1^{-1}, \ldots, \boldsymbol{\Sigma}_c^{-1}, \boldsymbol{B} \mid \boldsymbol{Y}) \propto \mid \boldsymbol{\Sigma}^{-1} \mid^{0.5(n+d-p-q-1)} \cdot \exp\left\{-0.5 \operatorname{tr} \boldsymbol{T}^*\boldsymbol{\Sigma}^{-1}\right\} \prod_{i=1}^{c} \boldsymbol{\Sigma}_i^{-1} \mid^{0.5(m_i+\delta_i-p-1)} \cdot \exp\left\{-0.5 \sum_{i=1}^{c} \operatorname{tr}(\boldsymbol{V}_i + \boldsymbol{B}_i'\boldsymbol{B}_i)\boldsymbol{\Sigma}_i^{-1}\right\}, \qquad (12.12)$$

so that given $(\boldsymbol{B}, \boldsymbol{Y})$, $(\boldsymbol{\Sigma}, \boldsymbol{\Sigma}_1, \ldots, \boldsymbol{\Sigma}_c)$ are mutually independent.

This result was obtained by integrating Equation (12.10) with respect to $\boldsymbol{\theta}$. Since $\boldsymbol{\Sigma}^{-1}$ follows a Wishart distribution, and so do the $\boldsymbol{\Sigma}_i^{-1}$s, we obtain the following theorem.

Theorem 12.2

(a) The posterior distribution of $\boldsymbol{\Sigma}$ conditional on $(\boldsymbol{B}, \boldsymbol{Y})$, has an inverted Wishart distribution with $\alpha = n + d - q + p + 1$ degrees of freedom and scale matrix $\boldsymbol{T}^*$.

(b) The posterior distribution of $\boldsymbol{\Sigma}_i$, conditional on $(\boldsymbol{B}_i, \boldsymbol{Y})$, has an inverted Wishart distribution with $\alpha_i = m_i + \delta_i + p + 1$ degrees of freedom and scale matrix $\boldsymbol{T}_i^* = \boldsymbol{V}_i + \boldsymbol{B}_i'\boldsymbol{B}_i$.

Now, to obtain the marginal posterior density of $\boldsymbol{B}$, one needs to integrate Equation (12.12) with respect to $\boldsymbol{\Sigma}^{-1}, \boldsymbol{\Sigma}_1^{-1}, \ldots, \boldsymbol{\Sigma}_c^{-1}$. This gives

$$p(\boldsymbol{B} \mid \boldsymbol{Y}) \propto \mid \boldsymbol{A} + (\boldsymbol{B} - \hat{\boldsymbol{B}})'\boldsymbol{U}'\boldsymbol{R}\boldsymbol{U}(\boldsymbol{B} - \hat{\boldsymbol{B}}) \mid^{-0.5(n+d-q)} \prod_{i=1}^{c} \mid \boldsymbol{V}_i + \boldsymbol{B}_i'\boldsymbol{B}_i \mid^{-0.5(m_i+\delta_i)}. \qquad (12.13)$$

That is, $(\boldsymbol{B} \mid \boldsymbol{Y})$ follows a multiple matrix $\boldsymbol{T}$-distribution. This distribution is a generalization of the matrix $\boldsymbol{T}$-distribution of Kshirsagar (1960); see also Dickey, 1967a, 1967b, 1968; Tiao and Zellner, 1964. This leads to the following theorem.

Theorem 12.3 The marginal posterior density of $\boldsymbol{B}$ is the multiple matrix $\boldsymbol{T}$-density, given by Equation (12.13).

For the usual Bayesian inferences on the variance components, that is, the covariance matrices $\boldsymbol{\Sigma}, \boldsymbol{\Sigma}_1, \ldots, \boldsymbol{\Sigma}_c$, we need their marginal posterior distributions. Their joint posterior density can be obtained from Equation (12.12) by integration with respect to $\boldsymbol{B}$. After this integration and simplification we find the following theorem.

Theorem 12.4 The joint posterior distribution of $(\boldsymbol{\Sigma}^{-1}, \boldsymbol{\Sigma}_1^{-1}, \ldots, \boldsymbol{\Sigma}_c^{-1} \mid Y)$ is given by

$$
\begin{aligned}
p(\boldsymbol{\Sigma}^{-1}, \boldsymbol{\Sigma}_1^{-1}, \ldots, \boldsymbol{\Sigma}_c^{-1} \mid Y) \propto \mid \boldsymbol{\Sigma}^{-1} \mid^{0.5(n+d-p-q)-1} \\
\cdot \left\{-0.5 \operatorname{tr}(\boldsymbol{A}\boldsymbol{\Sigma}^{-1})\right\} \prod_{i=1}^{c} \mid \boldsymbol{\Sigma}_i^{-1} \mid^{0.5(m_i+\delta_i-p)-1} \\
\cdot \exp\left\{-0.5 \operatorname{tr}(\boldsymbol{V}_i\boldsymbol{\Sigma}_i^{-1}) \mid \boldsymbol{A}_1 + \boldsymbol{A}_2 \mid^{-0.5}\right\} \\
\cdot \exp\left\{-0.5 \operatorname{tr} \hat{\boldsymbol{B}}'\boldsymbol{A}_1'[\boldsymbol{A}_1^{-1} - (\boldsymbol{A}_1 + \boldsymbol{A}_2)^{-1}\boldsymbol{A}_1\hat{\boldsymbol{\beta}}]\right\}, \quad (12.14)
\end{aligned}
$$

where $\boldsymbol{A}_1 = \boldsymbol{\Sigma}^{-1} \otimes \boldsymbol{U}'\boldsymbol{R}\boldsymbol{U}$, and $\boldsymbol{A}_2 = \operatorname{diag}(I_m \otimes \boldsymbol{\Sigma}_1^{-1}, \ldots, I_{m_c} \otimes \boldsymbol{\Sigma}_c^{-1})$.

It is clear that the joint posterior density of $\boldsymbol{\Sigma}^{-1}, \boldsymbol{\Sigma}_1^{-1}, \ldots, \boldsymbol{\Sigma}_c^{-1}$ is analytically intractable, although it could be evaluated numerically. What we would really like is the separate marginal posterior distribution of these quantities. An alternative is to look at the conditional distributions of these covariance matrices using the marginal posterior distributions of $\boldsymbol{B}$. However, here again, there is a problem because the marginal posterior distribution of $\boldsymbol{B}$ is multiple matrix $\boldsymbol{T}$, which is not very tractable (but see Dreze, 1977; Press and Davis, 1987). Alternatively, in the next section we develop a normal approximation to the marginal posterior distribution of $\boldsymbol{B}$. Then we use the normal distribution approximation to approximate the posterior means of both $\boldsymbol{\Sigma}$ and the $\boldsymbol{\Sigma}_i$s.

12.5.5 Approximation to the Posterior Distribution of B

Now we are concerned with inferences about the random effects, that is, the $\boldsymbol{B}_i$s or $\boldsymbol{B}$. We have already found the marginal posterior density of $\boldsymbol{B}$ (Theorem 12.3, Equation (12.13)) as a multiple matrix $\boldsymbol{T}$-distribution. But such a distribution is very complicated. Its proportionality constant cannot even be expressed in closed form (except in very special cases). Below we develop a normal distribution approximation to the distribution of $\boldsymbol{B}$.

We begin with a theorem on the first two moments of a matrix $\boldsymbol{T}$-distribution.

Theorem 12.5 Suppose $\boldsymbol{T} : (p \times q)$ denotes a random matrix with matrix $\boldsymbol{T}$-density function

$$
p(\boldsymbol{T}) \propto \frac{1}{\mid \boldsymbol{Q} + (\boldsymbol{T} - \boldsymbol{T}_0)'\boldsymbol{P}(\boldsymbol{T} - \boldsymbol{T}_0)^{0.5\tilde{m}}},
$$

for $\boldsymbol{P}$: $(p \times p)$, $\boldsymbol{P} > 0$, $\boldsymbol{Q}$: $(q \times q)$, $\boldsymbol{Q} > 0$, and $\boldsymbol{T}_0$: $(p \times q)$; $\tilde{m} > p + q - 1$. Then

1. $E(\boldsymbol{T}) = \boldsymbol{T}_0$;
2. $\operatorname{var}(\boldsymbol{T}) = [(\tilde{m} - q)/(\tilde{m} - q - p - 1)](\boldsymbol{P}^{-1} \otimes \boldsymbol{Q})$, where $\operatorname{var}(\boldsymbol{T}) = \operatorname{var}[\operatorname{vec}(\boldsymbol{T})]$.

Proof
See Marx and Nel, 1985.

Next we generalize and adopt an approximation used previously for the case of a vector multiple $\boldsymbol{t}$-distribution. Rajagopalan and Broemeling (1983, p. 708) replaced a multivariate Student $\boldsymbol{t}$ kernel by a multivariate normal kernel corresponding to a multivariate normal distribution having the same first two moments, and they did this for each kernel in a multiple $\boldsymbol{T}$-distribution.

First consider the product of the c kernels of matrix $\boldsymbol{t}$-distributions in Equation (12.13). Using Theorem 12.5, the kernels in the product in Equation (12.13) are replaced by

$$\exp\left\{-0.5\boldsymbol{\beta}^{(i)'}\tilde{\boldsymbol{\Delta}}_i^{-1}\boldsymbol{\beta}^{(i)}\right\},$$

where

$$\tilde{\boldsymbol{\Delta}}_i \equiv \boldsymbol{B}_i = \left(\frac{m_i + \delta_i - p}{\delta_i - p - 1}\right)(\boldsymbol{I}_{m_i} \otimes \boldsymbol{V}_i)$$

and

$$\boldsymbol{\beta}^{(i)} \equiv \mathrm{vec}(\boldsymbol{B}_i); \qquad i = 1, \ldots, c.$$

Defining

$$\boldsymbol{\beta} \equiv \begin{pmatrix} \boldsymbol{\beta}^{(1)} \\ \vdots \\ \boldsymbol{\beta}^{(c)} \end{pmatrix}, \qquad \tilde{\boldsymbol{\Delta}} = \begin{pmatrix} \tilde{\boldsymbol{\Delta}}_1 & & 0 \\ & \ddots & \\ 0 & & \tilde{\boldsymbol{\Delta}}_c \end{pmatrix},$$

we can write the approximation as

$$\prod_{i=1}^{c} |\, \boldsymbol{V}_i + \boldsymbol{B}_i\boldsymbol{B}_i \,|^{-0.5(m_i+\delta_i)} \sim \exp\left\{-0.5\sum_{i=1}^{c}\boldsymbol{\beta}^{(i)'}\tilde{\boldsymbol{\Delta}}_i^{-1}\boldsymbol{\beta}^{(i)}\right\} = \exp\left\{-0.5\boldsymbol{\beta}'\tilde{\boldsymbol{\Delta}}^{-1}\boldsymbol{\beta}\right\}. \tag{12.15}$$

Next consider the matrix Student $\boldsymbol{T}$ kernel in Equation (12.13) centered at $\hat{\boldsymbol{B}}$. Using Theorem 12.5, we replace the matrix $\boldsymbol{T}$ kernel by a normal kernel with the same mean vector, and the same covariance matrix, to obtain $\exp\{-0.5(\boldsymbol{\beta} - \hat{\boldsymbol{\beta}})'\boldsymbol{\Delta}^{-1}(\boldsymbol{\beta} - \hat{\boldsymbol{\beta}})\}$, where

$$\boldsymbol{\Delta} \equiv \left(\frac{n + d - q - p}{n + d - q - p - m - 1}\right)(\boldsymbol{U}'\boldsymbol{R}\boldsymbol{U})^{-} \otimes \boldsymbol{A},$$

and $\hat{\boldsymbol{\beta}}$ is partitioned as $\boldsymbol{\beta}$. Equation (12.13) now becomes, approximately,

$$p(\boldsymbol{\beta} \mid \boldsymbol{Y}) \propto \exp\left\{-0.5(\boldsymbol{\beta} - \hat{\boldsymbol{\beta}})'\boldsymbol{\Delta}^{-1}(\boldsymbol{\beta} - \hat{\boldsymbol{\beta}})\right\} \cdot \exp\left\{-0.5(\boldsymbol{\beta}'\tilde{\boldsymbol{\Delta}}^{-1}\boldsymbol{\beta})\right\}. \tag{12.16}$$

Now combine the two quadratic forms in the exponents in Equation (12.16) by completing the square, to obtain the final normal approximation:

$$(\boldsymbol{\beta} \mid \boldsymbol{Y}) \sim N(\tilde{\boldsymbol{\beta}}, \tilde{\boldsymbol{A}}), \tag{12.17}$$

where $\tilde{\boldsymbol{A}}^{-1} \equiv \boldsymbol{\Delta}^{-1} - \tilde{\boldsymbol{\Delta}}^{-1}, \tilde{\boldsymbol{\beta}} = \tilde{\boldsymbol{A}}\boldsymbol{\Delta}^{-1}\hat{\boldsymbol{\beta}}$. The result is summarized in the following theorem.

Theorem 12.6 The multiple matrix $\boldsymbol{T}$ posterior density for the random effects in a multivariate mixed linear model, as given in Equation (12.13), is given approximately by the normal distribution in Equation (12.16).

REMARK: Conditions under which the approximation is good were studied by Rajagopalan and Broemeling (1983). They concluded that the approximation is best when the degrees of freedom in the matrix $\boldsymbol{T}$ kernels are large. In Equation (12.13), this condition corresponds to moderate sample size and moderate degrees of freedom for the prior information.

12.5.6 Posterior Means for $\boldsymbol{\Sigma}, \boldsymbol{\Sigma}_1, \ldots, \boldsymbol{\Sigma}_c$

Recall that $\boldsymbol{\Sigma}$ and the $\boldsymbol{\Sigma}_i$s are often the quantities of greatest interest in the mixed linear model. The exact posterior distribution of $(\boldsymbol{\Sigma}, \boldsymbol{\Sigma}_1, \ldots, \boldsymbol{\Sigma}_c)$ is given in Equation (12.14), but is quite complicated. The marginal posteriors for $\boldsymbol{\Sigma}$ and the $\boldsymbol{\Sigma}_i$s are unknown. Short of evaluating Equation (12.14) numerically, we might, on some occasions, settle for the approximate marginal posterior means of $\boldsymbol{\Sigma}$ and the $\boldsymbol{\Sigma}_i$s. These approximate posterior means are derived below.

Approximate posterior means of $\boldsymbol{\Sigma}, \boldsymbol{\Sigma}_1, \ldots, \boldsymbol{\Sigma}_c$ can be computed by noting from Theorem 12.2 that $\boldsymbol{\Sigma}$ given $\boldsymbol{B}$, and the data, and $\boldsymbol{\Sigma}_i$ conditional on $\boldsymbol{B}_i$ and the data have inverted Wishart distributions. Moreover, $\boldsymbol{B}$ given the data has the approximate normal distribution given in Theorem 12.6. We obtain approximate posterior means for $\boldsymbol{\Sigma}$, and the $\boldsymbol{\Sigma}_i$s under these conditions, by conditioning, as in:

$$E(\boldsymbol{\Sigma}) = E\{E(\boldsymbol{\Sigma} \mid \boldsymbol{B})\}. \tag{12.18}$$

The marginal posterior mean $\boldsymbol{\Sigma}$ is obtained from Equation (12.18) as

$$E(\boldsymbol{\Sigma} \mid \boldsymbol{Y}) = E\{E(\boldsymbol{\Sigma} \mid \boldsymbol{B}, \boldsymbol{Y})\}.$$

From Theorem 12.2 $(\boldsymbol{\Sigma} \mid \boldsymbol{B}, \boldsymbol{Y})$ has an inverted Wishart distribution, so using, for example, Press, 1982, Equation (5.2.4) for the first moment of an inverted Wishart distribution,

$$E(\boldsymbol{\Sigma} \mid \boldsymbol{B}, \boldsymbol{Y}) = \frac{\boldsymbol{T}^*}{\alpha - 2p - 2}, \qquad \alpha - 2p > 2, \qquad \alpha = n + d - q + p + 1.$$

So

$$E(\boldsymbol{\Sigma} \mid \boldsymbol{Y}) = \frac{1}{\alpha - 2p - 2} E(\boldsymbol{T}^* \mid \boldsymbol{Y}).$$

Note that $\boldsymbol{T}^*$ is defined in Theorem 12.1. So

$$E(\boldsymbol{T}^* \mid \boldsymbol{Y}) = \boldsymbol{A} + \boldsymbol{E}[(\boldsymbol{B} - \hat{\boldsymbol{B}})'(\boldsymbol{U}'\boldsymbol{R}\boldsymbol{U})(\boldsymbol{B} - \hat{\boldsymbol{B}}) \mid \boldsymbol{Y}].$$

Define $\boldsymbol{B}$: $(m \times p)$ in terms of its columns:

$$\boldsymbol{B}_{(m\times p)} = (\boldsymbol{\beta}_1, \ldots, \boldsymbol{\beta}_p), \boldsymbol{\beta}_j: (m \times 1), \qquad \text{for} j = 1, \ldots, p.$$

Similarly, define $\hat{\boldsymbol{B}} = (\hat{\boldsymbol{\beta}}, \ldots, \hat{\boldsymbol{\beta}}_p)$. Then, if $\boldsymbol{T}^* \equiv (\alpha_{ij})$,

$$E(t^*_{ij} \mid \boldsymbol{Y}) = \alpha_{ij} + E[(\boldsymbol{\beta}_i - \hat{\boldsymbol{\beta}}_i)'(\boldsymbol{U}'\boldsymbol{R}\boldsymbol{U})(\boldsymbol{\beta}_j - \hat{\boldsymbol{\beta}}_j) \mid \boldsymbol{Y}].$$

So we must now evaluate the last expectation:

$$E[(\boldsymbol{\beta}_i - \hat{\boldsymbol{\beta}}_i)'(\boldsymbol{U}'\boldsymbol{R}\boldsymbol{U})(\boldsymbol{\beta}_j - \hat{\boldsymbol{\beta}}) \mid \boldsymbol{Y}] = \text{tr}\{(\boldsymbol{U}'\boldsymbol{R}\boldsymbol{U})E[(\boldsymbol{\beta}_j - \hat{\boldsymbol{\beta}}_j)(\boldsymbol{\beta}_i - \hat{\boldsymbol{\beta}}_i)'\boldsymbol{Y}]\}$$

Since $(\boldsymbol{\beta}_j - \hat{\boldsymbol{\beta}}_j) = (\boldsymbol{\beta}_j - \tilde{\boldsymbol{\beta}}_j) + (\tilde{\boldsymbol{\beta}}_j - \hat{\boldsymbol{\beta}}_j)$,

$$\begin{aligned} E[(\boldsymbol{\beta}_j - \hat{\boldsymbol{\beta}}_j)(\boldsymbol{\beta}_i - \hat{\boldsymbol{\beta}}_i)' \mid \boldsymbol{Y}] &= E[(\boldsymbol{\beta}_j - \tilde{\boldsymbol{\beta}}_j)(\boldsymbol{\beta}_i - \tilde{\boldsymbol{\beta}}_i)' \mid \boldsymbol{Y}] + E[(\boldsymbol{\beta}_j - \tilde{\boldsymbol{\beta}}_j)(\tilde{\boldsymbol{\beta}}_i - \hat{\boldsymbol{\beta}}_i)' \mid \boldsymbol{Y}] \\ &\quad + E[(\tilde{\boldsymbol{\beta}}_j - \hat{\boldsymbol{\beta}}_j)(\boldsymbol{\beta}_i - \tilde{\boldsymbol{\beta}}_i)' \mid \boldsymbol{Y}] + E[(\tilde{\boldsymbol{\beta}}_j - \hat{\boldsymbol{\beta}}_j)(\tilde{\boldsymbol{\beta}}_i - \hat{\boldsymbol{\beta}}_i)' \mid \boldsymbol{Y}]. \end{aligned}$$

From Theorem 12.6 it is approximately true that

$$[(\boldsymbol{\beta} - \tilde{\boldsymbol{\beta}}) \mid \boldsymbol{Y}] \sim N(\boldsymbol{0}, \tilde{\boldsymbol{A}}).$$

The marginal distribution of the subvector $\boldsymbol{\beta}_i$ is therefore $[(\boldsymbol{\beta}_i - \tilde{\boldsymbol{\beta}}_i) \mid \boldsymbol{Y}] \sim N(\boldsymbol{0}, \tilde{\boldsymbol{A}}_{ii})$; $\tilde{\boldsymbol{A}} = (\tilde{\boldsymbol{A}}_{ij})$, $i, j = 1, \ldots, p$. Moreover,

$$\text{cov}[(\boldsymbol{\beta}_i, \boldsymbol{\beta}_j) \mid \boldsymbol{Y}] = \tilde{\boldsymbol{A}}_{ij}.$$

So it is approximately true that

$$E[(\boldsymbol{\beta}_j - \hat{\boldsymbol{\beta}}_j)(\boldsymbol{\beta}_i - \hat{\boldsymbol{\beta}}_j)' \mid \boldsymbol{Y}] = \tilde{\mathbf{A}}_{ij} + (\tilde{\boldsymbol{\beta}}_j - \hat{\boldsymbol{\beta}}_j)(\tilde{\boldsymbol{\beta}}_i - \hat{\boldsymbol{\beta}}_i)'.$$

Combining terms gives

$$E(t^*_{ij} \mid \boldsymbol{Y}) = \alpha_{ij} + \text{tr}(\tilde{\boldsymbol{A}}_{ij}\boldsymbol{U}'\boldsymbol{R}\boldsymbol{U}) + (\tilde{\boldsymbol{\beta}}_i - \hat{\boldsymbol{\beta}}_i)'(\boldsymbol{U}'\boldsymbol{R}\boldsymbol{U})(\tilde{\boldsymbol{\beta}}_j - \hat{\boldsymbol{\beta}}_j).$$

So if $\boldsymbol{\Sigma} \equiv (\sigma_{ij})$, the approximate posterior mean is

$$\begin{aligned} E(\sigma_{ij} \mid \boldsymbol{Y}) = \left(\frac{1}{n + d - q - p - 1}\right)&\{a_{ij} + \text{tr}(\tilde{\boldsymbol{A}}_{ij}\boldsymbol{U}'\boldsymbol{R}\boldsymbol{U}) \\ &+ (\tilde{\boldsymbol{\beta}}_i - \hat{\boldsymbol{\beta}}_i)'(\boldsymbol{U}'\boldsymbol{R}\boldsymbol{U})(\tilde{\boldsymbol{\beta}}_j - \hat{\boldsymbol{\beta}}_j)\} \end{aligned} \tag{12.19}$$

The result is summarized in the following theorem.

Theorem 12.7 The marginal posterior mean of the elements of the model error covariance matrix, $\boldsymbol{\Sigma} \equiv (\sigma_{ij})$, is given approximately in Equation (12.19).

Now examine the posterior means of the $\boldsymbol{\Sigma}_i \equiv (\sigma_{uv}^{(i)})$; $i = 1, \ldots, c$. By analogy with $\boldsymbol{\Sigma}$,

$$E(\boldsymbol{\Sigma}_i \mid \boldsymbol{Y}) = \left(\frac{1}{\alpha_i - 2p - 2}\right) E(\boldsymbol{T}_i^* \mid \boldsymbol{Y}) + \left(\frac{1}{\alpha_i - 2p - 2}\right)[\boldsymbol{V}_i + E(\boldsymbol{B}_i'\boldsymbol{B}_i) \mid \boldsymbol{Y}],$$

for $\alpha_i = m_i + \delta_i + p + 1$.

Defining $\boldsymbol{B}_i$ in terms of its columns gives

$$\underset{(m_i \times p)}{\boldsymbol{B}_i} = (\boldsymbol{\beta}_1(i), \ldots, \boldsymbol{\beta}_p(i)),$$

$$\boldsymbol{B}_i'\boldsymbol{B}_i = (\boldsymbol{\beta}_u'(i)\boldsymbol{\beta}_v(i)), (\mu, v) = 1, \ldots, p.$$

Let $\boldsymbol{V}_i \equiv (V_{uv}(i))$. Then,

$$E[(\sigma_{uv}^{(i)}) \mid \boldsymbol{Y}] = \left(\frac{1}{\alpha_i - 2p - 2}\right)[V_{uv}(i) + E(\boldsymbol{\beta}_u'(i)\boldsymbol{\beta}_v(i)) \mid \boldsymbol{Y}].$$

From Theorem 12.6, the marginal posterior distribution of $\boldsymbol{\beta}_u(i)$ is, approximately,

$$[\boldsymbol{\beta}_u(i) \mid \boldsymbol{Y}] \sim N[\tilde{\boldsymbol{\beta}}_\mu(i), \tilde{\boldsymbol{A}}_{uu}(i)],$$

and

$$\begin{aligned}\operatorname{cov}\{[\boldsymbol{\beta}_v(i), \boldsymbol{\beta}_u(i)] \mid \boldsymbol{Y}\} &= E\{[\boldsymbol{\beta}_u(i) - \tilde{\boldsymbol{\beta}}_u(i)][\boldsymbol{\beta}_\mu(i) - \tilde{\boldsymbol{\beta}}_u(i)]' \mid \boldsymbol{Y}\} \\ &= \tilde{\boldsymbol{A}}_{uv}(i).\end{aligned}$$

So

$$\begin{aligned}E[\boldsymbol{\beta}_u'(i)\boldsymbol{\beta}_v(i) \mid \boldsymbol{Y}] &= \operatorname{tr}\{E[\boldsymbol{\beta}_v(i)\boldsymbol{\beta}_u'(i) \mid \boldsymbol{Y}]\} \\ &= \operatorname{tr}\{\tilde{\boldsymbol{A}}_{vu}(i) + \tilde{\boldsymbol{\beta}}_v(i)\tilde{\boldsymbol{\beta}}u(i)\} \\ &= \tilde{\boldsymbol{\beta}}_u(i)\tilde{\boldsymbol{\beta}}_v(i) + \operatorname{tr}[\tilde{\boldsymbol{A}}_{vu}(i)].\end{aligned}$$

Substituting gives the final result:

$$E[\sigma_{uv}^{(i)} \mid \boldsymbol{Y}] = \left(\frac{1}{m_i + \delta_i - p - 1}\right)\{V_{uv}(i) + \tilde{\boldsymbol{\beta}}_u'(i)\tilde{\boldsymbol{\beta}}_v(i) + \operatorname{tr}\tilde{\boldsymbol{A}}_{uv}(i)\}. \tag{12.20}$$

Theorem 12.8 The marginal posterior means of the variances and covariances of effects are given in Equation (12.20).

It would also be useful in some situations to evaluate the variances and covariances of the elements of $\boldsymbol{\Sigma}$ and the $\boldsymbol{\Sigma}_i$s. The required calculations have proven to be very difficult, however, involving multivariate fourth moments.

12.5.7 Numerical Example

The data for this example are taken from Morrison, 1967, p. 179. Three drugs A_1, A_2, A_3 are known to have toxic effects that are apparent in several measurable features of an organism ingesting them. Let these characteristics be the weight losses in the first and second weeks of the trial. It is also thought that the effects of the drugs might be different in male animals from those in females. For the trial 12 male rats of essentially equal ages and weights are drawn from a common laboratory strain and divided randomly and equally among the three drugs. Twelve female rats of comparable ages and weights are obtained from litters of the original strain and similarly assigned drugs.

For the analysis of this problem we adopt a two-dimensional, two-way, mixed model, bivariate analysis of variance. The model is also balanced. We will use the approximation to the posterior moments of the appropriate covariance matrices, as described above, to determine whether we can reduce the mixed model to a fixed effects model for further analysis.

There is one fixed effect (gender), and there are two random effects (effects due to drugs, and interaction effects between drugs and gender); so $c = 2$. The dependent variables are weight loss observations taken after the first week, and then again after the second week ($p = 2$). We also have $m_1 = 3, m_2 = 6, m = 9, q = 3$. The two random effects are each assumed to be normally distributed with mean zero with covariance matrices Σ_1 for drugs and Σ_2 for interactions between gender and drugs. The experimental errors are also assumed to be normal with mean zero, but with covariance matrix Σ. The weight losses in grams, for the two weeks for each rat, are shown in Table 12.2.

The model is defined by

$$\underset{(1\times p)}{\boldsymbol{y}_{ijk}} = \underset{(1\times p)}{\boldsymbol{\mu}} + \underset{(1\times p)}{\boldsymbol{a}_i} + \underset{(1\times p)}{\boldsymbol{b}_j} + \underset{(1\times p)}{(\boldsymbol{ab})_{ij}} + \underset{(1\times p)}{\boldsymbol{e}_{ijk}}$$

$i = 1, \ldots, I, j = 1, \ldots, J, k = 1, \ldots, K$, where the main effects, interaction effects, and disturbances are interpreted as in the conventional ANOVA model using the identifications above; this model may readily be placed into the format of Equation (12.1) where:

$$\underset{(24\times 3)}{\boldsymbol{X}} = \begin{bmatrix} 1 & 1 & 0 \\ \vdots & \vdots & \vdots \\ 1 & 1 & 0 \\ \hline 1 & 0 & 1 \\ \vdots & \vdots & \vdots \\ 1 & 0 & 1 \end{bmatrix}, \quad \underset{(4\times 1)}{\boldsymbol{1}} = \begin{bmatrix} 1 \\ 1 \\ 1 \\ 1 \end{bmatrix}, \quad \underset{(3\times 2)}{\Theta} = \begin{bmatrix} \mu' \\ \alpha_1' \\ \alpha_2' \end{bmatrix}, \quad \underset{(24\times 9)}{\boldsymbol{U}} = \begin{bmatrix} \underset{(24\times 3)}{\boldsymbol{U}_1} & \underset{(24\times 6)}{\boldsymbol{U}_2} \end{bmatrix},$$

Table 12.2 Weight Loss Each Week by Drug and Gender

	Drug		
	A_1	A_2	A_3
Male	[5, 6]	[7, 6]	[21, 15]
	[5, 4]	[7, 7]	[14, 11]
	[9, 9]	[9, 12]	[17, 12]
	[7, 6]	[6, 8]	[12, 10]
Female	[7, 10]	[10, 13]	[16, 12]
	[6, 6]	[8, 7]	[14, 9]
	[9, 7]	[7, 6]	[14, 8]
	[8, 10]	[6, 9]	[10, 5]

the top half of $\boldsymbol{X}$ are indicator variables for males, and the lower half for females,

$$\boldsymbol{U}_1 = \begin{bmatrix} \boldsymbol{I}_3 \\ \boldsymbol{I}_3 \end{bmatrix} \otimes 1, \qquad \boldsymbol{U}_2 = \boldsymbol{I}_6 \otimes 1,$$

$$\underset{(9\times 12)}{\boldsymbol{B}} = \begin{bmatrix} \boldsymbol{B}_1 \\ \boldsymbol{B}_1 \end{bmatrix}, \qquad \underset{(3\times 2)}{\boldsymbol{B}_1} = \begin{bmatrix} \underset{(1\times 2)}{\boldsymbol{b}_1}, & \underset{(1\times 2)}{\boldsymbol{b}_2}, & \underset{(1\times 2)}{\boldsymbol{b}_3} \end{bmatrix}', \quad \underset{(6\times 2)}{\boldsymbol{B}_2} = \begin{bmatrix} \underset{(1\times 2)}{(\boldsymbol{ab})_{11}} \cdots \underset{(1\times 2)}{(\boldsymbol{ab})_{23}} \end{bmatrix}'$$

Now we are concerned with the associated covariances $\boldsymbol{\Sigma}_1, \boldsymbol{\Sigma}_2$, and $\boldsymbol{\Sigma}$.

The hyperparameters of the prior Wishart distributions are assessed to be

$$\underset{(2\times 2)}{\boldsymbol{V}} = \begin{pmatrix} 1 & 0 \\ 0 & 1 \end{pmatrix}, \qquad \underset{(2\times 2)}{\boldsymbol{V}_1} = \begin{pmatrix} 2 & 0 \\ 0 & 2 \end{pmatrix}, \qquad \boldsymbol{V}_2 = \begin{pmatrix} 1 & 0 \\ 0 & 1 \end{pmatrix},$$

$$d = 50, \qquad \delta_1 = 75, \qquad \delta_2 = 100.$$

A computer program has been written in the Turbo Pascal 7.0 programming language which implements the multivariate covariance component estimation technique developed above. The program is designed so that one can specify an arbitrary experimental design.

Using posterior means as point estimates of the covariance matrices,

$$\hat{\boldsymbol{\Sigma}}_1 = \begin{bmatrix} 0.1272 & 0.0087 \\ 0.0087 & 0.1236 \end{bmatrix}, \qquad \hat{\boldsymbol{\Sigma}}_2 = \begin{bmatrix} 0.0715 & 0.0003 \\ 0.0003 & 0.0715 \end{bmatrix},$$

$$\hat{\boldsymbol{\Sigma}} = \begin{bmatrix} 6.7182 & 2.8266 \\ 2.8266 & 3.1842 \end{bmatrix}.$$

The posterior means for the experimental error covariance indicate that about 99 percent of the total covariance can be ascribed to the experimental errors. Moreover, since the elements of $\hat{\boldsymbol{\Sigma}}_1$ and $\hat{\boldsymbol{\Sigma}}_2$ are near zero, the random effects model can be replaced by a simple fixed effects model.

SUMMARY

This chapter has been devoted to a study of Bayesian inference in the general linear model. We have examined univariate and multivariate models of regression and the analysis of variance and covariance, and have addressed problems involving fixed effects models, random effects models, and mixed effect models. We have invoked the principle of exchangeability for the distribution of population means and have found that by assuming exchangeability, there results a substantial reduction in the number of hyperparameters that needs to be assessed.

EXERCISES

12.1* Consider the univariate multiple regression model

$$y_i \mid \boldsymbol{x}_i = \beta_0 + \beta_1 x_n + \cdots + \beta_p x_{ip} + u_i, \qquad i = 1, \ldots, n,$$

where y_i denotes yield for a crop in year i, x_{ij} denotes the observed value of explanatory variable j during year i, u_i denotes a disturbance term in year i, $\boldsymbol{\beta} \equiv (\beta_i)$: $(p+1) \times 1$ is a vector of coefficients to be estimated, and $x_i \equiv (\boldsymbol{X}_{ij})$ is a vector of observed explanatory variables during year i. Assume the u_is are independent $N(0, \sigma^2)$. Introduce a vague prior for $(\boldsymbol{\beta}, \sigma^2)$.

(a) Find the joint posterior density for $(\boldsymbol{\beta}, \sigma^2)$.
(b) Find the marginal posterior density for $(\boldsymbol{\beta})$.
(c) Find the marginal posterior density for $(\boldsymbol{\sigma}^2)$.
(d) Find the marginal posterior density for (β_1).
(e) Find the predictive density for a new observation y^*, based upon a new observation vector $\boldsymbol{x}^* \equiv (\boldsymbol{x}_j^*)$.

12.2* Consider the simple univariate regression model of Section 12.2.1. Introduce a natural conjugate prior for $(\beta_1, \beta_2, \sigma^2)$.

(a) Find the joint posterior density for $(\beta_1, \beta_2, \sigma^2)$.
(b) Find the marginal posterior density for (β_1, β_2).
(c) Find the marginal posterior density for (σ^2).
(d) Find the marginal posterior density for (β_2).
(e) Find the predictive density for a new observation y^*, based upon a new observation x^*.

12.3* For the simple univariate regression model of Section 12.2.1, suppose that instead of assuming the u_is are independent $N(0, \sigma^2)$, we assume that they are independent and that $u_i \sim N(0, \sigma_i^2)$, where $\sigma_i^2 = \sigma^2 x_i$ (heteroscedasticity). Answer parts (a)–(e) of Exercise 12.2 using the interpretation of σ_i^2 in this problem.

* Solutions for asterisked exercises may be found in Appendix 7.

12.4* For the simple univariate regression model of Section 12.2.1 suppose that instead of assuming the u_is are independent, we assume serially correlated errors with $\text{corr}(u_i u_j) = \rho, i \neq j, \text{var}(u_i) = \sigma^2, 0 < \rho < 1$. Adopt a vague prior for $(\beta_1, \beta_2, \sigma^2)$, assume ρ is *a priori* independent of $(\beta_1, \beta_2, \sigma^2)$, and assume a uniform prior for ρ on the unit interval.

(a) Find the joint posterior density for $(\beta_1, \beta_2, \sigma^2, \rho)$.

(b) Find the marginal posterior density for ρ.

(c) Find the marginal posterior density for β_2.

(d) Find the marginal posterior density for σ^2.

(e) Find the predictive density for a new observation y^* based upon a new observation x^*.

12.5 For the model in Exercise 12.2 give a highest posterior density credibility interval for β_2, of 95 percent probability.

12.6 For the model in Exercise 12.2 give:

(a) The model Bayesian estimator for β_2;

(b) The minimum-risk Bayesian estimator with respect to a quadratic loss function for β_2;

(c) The minimum-risk Bayesian estimator with respect to an absolute error loss function for β_2.

12.7 Suppose $\boldsymbol{X}: p \times 1$ and $\mathcal{L}(\boldsymbol{X} \mid \boldsymbol{\theta}) = N(\boldsymbol{\theta}, \boldsymbol{I})$. Give a natural conjugate prior family for $\boldsymbol{\theta}$.

12.8 Suppose $\boldsymbol{X}: p \times 1$ and $\mathcal{L}(\boldsymbol{X} \mid \boldsymbol{\theta}) = N(\boldsymbol{\theta}, I)$, for $\Lambda \equiv \boldsymbol{\Sigma}^{-1}$. Give a natural conjugate prior family for $\boldsymbol{\Sigma}$ is a positive definite matrix.

12.9 Suppose $\boldsymbol{X}: p \times 1$ and $\mathcal{L}(\boldsymbol{X} \mid \boldsymbol{\theta}, \boldsymbol{\Sigma}) = N(\boldsymbol{\theta}, \boldsymbol{\Sigma})$ for $\boldsymbol{\Sigma}$ a positive definite matrix. Find a natural conjugate prior family for $(\boldsymbol{\theta}$, by noting that $p(\boldsymbol{\theta}, \boldsymbol{\Lambda}) = p_1(\boldsymbol{\theta} \mid \boldsymbol{\Lambda}) p_2(\boldsymbol{\Lambda})$ and $\boldsymbol{\Lambda} \equiv \boldsymbol{\Sigma}^{-1}$ and by assuming that *a priori* $\text{var}(\boldsymbol{\theta} \mid \boldsymbol{\Sigma}) \propto \boldsymbol{\Sigma}$.

12.10 Use the example in Section 12.2.6 involving average microprocessor speeds in year i, y_i, and average RAM memory in year i, x_i, and test the hypothesis (using Bayesian procedures) that H: $\beta_2 = 3$, versus the alternative hypothesis that A: $\beta_2 \neq 3$ at the 5 percent level of credibility.

(a) Use the Jeffreys procedure for testing, and assume $P(H) = p(A) = 0.5$. (*Hint*: see Chapter 9.)

(b) Use the Lindley testing procedure (Chapter 9) of rejecting H if the 95 percent credibility interval for $(\beta_2 - 3)$ does not include the origin.

12.11* Do Exercise 12.1 assuming a g-prior for $(\boldsymbol{\beta}, \sigma^2)$ instead of the vague prior.

12.12 Provide conditions under which a set of exchangeable random variables might be treated approximately as if they were independent.

12.13 What can be said about de Finetti's theorem for finitely exchangeable events? (*Hint*: see Chapter 10.)

12.14 Assume that $\boldsymbol{\mu}, \boldsymbol{\alpha}_i$ are independent *a priori*, and assume that the $\boldsymbol{\alpha}_i$s are exchangeable and that $\boldsymbol{\alpha}_i \sim N(\xi^*, \Phi^*)$ for all i.

(a) Find the prior density for $\boldsymbol{\theta}_i$.
(b) Find the prior density for $\boldsymbol{B}$.
(c) Find the joint posterior density for $(\boldsymbol{B}, \boldsymbol{\Sigma})$.

12.15* Adopt the two-way layout

$$\underset{(p\times 1)}{y_{ijk}} = \underset{(p\times 1)}{\boldsymbol{\mu}} + \underset{(p\times 1)}{\boldsymbol{\alpha}_i} + \underset{(p\times 1)}{\boldsymbol{\beta}_j} + \underset{(p\times 1)}{\gamma_{ij}} + \underset{(p\times 1)}{\boldsymbol{\mu}_{ijk}},$$

where $i = 1, \ldots, I, j = 1, \ldots, J, k = 1, \ldots, K$, and the main effects, interaction effects, and disturbances are interpreted as in the conventional ANOVA model. Explain how you would generalize the results of this chapter for the one-way layout. (*Hint*: See Press and Shigemasu, 1985.)

12.16* How would you generalize the results of this chapter for the one-way layout to a one-way layout with covariates (MANOCOVA)?

12.17* Explain why the Bayesian estimators in the example involving test scores in Section 12.4.8 are so close to the MLEs.

12.18 Explain the relationship between a random effects model and a Bayesian analysis of a fixed effects model.

12.19 Explain why an analysis of variance and an analysis of covariance can be handled analytically in the same way.

FURTHER READING

Berger, J. O. (1980). *Statistical Decision Theory.* New York, Springer-Verlag.

Box, G. E. P. and Tiao, G. C. (1973). *Bayesian Inference in Statistical Analysis*, Massachusetts, Addison-Wesley.

Broemeling, L. (1985). *Bayesian Analysis of Linear Models*, New York, Marcel Dekker, Inc.

Dickey, J. M. (1967a). "Expansions of t-Densities and Related Complete Integrals," *Annals Math. Statist.*, **38**(2), 503–510.

Dickey, J. M. (1967b). "Multivariate Generalizations of the Multivariate t-Distribution and the Inverted Multivariate t-Distribution," *Annals Math. Statist.*, **38**, 511–518.

Dickey, J. M. (1968). "Three Dimensional Integral-Identities with Bayesian Applications," *Annals Math. Statist.*, **39**(5), 1615–1627.

Dreze, J. H. (1977). "Bayesian Regression Analysis Using the Poly-t Distribution," in A. Aykac and C. Brumat, Eds., *New Developments in the Application of Bayesian Methods*, New York, North-Holland Publishing Co.

Efron, B. and Morris, C. (1973). "Stein's Estimation Rule and Its Competitors—An Empirical Bayes Approach," *J. Am. Statist. Assoc.*, **68**, 117–130.

Gelfand, A. E., Hills, S. E., Racine-Poon, A. and Smith, A. F. M. (1990). "Illustration of Bayesian Inference in Normal Data Models Using Gibbs Sampling," *J. Am. Statist. Assoc.*, **85**, 972–985.

Haavelmo, T. (1947). "Methods of Measuring the Marginal Propensity to Consume," *J. Am. Statist. Assoc.*, **42**, 88.

Hulting, F. L. and Harville, D. A. (1991). "Some Bayesian and Non-Bayesian Procedures for the Analysis of Comparative Experiments and for Small-Area Estimation: Computational Aspects, Frequentist Properties, and Relationships," *J. Am. Statist. Assoc.*, **86**, 557–568.

James, W. and Stein, C. (1961). "Estimation with Quadratic Loss," in *Fourth Berkeley Symposium on Mathematical Statistics and Probability*, Berkeley, University of California Press, 361–379.

Jelenkowska, T. H. (1989), *Bayesian Multivariate Analysis*, 19 Coll. Metodl. z Agrobiom. PAN, Warszawa.

Jelenkowska, T. H. and Press, S. J. (1997). "Bayesian Inference in the Multivariate Mixed Model MANOVA," in Multivariate Statistical Analysis in Honor of Minoru Siotani on His 70th Birthday, *Am. J. Math. Mgmt Sci.*, 97–116, Vol. III.

Kshirsagar, A. M. (1960). "Some Extensions of the Multivariate t-Distribution and the Multivariate Generalization of the Distribution of the Regression Coefficient," *Proc. Cambridge Phil. Soc.*, **57**, 80–85.

Leamer, E. E. (1978). *Specification Searches*, New York, John Wiley and Sons, Inc.

Marx, D. G. and Nel, D. G. (1985). "Matric-t Distribution," in S. Kotz, N. L. Johnson, C. B, Read, Eds., *Encyclopedia of Statistical Sciences*, Read, **5**, 316–320.

Morrison, D. F. (1967). *Multivariate Statistical Methods*, New York, McGraw-Hill.

Press, S. J. (1979). "Bayesian Computer Programs," in A. Zellner, Ed., *Studies in Bayesian Econometrics and Statistics in Honor of Harold Jeffreys*, Amsterdam, North-Holland Publishing Co.

Press, S. J. (1980). "Bayesian Inference in MANOVA," in P. R. Krishnaiah, Ed., *Handbook of Statistics*, Vol. I, New York, North-Holland Publishing Co.

Press, S. J. (1982). *Applied Multivariate Analysis: Using Bayesian and Frequentist Methods of Inference*, Second Edition, Malabar, Florida, Robert E. Krieger Publishing Co.

Press, S. J., and Shigemasu, K. (1985). "Bayesian MANOVA and MANOCOVA Under Exchangeability," *Communications in Statistics: Theory and Methods, Vol. A*, **14**(5), 1053–1078.

Press, S. J. and Davis, A. W. (1987). "Asymptotics for the Ratio of Multiple t-Densities," in A. Gelfand, Ed., *Contributions to the Theory and Application of Statistics: Volume in Honor of Herbert Solomon*, New York, Academic Press, 155–177.

Rajagopalan, M. and Broemeling, L. (1983). "Bayesian Inference for the Variance Components of the General Mixed Models," *Comm. in Statist.*, **12**(6), 701–723.

Rao, C. R. and Kleffe, J. (1980). "Estimation of Variance Components," in P. R. Krishnaiah, Ed., *Handbook of Statistics 1: Analysis of Variance*, New York, North Holland Publishing Co., 1–40.

Robinson, G. K. (1991). "That BLUP is a Good Thing: The Estimation of Random Effects (with discussion)," *Statist. Sci.*, **6**, 15–51.

Scheffe, H. (1959). *The Analysis of Variance*, New York, John Wiley and Sons, Inc.

Tiao, G. C. and Zellner, A. (1964). "Bayes' Theorem and the Use of Prior Knowledge in Regression Analysis," *Biometrics*, **51**, 219–230.

Zellner, A. (1971). *An Introduction to Bayesian Inference in Econometrics*, New York, John Wiley and Sons, Inc.

Zellner, A. (1985). "Bayesian Econometrics," *Econometrica*, **53**(2), 253–269.

CHAPTER 13

Model Averaging

*Merlise Clyde**

13.1 INTRODUCTION

In Chapter 12, we considered inference in a normal linear regression model with q predictors. In many instances, the set of predictor variables $\mathbf{X}$ can be quite large, as one considers many potential variables and possibly transformations and interactions of these variables that may be relevant to modeling the response $\mathbf{Y}$. One may start with a large set to reduce chances that an important predictor has been omitted, but employ variable selection to eliminate variables that do not appear to be necessary and to avoid overfitting. Historically, variable selection methods, such as forwards, backwards, and stepwise selection, maximum adjusted R^2, AIC, Cp, and so on, have been used, and, as is well known, these can each lead to selection of a different final model (Weisberg, 1985). Other modeling decisions that may arise in practice include specifying the structural form of the model, including choice of transformation of the response, error distribution, or choice of functional form that relates the mean to the predictors. Decisions on how to handle outliers may involve multiple tests with a somewhat arbitrary cut-off for p-values or the use of robust outlier-resistant methods. Many of the modeling decisions are made conditional on previous choices, and final measures of significance may be questionable.

While one may not be surprised that approaches for selection of a model reach different conclusions, a major problem with such analyses is that often only a "best" model and its associated summaries are presented, giving a false impression that this is the only model that explains the data. This standard practice ignores uncertainty due to model choice, and can lead to overconfident inferences and predictions, and decisions that are riskier than one believes (Draper, 1995; Hodges, 1987).

In this chapter, we review a Bayesian approach to address model uncertainty known as Bayesian model averaging, and describe its use in linear regression models for the choice of covariates and generalized linear models for the choice of link function and covariates.

*Institute of Statistics and Decision Sciences, Duke University, Durham, NC;
E-mail: clyde@stat.duke.edu.

13.2 MODEL AVERAGING AND SUBSET SELECTION IN LINEAR REGRESSION

Let **X** denote the $n \times q$ matrix of all predictors under consideration. Under the full model (all predictors), the univariate multiple regression model is represented as

$$\mathbf{Y} = \mathbf{X}\boldsymbol{\beta} + \mathbf{u}$$

where $\mathbf{u} \sim N(0, \sigma^2 I)$. In the problem of subset or variable selection among the q predictor variables, models under consideration correspond to potentially all possible subsets of the q variables, leading to a model space $\mathcal{M} = \{\mathcal{M}_1, \ldots, \mathcal{M}_K\}$ where $K = 2^q$ and includes the model with no predictor variables at all. Models for different subsets may be represented by a vector of binary variables, $\gamma = (\gamma_1, \ldots, \gamma_q)'$ where γ_j is an indicator for inclusion of variable $\mathbf{X}_j$ under model $\mathcal{M}_k$. A convenient indexing of models in the all subset regression problem is to let γ denote the binary representation of k for model $\mathcal{M}_k$. Under $\mathcal{M}_k$ there are $q_\gamma = \sum_{j=1}^{q} \gamma_j$ nonzero parameters, $\boldsymbol{\beta}_\gamma$, with a $q_\gamma \times n$ design matrix $\mathbf{X}_\gamma$.

To incorporate model uncertainty regarding the choice of variables in the linear regression model, we build a hierarchical model (George and McCulloch, 1993, 1997, Raftery et al. 1997):

$$\mathbf{Y} \mid \boldsymbol{\beta}, \sigma^2, \mathcal{M}_k \sim N(\mathbf{X}_\gamma \boldsymbol{\beta}_\gamma, \sigma^2 I_n) \tag{13.1}$$

$$\boldsymbol{\beta}_\gamma \mid \sigma^2, \mathcal{M}_k \sim p(\boldsymbol{\beta}_\gamma \mid \mathcal{M}_k, \sigma^2) \tag{13.2}$$

$$\sigma^2 \mid \mathcal{M}_k \sim p(\sigma^2 \mid \mathcal{M}_k) \tag{13.3}$$

$$\mathcal{M}_k \sim p(\mathcal{M}_k) \tag{13.4}$$

where the first stage is based on the normal probability model of Chapter 12 using $\boldsymbol{\beta}_\gamma$ and the design matrix $\mathbf{X}_\gamma$ under model $\mathcal{M}_k$. Variables are excluded in model $\mathcal{M}_k$ by setting elements of the full parameter vector $\boldsymbol{\beta}$ to zero. The second stage corresponds to a prior distribution for $\boldsymbol{\beta}_\gamma$, the nonzero elements of $\boldsymbol{\beta}$, under each model. The resulting distribution for the q-dimensional vector of regression parameters $\boldsymbol{\beta}$ can be viewed as a mixture of point masses at zero and continuous distributions for $\boldsymbol{\beta}_\gamma$. The last stage of the hierarchical model assigns prior probabilities over all models under consideration, conditional on the model space $\mathcal{M}$.

The posterior distribution over models in $\mathcal{M}$ is

$$p(\mathcal{M}_k \mid \mathbf{Y}) = \frac{m(\mathbf{Y} \mid \mathcal{M}_k) p(\mathcal{M}_k)}{\sum_k m(\mathbf{Y} \mid \mathcal{M}_k) p(\mathcal{M}_k)}$$

where $m(\mathbf{Y} \mid \mathcal{M}_k)$ is the marginal distribution of the data under model $\mathcal{M}_k$,

$$m(\mathbf{Y} \mid \mathcal{M}_k) = \iint p(\mathbf{Y} \mid \boldsymbol{\beta}_\gamma, \sigma^2, \mathcal{M}_k) p(\boldsymbol{\beta}_\gamma \mid \sigma^2, \mathcal{M}_k) p(\sigma^2 \mid \mathcal{M}_k) d\boldsymbol{\beta}_\gamma d\sigma^2$$

obtained by integrating over the prior distributions for model-specific parameters. The posterior distribution $p(\mathcal{M}_k \mid \mathbf{Y})$ provides a summary of model uncertainty after observing the data $\mathbf{Y}$. For the more general problem of model selection, the marginals are obtained similarly by integrating over all model-specific parameters.

If Δ is a quantity of interest, say predicted values at a point x, then the expected value of Δ given the data $\mathbf{Y}$ is obtained by first finding the posterior expectation of Δ under each model, and then weighting each expectation by the posterior probability of the model:

$$E(\Delta \mid \mathbf{Y}) = \sum_k p(\mathcal{M}_k \mid \mathbf{Y}) E(\Delta \mid \mathcal{M}_k, \mathbf{Y}). \tag{13.5}$$

Similarly, the posterior distribution for Δ can be represented as a mixture distribution over all models,

$$p(\Delta \mid \mathbf{Y}) = \sum_k p(\mathcal{M}_k \mid \mathbf{Y}) p(\Delta \mid \mathbf{Y}, \mathcal{M}_k) \tag{13.6}$$

where $p(\Delta \mid \mathbf{Y}, \mathcal{M}_k)$ is the posterior distribution of Δ under model $\mathcal{M}_k$.

The Bayesian solution for incorporating model uncertainty has become known as Bayesian model averaging (BMA) (Hoeting et al., 1999) as quantities of interest can often be expressed as a weighted average of model-specific quantities, where the weights depend on how much the data support each model (as measured by the posterior probabilities on models). If the posterior probability is concentrated on a single model, then model uncertainty is not an issue and both model selection and model averaging will lead to similar results. In many cases, model uncertainty dominates other forms of uncertainty, such as parameter uncertainty and sampling variation, and BMA can lead to real improvements in predictive performance (see Hoeting et al. (1999) for several examples illustrating BMA).

On the surface, Bayesian model averaging and model selection are straightforward to implement: one specifies the distribution of the data under each model, and the prior probabilities of models and model-specific parameters; Bayes' theorem provides the rest. The two major challenges confronting the practical implementation of Bayesian model averaging are choosing prior distributions and calculating posterior distributions. The latter problem requires that one can actually carry out the integration necessary for obtaining the marginal distribution of the data $m(\mathbf{Y} \mid \mathcal{M}_k)$, as well as determining the normalizing constant in the denominator of the posterior probability if M_k. In the normal linear regression model, there are many choices of prior distributions that lead to closed-form solutions for the marginal distribution (George and McCulloch, 1997). If the number of models in $\mathcal{M}$ is too large to permit enumeration of all models, one can approximate BMA using a subset of models. Where integrals cannot be carried out analytically, there are a range of methods that can be used to implement BMA, from asymptotic approximations to reversible jump Markov chain Monte Carlo sampling (see review articles by Hoeting et al. (1999) and Chipman et al. (2001) for an overview).

13.3 PRIOR DISTRIBUTIONS

The specification of prior distributions is often broken down into two parts: (1) elicitation of distributions for parameters specific to each model, such as the distribution for regression coefficients in linear models, $p(\boldsymbol{\beta}_k \mid \mathcal{M}_k, \sigma^2)$, and (2) selection of a prior distribution over models $p(\mathcal{M}_k)$. For moderate to high dimensional problems, it is difficult to consistently specify separate prior distributions for parameters under each model and elicit probabilities of each $\mathcal{M}_k$ directly. Practical implementations of BMA have usually made prior assumptions to simplify prior elicitation and allow tractable computations using conjugate prior distributions.

13.3.1 Prior Distributions on Models

In the context of variable selection, one may specify a prior distribution over $\mathcal{M}$ by a probability distribution on the indicator variables $\boldsymbol{\gamma}$. In most applications to date, the indicator variables are taken to be independent *a priori*, using a product of independent Bernoulli distributions,

$$P(\gamma_k) = \prod_q \omega_j^{\gamma_{jk}} (1 - \omega_j)^{1-\gamma_{jk}}$$

for the prior distribution of γ. The hyperparameter ω_j corresponds to the prior probability that variable $\mathbf{X}_j$ is included. The choice of ω_j could be based on subjective information, or could be assigned a prior distribution, such as a beta distribution, reflecting additional uncertainty. As a special case of the independent prior distribution over models, the uniform distribution over models ($\omega_j = 0.5$) is often recommended as a default choice. The uniform distribution is appealing in that posterior probabilities of models depend only on the marginal likelihood of the data, and not prior probabilities of models. One should note, however, that the uniform distribution implies that the model size has a prior distribution that is Binomial(q, 0.5) and a prior belief that half of the variables are expected to be included, which may not be realistic in problems where q is large. For models that contain interactions, polynomial terms, or highly correlated variables, independent prior distributions may not be suitable, and a prior that takes into account the dependence structure may be preferred (Chipman, 1996). Prior distributions over models that account for expected correlations between variables form an area of ongoing research that should lead to more realistic prior distributions over models (see Chipman et al., 2001, for new directions).

13.3.2 Prior Distributions for Model-Specific Parameters

In the context of the linear regression model, by far the most common choice for prior distributions on parameters within models, $\boldsymbol{\beta}_\gamma$, is a conjugate normal prior distribution

$$\beta_\gamma \mid \gamma \sim N(0, \sigma^2 \boldsymbol{\Sigma}_\gamma)$$

(Chipman et al., 2001). As one cannot typically specify a separate prior distribution for $\boldsymbol{\beta}$ under each model, any practical implementation for BMA usually resorts to structured families of prior distributions. This also ensures that prior specifications for $\boldsymbol{\beta}_\gamma$ are compatible across models (Dawid and Lauritzen, 2001). To avoid incompatibilities between nested models, choices for $\boldsymbol{\Sigma}_\gamma$ are obtained from the prior distribution for $\boldsymbol{\beta}$ under the full model and finding the conditional distribution for $\boldsymbol{\beta}$ given that a subset of $\boldsymbol{\beta} = 0$. For example, Zellner's g-prior (Zellner, 1986) is commonly used, which leads to $\boldsymbol{\Sigma} = g(\mathbf{X}'\mathbf{X})^{-1}$ for the full model and $\boldsymbol{\Sigma}_\gamma = g(\mathbf{X}_\gamma'\mathbf{X}_\gamma)^{-1}$ for the coefficients in model γ. This reduces the choice of hyperparameters down to just g, a scale parameter in the prior variance.

In order to obtain closed-form marginal distributions for the data, a conjugate inverse gamma prior distribution for σ^2 is commonly used. This is equivalent to taking

$$\frac{\nu\lambda}{\sigma^2} \sim \chi^2_\nu \tag{13.7}$$

where ν and λ are fixed hyperparameters.

13.4 POSTERIOR DISTRIBUTIONS

Using a conjugate prior distributions for $\boldsymbol{\beta}_\gamma$ combined with the inverse gamma prior for σ^2, the marginal likelihood for $\mathbf{Y}$ is

$$m(\mathbf{Y} \mid \gamma) \propto \mid \mathbf{X}_\gamma'\mathbf{X}_\gamma + \boldsymbol{\Sigma}_\gamma^{-1} \mid^{0.5} \mid \boldsymbol{\Sigma}_\gamma \mid^{-0.5} (\lambda\nu + \mathbf{S}_\gamma^2)^{-0.5(n+\nu)} \tag{13.8}$$

where

$$S_\gamma^2 = \mathbf{Y}'\mathbf{Y} - \mathbf{Y}'\mathbf{X}_\gamma\left(\mathbf{X}_\gamma'\mathbf{X}_\gamma + \boldsymbol{\Sigma}_\gamma^{-1}\right)^{-1}\mathbf{X}_\gamma'\mathbf{Y}$$

is a Bayesian analog to the residual sum of squares. Under Zellner's g-prior the marginal simplifies further, leading to

$$m(\mathbf{Y} \mid \gamma) \propto (1+g)^{-0.5q\gamma}\left(\nu\lambda + \mathbf{Y}'\mathbf{Y} - \frac{g}{1+g}(\hat{\boldsymbol{\beta}}_\gamma'(\mathbf{X}_\gamma'\mathbf{X}_\gamma)\hat{\boldsymbol{\beta}}_\gamma\right)^{-0.5(n+\nu)} \tag{13.9}$$

where $\hat{\boldsymbol{\beta}}_\gamma$ is the ordinary least-squares estimate of $\boldsymbol{\beta}_\gamma$ under model γ and q_γ is the number of regression parameters (including the intercept) in model γ. This can be normalized

$$p(\gamma \mid \mathbf{Y}) = \frac{m(\mathbf{Y} \mid \gamma)}{\sum_{\gamma' \in M} m(\mathbf{Y} \mid \gamma\prime)} \tag{13.10}$$

to obtain the posterior model probabilities, which can be easily calculated from summary statistics using most regression packages.

13.5 CHOICE OF HYPERPARAMETERS

In order to implement BMA in the linear regression problem, even under structured prior distributions, there are still choices that must be made regarding the hyperparameters g, v, and λ, in addition to the parameters in $p(\gamma)$. Where possible, subjective considerations should be used, but one may often want to present an objective summary in addition to subjective analyses. As a default choice, both Smith and Kohn (1996) and Fernández et al., 2001, have recommended using a uniform prior distribution for $\log(\sigma^2)$, corresponding to the limiting case of the conjugate prior as $v \to 0$. This is invariant to scale changes in **Y** and while improper, leads to proper posterior distributions.

Vague choices for g in Zellner's g-prior should be avoided, since being too noninformative about $\boldsymbol{\beta}$ by taking g large can have the unintended consequence of favoring the null model *a posteriori* (Kass and Raftery, 1995), as g has an influential role in the posterior model probabilities, as inspection of Equation (13.8) reveals. Default choices for g can be calibrated based on information criteria such as AIC (Akaike Information Criterion; Akaike, 1973, see also Chapter 15), BIC (Bayes Information Criterion; Schwarz, 1978, see also Chapter 15), or RIC (Risk Inflation Criterion; Foster and George, 1994). Based on simulation studies, Fernández et al., 2001, prefer RIC-like prior distributions when $n < p^2$ and BIC-like prior distributions otherwise, and recommend using $g = \max(n, p^2)$.

BIC can be rapidly calculated for a wide class of models, including generalized linear models where analytic integration to obtain the marginal likelihood is not possible,

$$\text{BIC}(\mathcal{M}_k) = -2\log(\text{maximized likelihood} \mid \mathcal{M}_k) + q_k \log(n) \qquad (13.11)$$

where q_k is the dimension of model $\mathcal{M}_k$. This can be used to obtain approximate posterior model probabilities where

$$p(\mathcal{M}_k \mid \mathbf{Y}) = \frac{p(\mathcal{M}_k)\exp(-0.5\text{BIC}(\mathcal{M}_k))}{\sum\limits_k p(M_k)\exp(-0.5\text{BIC}(\mathcal{M}_k))}.$$

Model probabilities based on BIC permit a default objective analysis. They have provided improved out-of-sample predictive performance in a wide variety of settings (Hoeting et al., 1999) and avoid the issue of hyperparameter specification. BIC may be conservative in problems with small to moderate sample sizes.

Empirical Bayes (EB) approaches provide an adaptive (but data-based) choice for g (Clyde and George, 2000; George and Foster, 2000; Clyde, 2001; Chipman et al., 2001), which typically have better out-of-sample performance than fixed hyperparameter specifications or using BIC, but with additional computational complexity. EB estimates of g can be obtained iteratively, and still permit analytic expressions for BMA without using Monte Carlo sampling in small to moderate problems where the model space can be enumerated or a subset of models can be identified using leaps and bounds (see below). This provides a compromise between fixed specifications for g and using a fully Bayesian approach with an additional stage in the hierarchical model for a prior distribution on g. While more computationally demanding, the use of a prior distribution on g, such as an inverse gamma distribution, provides additional prior robustness by inducing heavier tails in the prior distribution.

13.6 IMPLEMENTING BMA

In the variable selection problem for linear regression, marginal likelihoods are available in closed form (at least for nice conjugate prior distributions); for generalized linear models and many other models, Laplace's method of integration or BIC can provide accurate approximations to marginal distributions. The next major problem is that the model space is often too large to allow enumeration of all models, and beyond 20–25 variables, estimation of posterior model probabilities, model selection, and BMA must be based on a sample of models.

Deterministic search for models using branch and bounds or leaps and bounds algorithms (Furnival and Wilson, 1974) is efficient for problems with typically fewer than 30 variables and is used in several programs such as BICREG and BICGLM (Hoeting et al., 1999). For larger problems, these methods are too expensive computationally or do not explore a large enough region of the model space. Gibbs and Metropolis Hastings MCMC algorithms (see Chapter 6) work well in larger problems for stochastically sampling models (Hoeting et al., 1999; George and McCulloch, 1993, 1997, for details on algorithms).

Given a subset of models obtained by deterministic or stochastic search, BMA is carried out by summing over the sampled models ($\mathcal{S}$) rather than the entire space $\mathcal{M}$ in Equations (13.5), (13.6), (13.10) and (13.11).

13.7 EXAMPLES

We now look at two examples to illustrate model averaging. In both examples, the model spaces can be enumerated and probabilities were approximated by BIC using the output from linear model and generalized linear model software. Software for BMA using BIC and Gibbs sampling is available on the BMA website, www.research.att.com/~volinsky/bma.html, maintained by Chris Volinsky.

13.7.1 Pollution and Mortality

There have been ongoing discussions about the effect of air pollution on mortality. Data in one early study (1960) was based on a cross-sectional sample of 60 metropolitan areas (data available in Ramsey and Schafer, 2002). The response variable is age-adjusted mortality from all causes (deaths per 100,000 population). Pollution indices include "relative pollution potential" for NOX (nitrogen dioxides), SO_2 (sulfur dioxide), and HC (hydrocarbons), where the relative pollution potential is the product of tonnes emitted per day per square kilometer and a factor used to correct for the area and exposure in the metropolitan areas. Twelve potential confounding variables under consideration include PRECIP (mean annual precipitation), JANTEMP (mean January temperature), JULYTEMP (mean annual temperature in July), OVER65 (percentage of population over age 65), HOUSE (population per household), EDUC (median number of years of school completed), SOUND (percentage of sound housing), DENSITY (population density), NONWHITE (percentage of population in 1960 that was nonwhite), WHITECOL (percentage of white collar occupations), and HUMIDITY (annual average relative humidity). Pollution concentrations were transformed using natural logarithms.

If uncertainty about whether all confounding variables and which of the pollution variables is related to mortality is taken into consideration, there are $2^{15} = 32,768$ potential models. All possible models were fit using ordinary least squares and approximate model probabilities based on BIC (Equation 13.11) were calculated for each model. Figure 13.1 illustrate the top 25 best models in terms of BIC. The columns in the image correspond to variables, with rows corresponding to models. White indicates variable exclusion, while dark rectangles indicate that the variable for that column is included for the model in that row. The y-axis corresponds to the intensity, which is scaled to the log(Bayes' Factor), the difference in BIC for comparing each model to the model with the largest BIC value. The best model in terms of BIC is at the top of the image. While log(NOX) is included in all of the top 25 models, there is more uncertainty about whether HC and SO_2, and several of the other potential explanatory variables need to be accounted for.

The log posterior odds that there is a pollution effect (one or more of the three indices is included) after adjusting for all potential confounders is 4.44. Using Jeffreys' scale of evidence (Kass and Raftery, 1995), applied to log posterior odds rather than Bayes' factors, this suggests positive evidence in favor of the hypothesis that pollution is associated with mortality. Posterior probabilities are preferable to Bayes' factors, as Bayes' factors are not necessarily monotonic for nested models (Lavine and Schervish, 1999).

Unlike p-values, marginal posterior probabilities can provide evidence of whether the coefficient is zero or nonzero. Overall, the marginal probability that the coefficient for log(NOX) is nonzero is 0.985, while the posterior probabilities for inclusion of log(HC) and log(SO_2) are 0.265 and 0.126, respectively. These are obtained by summing model probabilities over models that include each of the variables, and provide an overall measure for how likely it is that β_j equals zero. Caution should be used in interpreting the marginal posterior probabilities with many (highly) correlated predictors, as the variables may split the posterior mass among themselves and individually receive small (less than 0.5) marginal probabilities of inclusion, while the overall probability that at least one should be included is quite large (the "dilution" effect; Chipman et al., 2001). In this example, there is overall positive evidence after model averaging for a pollution effect, but the posterior probabilities are spread out over many models.

This analysis only incorporates uncertainty regarding covariates, but issues of outliers and transformations of variables are also important here. While log transformations had been suggested by other authors, the choice of which transformation

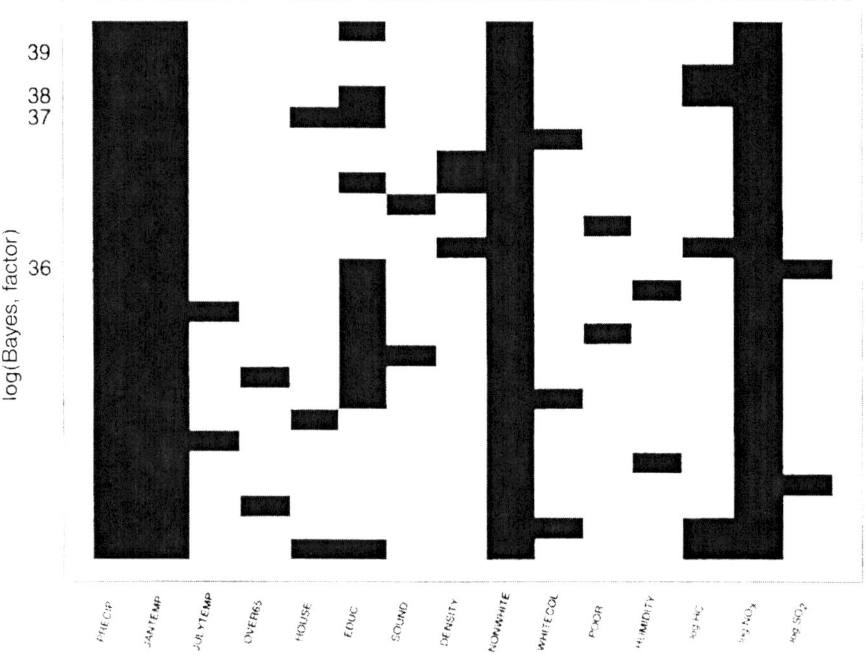

Figure 13.1 Top 25 models based on BIC for the pollution and mortality example

to use, outlier detection, and accommodation are other aspects of model uncertainty that can by addressed by BMA (Hoeting et al., 1996).

13.7.2 O-Ring Failures

On January 27, 1986, engineers who had built the space shuttles warned The National Aeronautics and Space Administration (NASA) that the *Challenger* should not be launched the next day because of temperatures predicted to be 31°F and the risk of fuel-seal failures at cold temperatures. Because the evidence regarding failures seemed inconclusive, the decision was made to proceed with the launch. Statistical models for the number of O-ring failures in each launch as a function of temperature (and pressure) have been explored by Dalal et al. (1989), Lavine (1991), and Draper (1995). The analysis below uses an indicator of at least one O-ring failure in a launch as a response (rather than the number of failures out of the six to avoid issues of dependence) with temperature and pressure as potential predictor variables.

Generalized linear models provide a natural extension of linear regression models for binary and other non-normal data. For O-ring failures, it is natural to model the failure indicator Y_i as Bernoulli with probability π_i. As in linear models, we may wish to explore how the mean π_i may depend on other covariates, $X\nu$. If we let the linear combination of the predictors be denoted by $\boldsymbol{\eta}$, ($\boldsymbol{\eta} = \mathbf{X}\boldsymbol{\beta}$) then in the normal linear regression, a natural starting point is to assume that the mean μ is identical to η, the linear predictor. For binary regression, the mean is a probability and so any posterior distributions on π_i should be constrained to the interval (0, 1). While such constraints are difficult to impose using an "identity link" where the mean π_i is equal to η_i, other functions are more natural for linking the mean of binary data to the linear predictor and enforcing the constraints on the probabilities. A widely used choice, the logit link corresponds to modeling the log odds of seal failure as a linear function of the predictors ($\text{logit}(\pi_i) = \log(\pi_i/(1 - \pi_i)) = \eta_i$). The probit link function is based on the inverse cdf of the normal distribution, $\Phi^{-1}(\pi_i) = \eta_i$, and is very similar to the logit model except in the extreme tails (i.e., at low temperatures). The complementary log–log link, $\log(-\log(1 - \pi_i))$, is another choice. Figure 13.2 shows the fitted probabilities as a function of temperature for the three link functions. It also shows the data points for failures and successes. For extrapolation, in the case of prediction at 31°F, the models may lead to different predictions. Rather than using a single link function, in BMA, the predictions are weighted by how much weight (*a posteriori*) is given to each link function and incorporates uncertainty regarding the structural form of the model.

Draper (1995) models the number of O-ring failures and considers structural (model) uncertainty in predicting failure at 31°F due to choice of link function and predictors (temperature, pressure, and quadratic terms in temperature). For illustration, we consider the choice of link function (logit, probit, and complemen-

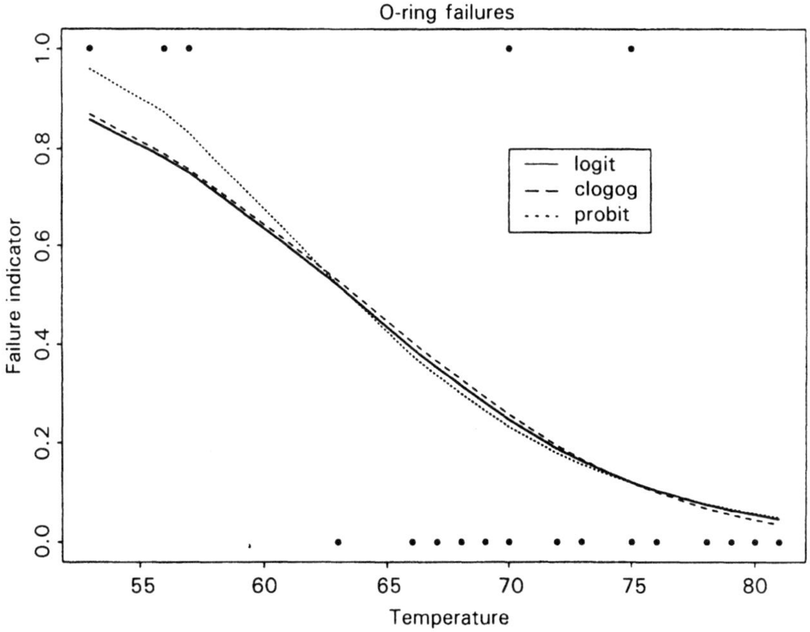

Figure 13.2 Fitted probabilities of O-ring failure using the logit, probit, and complementary log–log link functions

tary log–log) and within each link function, the inclusion/exclusion of temperature and pressure (linear terms only) leading to 12 possible models (3 link $\times$ 2^2 covariates). The models under the three link functions but with the intercept only, are all reparameterizations of the same model (the null model), so that in reality there are only 10 unique models under consideration. Assignment of uniform model probabilities in the latter case leads to a prior probability that temperature should be included of 0.6 and less weight on the null model (1/10); assigning uniform prior probabilities in the model space with 12 models (realizing that three are equivalent) leads to equal odds *a priori* that temperature (or pressure) should be included and probability of 1/4 on the null model. While prior odds, other than 1, for temperature (or pressure) inclusion could be based on subjective knowledge, even odds will be used for the rest of the example as a default or reference analysis.

Using BIC (Equation 13.11) the model probabilities can be calculated from deviances in the output of any GLM software. Using prior probabilities $p(\mathcal{M}_k) = 1/12$, the posterior probabilities for each of the 12 models based on choice of link function and covariates are given in the 12 cells of Table 13.1. The prior probability that temperature is related to failure is 1/2. We can readily find the average posterior probability that temperature is related to failure, averaged over the 12 models under consideration, in the following way.

Let $\delta(X)$ denote an indicator variable for whether the variable X is related to failure, or not. Then, if $P(X)$ denotes the Bayesian model average for including variable X,

$$\begin{aligned}P(X) &= E_{\text{posterior}}\{\delta(X) \mid \text{data}\}\\ &= E_{\text{posterior}}\{E_{\text{models}}[\delta(X) \mid \text{model, data}]\}\\ &= E_{\text{models}}\{E_{\text{posterior}}[\delta(X) \mid \text{model, data}]\}\\ &= \text{sum}_{\text{models}}\{(\text{models that include } X) \mid \text{data}]\}.\end{aligned}$$

If X denotes temperature, we find from Table 13.1 that

$$\begin{aligned}P\{\text{temperature}\} &= (0.1861 + 0.1803 + 0.2753 + 0.0835 + 0.0853 + 0.1143)\\ &= 0.9249 \cong 92 \text{ percent.}\end{aligned}$$

Table 13.1 Posterior Model Probabilities Based on the BIC Approximation for Choice of Link Function and Inclusion of Temperature and Pressure

	logit	probit	c log–log
Intercept only	0.0167	0.0167	0.0167
Temperature	0.1861	0.1803	0.2753
Pressure	0.0083	0.0082	0.0084
Temperature and Pressure	0.0835	0.0853	0.1143

and

$$
\begin{aligned}
P\{\text{pressure}\} &= (0.0083 + 0.0082 + 0.0084 + 0.0835 + 0.0853 + 0.1143) \\
&= 0.3081 \cong 31 \text{ percent.}
\end{aligned}
$$

There is a slightly higher probability in favor of the complementary log–log link, but otherwise the data are not strongly conclusive about link choice, which is not totally surprising given the small differences in probabilities over the range of observed data. For the problem of predicting failure at low temperatures (an extrapolation), the choice of link function can be influential, and in this case model averaging reflects the uncertainty regarding model form. While all models with temperature predict that the probability of failure at 31°F is above 0.99, the posterior mean with model averaging for the failure probability is 0.92, which incorporates uncertainty about whether temperature has an effect and choice of link function.

Using the normal approximation to the posterior distribution for the coefficients under each model, the posterior distribution and 95 percent posterior probability intervals for the failure probability under BMA can be obtained by Monte Carlo samples, by first drawing a model with replacement according to the posterior model probabilities, and then drawing the linear predictor $\boldsymbol{\eta}$ based on the asymptotic normal approximation to the posterior distribution for $\boldsymbol{\eta}$ given that model. Applying the inverse link to the linear predictor under the sampled model leads to one draw from the (approximate) posterior distribution of the probability of failure at 31°F. This is repeated to provide a Monte Carlo sample from the (approximate) posterior distribution (bottom histogram in Fig. 13.3). Using the mixture model under model averaging and drawing Monte Carlo samples, the 95 percent posterior probability interval for the failure probability is 0.2568 to 1, which is similar to results of Lavine (1991) and Draper (1995), but slightly wider (the other analyses were based on modeling the number of O-ring failures while here, the indicator of at least one failure was the response). The uncertainty about whether temperature and/or pressure should be included leads to bimodality in the posterior distribution of the failure probability under BMA (bottom plot in Fig. 13.3) and a long left tail.

SUMMARY

In problems where there is a plethora of models and no scientific rationale that requires that a single model be used, Bayesian model averaging addresses many of the problems associated with model selection. Model uncertainty is almost always an issue in statistical analysis of data. Bayesian model averaging provides a coherent approach for incorporating uncertainty due to predictor selection, transformations, outliers, model form, and much more. The examples illustrate how BMA can be carried out without using any specialized software using objective prior distributions based on BIC. While posterior model probabilities can often be approximated using BIC, there is an extensive literature on other approaches for implementing BMA in a wide variety of applications, using both subjective and objective prior distributions.

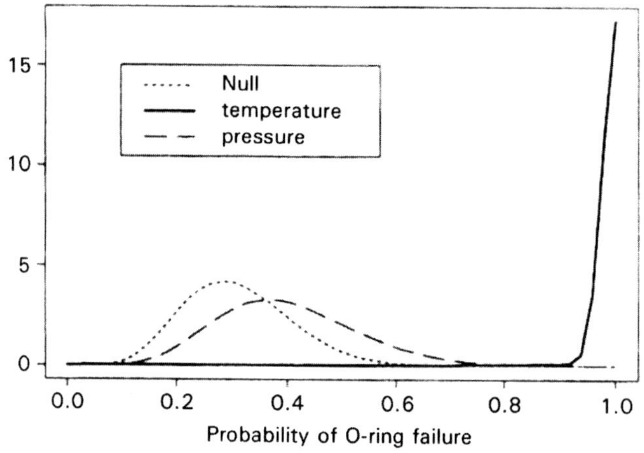

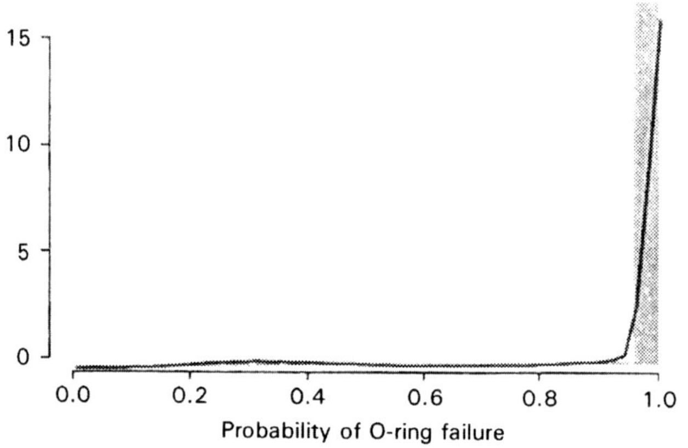

Figure 13.3 Posterior distributions for probability of O-ring failure at 31°F and pressure = 200. The top plot shows the posterior distributions of the probability of O-ring failure using the complementary log–log link functions for the null model (intercept only), and models with temperature alone and pressure alone. The smooth curve in the bottom plot shows the posterior distribution of the probability of O-ring failure under model averaging. The histogram is based on the Monte Carlo sample which was used to construct the density

Articles by Berger and Pericchi (2001), Chipman et al. (2001), Hoeting et al. (1999), and Clyde (1999) provide a description of current practice and open issues in model selection and model averaging, as well as additional history and references for the interested reader.

EXERCISES

13.1 Show that conditional distribution for γ_j given the other γ_k, $k \neq j$ and $\mathbf{Y}$ is Bernoulli. Describe how to implement a Gibbs sampler to explore the posterior distribution of models.

13.2 Consider the linear regression model, with Zellner's g-prior and the noninformative prior on σ^2, $p(\sigma^2) = 1/\sigma^2$.

(a) Find the predictive distribution of Y at $X = X_0$ under model averaging in the linear regression model.

(b) What is the mean of the predictive distribution?

(c) What is the variance of Y in the predictive distribution under model averaging?

(d) Rather than using Monte Carlo sampling, describe how one could construct a 95 percent prediction interval using the cumulative distribution functions under model averaging.

13.3 Let $\mathcal{M}$ denote the model $\mathbf{Y} = \boldsymbol{\beta}\mathbf{X} + \mathbf{u}$ and consider comparing $\mathcal{M}$ to the model where $\mathcal{M}^* : \boldsymbol{\beta} = 0$.

(a) Using a noninformative prior distribution for σ^2, $p(\sigma^2) = 1/\sigma^2$, and Zellner's g-prior for $\boldsymbol{\beta}$, $\boldsymbol{\beta} \sim N(0, \sigma^2 g(\mathbf{X}'\mathbf{X})^{-1})$, find the marginal likelihood for Y under model $\mathcal{M}$.

(b) Find the marginal likelihood for $\mathbf{Y}$ under the model $\mathcal{M}^* : \boldsymbol{\beta} = 0$ (use the same noninformative prior for σ^2).

(c) Using equal prior probabilities on $\mathcal{M}$ and $\mathcal{M}^*$, show that the posterior odds of $\mathcal{M}^*$ to $\mathcal{M}$ goes to a nonzero constant $(1+g)^{0.5(k-n)}$ as $\hat{\boldsymbol{\beta}}$ goes to infinity. (Even though one becomes certain that $\mathcal{M}^*$ is wrong, the posterior odds or Bayes' factor does not go to zero, which has led many to reject g-priors for model selection; this criticism applies equally for model averaging; see Berger and Pericchi (2001) for alternatives.)

(d) Using the marginal likelihood under $\mathcal{M}$, find the maximum likelihood estimate of g, $\hat{g}$. Using the Empirical Bayes g-prior ($\hat{g}$ substituted for g in the g-prior), what happens with the limit?

(e) For model $\mathcal{M}$, consider a Cauchy prior for $\boldsymbol{\beta} \mid \sigma^2$ Zellner and Siow, 1980) of the form

$$p(\boldsymbol{\beta} \mid \sigma^2) = \Gamma\left(\frac{q+1}{2}\right)\frac{\mid \mathbf{X}'\mathbf{X} \mid^{0.5}}{(n\sigma^2)^{0.5q}}\left(1 + \frac{\boldsymbol{\beta}'\mathbf{X}'\mathbf{X}\boldsymbol{\beta}}{n\sigma^2}\right)^{-0.5(q+1)}$$

rather than the conjugate normal. Show that the Cauchy distribution can be written as a scale mixture of normal distributions, where $\boldsymbol{\beta} \mid \sigma^2, \lambda$ is $N(0, \sigma^2/\lambda(\mathbf{X}'\mathbf{X})^{-1})$ and λ has a gamma(0.5, 2) distribution (parameterized so that the mean is 1).

(f) Investigate the limiting behavior of the posterior odds under the Cauchy prior distribution. *Using the scale mixture of normals representation, the marginal can be obtained using one-dimensional numerical integration.*

13.4 Consider an improper prior distribution for $\boldsymbol{\beta}$ and γ given σ^2: $p(\boldsymbol{\beta}_\gamma, \gamma)$ proportional to $\mid g\mathbf{X}_\gamma'\mathbf{X}_\gamma \mid^{0.5}$.

(a) What happens to the prior distribution of $\boldsymbol{\beta}_\gamma$ if g is multiplied by an arbitrary constant?

(b) Find the posterior distribution for $\boldsymbol{\beta}_\gamma$. What happens to the posterior distribution for $\boldsymbol{\beta}_\gamma$ if g is multiplied by an arbitrary constant?

(c) Find the posterior distribution of γ. What happens to the posterior distribution of γ when g is multiplied by an arbitrary constant?

(d) If the prior distribution is $p(\boldsymbol{\beta}_\gamma, \gamma) \propto c$, where c does not depend on the data, show that the posterior model probabilities are not invariant to scale changes in $\mathbf{X}$.

FURTHER READING

Akaike, H. (1973). "Information Theory and an Extension of the Maximum Likelihood Principle," in B. Petrox and F. Caski, Eds., *Second International Symposium on Information Theory*, 267.

Berger, J. O. and Pericchi, L. R. (2001). "Objective Bayesian Methods for Model Selection: Introduction and Comparison," in *IMS Lecture Notes—Monograph Series*, Volume 38, 135–193.

Chipman, H. (1996). "Bayesian Variable Selection with Related Predictors," *Canadian J. Statist.* **24**, 17–36.

Chipman, H., George, E. and McCulloch, R. (2001a). "The Practical Implementation of Bayesian Model Selection," in *IMS Lecture Notes—Monograph Series*, Volume 38, 65–134.

Clyde, M. (1999). "Bayesian Model Averaging and Model Search Strategies (with discussion)," in J. M. Bernardo, J. O. Berger, A. P. Dawid and A. F. M. Smith, Eds., *Bayesian Statistics 6*, Oxford University Press, Oxford, 157–185.

Clyde, M. (2001). "Discussion of 'the Practical Implementation of Bayesian Model Selection'," in *IMS Lecture Notes—Monograph Series*, Volume 38, 117–124.

Clyde, M. and George, E. (2000). "Flexible Empirical Bayes Estimation for Wavelets," *J. Royal Statist. Soc. B*, **62**, 681–698.

Dalal, S. R., Fowlkes, E. B. and Hoadley, B. (1989). "Risk Analysis of the Space Shuttle: Pre-Challenger Prediction of Failure," *J. Am. Statist. Assoc.* **84**, 945–957.

Dawid, A. and Lauritzen, S. (2001). "Compatible Prior Distributions," in E. I. George, *Bayesian Methods with Applications to Science, Policy, and Official Statistics, Selected Papers from ISBA 2000: The Sixth World Meeting of the International Society for Bayesian Analysis*, 109–118.

Draper, D. (1995). "Assessment and Propagation of Model Uncertainty (discussion, 71–97)," *J. Royal Statist. Soc. B*, **57**, 45–70.

Fernández, C., Ley, E. and Steel, M. F. (2001). "Benchmark Priors for Bayesian Model Averaging," *J. Econometrics*, 381–427.

Foster, D. P. and George, E. I. (1994). "The Risk Inflation Criterion for Multiple Regression," *Annals Statist.*, **22**, 1947–1975.

Furnival, G. M. and Wilson, R. W. J. (1974). "Regression by Leaps and Bounds," *Technometrics*, **16**, 499–511.

George, E. I. and Foster, D. P. (2000). "Calibration and Empirical Bayes Variable Selection," To appear in *Biometrika*.

George, E. I. and McCulloch, R. E. (1993). "Variable Selection via Gibbs Sampling," *J. Am. Statist. Assoc.*, **88**, 881–889.

George, E. I. and McCulloch, R. E. (1997). "Approaches for Bayesian Variable Selection," *Statistica Sinica*, **7**, 339–374.

Hodges, J. S. (1987). "Uncertainty, Policy Analysis and Statistics (c/r: P276–291)," *Statist. Sci.*, **2**, 259–275.

Hoeting, H. A., Madigan, D., Raftery, A. E. and Volinsky, C. T. (1999). "Bayesian Model Averaging: A Tutorial (with discussion)," *Statist. Sci.*, **14**, 382–417.

Hoeting, J., Raftery, A. E. and Madigan, D. (1996). "A Method for Simultaneous Variable Selection and Outlier Identification in Linear Regression," *Comput. Statist. Data Anal.*, **22**, 251–270.

Kass, R. E. and Raftery, A. E. (1995). "Bayes Factors," *J. Am. Statist. Assoc.*, **90**, 773–795.

Lavine, M. (1991). "Problems in Extrapolation Illustrated with Space Shuttle O-Ring Data (com: P921–922)," *J. Am. Statist. Assoc.*, **86**, 919–921.

Lavine, M. and Schervish, M. J. (1999). "Bayes Factors: What They Are and What They Are Not," *Am. Statist.*, **53**, 119–122.

Raftery, A. E., Madigan, D. and Hoeting, J. A. (1997). "Bayesian Model Averaging for Linear Regression Models," *J. Am. Statist. Assoc.* **92**, 179–191.

Ramsey, F. and Schafer, D. (2002). *The Statistical Sleuth: A Course in Methods of Data Analysis*.

Schwarz, G. (1978). "Estimating the Dimension of a Model," *Annals Statist.*, **6**, 461–464.

Smith, M. and Kohn, R. (1996). "Nonparametric Regression using Bayesian Variable Selection," *J. Econometrics*, **75**, 317–343.

Weisberg, S. (1985). *Applied Linear Regression* (Second Edition).

Zellner, A. (1986). "On assessing Prior Distributions and Bayesian Regression Analysis with *g*-Prior Distributions," in *Bayesian Inference and Decision Techniques: Essays in Honor of Bruno de Finetti*, 233–243.

Zellner, A. and Siow, A. (1980). "Posterior Odds Ratios for Selected Regression Hypotheses," in *Bayesian Statistics: Proceedings of the First International Meeting held in Valencia* (Spain), 585–603.

CHAPTER 14

Hierarchical Bayesian Modeling

*Alan M. Zaslavsky**

14.1 INTRODUCTION

Hierarchical modeling is a widely used approach to building complex models by specifying a series of more simple conditional distributions. It naturally lends itself to Bayesian inference, especially using modern tools for Bayesian computation. In this chapter we first present essential concepts of hierarchical modeling, and then suggest its generality by presenting a series of widely used specific models.

14.2 FUNDAMENTAL CONCEPTS AND NOMENCLATURE

14.2.1 Motivating Example

Suppose that for each of a collection of parameters $\boldsymbol{\theta} = \{\theta_i, i = 1, 2, \ldots, n\}$, a single observation $y_i \sim N(\theta_i, \sigma^2)$ is obtained. For concreteness, θ_i might be the mass of the i-th pill coming out of a pharmaceutical production process and y_i might be the measurement of its mass obtained using a scale with known error variance.

A naïve approach might estimate θ_i by the maximum likelihood estimator y_i. From a Bayesian standpoint, however, our inference for θ_i should combine a prior distribution $\pi_{\theta i}$ with the likelihood. Lacking additional information to differentiate among the indices i, we might assume that *a priori* the parameters $\{\theta_i\}$ are independent and identically distributed conditional on some other parameters, for example $\theta_i \sim N(\mu, \tau^2)$. In our example, μ and τ^2 are the mean and variance of the masses of the pills (assumed to be infinitely numerous if we allow the production process to go on indefinitely). Note that if we let $\tau^2 \to \infty$, that is, assume a uniform prior on $(-\infty, \infty)$ for θ_i, then y_i is also the posterior mean and mode of θ_i, but this assumption is physically implausible since we are fairly sure that pills always fall within some finite range of sizes. Furthermore, it is statistically implausible once we have observed a collection of values of $\{y_i\}$, with some finite sample variance. Instead,

* Department of Health Care Policy, Harvard Medical School, Boston, Massachusetts; E-mail: zaslavsk@hcp.med.harvard.edu

assume for the moment that σ^2 and τ^2 are known finite quantities. The unknown parameter μ itself has some prior distribution with density π_μ, so the joint prior density for $(\boldsymbol{\theta}, \mu)$ is $\pi_\mu(\mu) \prod_i \pi_\theta(\theta_i \mid \mu)$ and the joint prior density of $\boldsymbol{\theta}$, the masses of the items, is obtained by integrating out μ,

$$p(\boldsymbol{\theta}) = \int \pi_\mu(\mu) \prod_i \pi_\theta(\theta_i \mid \mu)\, d\mu \tag{14.1}$$

If we assume that μ itself has a uniform (improper) prior distribution, then the posterior distribution for θ_i can be readily obtained as

$$\theta_i \sim N(\tilde{\theta}_i, v_i) \tag{14.2}$$

where

$$\tilde{\theta}_i = (y_i/\sigma^2 + \bar{y}/\tau^2)/(1/\sigma^2 + 1/\tau^2),$$

and

$$v_i = \sigma^2(\sigma^2/n + \tau^2)/(\sigma^2 + \tau^2),$$

and $\bar{y} = \sum y_i/n$ is the sample mean. The posterior distribution combines information from the individual item with information about the entire collection of (related) items. Because each posterior mean $E[\theta_i \mid \mathbf{y}]$ is closer to the grand mean than the corresponding observation y_i, estimators like this one are called "shrinkage estimators." If the scale is accurate relative to the variation among the objects, $\sigma^2 \ll \tau^2$, then more weight is given to the measure for the individual object; conversely if $\tau^2 \ll \sigma^2$, more weight is given to the information about the mean for the collection (more "shrinkage"). Note that $\text{var}[\theta_i \mid y_i] = \sigma^2$, but $\text{var}[\theta_i \mid \mathbf{y}] = v_i < \sigma^2$ for $n > 1$, so using the additional information from the rest of the collection always reduces the posterior variance of the parameter.

This example, presented in the seminal article by Lindley and Smith (1972), illustrates some of the key features of hierarchical models: parameters organized into a hierarchy of levels, association of specific parameters with units of a population, and use of the entire collection of data to make estimates for each unit. Hierarchical modeling has advanced from being a theoretical research topic to being an extremely useful tool for statistical inference and the context for some of the most common applications of Bayesian methods.

14.2.2 What Makes a Hierarchical Model?

While it is difficult to formulate a definition that includes all of the many applications of the notion of hierarchical modeling, the key ideas are captured in the following common features.

Multilevel Parametrization

Hierarchical models are generally characterized by expression of a *marginal* model $P(\mathbf{y})$ (where $\mathbf{y}$ represents the entire data vector) through a sequence of *conditional* models involving *latent variables* (see also Section 10.8). Thus, in the motivating example (if σ^2 is a fixed known quantity), there are two conditional distributions: that of the data, $\mathbf{y} \mid \boldsymbol{\theta}$, and that of the parameters $\boldsymbol{\theta} \mid (\mu, \tau^2)$.

The sequence of conditional models can be represented graphically as in Figure 14.1(a), with arrows pointing to each variable from the conditioning variables (parents) of its prior model. In this representation, variables can be arranged in a series of *levels*, with data at the bottom and hyperparameters at the top, so all the arrows representing dependencies point downwards. Variables are assumed to be independent conditional on their common "ancestors" or on intervening variables in the diagram; thus, $\mathbf{y}$ is independent of (μ, τ^2) conditional on θ. It is common to simplify presentation of the model by explicitly conditioning only on the immediate parents and implicitly assuming the conditional independence relationships implied by the diagram.

Hierarchically Structured Data

Typically, data that are modeled hierarchically have a structure in which the basic observational units are grouped into larger units, possibly grouped into a hierarchy of larger units. For example:

1. In a biological experiment, rats are grouped into litters.
2. In a longitudinal analysis, observations over time are grouped by subject.
3. Students are grouped into classes, which are grouped into schools, which are grouped by districts.

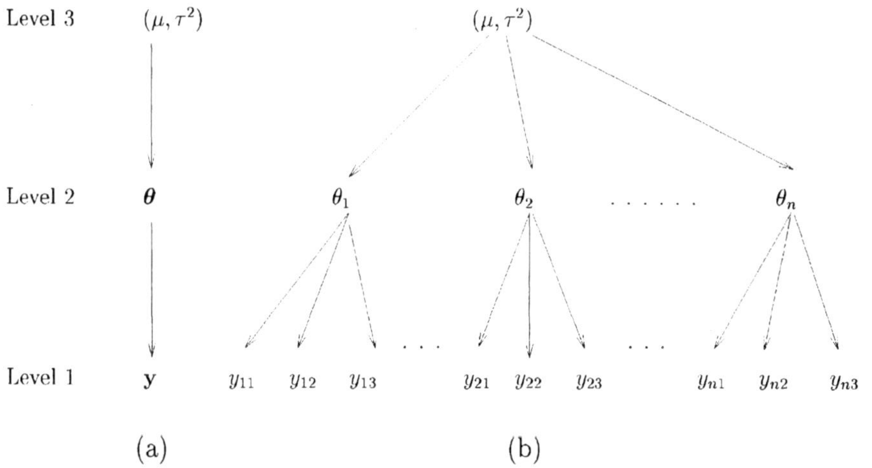

Figure 14.1 Graphical representations of a simple hierarchical model: (a) condensed and (b) expanded showing conditional independence relationships

Such data are commonly labeled using a vector of indices, from the highest level (largest units) to the lowest (single observations). Thus, in Example (1), above, the observation on the jth rat in the ith litter is y_{ij} while in Example (3), the observation on the lth student in classroom k of school j in district i is y_{ijkl}.

In these examples, the population structure is entirely nested. In that case, the units unambiguously can be arranged in levels that are related by a branching structure. Other structures are also common. For example, suppose each patient resides in a city that may have many health plans (in the health care model of the United States with multiple, competing, health plans), and is served by a health plan that operates across many cities. Then y_{ijk} might be the observation on the kth patient of those served by plan i in city j. In this case patients are nested within the cross-classification of plan and area. Other designs might combine crossing and nesting in different ways, as well as more complex structures such as multiple membership.

Correspondence of Parameters to Population Structures, and Conditional Independence

These population structures should be reflected in the parametrization of the model if they are relevant to the outcomes of interest. Typically, there is a parameter (scalar or vector) corresponding to each unit of the population structure, and the prior distribution of the parameter (or data, for the unit of observation) is conditioned on the parameters of the higher-level units in which the unit of interest is included (including, possibly, some parameters related to the entire population). Thus, the simple diagram Figure 14.1(a) for all the vectors of parameters can be expanded to one that more nearly corresponds to the population structure (Fig. 14.1(b)). The expanded diagram makes more explicit the conditional independence relationships among the parameters.

Figure 14.2 represents a mixed model (See Section 14.3.3) with a single random intercept $\theta_i \sim N(\mu, \tau^2)$ and regression in the bottom level, $y_{ij} \sim N(\theta_i + x'_{ij}\boldsymbol{\beta}, \sigma^2)$. In this case the conditional distribution of y_{ij} depends on $\boldsymbol{\beta}$, a general (population-level or Level 3) parameter, as well as the Level 2 parameter θ_i.

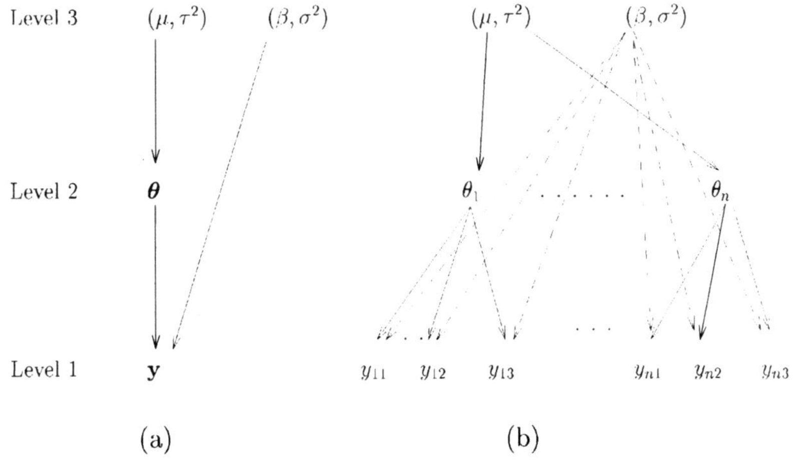

Figure 14.2 Graphical representations of a mixed model

As before, distributions of parameters or data within any unit are independent of those in another unit at the same level, conditional on the parameters of the unit's ancestor units at higher levels. Thus, in our initial example, $\theta_i \perp \theta_{i'} \mid \mu, \tau^2, y_{ij} \perp y_{i',j'} \mid \mu, \tau^2$, and $y_{ij} \perp y_{i,j'} \mid \theta_i$, where $A \perp B \mid C$ should be read "A is independent of B, conditional on C." Such conditional independence is crucial to the interpretation of hierarchical models.

14.2.3 Marginalization, Data Augmentation and Collapsing

Many hierarchical models can be expressed in several different forms, using different sets of parameters. Such alternate forms may be derived through marginalization, data augmentation and collapsing.

Marginalization involves integration over some or all parameters of the set of conditional models. For example, in the hierarchical model $y_i \sim N(\theta_i, 1)$, $\theta_i \sim N(\mu, \tau^2)$, we can integrate over θ_i to obtain the marginal model $y_i \sim N(\mu, \tau^2 + 1)$. Marginalization might be accompanied by reduction of the data to sufficient statistics for the marginal model. If the marginalized distribution is tractable (as in this case), then it can be used directly in inference. (See Section 14.3.2 for a more complex example.)

Data augmentation is the converse operation to marginalization, and involves *adding* parameters to the model to make the distributions involved more tractable (Tanner and Wong, 1987). It is useful when the benefits of simplifying the distributions outweigh the burden of adding parameters (Section 14.6.1).

Collapsing combines two or more levels of a model into a single distribution. It is most commonly applied with linear models in which the mean and variance structures are independent of each other. For example, consider the model $y_{ij} \sim N(\theta_i, \sigma^2)$, $\theta_i \sim N(\mu, \tau^2)$. This model could be rewritten as $\varepsilon_{ij} \sim N(0, \sigma^2)$, $\theta_i^* \sim N(\mu, \tau^2)$ and a final (deterministic) relationship $y_{ij} = \mu + \theta_i^* + \varepsilon_{ij}$. Thus the parameters $\boldsymbol{\theta}^*, \boldsymbol{\varepsilon}$ have independent distributions conditional only on the variance components, and therefore could be placed on a single level. (The same operation could be applied to models with any number of levels of nesting.)

Furthermore, the terms $\theta_i^*, \varepsilon_{ij}$ could be combined by a marginalization step into a single term $\varepsilon_{ij}^* = \theta_i^* + \varepsilon_{ij}$; then $\boldsymbol{\varepsilon}^*$ has covariance matrix $\tau^2\mathbf{J} + \sigma^2\mathbf{I}$, where $\mathbf{I}$ is an identity matrix and $\mathbf{J}$ is a block-diagonal matrix of 1's and 0's. For example, with two Level 2 clusters, $i = 1, 2$, and two Level 1 units in each cluster, $j = 1, 2$, we would write

$$\mathbf{J} = \begin{pmatrix} 1 & 1 & 0 & 0 \\ 1 & 1 & 0 & 0 \\ 0 & 0 & 1 & 1 \\ 0 & 0 & 1 & 1 \end{pmatrix}, \tag{14.3}$$

indicating that the corresponding effects have correlation 1, that is, they are identical, for observations (1, 1) and (1, 2), but these are independent of the effects for (2, 1) and (2, 2). This formulation reduces the number of levels in the model at the cost of making the covariance structure for one of the parameters more complex.

The choice among these three ways of expressing this model is largely a matter of convenience and interpretability. For some purposes, the parameters θ_i might have an interesting interpretation. For others, it might be more convenient to estimate the two parameters of the marginalized covariance structure than to deal with a model with a large number of random effects.

14.2.4 Hierarchical Models, Exchangeability, and De Finetti's Theorem

The alert reader will have noted the resemblance of Equation (14.1) to de Finetti's results on exchangeable distributions (Section 10.5); Lindley and Smith (1972) take these results as a starting point of their exposition. A two-level hierarchical model might be regarded simply as a device for characterizing an exchangeable distribution. From a predictivist standpoint (Chapter 10), no physical or scientific meaning need be attached to the Level 2 parameters $\boldsymbol{\theta}$; inference for these parameters can be simply a means to predict future observations. On the other hand, often an intuitive meaning can be attached to these latent parameters.

More complex hierarchical models extend de Finetti's results to more complex relationships of exchangeability. In the three-level model $y_{ij} \sim N(\theta_i, \sigma^2)$, $\theta_i \sim N(\mu, \tau^2)$, the observations are only *partially exchangeable*. (Note that there is no consensus on how the number of levels are counted, as with many aspects of notation for hierarchical models.) The joint distribution is *not* invariant under arbitrary permutation of all observations, since this would break up the clusters, which are meaningful units. The distribution is exchangeable, however, under permutation of the labels j for any fixed i (exchangeability of observations within a cluster), and also under permutation of the labels i (exchangeability of clusters).

Other model specifications correspond to different exchangeability assumptions. An alternative model for doubly subscripted data is $y_{ij} \sim N(\theta_i + \lambda_j, \sigma^2)$, $\theta_i \sim N(\mu, \tau_\theta^2)$, $\lambda_j \sim N(0, \tau_\lambda^2)$. This model is invariant under permutation of the indices i or of the indices j, but not under independent permutation of the indices j for different values of i. Thus this model represents a cross-classification rather than nesting.

In some models (for example, regression models with arbitrary covariate values associated with each set of indices), exchangeability does not apply directly. Nonetheless, understanding the structure of exchangeability is crucial to specification and interpretation of hierarchical models.

14.3 APPLICATIONS AND EXAMPLES

14.3.1 Generality of Hierarchical Models

Hierarchical models, like most good ideas, have been invented many times. Consequently, there are many names for specific models within this framework. Table 14.1 names and describes some of the commonly used models from the important class of hierarchical linear models. Other hierarchical models, more broadly conceived, include Bayesian neural networks (Section 10.8).

Table 14.1 Summary of Some Common Hierarchical Linear Models

Name of Model	Description
Hierarchical linear models	Specification at each stage is linear; usually, distributions are normal.
Random effects models	All parameters are random, associated with units at various levels.
Mixed models	Both random effects and fixed (not unit-specific) coefficients appear in the same model.
Variance component models	Additive random effects, typically modeled as independent and normally distributed, for nested and/or crossed units.
Generalized linear mixed models	Mixed model in which Level 1 (observational) is a generalized linear model.
Random coefficient models	At least some regression coefficients at a level are random, associated with units at a higher level.
Growth curve models	Random coefficient models in which Level 2 clusters are groups of observations over time on a subject.

14.3.2 Variance Component Models

The variance component model is one of the simplest hierarchical models. A two-level nested model can be specified as follows:

$$\begin{aligned}
&\text{Level 1: } y_{ij} \sim N(\theta_i, \sigma^2)\\
&\text{Level 2: } \theta_i \sim N(\mu, \tau^2)\\
&\text{Level 3: } \mu \sim \pi_\mu, \sigma^2 \sim \pi_\sigma, \tau^2 \sim \pi_\tau
\end{aligned} \tag{14.4}$$

This is a natural model for clustered data with no covariates. It characterizes the observation for unit j of cluster i as the sum of independent contributions from the unit and the cluster. With more than two nested levels, the generalization to additional variance components is obvious.

Variance component models can also be defined for non-nested data structures. Suppose that patients rate the quality of care they receive in their city from their health plan, with a crossed structure as described in the example in Section 14.2.2. A model for a variable defined on cross-classified categories indexed by i (city) and j (health plan), with individual units k (patients) nested within the cross-classification, is:

$$\begin{aligned}
&\text{Level 1: } y_{ijk} \sim N(\theta_{ij}, \sigma^2)\\
&\text{Level 2: } \theta_{ij} \sim N(\mu + \alpha_i + \gamma_j, \tau_\theta^2)\\
&\text{Level 3A: } \alpha_i \sim N(0, \tau_\alpha^2)\\
&\text{Level 3B: } \gamma_j \sim N(0, \tau_\gamma^2)\\
&\text{Level 4: } \mu \sim \pi_\mu, \sigma^2 \sim \pi_\sigma, \tau_\theta^2 \sim \pi_{\tau_\theta}, \tau_\alpha^2 \sim \pi_{\tau_\alpha}, \tau_\gamma^2 \sim \pi_{\tau_\gamma}
\end{aligned} \tag{14.5}$$

Variance component models can be collapsed to three levels by explicitly writing the outcome as a sum of random effects. Thus, an alternative specification of Equation (14.5) is:

$$\begin{aligned}
&\text{Level 1: } y_{ijk} = \mu + \alpha_i + \gamma_j + \delta_{ij} + \varepsilon_{ijk} \\
&\text{Level 2A: } \alpha_i \sim N(0, \tau_\alpha^2) \\
&\text{Level 2B: } \gamma_j \sim N(0, \tau_\gamma^2) \\
&\text{Level 2C: } \delta_{ij} \sim N(0, \tau_\delta^2) \\
&\text{Level 2D: } \varepsilon_{ijk} \sim N(0, \sigma^2) \\
&\text{Level 3: } \mu \sim \pi_\mu, \sigma^2 \sim \pi_\sigma, \tau_\alpha^2 \sim \pi_{\tau_\alpha}, \tau_\gamma^2 \sim \pi_{\tau_\gamma}, \tau_\delta^2 \sim \pi_{\tau_\delta}
\end{aligned}$$

In this formulation, Level 1 is deterministic. In our example, α_i might represent the general quality of care in city i, γ_j the general quality of health plan j, δ_{ij} the interaction of the health plan and city (indicating whether plan j does better or worse than expected in city i), and ε_{ijk} the specific experiences of a single patient.

As noted above, this model can be further collapsed into a two-level model with a structured covariance matrix determined by the variance component parameters. In the latter formulation, the relationships of exchangeability and conditional independence are no longer explicit in the hierarchical diagram but rather must be deduced from the symmetries of the covariance matrix. Alternatively, the same model could be expressed in three levels by collapsing levels 2A, 2B, and 2C to express the joint distribution of the entire vector $\boldsymbol{\theta}$,

$$\begin{aligned}
&\text{Level 1: } y_{ij} \sim N(\theta_{ij}, \sigma^2) \\
&\text{Level 2: } \boldsymbol{\theta} \sim N(\mu\mathbf{1}, \tau_\alpha^2\mathbf{J} + \tau_\gamma^2\mathbf{K} + \tau_\delta^2\mathbf{L}),
\end{aligned} \tag{14.6}$$

with Level 3 as before, where $\mathbf{J}, \mathbf{K}, \mathbf{L}$ are block-structured matrices.

14.3.3 Random Coefficient Models, Mixed Models, Longitudinal Data

In a random coefficient model, regression coefficients at some level vary across units at a higher level. A simple three-level example is:

$$\begin{aligned}
&\text{Level 1: } y_{ij} \sim N(\mathbf{z}_{ij}'\boldsymbol{\beta}_i, \sigma^2) \\
&\text{Level 2: } \boldsymbol{\beta}_i \sim N(\mathbf{x}_i\boldsymbol{\lambda}, \mathbf{T}) \\
&\text{Level 3: } \boldsymbol{\lambda} \sim \pi_\lambda, \sigma^2 \sim \pi_\sigma, \mathbf{T} \sim \pi_T
\end{aligned} \tag{14.7}$$

where $\mathbf{z}_{ij}$ is a covariate vector associated with a Level 1 unit, and $\mathbf{x}_i$ is a covariate *matrix* associated with a Level 2 unit. This formulation can be extended to any number of levels.

Like the variance component model, this one can be flattened into two stochastic levels, with an additional deterministic level at the bottom to combine the components; in our simple example, the latter has the form $y_{ij} = \mathbf{z}_{ij}'\boldsymbol{\beta}_i + \varepsilon_{ij}$.

This model can also be written using a structured covariance matrix; for example

$$\mathbf{y}_i \sim N(\mathbf{z}_i\mathbf{x}_i\lambda, \ \mathbf{S}_i)$$

where $\mathbf{z}_i$ is the matrix whose rows are the vectors $\mathbf{z}_{ij}$ and $\mathbf{S}_i = \mathbf{S}_i(\sigma^2, \mathbf{T}) = \mathbf{z}_i\mathbf{T}\mathbf{z}_i' + \sigma^2\mathbf{I}$. Often, however, the Level 2 unit-specific regression coefficients are of intrinsic scientific interest, so a parametrization that makes them explicit is useful.

In the analysis of longitudinal data, Level 2 represents subjects and Level 1 the observations on subjects at different times. Then in a growth curve model, $\boldsymbol{\beta}_i$ might describe the trajectory of values (constant and linear and quadratic coefficients of time). It is common in these analyses to combine the Level 1 regression model with a structured covariance matrix for ε_i representing a plausible process for consecutive residuals, such as an autoregressive or random-walk model.

A model that includes both unit-specific and nonunit-specific regression coefficients is called a "mixed model." (The multivariate mixed model is discussed in Section 12.5.) In the "flattened" notation, we might have

$$y_{ij} = \mathbf{z}_{ij}'\boldsymbol{\beta}_i + \mathbf{w}_{ij}'\boldsymbol{\eta} + \varepsilon_{ij}. \tag{14.8}$$

14.3.4 Models with Normal Priors and Non-Normal Observations

Normal models can be extended by letting Level 2 specify the distributions of the parameters of a non-normal Level 1 (observational) model. The normal linear model is a powerful and flexible tool for describing multivariate relationships, while a non-normal observational model can be selected to correspond to the specific structure of the data at hand, such as count data (commonly modeled using Poisson regression) or binomial data (logistic or probit regression).

In an important subclass of models, the Level 2 parameter is the linear predictor in a generalized linear model (McCullagh and Nelder, 1989), yielding a generalized linear mixed model (GLMM). For binary data, we might replace Level 1 of Equation (14.4) with a logistic regression model

$$\operatorname{logit} P(y_{ij} = 1 \mid \theta_i) = \theta_i$$

(where logit represents the logistic transformation, $\operatorname{logit} p = \log[p/(1-p)]$) or a probit model,

$$\Phi(P(y_{ij} = 1 \mid \theta_i)) = \theta_i \tag{14.9}$$

(where Φ is the standard normal cdf). Parameters of other types of nonlinear models can also be modeled in a similar fashion (Pinheiro and Bates, 2000). The normal priors are not conjugate to the Level 1 distributions for these models.

14.3.5 Non-Normal Conjugate Models

A class of three-level models is obtained by assuming any exponential family distribution at Level 1 and the conjugate family at Level 2. In such cases the posterior distribution of the Level 2 parameters conditional on data and Level 3 parameters is also a member of the conjugate family, and the marginal distribution of the Level 1 sufficient statistics is often a member of a familiar family. One commonly used representative of this class is the Gamma–Poisson regression model (Christiansen and Morris, 1997a), with

$$\begin{aligned} &\text{Level 1: } y_{ij} \sim \text{Poisson}(\lambda_i) \\ &\text{Level 2: } \lambda_i \sim \text{Gamma}(a, b_i) \end{aligned} \tag{14.10}$$

where $b_i = a \exp(\mathbf{x}_i'\beta)$ and the Gamma(a, b) density is $(\lambda^{a-1}/b^a)\exp(-\lambda/b)$. Then $\lambda_i \mid \mathbf{y}, a, b \sim \text{Gamma}(a + n_i, b_i + y_{i+})$, where n_i is the number of observations in cluster i and $y_{i+} = \sum_{j=1}^{n_i} y_{ij}$. The marginal distribution of the sufficient statistic y_{i+} is negative binomial. Similarly, a Beta–Binomial model has Beta conditionals and a beta–binomial or Pólya–Eggenburger marginal distribution (Johnson and Kotz, 1969, pp. 78–79, 229–231).

Because the posterior expectation with a conjugate prior (conditional on the hyperparameters) is a linear function of the data (Diaconis and Ylvisaker, 1979), these conjugate models have a simple "uniform shrinkage" interpretation.

14.4 INFERENCE IN HIERARCHICAL MODELS

14.4.1 Levels of Inference

Depending on the scientific questions of interest, the analyst might be interested in inferences concerning parameters at one or more levels of a hierarchical model. The hyperparameters (those at the top level) are general, characterizing the entire population or process of interest. Thus, for example, inferences about hyperparameters in a multicenter clinical experiment might concern the average magnitude of a treatment effect, the amount of variation in that effect across centers, and the amount of residual variation in the clinical outcomes. In an analysis of educational outcomes using a multilevel mixed model, inferences for hyperparameters might characterize the amount of variation attributable to schools, teachers, and individual students as well as the systematic influence of observable characteristics at each level on the outcomes.

Parameters at levels above the observational Level 1 but below the top level characterize specific units (possibly at several different levels). In a wide variety of applications, inference about specific units is essential to guide action. Parameters for such units define predictive distributions for future observations from that unit; thus in Equation (14.4), θ_i is the expected value of a future observation from unit i and we would integrate over the posterior distribution of (θ_i, σ^2) to obtain the predictive distribution of a new observation.

Inference for these variables, conditional on variables at higher and lower levels of the model, involves a straightforward application of Bayes' theorem. The prior is determined by the immediate parents of the variable and the likelihood by the immediate children. This calculation does not require acceptance of a Bayesian statistical philosophy, since the prior distribution is in some way estimated from the data, and hence such inferences are widely acceptable.

An early illustration of "shrinkage" estimation by Efron and Morris (1975) pointed out how relative standings of baseball players might be greatly affected by shrinkage under a hierarchical model, because the number of at-bats on which they are based varied greatly. Thus, a naïve interpretation of the sample proportion of hits would suggest that the best and worst batters in the major leagues are those with one or zero hits, respectively, on one at-bat. These estimates would be shrunk strongly toward the overall mean, however, and a player whose batting average is above average based on a large number of trials emerges with the highest predicted rate for the remainder of the season.

In analysis of data from sample surveys, estimates are often needed for domains (geographic areas, population subgroups) for which samples are inadequate to provide direct estimates based only on data from the area, such as the classical unbiased survey estimates, of adequate precision. Small-area estimation for such domains is increasingly conducted using hierarchical models, including the simple three-level model of Fay and Herriot (1979) as well as more complex models that include covariates at both Levels 1 and 2; see the review by Ghosh and Rao (1994), or Citro and Kalton (2000) for a recent example.

Similar methods are used for profiling of the outcomes obtained by providers of medical care (Normand et al., 1997; Christiansen and Morris, 1997b). The units being compared (physicians or hospitals) have very different numbers of cases undergoing a particular medical procedure. Hence the posterior variances of estimates, and the amount of shrinkage of posterior means, also differs greatly. Various methods of competitively ranking the providers using inferences from a hierarchical model (posterior mean, probability of being in worst 10 percent of units, probability of being below a fixed threshold) might then be substantially different from each other as well as from sample proportions of successful outcomes. Considerable attention has been given to methods for summarizing and presenting such inferences in a variety of contexts (Goldstein and Spiegelhalter, 1996).

In some settings, predictions might be required for Level 1 variables. Although conventionally the observed values are at Level 1, there might be other, unobserved, values that are of interest. For example, a hierarchical model might be used to fill in missing values in a data set (Schafer and Yucel, 2002), or to make predictions for future values.

14.4.2 Full Bayes' Inference

The general arguments for full Bayes inference appear throughout this volume and do not require reiteration here. Several specific arguments, however, are worth mentioning. First, inferences at several levels of the model might be required.

Because inferences at intermediate levels are inherently Bayesian, it is natural to use a fully Bayesian inference.

Secondly, fully Bayesian computations in a hierarchical model can often be broken down into simple components using the conditional independence relationships of the model. These strategies are less readily applicable in the maximum likelihood framework.

Thirdly, asymptotic methods are difficult to apply in hierarchical models, owing to the difficulty of expressing the marginal likelihood in closed form (Section 14.5.1).

14.4.3 Priors for Hyperparameters of Hierarchical Models

Selection of priors for hierarchical models is somewhat different than in superficially similar nonhierarchical models. Consider the simple variance components model

$$y_i|\theta_i \sim N(\theta_i, 1), \quad \theta_i|\tau \sim N(0, \tau^2), \tag{14.11}$$

with τ^2 the single unknown hyperparameter. The integrated model is $y_i \sim N(0, \tau^2 + 1)$. The likelihood is $(1 + \tau^2)^{-0.5n} \exp(-ns^2/2(\tau^2 + 1))$, where $s^2 = n^{-1} \sum y_i^2$, and is bounded away from zero in the neighborhood of $\tau^2 = 0$. Hence any prior with an infinite mass in the neighborhood of zero will yield a posterior that also has an infinite mass, that is, a degenerate distribution. This excludes some of the popular default priors used in the nonhierarchical normal model. On the other hand, an (improper) uniform prior for τ^2 yields a proper posterior when there are three or more observations.

Similarly, consider the categorical errors-in-variables model with θ_i, y_i belonging to $\{0, 1\}$ and $y_i = \theta_i$ with probability p and $1 - \theta_i$ otherwise, with a (constant) logistic regression at Level 2, $\text{logit}\, P(\theta_i = 1 \mid \beta) = \beta$. The likelihood $L(\beta \mid \mathbf{y}) \to p^{y^+}(1-p)^{n-y^+} > 0$ as $\beta \to +\infty$, while $L(\beta \mid \mathbf{y}) \to p^{n-y^+}(1-p)^{y^+} > 0$ as $\beta \to -\infty$, where there are n observations and $y^+ = \sum y_i$. Hence with a uniform prior the posterior is improper. If we reparameterize the Level 3 parameter as a probability $q = (1 + \exp(-\beta))^{-1}$ instead of a logit, the prior is Beta(0,0) and the posterior, like the prior, concentrates all its mass around $q = 0$ and $q = 1$. The difference between the degenerate posterior (all mass concentrated at one or a few points) and the improper one is simply a transformation of scale.

In simple models, the propriety of the posterior with an improper prior may be easy to check. As the examples above illustrate, even with moderately complex hierarchical models, this becomes more difficult, and proving the appropriate conditions is an active research topic (Ibrahim and Laud, 1991; Ghosh et al., 1998). It is easy to make mistakes in this area because a hierarchical model is specified through several conditional distributions, each of which may lead to conditionally proper posteriors although the posterior for the entire model is improper or degenerate. In either of our examples, the posterior of the general parameter (variance component τ^2 or regression coefficient β) is proper conditional on the Level 2 parameters $\{\theta_i\}$, but not conditional only on the data.

If only proper prior distributions are used, the propriety of the posterior follows. This suggests the use of almost vague, but proper prior distributions, such as Gamma distributions with fractional degrees of freedom for variance components. Even these distributions, which approach impropriety, may give undesirable results in modeling (Natarajan and McCulloch, 1998). The sensitivity of inferences to the choice among reasonable modeling alternatives depends on the specifics of the data, and sensitivity analyses are therefore particularly important in hierarchical modeling.

14.5 RELATIONSHIP TO NON-BAYESIAN APPROACHES

14.5.1 Maximum Likelihood Empirical Bayes and Related Approaches

Some popular estimation strategies use Bayes' theorem for inference about random effects, but non-Bayesian methods for estimation of the hyperparameters. Estimation of the latter may be by the method of moments (Swallow and Searle, 1978) or by maximum likelihood; the latter is most general and generally more efficient. The term "Empirical Bayes" is commonly used to describe these approaches (Robbins, 1955), because the prior distribution of the random effects is estimated, in contrast to methods in which the hyperparameters are known *a priori* (See also Section 8.5). Because this term is also applied to any application of Bayes methods to real data, more precise terms, such as maximum likelihood Empirical Bayes (MLEB) and Bayes Empirical Bayes (Deely and Lindley, 1981), should be used.

Because the full likelihood includes parameters whose number grows in proportion to the number of units at Levels 2 and above, estimates based on maximization of this likelihood are not typically consistent as the number of such parameters goes to infinity. For example, in the simple model $\theta_i \sim N(0, \tau^2)$, $i = 1, 2, \ldots, n$, $y_{ij} \sim N(\theta_i, \sigma^2)$, $j = 1, 2$, with $\sigma^2 \ll \tau^2$, maximization of the full likelihood yields $\hat{\sigma}^2 \to \sigma^2/2$ (the Neyman–Scott problem; Neyman and Scott, 1948). Hence, the appropriate likelihood to maximize for MLEB is the marginal likelihood, after integrating out the random effects (Section 14.2.3). Commonly, the fixed regression coefficients are also integrated out under a uniform prior to obtain Restricted Maximum Likelihood (REML) estimates (Corbeil and Searle, 1976). In normal linear models, the remaining parameters are variance components. Bayesian estimation, of course, usually integrates over all parameters.

Inferences for random effects under MLEB typically underestimate the variability of the prediction, because they plug in estimates of the hyperparameters rather than drawing from their posterior distribution. More sophisticated estimators of predictive variance include a component for estimation of the hyperparameters, based on an asymptotic calculation (Prasad and Rao, 1990) or a resampling variance estimation technique (Laird and Louis, 1987). These methods may fairly well approximate a full Bayesian inference when the data support precise estimates of the hyperparameters (sufficiently so that substituting a point estimate is almost equivalent to integrating over a posterior distribution) or when asymptotic approximations are adequate. These approximations break down, however, when maximum likelihood

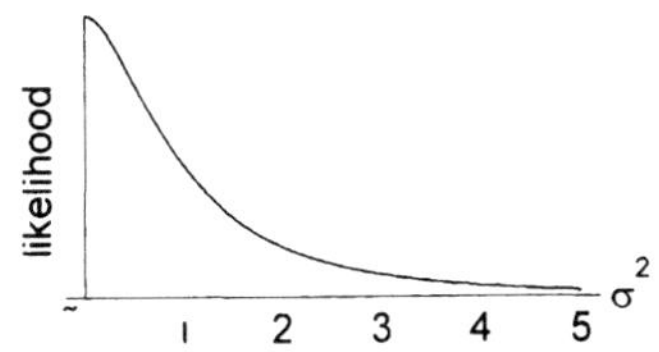

Figure 14.3 Likelihood for model (14.11) with $n = 8, s^2 = 1$

estimates fall on the boundary of the parameter space. In the model (14.11), if $s^2 = 1$, the likelihood (and posterior distribution under a uniform prior on $(0, \infty)$ for τ^2) appears as in Figure 14.3, with MLE $\hat{\tau}^2 = 0$; hence MLEB estimates all $\theta_i = 0$ with probability 1. A Bayesian analysis, by integrating over the posterior distribution of τ^2, allows for uncertainty about θ_i. The same phenomenon occurs with more realistic regression models, in which case MLEB treats the regression predictions as certain; see Rubin (1980) for an example.

14.5.2 Non-Bayesian Theoretical Approaches: Stein Estimation, Best Linear Unbiased Predictor

Estimators such as (14.2) were first justified on purely frequentist grounds. A classical result by Stein (1955) demonstrated that under an observational model $y_i \sim N(\theta_i, 1)$ as in Section 14.2.1, but with no assumed relationship of the parameters $\boldsymbol{\theta} = \{\theta_i\}$, the sample mean vector $\mathbf{y}$ is inadmissable as an estimator of $\boldsymbol{\theta}$, because another estimator, similar to Equation (14.2), has smaller expected mean squared error regardless of the value of $\boldsymbol{\theta}$. Hence, these estimators are sometimes known as Stein estimators (Efron and Morris, 1975). (See also James and Stein, 1961, and Chapter 11, and extensions to more complex models by Brown (1966) and others.)

Despite the similarity of the Stein and hierarchical Bayes' estimators, however, their philosophical bases are quite different, because the hierarchical model makes explicit assumptions about the relatedness and exchangeability of the parameters that are absent from the frequentist derivation. The Bayesian's starting point includes a model for the prior distribution of $\boldsymbol{\theta}$, as well as the observational model. Therefore, the thoughtful Bayesian modeler is defended against common criticisms of Stein estimators. For example, paradoxically, estimates of the masses of six fleas and an elephant can be improved in the frequentist sense by shrinking toward a common mean, but a Bayesian modeler aware of the heterogeneity of this collection would be unlikely to make the assumption of exchangeability necessary to perform such an operation. On the other hand, when sensible assumptions can be made relating the components of $\boldsymbol{\theta}$, deriving the corresponding estimators (or full posterior distributions) under the Bayesian hierarchical model is essentially a computational problem, which can be addressed using standard tools and without developing new theory for each case.

Another non-Bayesian approach to estimation of linear functions of random effects in linear hierarchical models assumes that the variance components are known, and then derives the best linear (in the data) unbiased predictor of the desired functions, the BLUP (Henderson, 1975). When an estimate of the variance components is substituted, the estimator is an Empirical BLUP or EBLUP, by analogy to Empirical Bayes. In linear models with normal random effects, the results of this approach generally agree with those obtained using MLEB (Hulting and Harville, 1991).

14.5.3 Contrast to Marginal Modeling Approaches with Clustered Data

An alternative approach to modeling clustered data is to fit a marginal model (one that ignores the clustering) but to estimate variances taking into account clustering of the data. Such approaches are standard in analysis of data from surveys with clustered observations (see, for example, Cochran, 1977, Chapter 9, 9A; Särndal et al., 1992, Chapter 4) and under the rubric of "generalized estimating equations" (GEE) are also widely used with longitudinal data and other clustered experimental data (Zeger et al., 1988).

Suppose that a hierarchical modeler fits a three-level mixed model as in Equation (14.8) to a clustered data set. Variances are calculated under the model; they may be incorrect if the model is wrong. In particular, if the modeler assumes that a certain coefficient is fixed across Level 2 units (by including it in $\mathbf{w}$ when it actually varies across units and therefore should have been included in $\mathbf{z}$), a meaningful variance estimate will not be obtained. With an adequate model, however, the hierarchical modeler can report inferences about how much the within-unit regressions vary across units, as well as about random effects for specific units.

A marginal analyst fits a simple regression model, $y_{ij} \sim N(\mathbf{x}_{ij}\boldsymbol{\beta}, \sigma^2)$, to the same data, where $\mathbf{x}$ includes variables in $\mathbf{z}$ and $\mathbf{w}$. This analyst estimates variances using a robust estimator that is consistent even if the regression model is wrong. In fact, if the robust variance estimate differs from the ordinary estimate, the model must be wrong in some respect, possibly related to clustering. This need not be a problem to this analyst, whose estimand is *defined* as the regression coefficient $\hat{\boldsymbol{\beta}}$ obtained by applying the usual ordinary least-squares estimator to the data. Estimation of random effects plays no role in this context, and there is no distinction between the within- and between-cluster components of the regression.

The two approaches proceed from different statistical philosophies and their results have different interpretations. The hierarchical model provides a rich summary of the data, treating clustering as an important feature of the population structure, but is also more sensitive to errors in characterizing that structure. The marginal model ignores the clustering structure but focuses on robustness of variance estimation for a relatively simple summary of the data. The choice of modeling strategy ultimately depends on the objectives of the study.

14.6 COMPUTATION FOR HIERARCHICAL MODELS

Hierarchical models provide excellent opportunities for application of the standard tools for Bayesian inference, particularly those based on simulation. Often several techniques can be combined to handle the distributions that arise at the various levels of the model. Because of the huge variety of hierarchical models, only an overview appears here; see Chapter 6 for details of these methods and further examples. Other techniques provide MLEB estimates of top-level parameters. Both the fully Bayesian and the maximum likelihood approaches integrate over the unit-specific parameters, differing in the treatment of those at the top level (integration versus maximization). Consequently, largely identical computer code may be used for Bayesian inference with a Gibbs sampler and for maximum likelihood inference using a Monte Carlo EM (expectation–maximization) algorithm for a multilevel model (Zaslavsky, 1997).

14.6.1 Techniques Based on Conditional Distributions: Gibbs Samplers and Data Augmentation

Because of their specification as a series of conditional distributions, hierarchical models naturally lend themselves to simulation using a Gibbs sampler (see Chapter 6 and its Complement). Conditional on all other parameters, the parameters corresponding to a given level of the model have posterior distributions that are independent (for different units), with priors determined by parameters at higher levels and likelihoods reflecting the distributions of data and/or lower-level parameters. A Gibbs sampler can cycle through the levels of the model, drawing the corresponding sets of parameters in turn. Because the Gibbs sampler draws parameters at all levels of the model, it is well adapted to performing several inferences simultaneously as described in Section 14.4.1.

A variety of Gibbs samplers may be possible corresponding to different ways of expressing the same model. For example, the flattened version (14.6) of the three-level model reduces the number of steps in the Gibbs sampler because fewer vectors of random-effects parameters must be drawn: only $\boldsymbol{\theta}$, rather than $\boldsymbol{\alpha}, \boldsymbol{\beta}, \boldsymbol{\gamma}$ as in Equation (14.5). The cost of this simplification is that the simple independence structure of the draws in Equation (14.5) is lost. The likelihood for the variance components $\sigma_\alpha^2, \sigma_\gamma^2, \sigma_\delta^2$ becomes complex and nonstandard, and drawing $\boldsymbol{\theta}$ involves calculations with the entire covariance matrix $\tau_\alpha^2\mathbf{J} + \tau_\gamma^2\mathbf{K} + \tau_\delta^2\mathbf{I}$, which becomes large and intractable if the number of units is large.

Our warnings about use of improper posterior distributions in hierarchical modeling bear repetition in the context of computing. In particular, the propriety of all the conditional distributions of the Gibbs sampler does *not* imply the existence of a proper joint distribution. In such cases the sampler does not converge to any probability distribution, but this might not be at all apparent from examination of the sampler's output (Hobert and Casella, 1996). Therefore, propriety must be established on theoretical grounds (if the model specification includes improper distributions) before attempting to interpret draws from a sampler.

In nonconjugate models, another technique such as a Metropolis–Hastings or rejection sampling step must often be used to draw from the conditional distributions (Section 6.3). Such techniques are common, for example, in fitting GLMMs.

In some cases data augmentation offers a simple alternative (Seçtion 6.4). Augmentation simply adds another level to the hierarchy; in some cases, the additional level has a scientific interpretation and facilitates model specification as well as computation (Bradlow and Zaslavsky, 1999).

14.6.2 Techniques Based on Marginal Likelihoods

As noted in Section 14.3.5, marginal models for simple two-stage conjugate models often have a tractable form. For example, the two-stage model (14.10) can be fitted as a negative binomial model.

The complex crossed model (14.5) can be written in the fully collapsed form $\mathbf{y} \sim N(\mu\mathbf{1}, \tau_\alpha^2\mathbf{J} + \tau_\gamma^2\mathbf{K} + \tau_\delta^2\mathbf{L} + \sigma^2\mathbf{I})$. The dimension of the parameter of this integrated model is the number of variance components (four) plus one for μ. Hence it might be fitted using integration over a grid or an adaptive Metropolis jumping sampler for the variance components. If the observational model is non-normal, the likelihood can be approximated using numerical integration methods such as quadrature (Gibbons et al., 1994).

14.7 SOFTWARE FOR HIERARCHICAL MODELS

An increasing number of statistical software packages can fit hierarchical models. Among these are modules of general-purpose packages including SAS (PROC MIXED), S-Plus/R (`nlme` and `varcomp` functions and the OSWALD library), and Stata (`gllamm` module). Special-purpose software packages include MLwiN (from the Institute of Education, University of London, UK), and HLM (Raudenbush, et al., 2000), and the free packages MIXOR/MIXREG (Hedeker and Gibbons) and MLA (Busing, van der Leeden and Meijer). In addition, WinBUGS (described in Complement A to Chapter 6) is well adapted to specifying a wide range of hierarchical models.

Most of the listed packages fit hierarchical normal models, and some also fit GLMMs (particularly the widely used logistic and Poisson regression models) or other nonlinear models. Other than WinBUGS, only MLwiN has a fully Bayesian option, using a limited set of default priors and a MCMC algorithm. The other packages use MLEB and give asymptotic standard errors that are of dubious validity in some applications. The packages have various restrictions and capabilities in terms of number of levels allowed in the model, ability to accomodate random coefficients and crossed effects, generality of the Level 1 specification, and user interface. Furthermore, the price of greater generality of models is often loss of efficiency in computation; thus, for example, MLEB with a specialized package is likely to be much faster than full Bayes' inference with WinBUGS, especially for large data sets.

A useful web site for up-to-date software information is maintained by the Centre for Multilevel Modeling, also at the Institute of Education, University of London (http://multilevel.ioe.ac.uk/); they also publish a free newsletter. Information on an e-mail discussion list for hierarchical modeling appears at http://www.jiscmail.ac.uk/lists/multilevel.html.

SUMMARY

In this chapter we have introduced hierarchical modeling as a very general approach to specifying complex models through a sequence of more simple stages. Hierarchical models are useful for modeling collections of observations with a simple or complex exchangeability structure. Using fully Bayesian approaches, both general parameters (characterizing the entire population) and parameters specific to individual units (as in small-area estimation or profiling) can be estimated. Modern Bayesian computational approaches are well adapted to estimation of such models.

Introductory texts on hierarchical modeling for the applied statistician include Raudenbush and Bryk (2001), Goldstein (1995), and Snijders and Bosker (1999). More advanced treatments include Carlin and Louis (2000), Gelman et al. (1995), and Longford (1993).

EXERCISES

14.1* Prove expression (14.2), with the associated definition of $\tilde{\theta}_i$, v_i. (Recall that σ^2 and τ^2 are assumed known. *Hint*: first derive the distributions of $\mu \mid \mathbf{y}$ and $\theta_i \mid \mu, \mathbf{y}$ and then integrate out μ using the formula for a normal mixture of normals.) Show that this equation can be written in the form $\tilde{\theta}_i = \bar{y} + (1 - B)(y_i - \bar{y}_i)$, where B is the shrinkage factor, and give an expression for B.

14.2 Derive all of the conditional distributions required to implement a Gibbs sampler for the model (14.4).

14.3* Consider the hierarchical model (14.5).

(a) Draw a summary version of the hierarchical diagram as in Figure 14.1(a).

(b) Assume that i, j, and k each have two levels and show the detailed diagram as in Figure 14.1(b).

(c) Describe (in words) the exchangeability relationships that characterize this model (i.e., which indices can be permuted without changing the joint distribution).

* Solutions for asterisked exercises may be found in Appendix 7.

14.4 Consider the collapsed hierarchical model (14.6).

(a) Write out the matrices **J**, **K**, **L**, as in Equation (14.3). Assume that i, j, and k each have two levels; make clear the ordering of the indices (i, j, k) labeling each row and column of these matrices.

(b) Write the marginal covariance matrix of **y** given μ and the variance components (with entries that are combinations of the variance components).

(c) Verify visually (without writing out the details) that the exchangeability relationships described in Exercise 14.3(c) hold for the covariance matrix you wrote.

(d) For a more notationally challenging version, write the same matrices **J**, **K**, **L** for a general number of levels of each index using Kronecker product notation.

14.5* Each coin in a collection of bent coins has a different probability of coming up heads when tossed; the probability for each coin is drawn from a Beta distribution. Each coin is tossed several times with outcome recorded as 1 if heads, 0 if tails. Because the lighting is poor, the coin tosser sometimes records the outcome correctly (with probability p, a fixed known quantity) and sometimes records the opposite outcome (with probability $1 - p$).

(a) Define the variables needed to specify a hierarchical model for this situation, including the recorded outcome of a single toss, the true outcome, and the probability of heads for a single coin. Use subscripts appropriately.

(b) Draw a diagram showing how the variables are related in the style of Figure 14.1(b).

(c) Write out all the conditional distributions needed to complete the model.

(d) Derive an expression for the posterior probability that a particular toss was actually a head given that it was recorded as a head, conditional on the probability of heads for that coin.

(e) Derive an expression for the probability that the coin is recorded as a head given the parameter value for the probability of heads for that coin.

(f) Derive an expression for the posterior distribution of the parameter for the probability of heads for a particular coin, given the Beta prior parameters and the information that it was recorded as having two heads and two tails out of four tosses. (It may involve an integral that you cannot do.)

14.6 The genetic composition (genome) of a person is derived from that of the parents, but the way they are combined involves a random process. We will consider the process for a small group of related individuals. Elaine and Juan have two children, Fred and Rosalia. Shanti and Nur have a child, Gauri. Rosalia has a child with Gauri, named Billy. Later she has a child with Yun-fat, named Liz. Fred and Ludmilla (a woman unrelated to any of the other individuals listed) have a child named Mariko.

(a) Draw a diagram representing the conditional relationships among the genomes of the different people in this family, in the style of Figure 14.1(b). (*Hint*: the variable at each node represents the genome of one person, either the complete genome or a specific characteristic.)

(b) What individuals' genomes would you condition on to make the genomes for Liz and Bill conditionally independent? What about Liz and Mariko? (For which can you give more than one answer?)

(c) Compare this diagram to a standard family tree. How are they different or similar in terms of the information conveyed? (This is not a technical question, just a question of interpretation.)

(d) Comment on the extent to which the models associated with this diagram fit the informal characterization of hierarchical models in Section 14.2.2.

(e) An individual's observable characteristics (without DNA sequencing) reflect some combination of effects of the genome and of various kinds of environmental influences. Supposing that the environmental influences are independent for each individual, augment your diagram to show where variables for the *observable* characteristics of Mariko, Rosalia, and Bill would appear.

(f) How plausible is this assumption of independence, for example, for Fred and Rosalia? Explain your view, and suggest the possible consequences of assuming independence of environmental effects if they are not in fact independent.

Note: models of this general form are common in a number of fields concerned with inherited characteristics. In animal husbandry, the variable might be a single characteristic such as the animal's weight (corrected for sex differences), and a variance component model similar to (14.5) might be appropriate. In studies of inheritance of genetically linked diseases, the variable might represent the expression of a specific gene and the conditional distributions are based on the laws of heredity.

14.7 Consider the model (14.10) and suppose that there is a single observation $y_{i1} = y_i$ from each Level 2 unit i, and that $b_i = b$ is the same for every unit (i.e., the regression is a constant).

(a) Derive the negative binomial marginal distribution of $y_i \mid a, b$. Show that the probability parameter of this distribution is $p = 1/(b+1)$.

(b) For the remainder of this problem, assume that $a = 4$ is known. The following data were obtained from 25 units:

4, 2, 8, 11, 1, 3, 2, 4, 7, 5, 13, 12, 1, 8, 15, 8, 11, 10, 10, 10,
2, 8, 7, 2, 14

Obtain the maximum likelihood estimate of p and hence of b.

(c) What is the posterior distribution of $p \mid \mathbf{y}$ under a uniform, that is, Beta(1,1), prior distribution for p?

(d) What is the MLEB posterior distribution of λ_3 (posterior mean, mode, and variance), given that $y_3 = 8$?

(e) Use a random number generator to obtain 10 draws from the posterior distribution of p (and hence of b). Approximate the full Bayes' inference (mean and variance) by integrating over the distribution of b using your 10 draws. How good is the MLEB inference as an approximation to full Bayes inference?

14.8* Consider the random coefficient model (14.7).

(a) Write the Level 1 conditional likelihood (given the variance components) for $\boldsymbol{\beta}_i$, i.e., the density of the part of the data that depends on $\boldsymbol{\beta}_i$.

(b) What is the posterior distribution of $\boldsymbol{\beta}_i \mid \mathbf{y}_i, \lambda, \sigma^2, \mathbf{T}$? (*Hint*: combine a normal prior and a normal likelihood as in Section 14.2.1, but with a multivariate distribution.)

(c) Write the posterior mean conditional on $\lambda, \sigma^2, \mathbf{T}$ as $\tilde{\boldsymbol{\beta}}_i = \boldsymbol{\beta}_i^* + (\mathbf{I} - \mathbf{B})(\hat{\boldsymbol{\beta}}_i - \boldsymbol{\beta}_i^*)$, where $\boldsymbol{\beta}_i^* = \mathbf{x}_i\lambda$ is the Level 2 regression prediction and $\hat{\boldsymbol{\beta}}_i$ is the least-squares estimate of β_i based on $\mathbf{y}_i$. Note the similarity to the shrinkage factor in Exercise 14.8.

14.9 Consider the following three-level hierarchical model

$$
\begin{aligned}
y_{ij} &\sim N(\theta_{ij}, \operatorname{var}{}_i\theta_{ij}^2),\\
\theta_{ij} &\sim N(\mu_i, \tau^2),\\
\sigma_i^2 &\sim IG(\nu_1, \nu_2),\\
\mu_i &\sim N(\lambda, \omega^2),\\
\tau^2 &\sim IG(\eta_1, \eta_2),
\end{aligned}
$$

where $i \in \{1, \ldots, n\}, j \in \{1, \ldots, r_i\}$, and $(\nu_i, \nu_2, \lambda, \omega^2, \eta_1, \eta_2)$ is known.

(a) Derive the closed-form conditional posterior distributions (when possible) required by the Gibbs sampler.

(b) One conditional posterior does not have a closed-form solution. Which one? Describe how variates could be drawn from this distribution, thus enabling the Gibbs sampler to be implemented.

14.10 In a subject area that interests you, find a data set that has a hierarchical (clustered) structure and analyse it using any convenient software.

FURTHER READING

Bradlow, E. T. and Zaslavsky, A. M. (1999). "A Hierarchical Latent Variable Model for Ordinal Data from a Customer Satisfaction Survey with "No Answer" Responses," *J. Am. Statist. Assoc.*, **94**, 43–52.

Brown, L. D. (1966). "On the Admissibility of Invariant Estimators of One or More Location Parameters," *Annals Math. Statist.*, **37**, 1087–1136.

Carlin, B. P. and Louis, T. A. (2000). *Bayes and Empirical Bayes Methods for Data Analysis.* Boca Raton, Chapman and Hall/CRC.

Christiansen, C. L. and Morris, C. N. (1997a). "Hierarchical Poisson Regression Modeling." *J. Am. Statist. Assoc.*, **92**, 618–632.

Christiansen, C. L. and Morris, C. N. (1997b). "Improving the Statistical Approach to Health Care Provider Profiling." *Annals Internal Med.*, **127**, 764–768.

Citro, C. F. and Kalton, G. (Eds.) (2000). *Small-Area Income and Poverty Estimates: Evaluation of Current Methodology.* Washington, D.C., National Academy Press.

Cochran, W. G. (1977). *Sampling Techniques* (3rd edition). New York, John Wiley and Sons.

Corbeil, R. R. and Searle, S. R. (1976). "Restricted Maximum Likelihood (REML) Estimation of Variance Components in the Mixed Model." *Technometrics*, **18**, 31–38.

Deely, J. J. and Lindley, D. V. (1981). "Bayes Empirical Bayes." *J. Am. Statist. Assoc.*, **76**, 833–841.

Diaconis, P. and Ylvisaker, D. (1979). "Conjugate Priors for Exponential Families." *Annals Statist.*, **7**, 269–281.

Efron, B. and Morris, C. (1975). "Data Analysis Using Stein's Estimator and Its Generalizations." *J. Am. Statist. Assoc.* **70**, 311–319.

Fay, R. E. I. and Herriot, R. A. (1979). "Estimates of Income for Small Places: An Application of James–Stein Procedures to Census Data." *J. Am. Statist. Assoc.*, **74**, 269–277.

Gelman, A., Carlin, J. B., Stern, H. S. and Rubin, D. B. (1995). *Bayesian Data Analysis.* London, Chapman and Hall.

Ghosh, M., Natarajan, K., Stroud, T. W. F. and Carlin, B. P. (1998). "Generalized Linear Models for Small-Area Estimation." *J. Am. Statist. Assoc.*, **93**, 273–282.

Ghosh, M. and Rao, J. N. K. (1994). "Small Area Estimation: An Appraisal." *Statist. Sci.*, **9**, 55–76.

Gibbons, R. D., Hedeker, D., Charles, S. C. and Frisch, P. (1994). "A Random-Effects Probit Model for Predicting Medical Malpractice Claims." *J. Am. Statist. Assoc.*, **89**, 760–767.

Goldstein, H. (1995). *Multilevel Statistical Models* (Second edition), London, Edward Arnold Publishers Ltd.

Goldstein, H. and Spiegelhalter, D. J. (1996). "League Tables and Their Limitations: Statistical Issues in Comparisons of Institutional Performance." *J. Royal Statist. Soc., A*, **159**, 385–409.

Henderson, C. R. (1975). "Best Linear Unbiased Estimation and Prediction Under a Selection Model." *Biometrics*, **31**, 423–447.

Hobert, J.P. and Casella, G. (1996). "The Effect of Improper Priors on Gibbs Sampling in Hierarchical Linear Mixed Models." *J. Am. Statist. Assoc.*, **91**, 1461–1473.

Hulting, F. L. and Harville, D. A. (1991). "Some Bayesian and Non-Bayesian Procedures for the Analysis of Comparative Experiments and for Small-Area Estimation: Computational Aspects, Frequentist Properties, and Relationships." *J. Am. Statist. Assoc.*, **86**, 557–568.

Ibrahim, J. G. and Laud, P. W. (1991). "On Bayesian Analysis of Generalized Linear Models Using Jeffreys's Prior." *J. Am. Statist. Assoc.*, **86**, 981–986.

James, W. and Stein, C. (1961). "Estimation with Quadratic Loss." In Proceedings of the Fourth Berkeley Symposium on Mathematical Statistics and Probability, Berkeley, University of California Press, 361–380.

Johnson, N. L. and Kotz, S. (1969). *Discrete Distributions*, New York, John Wiley and Sons.

Laird, N. M. and Louis, T. A. (1987). "Empirical Bayes Confidence Intervals Based on Bootstrap Samples." *J. Am. Statist. Assoc.*, **82**, 739–750.

Lindley, D. V. and Smith, A. F. M. (1972). "Bayes Estimates for the Linear Model." *J. Royal Statist. Soc. B*, **34**, 1–41.

Longford, N. T. (1993). *Random Coefficient Models*, Oxford, Clarendon Press.

McCullagh, P. and Nelder, J. A. (1989). *Generalized Linear Models* (Second edition), London, Chapman and Hall Ltd.

Natarajan, R. and McCulloch, C. E. (1998). "Gibbs Sampling with Diffuse Proper Priors: A Valid Approach to Data-Driven Inference?" *J. Comput. Graphical Statist.*, **7**, 267–277.

Neyman, J. and Scott, E. L. (1948). "Consistent Estimates Based on Partially Consistent Observations." *Econometrica*, **16**, 1.

Normand, S.-L. T., Glickman, M. E. and Gatsonis, C. A. (1997). "Statistical Methods for Profiling Providers of Medical Care: Issues and Applications." *J. Am. Statist. Assoc.*, **92**, 803–814.

Pinheiro, J. C. and Bates, D. M. (2000). *Mixed Effects Models in S and S-plus*, New York, Springer.

Prasad, N. G. N. and Rao, J. N. K. (1990). "The Estimation of the Mean Squared Error of Small-Area Estimators." *J. Am. Statist. Assoc.*, **85**, 163–171.

Raudenbusch, S. W. and Bryck, A. S. (2001). *Hierarchical Linear Models: Applications and Data Analysis Methods*, Second Edition, Thousand Oaks, CA: Sage Publications, Inc.

Raudenbush, S. W., Bryk, A. S., Cheong, Y. F. and Congdon, R. T. Jr (2000). *HLM5: Hierarchical Linear and Nonlinear Models*, Lincolnwood, IL, Scientific Software International, Inc.

Robbins, H. (1955). "An Empirical Bayes Approach to Statistics." In *Proceedings of the Third Berkeley Symposium on Mathematical Statistics and Probability*, Berkeley, University of California Press, 157–163.

Rubin, D. B. (1980). "Using Empirical Bayes Techniques in the Law School Validity Studies." *J. Am. Statist. Assoc.*, **75**, 801–816.

Särndal, C.-E., Swensson, B. and Wretman, J. (1992). *Model Assisted Survey Sampling*, New York, Springer-Verlag.

Schafer, J. L. and Yucel, R. M. (2002). "Computational Strategies for Multivariate Linear Mixed-Effects Models with Missing Values." *J. Computat. Graphical Statist.*, **11**(2), 437–457, page 358.

Snijders, T. A. B. and Bosker, R. J. (1999). *Multilevel Analysis*. London, Sage Publications.

Stein, C. (1956). "Inadmissibility of the Usual Estimator for the Mean of a Multivariate Normal Distribution." In *Proceedings of the Third Berkeley Symposium on Mathematical Statistics and Probability*, Berkeley, University of California Press, 197–206.

Swallow, W. H. and Searle, S. R. (1978). "Minimum Variance Quadratic Unbiased Estimation (MIVQUE) of Variance Components." *Technometrics*, **20**, 265–272.

Tanner, M. A. and Wong, W. H. (1987). "The Calculation of Posterior Distributions by Data Augmentation." *J. Am. Statist. Assoc.*, **82**, 528–540.

Zaslavsky, A. M. (1997). "Comment on Meng and van Dyk, "The EM Algorithm—an Old Folk-Song Sung to a Fast New Tune". *J. Royal Statist. Soc., B*, **59**, 559.

Zeger, S. L., Liang, K.-Y. and Albert, P. S. (1988). "Models for Longitudinal Data: A Generalized Estimating Equation Approach." *Biometrics*, **44**, 1049–1060.

CHAPTER 15

Bayesian Factor Analysis

15.1 INTRODUCTION

This chapter presents a Bayesian approach to inference in the factor analysis model. We will present a model that differs somewhat from the classical (frequentist) model in that we will adopt a full disturbance covariance matrix, rather than a diagonal one. Accordingly, such a model provides the analyst with more opportunity to bring prior information to bear on the results, and will yield a unique solution to the factor analysis problem. We will assume throughout that data are normally distributed and that we have available a data matrix of covariances. We will first approach the problem by assuming that the data vectors are mutually independent, and that we have preassigned the number of underlying (latent) factors that satisfy the constraints of the model. Later we generalize the model to the case in which we use Bayesian inference to select the number of latent factors. Finally, we will briefly discuss some additional model considerations (such as generalizing the model to the case in which the data vectors are permitted to be correlated, and assessing the hyperparameters of the model). We will adopt vague and natural conjugate prior distributions, and show that the marginal posterior distribution of the factor scores follows a matrix T-distribution in large samples. As a consequence, we will be able to make simple interval estimates of the factor scores and test simple hypotheses about them.

15.2 BACKGROUND

An early formulation (see Press, 1972, 1982) of a Bayesian factor analysis model adopted the Wishart distribution as the distribution for the sample covariance matrix, and a vague prior distribution for the parameters. Only implicit numerical solutions could be obtained from this model. Kaufman and Press (1973a, b) proposed a new formulation of that model in terms of a model with more factors than observations (a characteristic of the model shared with that of Guttman, 1953), but the prior on most of the factors was centered at zero. This work was developed further in Kaufman and

Press (1976), who showed that the posterior distribution of the factor loading matrix was truncated multivariate normal in large samples.

Martin and McDonald (1975) approached the factor analysis problem looking for a Bayesian solution to Heywood cases (cases in which MLE gives negative or zero estimates for the variances of the disturbances). They adopted a diagonal disturbance covariance matrix and used a Jeffreys-type vague prior for the elements. They proposed finding posterior joint modal estimators of the factor loading and disturbance covariance matrices, and obtained an implicit numerical solution. A point estimate of the factor loading matrix was also obtained.

Wong (1980) addressed the factor analysis problem from the empirical Bayes point of view, adopting normal priors for the factor loadings. He suggested use of the EM algorithm (see Dempster et al., 1977) to find a posterior mode for the factor loading matrix, but an explicit algorithm was not obtained.

Lee (1981) adopted a hierarchical Bayesian approach to confirmatory factor analysis, starting from the assumption that the free parameters in the factor loading matrix were exchangeable and normally distributed. The disturbance covariance matrix was assumed to be diagonal. Joint modal estimates were found of the factor loading matrix, the variances of the disturbances, and the covariance matrix of the factors. This modal solution was implicit and numerical. A point estimate of the factor loading matrix was obtained.

Euverman and Vermulst (1983) studied the Bayesian factor analysis model with a diagonal disturbance covariance matrix, and a preassigned number of factors. A numerical computer routine was described for implicitly finding the posterior joint mode of the factor loadings and error variances.

Mayekawa (1985) studied the Bayesian factor analysis problem examining factor scores as well as factor loadings and error (specific) variances. The factor loadings were assumed to be normal, *a priori*. The author used the EM algorithm to find point estimates of the parameters as marginal modes of the posterior distributions. Unfortunately, however, there were no proofs about convergence of the EM algorithm used.

Shigemasu (1986) used a natural conjugate prior Bayesian approach to the factor analysis model and found implicit numerical solutions for the factor loading matrix and specific variances.

Akaike (1987) suggested that the AIC criterion could be used to select the appropriate number of factors to use in a Bayesian model (see also Press, 1982). He was motivated by the desire to deal with the problem of frequent occurrence of improper solutions in maximum likelihood factor analysis caused by overparameterization of the model. By minimizing, the AIC results can be used to test hypothesis about the appropriate number of factors. Bozdogan (1987) extended AIC by penalizing overparameterization more stringently and thereby developed a modified AIC criterion that was consistent.

Arminger and Muthen (1998) examined a nonlinear LISREL oriented factor analysis model using natural conjugate prior distributions for the parameters and a traditional factor analysis model with diagonal disturbance covariance matrix. They used the Markov chain Monte Carlo (MCMC) and the Metropolis–Hastings algorithms (see Chapter 6) to estimate the unknown parameters by sampling from the posterior distribution. Closed-form solutions could not be obtained. Shi and Lee (1998) investigated the Bayesian approach to the factor analysis model involving some jointly distributed continuous data and some polytomous variable data. They used MCMC and the Metropolis–Hastings algorithms to estimate the latent factor scores and the structural parameters.

In contrast to the earlier work on Bayesian factor analysis, which focused upon point estimation, in this chapter we also develop methods for obtaining large sample interval estimators of factor scores, factor loadings, and specific variances. Consequently, standard Bayesian hypothesis testing methods (see Chapter 9) can be used for testing hypothesis about all the fundamental quantities (apart from the number of factors) in the model. Because we develop exact (large sample) posterior distributions for these quantities, level curves, or contours, of the posterior distributions can be studied for sensitivity around the point estimators by examining the steepness and shape of the level curves. This development yields *explicit* analytical results for the distributions of the quantities of interest (as well as some general implicit solutions), whereas most earlier work focused only on *implicit* numerical solutions of the matrix equations.

15.3 BAYESIAN FACTOR ANALYSIS MODEL FOR FIXED NUMBER OF FACTORS[1]

In this section we develop the basic factor analysis model. We first define the likelihood function. Then we introduce prior distributions on the parameters, and calculate the joint probability density function of the parameters. Finally we find the marginal posterior probabilities for the parameters.

15.3.1 Likelihood Function

Define p-variate observation vectors, $(x_1, \ldots, x_N) \equiv X'$ on N subjects. The means are assumed to have been subtracted out, so that $E(X') = 0$. The prime denotes transposed matrix. The traditional factor analysis model is

$$\underset{(p \times 1)}{\boldsymbol{x}_j} = \underset{(p \times m)}{\boldsymbol{\Lambda}} \; \underset{(m \times 1)}{\boldsymbol{f}_j} + \underset{(p \times 1)}{\boldsymbol{\epsilon}_j}, \qquad m < p, \tag{15.1}$$

for $j = 1, \ldots, N$, where $\boldsymbol{\Lambda}$ denotes a matrix of constants called the factor loading matrix; $\boldsymbol{f}_j$ denotes the factor score vector for subject j; $\boldsymbol{F}' \equiv (\boldsymbol{f}_1, \ldots, \boldsymbol{f}_N)$. The $\boldsymbol{\epsilon}_j$s are assumed to be mutually uncorrelated and normally distributed as $N(\boldsymbol{0}, \boldsymbol{\Psi})$; in the traditional model, $\boldsymbol{\Psi}$ is taken to be diagonal, whereas here, it is permitted to be a

[1]The work described in Section 15.3 of this chapter is adapted from Press and Shigemasu (1989) with the permission of Springer-Verlag.

general symmetric positive definite matrix, that is, $\boldsymbol{\Psi} > \mathbf{0}$. While $\boldsymbol{\Psi}$ is not assumed to be diagonal, note from Equation (15.5b) that $E(\boldsymbol{\Psi})$ is diagonal.

We assume that $(\boldsymbol{\Lambda}, \boldsymbol{F}, \boldsymbol{\Psi})$ are unobserved and fixed quantities, and we assume that we can write the probability law of x_j as

$$\mathcal{L}(\boldsymbol{x}_j \mid \boldsymbol{\Lambda}, \boldsymbol{f}_j, \boldsymbol{\Psi}) = N(\boldsymbol{\Lambda f}_j, \boldsymbol{\Psi}), \tag{15.2}$$

where $\mathcal{L}(\cdot)$ denotes probability law. Equivalently, if "$\propto$" denotes proportionality, the likelihood for $(\boldsymbol{\Lambda}, \boldsymbol{F}, \boldsymbol{\Psi})$ is

$$p(\boldsymbol{X} \mid \boldsymbol{\Lambda}, \boldsymbol{F}, \boldsymbol{\Psi}) \propto |\Psi|^{-0.5N} e^{-0.5 \operatorname{tr} \Psi^{-1}(\boldsymbol{X}-\boldsymbol{F\Lambda}')'(\boldsymbol{X}-\boldsymbol{F\Lambda}')} \tag{15.3}$$

We will use $p(\cdot)$ generically to denote "density"; the ps will be distinguished by their arguments. This should not cause confusion. The proportionality constant in Equation (15.3) is numerical, depending only on (p, N) and not upon $(\boldsymbol{\Lambda}, \boldsymbol{F}, \boldsymbol{\Psi})$.

15.3.2 Priors

We use a generalized natural conjugate family (see Press, 1982) of prior distributions for $(\boldsymbol{\Lambda}, \boldsymbol{F})$. We take as prior density for the unobservables (to represent our state of uncertainty)

$$p(\boldsymbol{\Lambda}, \boldsymbol{F}, \boldsymbol{\Psi}) \propto p(\boldsymbol{\Lambda} \mid \boldsymbol{\Psi}) p(\boldsymbol{\Psi} p(\boldsymbol{F}), \tag{15.4}$$

where

$$p(\boldsymbol{\Lambda} \mid \boldsymbol{\Psi}) \propto |\boldsymbol{\Psi}|^{-0.5m} e^{-0.5 \operatorname{tr} \boldsymbol{\Psi}^{-1}(\boldsymbol{\Lambda}-\boldsymbol{\Lambda}_0)\boldsymbol{H}(\boldsymbol{\Lambda}-\boldsymbol{\Lambda}_0)'}, \tag{15.5a}$$

$$p(\boldsymbol{\Psi}) \propto |\boldsymbol{\Psi}|^{-0.5\nu} e^{-0.5 \operatorname{tr} \boldsymbol{\Psi}^{-1}\boldsymbol{B}}, \tag{15.5b}$$

with $\boldsymbol{B}$ a diagonal matrix, and $\boldsymbol{H} > 0$. (Choices for the prior density $p(\boldsymbol{F})$ will be discussed below.) Thus, $\boldsymbol{\Psi}^{-1}$ follows a Wishart distribution, $(\nu, \boldsymbol{B})$ are hyperparameters to be assessed; Λ conditional on $\boldsymbol{\Psi}$ has elements that are jointly normally distributed, and $(\boldsymbol{\Lambda}_0, \boldsymbol{H})$ are hyperparameters to be assessed. Note that $E(\boldsymbol{\Psi} \mid \boldsymbol{B})$ is diagonal, to represent traditional views of the factor model containing "common" and "specific" factors. Also, note that if $\boldsymbol{\Lambda} \equiv (\boldsymbol{\lambda}_1, \ldots, \boldsymbol{\lambda}_m)$, $\boldsymbol{\lambda} \equiv \operatorname{vec}(\boldsymbol{\Lambda}) = (\boldsymbol{\lambda}_1', \ldots, \boldsymbol{\lambda}_m')'$, then $\operatorname{var}(\boldsymbol{\lambda} \mid \boldsymbol{\Psi}) = \boldsymbol{H}^{-1} \otimes \boldsymbol{\Psi}$, $\operatorname{var}(\boldsymbol{\lambda}) = \boldsymbol{H}^{-1} \otimes (E\boldsymbol{\Psi})$, and $\operatorname{cov}[(\boldsymbol{\lambda}_i, \boldsymbol{\lambda}_j) \mid \boldsymbol{\Psi}] = \boldsymbol{H}_{ij}^{-1}\boldsymbol{\Psi}$. Moreover, we will often take $\boldsymbol{H} = n_0\boldsymbol{I}$, for some preassigned scalar n_0. These interpretations of the hyperparameters will simplify assessment.

15.3.3 Joint Posteriors

Combining Equations (15.2)–(15.5), the joint posterior density of the parameters becomes

$$p(\boldsymbol{\Lambda}, \boldsymbol{F}, \boldsymbol{\Psi} \mid \boldsymbol{X}) \propto p(\boldsymbol{F}) \mid \boldsymbol{\Psi} \mid^{-0.5(N+m+v)} e^{-0.5 \operatorname{tr} \boldsymbol{\Psi}^{-1} \boldsymbol{G}}, \tag{15.6}$$

where $\boldsymbol{G} = (\boldsymbol{X} - \boldsymbol{F}\boldsymbol{\Lambda}')'(\boldsymbol{X} - \boldsymbol{F}\boldsymbol{\Lambda}') + (\boldsymbol{\Lambda} - \boldsymbol{\Lambda}_0)\boldsymbol{H}(\boldsymbol{\Lambda} - \boldsymbol{\Lambda}_0)' + \boldsymbol{B}$.

15.3.4 Marginal Posteriors

Integrating with respect to Ψ, and using properties of the Inverted Wishart density, gives the marginal posterior density of (Λ, F):

$$p(\boldsymbol{\Lambda}, \boldsymbol{F} \mid \boldsymbol{X}) \propto p(\boldsymbol{F}) \mid \boldsymbol{G} \mid^{-0.5(N+m+v-p-1)}. \tag{15.7}$$

We next want to integrate Equation (15.7) with respect to $\boldsymbol{\Lambda}$, to obtain the marginal posterior density of $\boldsymbol{F}$. We accomplish this by factoring $\boldsymbol{G}$ into a form which makes it transparent that in terms of $\boldsymbol{\Lambda}$, the density is proportional to a matrix $\boldsymbol{T}$-density. Thus, completing the square in $\boldsymbol{\Lambda}$ in the $\boldsymbol{G}$ function defined in Equation (15.6), Equation (15.7) may be rewritten as

$$p(\boldsymbol{\Lambda}, \boldsymbol{F} \mid \boldsymbol{X}) \propto \frac{p(\boldsymbol{F})}{\mid \boldsymbol{R}_F + (\boldsymbol{\Lambda} - \boldsymbol{\Lambda}_F)\boldsymbol{Q}_F(\boldsymbol{\Lambda} - \boldsymbol{\Lambda}_F)' \mid^{0.5\gamma}}. \tag{15.8}$$

where $\boldsymbol{Q}_F = \boldsymbol{H} + \boldsymbol{F}'\boldsymbol{F}$,

$$\boldsymbol{R}_F = \boldsymbol{X}'\boldsymbol{X} + \boldsymbol{B} + \boldsymbol{\Lambda}_0\boldsymbol{H}\boldsymbol{\Lambda}_0' - (\boldsymbol{X}'\boldsymbol{F} + \boldsymbol{\Lambda}_0\boldsymbol{H})\boldsymbol{Q}_F^{-1}(\boldsymbol{X}'\boldsymbol{F} + \boldsymbol{\Lambda}_0\boldsymbol{H})', \tag{15.9}$$

$$\boldsymbol{\Lambda}_F = (\boldsymbol{X}'\boldsymbol{F} + \boldsymbol{\Lambda}_0\boldsymbol{H})(\boldsymbol{H} + \boldsymbol{F}'\boldsymbol{F})^{-1}, \tag{15.10}$$

$$\gamma = N + m + v - p - 1. \tag{15.11}$$

Equation (15.8) is readily integrated with respect to $\boldsymbol{\Lambda}$ (by using the normalizing constant of a matrix T-distribution) to give the marginal posterior density of $\boldsymbol{F}$,

$$p(\boldsymbol{F} \mid \boldsymbol{X}) \propto \frac{p(\boldsymbol{F})}{\mid \boldsymbol{R}_F \mid^{0.5(\gamma-m)} \mid \boldsymbol{Q}_F \mid^{0.5p}}. \tag{15.12}$$

After some algebra, the marginal posterior density of $\boldsymbol{F}$ in Equation (15.12) may be written in the form

$$p(\boldsymbol{F} \mid \boldsymbol{X}) \propto \frac{p(\boldsymbol{F}) \mid \boldsymbol{H} + \boldsymbol{F}'\boldsymbol{F} \mid^{0.5(\gamma-m-p)}}{\mid \boldsymbol{A} + (\boldsymbol{F} - \hat{\boldsymbol{F}})'(\boldsymbol{I}_N - \boldsymbol{X}\boldsymbol{W}^{-1}\boldsymbol{X}')(\boldsymbol{F} - \hat{\boldsymbol{F}}) \mid^{0.5(\gamma-m)}}. \tag{15.13}$$

where

$$\hat{F} \equiv (I_N - XW^{-1}X')^{-1}XW^{-1}\Lambda_0 H$$
$$= (I_N - X(X'X - W)^{-1}X')XW^{-1}\Lambda_0 H, \tag{15.14}$$

$$W \equiv X'X + B + \Lambda_0 H \Lambda_0', \tag{15.15}$$

$$A \equiv H - (\Lambda_0 H)'W^{-1}\Lambda_0 H$$
$$- (XW^{-1}\Lambda_0 H)'(I_N - XW^{-1}X')^{-1}(XW^{-1}\Lambda_0 H)$$
$$\equiv H - (\Lambda_0 H)'W^{-1}\Lambda_0 H$$
$$- (XW^{-1}\Lambda_0 H)'(I_N - X(X'X - W)^{-1}X')(XW^{-1}\Lambda_0 H). \tag{15.16}$$

Note 1: In Equations (15.14) and (15.16) the second representations of $\hat{F}$ and A are more convenient for numerical computation than the first ones, because we need only invert a matrix of order p, instead of one of order N.
Note 2: In Equation (15.15) the quantity $(X'X)/N$ is the sample covariance matrix of the observed data (since the data are assumed to have mean zero). If the data are scaled to have variance of unity, $(X'X)/N$ denotes the data correlation matrix.
Note 3: In Equation (15.16), $H = H'$, but we have left H' to preserve the symmetry of the representation.

15.3.5 Estimation of Factor Scores

Now examine Equation (15.13). There are several cases of immediate interest. We take factor scores of subjects to be independent, *a priori*, so we can think of $p(F)$ as $p(f_1) \cdot p(f_2) \cdot \ldots \cdot p(f_N)$.

15.3.6 Historical Data Assessment of F

Suppose that, on the basis of historical data that is similar to the current data set, we can assess $p(F)$. We can then evaluate $p(F \mid X)$ numerically from Equation (15.13) to construct point estimators, and we can make interval estimates from the cdf of $p(F \mid X)$.

15.3.7 Vague Prior Estimator of F

Suppose instead that we are uninformed about F, *a priori*, and we accordingly adopt a vague prior

$$p(F) \propto \text{constant}. \tag{15.17}$$

Then Equation (15.13) becomes

$$p(\boldsymbol{F} \mid \boldsymbol{X}) \propto \frac{\mid \boldsymbol{H} + \boldsymbol{F}'\boldsymbol{F} \mid^{0.5(\gamma-m-p)}}{\mid \boldsymbol{A} + (\boldsymbol{F} - \hat{\boldsymbol{F}})'(\boldsymbol{I}_N - \boldsymbol{X}\boldsymbol{W}^{-1}\boldsymbol{X}')(\boldsymbol{F} - \hat{\boldsymbol{F}}) \mid^{0.5(\gamma-m)}}. \tag{15.18}$$

Again, interval estimates of $\boldsymbol{F}$ can be made numerically from the cdf of $p(\boldsymbol{F} \mid \boldsymbol{X})$, and point estimates can also be obtained numerically from Equation (15.18). Such numerical evaluations are treated in Press and Davis (1987).

15.3.8 Large Sample Estimation of $\boldsymbol{F}$

We note that

$$\frac{\boldsymbol{F}'\boldsymbol{F}}{N} = \frac{1}{N} \sum_{j=1}^{N} f_j f_j'.$$

If we assume (without loss of generality) that $E(\boldsymbol{f}_j) = 0$, $\text{var}(\boldsymbol{f}_j) = \boldsymbol{I}_m$, then for large N, by the law of large numbers,

$$\frac{\boldsymbol{F}'\boldsymbol{F}}{N} \approx \boldsymbol{I}_m.$$

Thus, for large N, $\mid \boldsymbol{H} + \boldsymbol{F}'\boldsymbol{F} \mid \approx \mid \boldsymbol{H} + N\boldsymbol{I}_m \mid$, a term which can be incorporated into the proportionality constant in Equation (15.13), because it no longer depends on $\boldsymbol{F}$. Equation (15.13) may now be rewritten, for large N, as

$$p(\boldsymbol{F} \mid \boldsymbol{X}) \propto \frac{p(\boldsymbol{F})}{\mid \boldsymbol{A} + (\boldsymbol{F} - \hat{\boldsymbol{F}})'(\boldsymbol{I}_N - \boldsymbol{X}\boldsymbol{W}^{-1}\boldsymbol{X}')(\boldsymbol{F} - \hat{\boldsymbol{F}}) \mid^{0.5(\gamma-m)}},$$

where $\hat{\boldsymbol{F}}$ is defined by Equation (15.14).

Suppose $p(\boldsymbol{F}) \propto$ constant. Then, $(\boldsymbol{F} \mid \boldsymbol{X})$ follows a matrix $\boldsymbol{T}$-distribution with density

$$p(\boldsymbol{F} \mid \boldsymbol{X}) \propto \mid \boldsymbol{A} + (\boldsymbol{F} - \hat{\boldsymbol{F}})'(\boldsymbol{I}_N - \boldsymbol{X}\boldsymbol{W}^{-1}\boldsymbol{X}')(\boldsymbol{F} - \hat{\boldsymbol{F}}) \mid^{-0.5(\gamma-m)}. \tag{15.19}$$

Alternatively, suppose $\mathcal{L}(\boldsymbol{f}_j) = N(\boldsymbol{0}, \boldsymbol{I}_m)$, and the $\boldsymbol{f}_j$s are mutually independent. Then,

$$p(\boldsymbol{F}) \propto e^{-0.5\,\text{tr}(\boldsymbol{F}'\boldsymbol{F})}.$$

For large N, since $\boldsymbol{F}'\boldsymbol{F} \approx N\boldsymbol{I}_m$, $p(\boldsymbol{F})$ can be incorporated into the proportionality constant in Equation (15.13) to yield Equation (15.19). The same argument applies to any prior density for $\boldsymbol{F}$ which can depend upon $\boldsymbol{F}'\boldsymbol{F}$.

In summary, we conclude that for large N, and for a wide variety of important priors for $\boldsymbol{F}$ (a vague prior, or for any prior which depends on $\boldsymbol{F}$ only through $\boldsymbol{F}'\boldsymbol{F}$), the marginal posterior density of $\boldsymbol{F}$, given the observed data vectors, is approximately matrix $\boldsymbol{T}$, as given in Equation (15.19), centered at $\hat{\boldsymbol{F}}$. In particular, $E(F \mid X) \approx \hat{\boldsymbol{F}}$, for large N.

15.3.9 Large Sample Estimation of $\boldsymbol{f}_j$

Since $(\boldsymbol{F} \mid \boldsymbol{X})$ is approximately distributed as matrix $\boldsymbol{T}$, $(\boldsymbol{f}_j \mid \boldsymbol{X})$ is distributed as multivariate $\boldsymbol{t}$ (see Theorem 6.2.4 in Press, 1982, p. 140), where $\boldsymbol{f}_j$ is the jth column of $\boldsymbol{F}'$, with density given by

$$p(\boldsymbol{f}_N \mid \boldsymbol{X}) \propto \left(\boldsymbol{P}_{22.1}^{-1} + (\boldsymbol{f}_N - \hat{\boldsymbol{f}}_N)'\boldsymbol{A}^{-1}(\boldsymbol{f}_N - \hat{\boldsymbol{f}}_N)\right)^{-0.5(\gamma-m-N+1)} \tag{15.20}$$

where $\boldsymbol{P}_{22.1} = \boldsymbol{P}_{22} - \boldsymbol{P}_{21}\boldsymbol{P}_{11}^{-1}\boldsymbol{P}_{12}$ is obtained from

$$\boldsymbol{P} \equiv \boldsymbol{I}_N - \boldsymbol{X}\boldsymbol{W}^{-1}\boldsymbol{X}' = \begin{pmatrix} \boldsymbol{P}_{11} & \boldsymbol{P}_{12} \\ \boldsymbol{P}_{21} & \boldsymbol{P}_{22} \end{pmatrix}, \qquad \boldsymbol{P}_{11}: (N-1) \times (N-1).$$

By reindexing the subjects, Equation (15.4) gives the posterior density of the factor score vector for any of the N subjects. Equation (15.20) can readily be placed into canonical form

$$p(\boldsymbol{f}_N \mid \boldsymbol{X}) \propto \left(1 + (\boldsymbol{f}_N - \hat{\boldsymbol{f}}_N)'\left(\frac{\boldsymbol{A}}{P_{22.1}}\right)^{-1}(\boldsymbol{f}_N - \hat{\boldsymbol{f}}_N)\right)^{-0.5(\delta+m)}, \tag{15.21}$$

where $\delta \equiv \nu - p - m$.

15.3.10 Large Sample Estimation of the Elements of $\boldsymbol{f}_j$

Now suppose we wish to make posterior probability statements about a particular element of $\boldsymbol{f}_N \equiv (f_{kN})$, $k = 1, \ldots, m$, say f_{1N}. We use the posterior density of a Student t-variate obtained as the marginal of the multivariate t-density in Equation (15.21) (see, for example, Press, 1982, p. 137). It is given by

$$p(f_{1N} \mid \boldsymbol{X}) \propto \left(\delta + \left(\frac{f_{1N} - \hat{f}_{1N}}{\sigma_1}\right)^2\right)^{-0.5(\delta+1)}, \tag{15.22}$$

where σ_1^2 is the (1, 1) element of

$$\frac{\boldsymbol{A}}{\delta \boldsymbol{P}_{22.1}} = \begin{pmatrix} \sigma_1^2 & \boldsymbol{\Sigma}_{12} \\ \boldsymbol{\Sigma}_{21} & \boldsymbol{\Sigma}_{22} \end{pmatrix}.$$

$\hat{f}_{1N}$ is of course the (1, N) element of $\hat{\boldsymbol{F}}'$. From Equation (15.22) we can make posterior probability statements about any factor score for any subject; that is, we can obtain credibility (confidence) intervals for any factor score. For example,

$$\left.\left(\frac{f_{1N} - \hat{f}_{1N}}{\sigma_1}\right)\right| \boldsymbol{X} \sim t_\delta.$$

15.3.11 Estimation of the Factor Loadings Matrix

We now return to the joint posterior density of ($\boldsymbol{\Lambda}$, $\boldsymbol{F}$), given in Equation (15.8). One method of estimating $\boldsymbol{\Lambda}$ would be to integrate $\boldsymbol{F}$ out of Equation (15.8) to obtain the marginal posterior density of $\boldsymbol{\Lambda}$. Then, some measure of location of the distribution could be used as a point estimator of $\boldsymbol{\Lambda}$. Unfortunately, while the integration can be carried out, the resulting marginal density is extremely complicated, and it does not seem possible to obtain a mean or mode of the distribution for any realistic prior densities for $\boldsymbol{F}$, except numerically. The result is

$$p(\boldsymbol{\Lambda} \mid \boldsymbol{X}) \propto |\, \boldsymbol{P}_\Lambda \,|^{-0.5\gamma} |\, \boldsymbol{\Lambda}'\boldsymbol{P}_\Lambda^{-1}\boldsymbol{\Lambda} \,|^{-0.5N} |\, \boldsymbol{Z} \,|^{-0.5(\gamma - m)},$$

where

$$\boldsymbol{P}_\Lambda \equiv \boldsymbol{B} + (\boldsymbol{\Lambda} - \boldsymbol{\Lambda}_0)\boldsymbol{H}(\boldsymbol{\Lambda} - \boldsymbol{\Lambda}_0)',$$

and

$$\boldsymbol{Z} \equiv \boldsymbol{I}_N + \boldsymbol{X}\boldsymbol{P}_\Lambda^{-1}\boldsymbol{X}' - (\boldsymbol{X}\boldsymbol{P}_\Lambda^{-1}\boldsymbol{\Lambda})(\boldsymbol{\Lambda}'\boldsymbol{P}_\Lambda^{-1}\boldsymbol{\Lambda})^{-1}(\boldsymbol{X}\boldsymbol{P}_\Lambda^{-1}\boldsymbol{\Lambda})'.$$

Since this distribution is so complicated, we will alternatively estimate $\boldsymbol{\Lambda}$ for given $\boldsymbol{F} = \hat{\boldsymbol{F}}$. First note from Equation (15.8) that

$$p(\boldsymbol{\Lambda} \mid \boldsymbol{F}, \boldsymbol{X}) \propto |\, \boldsymbol{R}_F + (\boldsymbol{\Lambda} - \boldsymbol{\Lambda}_F)\boldsymbol{Q}_F(\boldsymbol{\Lambda} - \boldsymbol{\Lambda}_F)' \,|^{-0.5\gamma}. \tag{15.23}$$

That is, the conditional distribution of $\boldsymbol{\Lambda}$ for prespecified $\boldsymbol{F}$ is matrix $\boldsymbol{T}$. Our point estimator of $\boldsymbol{\Lambda}$ is $E(\boldsymbol{\Lambda} \mid \hat{\boldsymbol{F}}, \boldsymbol{X})$, or

$$\hat{\boldsymbol{\Lambda}} \equiv \boldsymbol{\Lambda}_{\hat{F}} = (\boldsymbol{X}'\hat{\boldsymbol{F}} + \boldsymbol{\Lambda}_0\boldsymbol{H})(\boldsymbol{H} + \hat{\boldsymbol{F}}'\hat{\boldsymbol{F}})^{-1} \tag{15.24}$$

Any scalar element of $\boldsymbol{\Lambda}$, conditional on $(\hat{\boldsymbol{F}}, \boldsymbol{X})$, follows a general Student t-distribution, analogous to the general univariate marginal Student t-density in Equation (15.22), which corresponds to the matrix $\boldsymbol{T}$-density in Equation (15.19).

We note also that $\hat{\boldsymbol{\Lambda}}$ in Equation (15.24) is both a mean and a modal estimator of the joint distribution of ($\boldsymbol{\Lambda}$, $\boldsymbol{F} \mid \boldsymbol{X}$) under a vague prior for $\boldsymbol{F}$ (see Equations (15.8) and (15.10)). This follows from the unimodality and the symmetry of the density in Equation (15.8). Thus, in this case, ($\hat{\boldsymbol{F}}$, $\hat{\boldsymbol{\Lambda}}$) is a joint modal estimator of ($\boldsymbol{F}$, $\boldsymbol{\Lambda}$).

15.3.12 Estimation of the Disturbance Covariance Matrix

The disturbance covariance matrix, $\boldsymbol{\Psi}$, is estimated conditional upon $(\boldsymbol{F}, \boldsymbol{\Lambda}) = (\hat{\boldsymbol{F}}, \hat{\boldsymbol{\Lambda}})$. The joint posterior density of $(\boldsymbol{\Psi}, \boldsymbol{\Lambda}, \boldsymbol{F} \mid \boldsymbol{X})$ is given in Equation (15.6). The conditional density of $(\boldsymbol{\Psi} \mid \boldsymbol{\Lambda}, \boldsymbol{F}, \boldsymbol{X})$ is obtained by dividing Equation (15.6) by (15.3.7) and setting $\boldsymbol{G} = \hat{\boldsymbol{G}}$ ($\hat{\boldsymbol{G}}$ depends only on the data $\boldsymbol{X}$ and the hyperparameters $\boldsymbol{H}$, $\boldsymbol{B}$, and $\boldsymbol{\Lambda}_0$). The result is

$$p(\boldsymbol{\Psi} \mid \hat{\boldsymbol{\Lambda}}, \hat{\boldsymbol{F}}, \boldsymbol{X}) \propto \frac{e^{-0.5 \operatorname{tr} \boldsymbol{\Psi}^{-} \hat{\boldsymbol{G}}}}{\mid \boldsymbol{\Psi} \mid^{0.5(N+m+v)}} \tag{15.25}$$

where

$$\hat{\boldsymbol{G}} = (\boldsymbol{X} - \hat{\boldsymbol{F}}\hat{\boldsymbol{\Lambda}}')'(\boldsymbol{X} - \hat{\boldsymbol{F}}\hat{\boldsymbol{\Lambda}}') + (\hat{\boldsymbol{\Lambda}} - \boldsymbol{\Lambda}_0)\boldsymbol{H}(\hat{\boldsymbol{\Lambda}} - \boldsymbol{\Lambda}_0)' + \boldsymbol{B}. \tag{15.26}$$

That is, the posterior conditional distribution of $\boldsymbol{\Psi}$ given $(\hat{\boldsymbol{F}}, \hat{\boldsymbol{\Lambda}}, \boldsymbol{X})$ is Inverted Wishart. A point estimator of $\boldsymbol{\Psi}$ is given by $\hat{\boldsymbol{\Psi}} = E(\boldsymbol{\Psi} \mid \hat{\boldsymbol{\Lambda}}, \hat{\boldsymbol{F}}, \boldsymbol{X})$. Equation (5.2.4) of Press (1982, p. 119) gives

$$\hat{\boldsymbol{\Psi}} = \frac{\hat{\boldsymbol{G}}}{N + m + v - 2p - 2}, \tag{15.27}$$

with $\hat{\boldsymbol{G}}$ given in Equation (15.26).

15.3.13 Example[2]

We have extracted some data from an illustrative example in Kendall (1980), and have analysed these data from the Bayesian viewpoint using our model. There are 48 applicants for a certain job, and they have been scored on 15 variables regarding their acceptability. The variables are:

1. Form of letter application
2. Appearance
3. Academic ability
4. Likeability
5. Self-confidence
6. Lucidity
7. Honesty
8. Salesmanship
9. Experience
10. Drive
11. Ambition

[2]I am grateful to Daniel Rowe for correcting some typographical errors in the data for this example.

12. Grasp
13. Potential
14. Keenness to join
15. Suitability

The raw scores of the applicants on these 15 variables, measured on the same scale, are presented in Table 15.1. The question is, is there an underlying subset of factors that explain the variation observed in the scores? If so, then the applicants could be compared more easily. The correlation matrix for the 15 variables is given in Table 15.2. (*Note*: we assume the sample size of 48 is large enough to estimate the mean well enough for it to be ignored after subtracting it out.)

Now we postulate a model with four factors. This choice is based upon our having carried out a principal components analysis and our having found that four factors accounted for 81.5 percent of the variance. This is therefore our first guess, a conclusion that might be modified if we were to do hypothesis testing to see how well a four-factor model fit the data. Based upon underlying beliefs we constructed the prior factor loading matrix

$$\Lambda_0 = \begin{matrix} 1 \\ 2 \\ 3 \\ 4 \\ 5 \\ 6 \\ 7 \\ 8 \\ 9 \\ 10 \\ 11 \\ 12 \\ 13 \\ 14 \\ 15 \end{matrix} \begin{pmatrix} 0 & 0 & 0.7 & 0 \\ 0 & 0 & 0 & 0 \\ 0 & 0.7 & 0 & 0 \\ 0 & 0 & 0 & 0.7 \\ 0.7 & 0 & 0 & 0 \\ 0.7 & 0 & 0 & 0 \\ 0 & 0 & 0 & 0.7 \\ 0.7 & 0 & 0 & 0 \\ 0 & 0 & 0.7 & 0 \\ 0.7 & 0 & 0 & 0 \\ 0.7 & 0 & 0 & 0 \\ 0.7 & 0 & 0 & 0 \\ 0.7 & 0 & 0 & 0 \\ 0 & 0 & 0 & 0 \\ 0 & 0 & 0.7 & 0 \end{pmatrix}$$

The hyperparameter $\boldsymbol{H}$ was assessed as $\boldsymbol{H} = 10\boldsymbol{I}_4$. The prior distribution for $\boldsymbol{\Psi}$ was assessed with $\boldsymbol{B} = 0.2\boldsymbol{I}_{15}$, and $\nu = 33$. Note that when our observational data are augmented by proper prior information, as in this example, the identification-of-parameters problem of classical factor analysis disappears. The factor scores, factor loadings, and disturbance variances and covariances may now be estimated from Equations (15.14), (15.24), and (15.27), respectively. Results are given in Tables 15.3, 15.4, and 15.5.

Note that since we used standardized scores, the elements in Table 15.2 may be interpreted as correlations. It may also be noted from Table 15.5 that most of the off-diagonal elements of the estimated disturbance matrix $\boldsymbol{\Psi}$ are very small relative to the diagonal elements (the variances). That is, $\boldsymbol{\Psi}$ is approximately diagonal. Tables 15.3, 15.4, and 15.5 give the Bayesian point estimates for $(\boldsymbol{F}, \boldsymbol{\Lambda}, \boldsymbol{\Psi})$. We next obtain

Table 15.1 Raw Scores of 48 Applicants Scaled on 15 Variables

Person	1	2	3	4	5	6	7	8	9	10	11	12	13	14	15
1	6	7	2	5	8	7	8	8	3	8	9	7	5	7	10
2	9	10	5	8	10	9	9	10	5	9	9	8	8	8	10
3	7	8	3	6	9	8	9	7	4	9	9	8	6	8	10
4	5	6	8	5	6	5	9	2	8	4	5	8	7	6	5
5	6	8	8	8	4	5	9	2	8	5	5	8	8	7	7
6	7	7	7	6	8	7	10	5	9	6	5	8	6	6	6
7	9	9	8	8	8	8	8	8	10	8	10	8	9	8	10
8	9	9	9	8	9	9	8	8	10	9	10	9	9	9	10
9	9	9	7	8	8	8	8	5	9	8	9	8	8	8	10
10	4	7	10	2	10	10	7	10	3	10	10	10	9	3	10
11	4	7	10	0	10	8	3	9	5	9	10	8	10	2	5
12	4	7	10	4	10	10	7	8	2	8	8	10	10	3	7
13	6	9	8	10	5	4	9	4	4	4	5	4	7	6	8
14	8	9	8	9	6	3	8	2	5	2	6	6	7	5	6
15	4	8	8	7	5	4	10	2	7	5	3	6	6	4	6
16	6	9	6	7	8	9	8	9	8	8	7	6	8	6	10
17	8	7	7	7	9	5	8	6	6	7	8	6	6	7	8
18	6	8	8	4	8	8	6	4	3	3	6	7	2	6	4
19	6	7	8	4	7	8	5	4	4	2	6	8	3	5	4
20	4	8	7	8	8	9	10	5	2	6	7	9	8	8	9
21	3	8	6	8	8	8	10	5	3	6	7	8	8	5	8
22	9	8	7	8	9	10	10	10	3	10	8	10	8	10	8
23	7	10	7	9	9	9	10	10	3	9	9	10	9	10	8
24	9	8	7	10	8	10	10	10	2	9	7	9	9	10	8
25	6	9	7	7	4	5	9	3	2	4	4	4	4	5	4
26	7	8	7	8	5	4	8	2	3	4	5	6	5	5	6
27	2	10	7	9	8	9	10	5	3	5	6	7	6	4	5
28	6	3	5	3	5	3	5	0	0	3	3	0	0	5	0
29	4	3	4	3	3	0	0	0	0	4	4	0	0	5	0
30	4	6	5	6	9	4	10	3	1	3	3	2	2	7	3
31	5	5	4	7	8	4	10	3	2	5	5	3	4	8	3
32	3	3	5	7	7	9	10	3	2	5	3	7	5	5	2
33	2	3	5	7	7	9	10	3	2	2	3	6	4	5	2
34	3	4	6	4	3	3	8	1	1	3	3	3	2	5	2
35	6	7	4	3	3	0	9	0	1	0	2	3	1	5	3
36	9	8	5	5	6	6	8	2	2	2	4	5	6	6	3
37	4	9	6	4	10	8	8	9	1	3	9	7	5	3	2
38	4	9	6	6	9	9	7	9	1	2	10	8	5	5	2
39	10	6	9	10	9	10	10	10	10	10	8	10	10	10	10
40	10	6	9	10	9	10	10	10	10	10	10	10	10	10	10
41	10	7	8	0	2	1	2	0	10	2	0	3	0	0	10
42	10	3	8	0	1	1	0	0	10	0	0	0	0	0	10
43	3	4	9	8	2	4	5	3	6	2	1	3	3	3	8
44	7	7	7	6	9	8	8	6	8	8	10	8	8	6	5
45	9	6	10	9	7	7	10	2	1	5	5	7	8	4	5
46	9	8	10	10	7	9	10	3	1	5	7	9	9	4	4
47	0	7	10	3	5	0	10	0	0	2	2	0	0	0	0
48	0	6	10	1	5	0	10	0	0	2	2	0	0	0	0

Table 15.2 Correlation Matrix of Variables 1 through 15

	1	2	3	4	5	6	7	8	9	10	11	12	13	14	15
1	1.000	0.239	0.044	0.306	0.092	0.229	−0.107	0.269	0.548	0.346	0.285	0.338	0.367	0.467	0.586
2		1.000	0.123	0.380	0.431	0.376	0.354	0.477	0.141	0.341	0.550	0.506	0.507	0.284	0.384
3			1.000	0.002	0.001	0.080	−0.030	0.046	0.266	0.094	0.044	0.198	0.290	−0.323	0.140
4				1.000	0.302	0.489	0.645	0.347	0.141	0.393	0.347	0.503	0.606	0.685	0.327
5					1.000	0.802	0.410	0.816	0.015	0.704	0.842	0.721	0.672	0.482	0.250
6						1.000	0.360	0.823	0.155	0.700	0.758	0.890	0.785	0.533	0.420
7							1.000	0.231	−0.156	0.280	0.215	0.386	0.416	0.448	0.003
8								1.000	0.233	0.811	0.860	0.766	0.735	0.549	0.548
9									1.000	0.337	0.195	0.299	0.348	0.215	0.693
10										1.000	0.780	0.714	0.788	0.613	0.623
11											1.000	0.784	0.769	0.547	0.435
12												1.000	0.876	0.549	0.528
13													1.000	0.539	0.574
14														1.000	0.396
15															1.000

two-tailed 95 percent credibility intervals for the 48th subject's factor scores, and for the last (15th) row of the factor loading matrix.

The factor scores for subject 48 are given in the last row of the matrix in Table 15.3 as

$$(-2.156, 2.035, -2.529, -0.751).$$

Now calculate the two-tailed credibility interval at the 95 percent level from Equation (15.22) and find the intervals

$$\begin{aligned}&[-3.160, -1.152],\\&[-0.578,\ 4.649],\\&[-4.058, -1.001],\\&[-2.617,\ 1.115].\end{aligned}$$

The factor loadings for row 15 of the factor loading matrix are obtained from Table 15.4 as

$$(0.128, -0.015, 0.677, 0.011).$$

Now calculate 95 percent two-tailed credibility intervals from the marginals of Equation (15.1), just as we obtained the result in Equation (15.22) from Equation (15.19). Results for the last row factor loadings are

$$\begin{aligned}&[-0.219, 0.475],\\&[-0.295,\ 0.266],\\&[0.341, 1.012],\\&[-0.314,\ 0.337].\end{aligned}$$

Hypothesis about the elements of $(\boldsymbol{F}, \boldsymbol{\Lambda}, \boldsymbol{\Psi})$ may be tested using the associated marginal posterior densities. These are quite simple, being Student t, Student t for given $\boldsymbol{F}$, and Inverted Wishart given $\boldsymbol{F}$, and $\boldsymbol{\Lambda}$, respectively. For example, note that the credibility intervals for the first, second, and fourth factor loadings corresponding to the last row of Table 15.4 include the origin. A commonly used Bayesian hypothesis testing procedure (see Section 9.4) suggests that we should therefore conclude that we cannot reject the hypothesis that these three factor loadings are zero.

15.4 CHOOSING THE NUMBER OF FACTORS[3]

15.4.1 Introduction

Akaike (1973, 1987) proposed the AIC criterion for selecting the appropriate number of factors to use in a MLE factor analysis model. He derived the statistic

Table 15.3 Bayes' Estimates of Factor Scores

Person	1	2	3	4
1	0.728	−3.548	0.405	−0.301
2	1.476	−1.454	1.225	0.735
3	1.020	−2.850	0.726	0.231
4	−0.288	0.640	0.226	−0.021
5	−0.324	0.640	0.691	0.735
6	0.263	−0.058	0.868	0.510
7	1.188	0.640	1.942	0.456
8	1.475	1.338	1.942	0.456
9	0.876	−0.058	1.799	0.456
10	1.880	2.035	0.050	−1.336
11	1.550	2.035	−0.382	−2.957
12	1.547	2.035	−0.525	−0.832
13	−0.590	0.640	0.261	1.239
14	−0.623	0.640	0.472	0.708
15	−0.708	0.640	0.049	0.762
16	0.903	−0.756	1.123	0.204
17	0.422	−0.058	0.903	0.204
18	−0.195	0.640	−0.458	−1.111
19	−0.217	0.640	−0.315	−1.390
20	0.729	−0.058	−0.237	1.014
21	0.597	−0.756	−0.415	1.014
22	1.591	−0.058	0.650	1.014
23	1.591	−0.058	0.295	1.266
24	1.364	−0.058	0.506	1.518
25	−0.932	−0.058	−0.602	0.483
26	−0.702	−0.058	0.007	0.456
27	0.326	−0.058	−1.024	1.266
28	−1.822	−1.454	−1.464	−1.642
29	−2.047	−2.152	−1.819	−3.038
30	−0.976	−1.454	−1.244	0.510
31	−0.586	−2.152	−0.923	0.762
32	−0.151	−1.454	−1.422	0.762
33	−0.492	−1.454	−1.600	0.762
34	−1.601	−0.756	−1.566	−0.553
35	−2.198	−2.152	−0.889	−0.526
36	−0.699	−1.454	−0.213	−0.301
37	0.674	−0.756	−1.388	−0.553
38	0.722	−0.756	−1.388	−0.327
39	1.720	1.338	2.120	1.518
40	1.859	1.338	2.120	1.518
41	−2.283	0.640	2.120	−3.236
42	−2.709	0.640	2.120	−3.794
43	−1.647	1.338	0.015	−0.381
44	1.090	−0.058	0.581	−0.048
45	−0.007	2.035	−0.069	1.266
46	0.521	2.035	−0.212	1.518
47	−2.156	2.035	−2.529	−0.247
48	−2.156	2.035	−2.529	−0.751

Table 15.4 Bayes' Estimates of Factor Loadings

Variables	1	2	3	4
1	−0.045	−0.065	0.711	0.028
2	0.241	0.046	0.094	0.175
3	0.000	0.726	0.000	0.000
4	−0.010	−0.010	0.152	0.703
5	0.775	−0.051	−0.192	−0.029
6	0.719	−0.011	−0.058	0.035
7	0.012	0.010	−0.152	0.722
8	0.754	−0.049	0.016	−0.091
9	−0.081	0.080	0.737	−0.040
10	0.653	−0.020	0.116	−0.021
11	0.771	−0.046	−0.028	−0.100
12	0.659	0.057	0.052	0.066
13	0.592	0.120	0.094	0.141
14	0.244	−0.263	0.221	0.311
15	0.128	−0.015	0.677	0.011

for the problem of selection of variables in nested linear regression models. The motivation was his desire to control the effect of overparameterization of the model, and therefore to move towards parsimony in statistical modeling. His concern was with the maximum likelihood approach to estimation and he believed that models should be selected by adjusting the MLE for the number of parameters being estimated, which is exactly what the AIC criterion does. By definition,

$$\begin{aligned}\text{AIC} = &-2\ \log(\text{maximized likelihood}) \\ &+ 2(\text{number of free, that is, independent, parameters}).\end{aligned}$$

For an example of the use of AIC in a factor analysis context, see Press, 1982, Section 10.4.

In Akaike (1987) the author points out (p. 11) that "the quantity that minimizes the AIC criterion is not a consistent estimate of the correct model." As sample size increases, AIC tends to choose models that are too complex. The procedure is derived as a finite sample approximation to an asymptotic result. Zellner (1978) pointed out that posterior odds ratio contains much more information than does AIC, and he states in his abstract: "It is found that the AIC is a truncated version of the POC (posterior odds criterion) that omits important factors."

[3]Section 15.4 is adapted from Press and Shigemasu, 1999, with the permission of Marcel Dekker, Inc.

Table 15.5 Bayes' Estimates of the Disturbance Covariance Matrix

	1	2	3	4	5	6	7	8	9	10	11	12	13	14	15
1	0.278	0.008	0.000	0.035	0.015	0.009	−0.035	−0.026	−0.148	−0.035	0.029	0.008	0.000	0.082	−0.127
2		0.650	0.000	−0.025	0.019	−0.077	0.026	0.035	−0.058	−0.099	0.109	0.016	−0.002	−0.061	0.050
3			0.004	0.000	0.000	0.000	0.000	0.000	0.000	0.000	0.000	0.000	0.001	−0.001	0.000
4				0.132	−0.055	0.026	−0.128	0.002	−0.024	−0.015	0.018	−0.001	0.026	0.049	−0.010
5					0.150	−0.017	0.055	0.001	0.013	−0.031	0.015	−0.052	−0.061	−0.018	−0.029
6						0.148	−0.026	−0.011	−0.002	−0.077	−0.071	0.055	−0.022	−0.025	−0.008
7							0.132	−0.002	0.024	0.015	−0.018	0.001	−0.025	−0.048	0.010
8								0.126	−0.011	0.001	−0.003	−0.055	−0.054	0.007	0.038
9									0.199	−0.004	−0.003	0.005	0.002	−0.026	−0.047
10										0.203	−0.018	−0.077	0.003	0.052	0.039
11											0.128	−0.033	−0.014	0.021	−0.026
12												0.141	0.025	−0.012	−0.012
13													0.125	−0.023	−0.002
14														0.308	−0.055
15															0.177

Akaike (1978) proposed a Bayesian version of model selection that he called BIC (Bayesian Information Criterion), which he adopted for use in (nested) regression; see also Section 13.5. His definition is that if BIC is defined as $-2\log$(model probability), then

$$\begin{aligned}\text{BIC} &= -2\ \log(\text{maximum marginal likelihood})\\ &\quad + (\log n)(\text{number of free parameters, including hyperparameters}).\end{aligned}$$

The BIC criterion is asymptotically consistent, but in finite samples tends to choose models that are too simple because it penalizes complex models. Schwarz (1978, p. 463) claimed that AIC cannot be asymptotically optimal, and also proposed essentially the BIC. Geisser and Eddy (1979) pointed out that the Schwarz criterion for determining the proper dimension of a (nested) model depends on sample size, but not upon the type of error made, so it is not very helpful for prediction (prediction of factor scores in this case). Schwarz assumed that the data come from a Koopman–Darmois (exponential) family. In a regression context, the maximum marginal likelihood is the mean sum of squares of the residuals. The marginal likelihood is the likelihood that remains after integrating out the parameters that are not of interest (Good, 1965). It is then maximized over the remaining parameters.

Geisser and Eddy (1979) proposed their own model selection procedure, and it was based upon the "sample reuse method." Berger and Pericchi (1996) pointed out that "BIC seems to overstate the evidence in favor of the complex model, which is a 'bias' that, if anything, goes in the wrong direction." Both AIC and BIC depend upon asymptotic expansions whose first couple of terms depend upon the likelihood function. Zellner and Min (1993, p. 202) suggest five different methods of comparing models, including: AIC, BIC, Mean Squared Error, the Schwarz Criterion, and the Mallows Cp Criterion, 1973. Results for the Bayesian factor analysis model using posterior odds will be compared below with AIC and BIC.

A fundamental assumption made in most factor analysis models (Bayesian and otherwise) is that the number of factors in the model is known or preassigned. While such an assumption may be satisfied in some confirmatory factor analysis problems, it is generally not a tenable assumption in exploratory factor analysis. In this section we relax this assumption and then develop the posterior probability and the posterior odds ratio for the number of factors in the model.

15.4.2 Posterior Odds for the Number of Factors: General Development

Since the number of factors in the model is now considered unknown, we assume it is random and denote it by M. We will evaluate the posterior probability, $P\{M = m \mid X\}$, and choose the number of factors for the model that maximizes this posterior probability. In a decision–theoretic context the analyst could alternatively introduce loss functions and use criteria appropriate for the given loss function. Sometimes, as an alternative to maximizing the posterior probability of m

factors, it might be appropriate to maximize the posterior probability of a weighted average of several models (with different m). We will proceed using the maximum posterior probability criterion.

By Bayes' theorem,

$$P\{M = m \mid X\} = C(X)P\{X \mid M = m\}P\{M = m\}. \tag{15.28}$$

The term $P\{M = m\}$ is the prior probability for the number of factors in the model; the term $P\{X \mid M = m\}$ is the unconditional likelihood function for the observed data (unconditional because it does not depend upon any unknown parameters, other than m); and $C(X)$ is the normalizing constant. The unconditional likelihood function is given by

$$\begin{aligned} p(X \mid M = m) &\propto I(m) \\ &= \iiint p(X \mid \Lambda, F, \Psi, M = m) p(\Lambda, F, \Psi \mid M = m) d\Lambda dF d\Psi, \end{aligned} \tag{15.29}$$

where we have used the notation $p(\cdot)$ generically to denote a density, and the various $p(\cdot)$s are distinguished by their arguments. Since $C(X)$ depends only upon X, and not upon M, we may rewrite Equation (15.28) as

$$P\{M = m \mid X\} \propto p(X \mid M = m)P\{M = m\},$$

where $\propto$ denotes proportionality, and combining with Equation (15.29) gives

$$P\{M = m \mid X\} \propto P\{M = m\}I(m). \tag{15.30}$$

Note that in Equation (15.29), the first integrand term is a "conditional likelihood specification," while the second integrand term is a "prior specification."

15.4.3 Likelihood Function

The conditional likelihood function is found readily by recalling that $(x_j \mid \Lambda, f_j, \Psi, m) \sim N(\Lambda f_j, \Psi)$, and the x_js are assumed to be independent. It is

$$\begin{aligned} p(X \mid \Lambda, F, \Psi, m) &= \prod_1^N (2\pi)^{-0.5p} \mid \Psi \mid^{-0.5} \exp\{-0.5(x_j - \Lambda f_j)' \Psi^{-1}(x_j - \Psi f_j)\} \\ &= (2\pi)^{-0.5Np} \mid \Psi \mid^{-0.5N} \\ &\quad \times \exp\left\{-0.5 \operatorname{tr} \Psi^{-1}\left[\sum_1^N (x_j - \Lambda f_j)(x_j - \Lambda f_j)'\right]\right\}. \end{aligned} \tag{15.31}$$

Noting in Equation (15.31) that the sum is $(X - F\Lambda')'(X - F\Lambda')$ gives the simplification

$$p(X \mid \Lambda, F, \Psi, m) = (2\pi)^{-0.5Np} \mid \Psi \mid^{-0.5N} \cdot \exp\{-0.5 \operatorname{tr}(X - F\Lambda')'(X - F\Lambda')\Psi^{-1}\}. \tag{15.32}$$

15.4.4 Prior Densities

We now develop the joint prior distribution for all of the factor analysis parameters conditional upon the number of factors; then we find the combined distribution of the data with the prior.

Assume the joint prior density of all the parameters, for given M, is given by

$$p(\Lambda, F, \Psi \mid M) = p(\Lambda \mid \Psi, M)p(\Psi \mid M)p(F \mid M) \tag{15.33}$$

We adopt the normal prior density for the factor loading matrix conditional on the matrix of disturbance covariances and the number of factors,

$$p(\Lambda \mid \Psi, M = m) \propto \frac{1}{\mid \Psi \mid^{0.5m}} e^{-0.5 \operatorname{tr}(\Lambda - \Lambda_0)H(\Lambda - \Lambda_0)'\Psi^{-1}}, \tag{15.34}$$

where (Λ_0, H) are hyperparameters. Using the identity

$$\operatorname{tr}[(\Lambda - \Lambda_0)H(\Lambda - \Lambda_0)'\Psi^{-1}] = (\lambda - \lambda_0)'Q^{-1}(\lambda - \lambda_0),$$

where

$$\begin{aligned} \underset{(m \times p)}{\Lambda'} &\equiv (\lambda_1, \ldots, \lambda_p), \qquad \lambda_j\colon (m \times 1), \qquad j = 1, \ldots, p, \\ \underset{(mp \times 1)}{\lambda} &\equiv \operatorname{vec}(\Lambda') = (\lambda_1', \ldots, \lambda_p')', \qquad Q \equiv \Psi \otimes H^{-1}, \\ p(\Lambda \mid \Psi, M = m) &\propto \frac{1}{\mid \Psi \mid^{0.5m}} e^{-0.5(\lambda - \lambda_0)'Q^{-1}(\lambda - \lambda_0)} \end{aligned} \tag{15.35}$$

where λ_0 is defined analogously to λ. Note next that $\mid Q \mid = \mid \Psi \otimes H^{-1} \mid = \mid \Psi \mid^m \mid H^{-1} \mid^p$, and H is just a constant to be assessed. Moreover, $\Psi\colon (p \times p), H\colon (m \times m)$. So

$$p(\Lambda \mid \Psi, M = m) = \frac{\exp\{-0.5 \operatorname{tr}(\Lambda - \Lambda_0)H(\Lambda - \Lambda_0)'\Psi^{-1}\}}{(2\pi)^{0.5mp} \mid \Psi \mid^{0.5m} \mid H \mid^{-0.5p}}. \tag{15.36}$$

Next, assume that $(\boldsymbol{\Psi} \mid M = m) \sim W^{-1}(\boldsymbol{B}, p, v)$, where $W^{-1}(\cdot)$ denotes an inverted Wishart distribution with scale matrix hyperparameter $\boldsymbol{B}$, of dimension p, and with v degrees of freedom (v is also a hyperparameter). Then,

$$p(\boldsymbol{\Psi} \mid M = m) = \frac{C_0(v, p) \mid \boldsymbol{B} \mid^{0.5(v-p-1)}}{\mid \boldsymbol{\Psi} \mid^{0.5v}} e^{-0.5 \operatorname{tr} \boldsymbol{B}\boldsymbol{\Psi}^{-1}}, \tag{15.37}$$

where $C_0(v, p)$ is a constant depending only upon (v, p), and not upon m, given by

$$C_0^{-1}(v, p) = 2^{(v-p-1)/1} \pi^{p(p-1)/4} \prod_{j=1}^{p} \Gamma[0.5(v - p - j)].$$

(Note that it will be important to have the constants for numerical evaluation.) Multiplying Equations (15.36) and (15.37) gives the joint prior density:

$$\begin{aligned} p(\boldsymbol{\Lambda}, \boldsymbol{F}, \boldsymbol{\Psi} \mid M = m) = & \frac{C_0(v, p) p(\boldsymbol{F} \mid M = m) \mid \boldsymbol{B} \mid^{0.5(v-p-1)}}{(2\pi)^{0.5mp} \mid \boldsymbol{\Psi} \mid^{0.5m} \mid \boldsymbol{H} \mid^{-0.5p} \mid \boldsymbol{\Psi} \mid^{0.5v}} \\ & \cdot \exp\{-0.5 \operatorname{tr}[\boldsymbol{B} + (\boldsymbol{\Lambda} - \boldsymbol{\Lambda}_0)\boldsymbol{H}(\boldsymbol{\Lambda} - \boldsymbol{\Lambda}_0)']\boldsymbol{\Psi}^{-1}\}. \end{aligned} \tag{15.38}$$

Now multiply the conditional likelihood function (15.32) by the joint prior density (15.38), to obtain

$$\begin{aligned} R \equiv p(\boldsymbol{X}, \boldsymbol{\Lambda}, \boldsymbol{F}, \boldsymbol{\Psi} \mid M = m) = & \frac{C_0(v, p) \mid \boldsymbol{B} \mid^{0.5(v-p-1)} \mid \boldsymbol{H} \mid^{0.5p} p(\boldsymbol{F} \mid M)}{(2\pi)^{0.5(Np+mp)} \mid \boldsymbol{\Psi} \mid^{0.5(N+m+v)}} \\ & \cdot \exp\{-0.5 \operatorname{tr}[\boldsymbol{B} + (\boldsymbol{\Lambda} - \boldsymbol{\Lambda}_0)\boldsymbol{H}(\boldsymbol{\Lambda} - \boldsymbol{\Lambda}_0)' \\ & + (\boldsymbol{X} - \boldsymbol{F}\boldsymbol{\Lambda}')'(\boldsymbol{X} - \boldsymbol{F}\boldsymbol{\Lambda}')]\boldsymbol{\Psi}^{-1}\}. \end{aligned} \tag{15.39}$$

15.4.5 Posterior Probability for the Number of Factors

The desired posterior probability for the number of factors can now be calculated, and is given in the following theorem.

Theorem 15.1 For moderate sample sizes, the posterior probability for the number of factors is given approximately by

$$P\{M = m \mid X\} \cong C(X) \left[\frac{\Gamma_m[0.5(N+m+v-2p-1)]\Gamma_N[0.5(N+v-p-m-1)] \prod_{j=1}^{p} \Gamma[0.5(N+m+v-p-j)]}{(2e)^{0.5NM}\pi^{0.5Np}\Gamma_m[0.5(N+m+v-p-1)] \Gamma_N[0.5(N+v-p-1)] \prod_{j=1}^{p} \Gamma[0.5(v-p-j)]} \right] \cdot \left[\frac{|\boldsymbol{B}|^{0.5(v-p-1)} |\boldsymbol{H}|^{0.5p} |\boldsymbol{H}+N\boldsymbol{I}_m|^{0.5(N+v-2p-1)}}{|\boldsymbol{W}|^{0.5(N+v-p-1)} |\boldsymbol{A}|^{0.5(v-p-1)} |\boldsymbol{P}_0|^{0.5m}} \right] P\{M = m\}, \tag{15.40}$$

where $\boldsymbol{W} = \boldsymbol{X}'\boldsymbol{X} + \boldsymbol{B} + \boldsymbol{\Lambda}_0\boldsymbol{H}\boldsymbol{\Lambda}_0'$, $\boldsymbol{P}_0 = \boldsymbol{I}_N - \boldsymbol{X}\boldsymbol{W}^{-1}\boldsymbol{X}'$, and $C(\boldsymbol{X})$ is a normalizing constant possibly depending upon $\boldsymbol{X}$, but that does not depend upon m, and for which $\sum_{m=1}^{p} P\{M = m \mid \boldsymbol{X}\} = 1$.

Proof
Given in Complement to Chapter 15.

REMARK 1: The second quantity in large brackets in Equation (15.40) is $I(m) \equiv p(\boldsymbol{X} \mid M = m)$, the unconditional likelihood function defined by Equation (15.29).

REMARK 2: The posterior odds ratio for m factors versus n factors is given by

$$\frac{P\{M = m \mid \boldsymbol{X}\}}{P\{M = n \mid \boldsymbol{X}\}} = \frac{P\{\boldsymbol{X} \mid M = m\}P\{M = m\}}{P\{\boldsymbol{X} \mid M = n\}P\{M = n\}}. \tag{15.41}$$

Note that the normalizing constant in Equation (15.40) disappears in Equation (15.41) since it cancels out in the ratio.

REMARK 3: Note from Equation (15.40) that to compute the optimum m it is only necessary to substitute the assessed hyperparameters $(\boldsymbol{H}, \boldsymbol{B}, v, \boldsymbol{\Lambda}_0)$, and the data-related quantities $(\boldsymbol{X}, N, p)$. That is, it is not necessary to actually carry out the corresponding Bayesian factor analysis, unless the estimates of $(\boldsymbol{\Lambda}, \boldsymbol{F}, \boldsymbol{\Psi})$ are desired. Moreover, to compute the optimum m it is not necessary to evaluate the normalizing constant since that constant does not depend upon m; it is only necessary to maximize Equation (15.40) over m without the constant $C(\boldsymbol{X})$.

15.4.6 Numerical Illustrations and Hyperparameter Assessment

Data Generation
To illustrate the efficacy of the maximum posterior probability selection criterion proposed in Theorem 15.1 we elected to study how well the criterion works with

simulated data where we would know the correct number of factors that should be selected by the criterion. By contrast, with real data, there is really no true value for the number of factors, although some solutions for the number of factors are more acceptable than others in most situations. We therefore generated many data sets by simulation. The data are 10-dimensional ($p = 10$). The factors $\boldsymbol{f}_j$: $(m \times 1)$ were generated randomly and independently from $N(0, I_m)$. The disturbances, $\boldsymbol{\epsilon}_j$, were generated randomly and independently from $N(0, 0.36I_p)$. For factor loadings we used the matrix $\boldsymbol{\Lambda}$, where its transpose is given by

$$\Lambda' = \begin{pmatrix} 0.8 & 0.8 & 0.8 & 0.8 & 0 & 0 & 0 & 0 & 0 & 0 \\ 0 & 0 & 0 & 0 & 0.8 & 0.8 & 0.8 & 0 & 0 & 0 \\ 0 & 0 & 0 & 0 & 0 & 0 & 0 & 0.8 & 0.8 & 0.8 \end{pmatrix}.$$

The observable data were then generated from Equation (15.1) by addition. A sample of size $N = 200$ data vectors was generated. This collection of data vectors constituted one data set. We adopted the vague or indifference prior probabilities: $P\{M = m\} = 1/p = 0.10$.

Recent research has shown (Lee and Press, 1998) that the posterior probabilities that result from the Bayesian factor analysis model of PS89 are robust with respect to variations in $\nu, \boldsymbol{B}$, and $\boldsymbol{H}$, but not with respect to the hyperparameter $\boldsymbol{\Lambda}_0$. We therefore took special care to assess $\boldsymbol{\Lambda}_0$, and were less concerned about our assessments for ν, $\boldsymbol{B}$, and $\boldsymbol{H}$. The hyperparameter $\boldsymbol{\Lambda}_0$ was assessed by dividing the data set of 200 vectors into two equal parts, and doing principal components analysis on the covariance matrix of the first 100 data vectors. $\boldsymbol{\Lambda}_0$ was then taken to be the matrix of latent vectors corresponding to the principal components. We refer to these first 100 data vectors as our calibration data, because they were used to assess $\boldsymbol{\Lambda}_0$, which can be thought of as a calibrating hyperparameter for the model. Assessments for the other hyperparameters were:

$$\nu = 30, \qquad \boldsymbol{B} = 0.2\boldsymbol{I}_p = 0.2\boldsymbol{I}_{10}, \qquad \boldsymbol{H} = 10\boldsymbol{I}_m.$$

We tried different numbers of factors (values for m).

We used the remaining 100 data vectors to evaluate the various estimates, and we selected m to be the number that maximized the posterior probability, as given in Theorem 15.1. The entire procedure was repeated 10 times for 10 distinct data sets of 200 data vectors each.

Results

Results are presented in Table 15.6, where the table entries are the logs of the posterior probability, excluding the normalizing constant $C(\boldsymbol{X})$, given in Equation (15.40) (see REMARK (3) following Theorem 15.1). It may readily be seen from Table 15.6 that use of the maximum posterior probability criterion would have correctly selected a three-factor model in every one of the ten cases (since three factors maximizes the posterior probability of M).

Table 15.6 Log of Posterior Probability of M (Within an Additive Constant) for $M = 2, 3, 4$

	Number of Factors		
Data Set	2	3	4
1	−1966.707	−1855.164	−1860.475
2	−2007.874	−1896.381	−1901.199
3	−1963.285	−1851.879	−1858.199
4	−1969.922	−1858.494	−1862.771
5	−1927.428	−1816.727	−1822.969
6	−1963.074	−1852.200	−1859.598
7	−2000.221	−1889.349	−1894.267
8	−1964.465	−1853.407	−1857.357
9	−1936.254	−1825.548	−1832.627
10	−1947.455	−1837.252	−1842.724

15.4.7 Comparison of the Maximum Posterior Probability Criterion with AIC and BIC

In Table 15.7 below, there are two entries in each cell. The upper entries in each cell are results for the AIC criterion, and the lower entries in each cell are results for the BIC criterion. Thus, for data set no. 1, for two factors, the AIC criterion gives a result of 1.625, whereas the BIC criterion gives a result of 2.126. These cell entries are the optimizing target functions that result from applying the modified maximized likelihoods specified by the AIC and BIC criteria. The last column on the right gives the selection of the number of factors that would be made by means of the AIC criterion (the selection is made by choosing the minimum value of AIC; the choice is stated in the upper part of the cell) and the BIC criterion (the choice is made by choosing the minimum of BIC; it is given in the lower part of the cell). Inspection of Table 15.7 shows that while use of the BIC criterion would have correctly selected three factors for all 10 data sets, use of the AIC criterion would have incorrectly selected four factors for data set no. 10. So we find that the maximum posterior probability criterion improves upon the AIC criterion. We expect, furthermore, that as the sample size increases, the difficulties with the BIC criterion in large samples will begin to show up in incorrect model selections, whereas the maximum posterior probability criterion will only improve.

15.5 ADDITIONAL MODEL CONSIDERATIONS

One study of the Bayesian factor analysis model described in this chapter has been to examine the robustness of the estimators with respect to the assessments of the hyperparameters (Lee, 1994; Lee and Press, 1998). In this research it was found that the Bayesian estimators are quite robust to variations in the values of most of the

Table 15.7 Comparison of Maximum Posterior Probability with AIC and BIC

Data Set	Number of Factors 2	3	4	Selected Number of Factors
1	1.625	0.478	0.487	3
	2.126	1.091	1.216	3
2	1.465	0.436	0.476	3
	1.946	1.049	1.207	3
3	1.840	0.467	0.479	3
	2.320	1.080	1.208	3
4	1.676	0.443	0.456	3
	2.157	1.057	1.194	3
5	1.715	0.479	0.485	3
	2.196	1.092	1.214	3
6	1.631	0.415	0.462	3
	2.112	1.028	1.191	3
7	1.466	0.473	0.482	3
	1.946	1.086	1.211	3
8	1.691	0.495	0.508	3
	2.171	1.108	1.238	3
9	1.552	0.426	0.463	3
	2.002	1.040	1.192	3
10	1.345	0.512	0.493	4
	1.825	1.125	1.223	3

hyperparameters, with the exception of $\mathbf{\Lambda}_0$, the location hyperparameter of the prior distribution of $\mathbf{\Lambda}$. Results have been found to be sensitive to values of $\mathbf{\Lambda}_0$, so care must be taken in assessing that hyperparameter.

The Bayesian factor analysis model considered here was extended to include the case of correlated data vectors, as in time series or spatial analysis structures (Rowe, 1998). Markov chain Monte Carlo sampling was used for evaluation of a number of models that reflected special data dependence structures, such as *separable models* in which the covariance matrix of the data vectors was expressible as a direct product:

$$\text{var}(x_1', \ldots, x_n') = \mathbf{\Phi} \otimes \mathbf{\Psi},$$

for p-dimensional column vectors $\mathbf{x}_1, \ldots, \mathbf{x}_n$, and for matrix intraclass covariance matrix structures.

Shigemasu, Nakamura, and Shinzaki (1993) examined the model from a Markov chain Monte Carlo (MCMC) point of view and derived first and second moments for the marginal posterior distributions of the parameters. An advantage of the MCMC approach is that Bayesian estimates can be found even for very small sample sizes. They used conditional modes as their starting values.

There have also been some major efforts to provide assessment procedures for assessing the hyperparameters of this model (see, for example, Hyashi, 1997; and Shera and Ibrahim, 1998). Hyashi adopted a frequentist approach and was concerned with bias of the model estimators. Shera and Ibrahim adopted a Markov chain Monte Carlo approach for estimation in this model, and used historical prior information.

Polasek (1997) was concerned with detecting multivariate outliers in the Bayesian factor analysis model. He found that while detecting multivariate outliers is computationally burdensome, by following Mori et al., 1998, using Markov chain Monte Carlo, the procedure simultaneously computes marginal likelihoods, easing the computational detection burden. Polasek (1997) extended the univariate Verdinelli and Wasserman (1991) location shift outlier model to the multivariate case.

SUMMARY

This chapter presented a model for carrying out Bayesian factor analysis. The model differs from the traditional factor model in that the matrix of disturbance variances and covariances in the traditional model is diagonal, whereas in the model presented in this chapter that matrix is permitted to be more general, but is taken to be diagonal on the average. Misspecification of the relationships among variables could easily lead to a nondiagonal covariance matrix, such as we suppose. The Bayesian approach to estimation in this more general covariance matrix framework, using subjective prior distributions, has led to a model that no longer suffers from the indeterminacies that have plagued traditional factor analysis models, and permits the factor analyst to bring prior information to bear on the problem. We have shown how closed-form solutions can be obtained in terms of well-known (Student t) distributions, and we have shown how the number of factors in the model can be determined by examining posterior probabilities, or posterior odds ratios. We have also pointed to extensions of the model to cases involving nonindependent data vectors, robustness studies for estimation of the hyperparameters, and the use of MCMC to estimate the hyperparameters of the model, and to obtain estimates of the factor loadings, factor scores, and disturbance variances and covariances in small samples.

EXERCISES

15.1 Suppose you have data that consist of $n = 10$ independent and joint normally distributed random vectors representing multivariate data intensities. The data might correspond to the output of micro arrays, as used in biology, gray levels, as used in spatial analysis and imaging, or any other collection of intensities. The vectors are $p = 100$ dimensions in length. How might Bayesian factor analysis be used to obtain a modified data set in which $p < n$?

* Solutions for asterisked exercises may be found in Appendix 7.

15.2 Explain how you might assess the hyperparameters of the Bayesian factor analysis model when sample sizes are large.

15.3* What might be the advantages of standardizing the data vectors (with respect to location and scale) before carrying out a Bayesian factor analysis?

15.4 Of what use might factor scores be in: marketing analysis; a political science voting context; an astronomy context; a sociology context; a biology context; and a medical context?

15.5* Suppose you carry out an exploratory Bayesian factor analysis (you had no idea in advance of collecting data of how many factors there might be) and you find there are four factors; but the four factors do not seem to correspond to anything real. What would you conclude?

15.6 Explain how you would go about finding interval estimates of factor loadings in a Bayesian factor analysis containing a large number of data vectors.

15.7* Why should you not be concerned about the possible biases of Bayesian estimators in the factor analysis model?

15.8* What is the main advantage of using MCMC in the Bayesian factor analysis model over the closed-form solutions proposed in Sections 15.3 and 15.4?

15.9 Suppose a Bayesian factor analysis is contemplated but some of the observable variables in the data vectors are jointly continuous, and others are jointly discrete. What ways might you explore to carry out the analysis?

15.10 Suppose that examination in each dimension, separately, of the data vectors shows that in each dimension beyond the first, the data are indeed approximately normally distributed, but that in the first dimension, the data tend to follow some other continuous distribution. What might you do in order to have the assumptions of the Bayesian factor analysis model apply, at least approximately?

FURTHER READING

Akaike, H. (1973). "Information Theory and an Extension of the Maximum Likelihood Principle," in B. Petrov and F. Csaki, Eds., *2nd International Symposium of Information Theory*, Akademiai Kiado, Budapest.

Akaike, H. (1978). "A Bayesian Analysis of the Minimum AIC Procedure," *Ann. Inst. Statist. Math., Part A*, **30**, 9–14.

Akaike, H. (1987). "Factor analysis and AIC," *Psychometrika*, **52**(3), 317–332.

Arminger, G. and Muthen, B. O. (1998). "A Bayesian Approach to Non-linear, Latent Variable Models Using the Gibbs Sampler and the Metropolis–Hastings Algorithm," *Psychometrika*, **63**(3), 271–300.

Berger, J. O. and Pericchi, L. R. (1996). "The Intrinsic Bayes Factor for Linear Models," in J. M. Bernardo, J. O. Berger, A. P. Dawid and Smith, A. F. M., Eds., *Proceedings of the Fifth International Meeting On Bayesian Statistics*, Oxford, Oxford University Press, 25–44.

Bozdogan, H. (1987). "Model Selection and Akaike's Information Criterion (AIC): The General Theory and Its Analytical Extensions," *Psychometrika*, **52**(3), 345–370.

Dempster, A. P., Laird, N. M. and Rubin, D. B. (1977). "Maximum Likelihood from Incomplete Data Via the EM Algorithm," *J. Roy. Statist. Soc. B*, **39**, 1–38.

Euverman, T. J. and Vermulst, A. A. (1983). *Bayesian Factor Analysis*, Department of Social Sciences, State University of Groningen, November 3, 1983.

Geisser, S. and Eddy, W. F. (1979). "A Predictive Approach to Model Selection," *J. Am. Statist. Assoc.*, **74**(365), 153–160.

Good, I. J. (1965). *The Estimation of Probabilities: An Essay On Modern Bayesian Methods*, Res. Monograph 30, Cambridge MA: The MIT Press.

Guttman, L. (1953). "Image Theory for the Structure of Quantitative Variates," *Psychometrika*, **18**(4), 277–296.

Hyashi, K. (1997). *The Press–Shigemasu Factor Analysis Model with Estimated Hyperparameters*, Ph.D. Thesis, Department of Psychology, University of North Carolina, Chapel Hill.

Kaufman, G. M. and Press, S. J. (1973a). "Bayesian Factor Analysis," Report No. 73t-25, *Bull. Inst. Math. Statist.*, **2**(2), Issue No. 7, March 1973.

Kaufman, G. M. and Press, S. J. (1973b). "Bayesian Factor Analysis," Report No. 7322, Center for Mathematical Studies in Business and Economics, University of Chicago, April 1973.

Kaufman, G. M. and Press, S. J. (1976), "Bayesian Factor Analysis," Working Paper No. 413, Department of Commerce and Business Administration, University of British Columbia, Vancouver, B.C. Canada, September 1983.

Kendall, M. (1980). *Multivariate Analysis*, Second Edition, Charles Griffin Publisher, 34.

Lee, S. Y. (1981). "A Bayesian Approach to Confirmatory Factor Analysis," *Psychometrika*, **46**(2), 153–159.

Lee, S. E. (1994). *Robustness of Bayesian Factor Analysis Estimates*, Ph.D. Thesis, Department of Statistics, University of California, Riverside.

Lee, S. E. and Press, S. J. (1998). "Robustness of Bayesian Factor Analysis Estimates," *Communications in Statistics—Theory and Methods*, **27**(8), 1871–1893.

Mallows, C. L. (1973). "Some Comments On Cp," *Technometrics*, **15**, 661–675.

Martin, J. K. and McDonald, R. P. (1975). "Bayesian Estimation in Unrestricted Factor Analysis: A Treatment for Heywood Cases," *Psychometrika*, **40**(4), 505–517.

Mayekawa, S. (1985). "Bayesian Factor Analysis," ONR Technical Report No 85–3, School of Education, The University of Iowa, Iowa City, Iowa.

Mori, M., Watadani, S., Tanrumi, T. and Tanaka, Y. (1998). "Development of Statistical Software SAMMIF for Sensitivity Analysis in Multivariate Method," *COMPSTAT*, 395–400, Physica-Verlag.

Polasek, W. (1997). "Factor Analysis and Outliers: A Bayesian Approach," Institute of Statistics and Econometrics, University of Basel Discussion Paper, February 24, 2000.

Press, S. J. (1972). *Applied Multivariate Analysis*, New York, Holt Rinehart, and Winston, Inc.

Press, S. J. (1982). *Applied Multivariate Analysis: Using Bayesian and Frequentist Methods of Inference*, Malabar, Florida, Robert E. Krieger Publishing Co.

Press, S. J. (1989). *Bayesian Statistics: Principles, Models, and Applications*, New York, John Wiley and Sons.

Press, S. J. and Davis, A. W. (1987). "Asymptotics for the Ratio of Multiple t-Densities," in A. Gelfand, Ed., *Contributions to the Theory and Applications of Statistics*, New York, Academic Press, 155–177.

Press, S. J. and Shigemasu, K. (1989). "Bayesian Inference in Factor Analysis," in L. J. Gleser, M.D. Perlman, S. J. Press, and A. R. Simpson, Eds., *Contributions to Probability and Statistics: Essays in Honor of Ingram Olkin*, New York, Springer-Verlag, 271–287.

Press, S. J. and Shigemasu, K. (1997). "Bayesian Inference in Factor Analysis-Revised," with an appendix by Daniel B. Rowe, Technical Report No. 243, Department of Statistics, University of California, Riverside, CA, May 1997.

Press, S. J. and Shigemasu, K. (1999). "A Note on Choosing the Number of Factors," *Comm. Statist.—Theory and Methods*, **28**(7), 1653–1670.

Rowe, D. B. (1998). *Correlated Bayesian Factor Analysis*, Ph.D. Thesis, Department of Statistics, University of California, Riverside.

Schwarz, G. (1978). "Estimating the Dimension of the Model," *Ann. Statist.*, **6**, 461–464.

Shera, D. M. and Ibrahim, J. G. (1998). "Prior Elicitation and Computation for Bayesian Factor Analysis," submitted manuscript, Department of Biostatistics, Harvard School of Public Health, September 28, 1998.

Shigemasu, K. (1986). "A Bayesian Estimation Procedure with Informative Prior in Factor Analysis Model." Tokyo Institute of Technology, Jinmon-Ronso, No. 11, 67–77.

Shigemasu, K., Nakamura, T. and Shinzaki, S. (1993). "A Bayesian Numerical Estimation Procedure in Factor Analysis Model," Educational Science and Technology Research Group, Report 93-6, Tokyo Institute of Technology, December 1993.

Shi, J.-Q. and Lee, S.-Y. (1998). "Bayesian Sampling-Based Approach for Factor Analysis Models with Continuous and Polytomous Data," *British Journal of Math. and Stat. Psychology*, **51**, 233–252.

Verdinelli, I. and Wasserman, L. (1991). "Bayesian Analysis of Outlier Problems Using the Gibbs Sampler," *Statistics and Computing*, **1991-1**, 105–117.

Wong, G. (1980). "A Bayesian Approach to Factor Analysis." Program in Statistical Technical Report, No. 80-9, Educational Testing Service, Princeton, N.J., August 1980.

Zellner, A. (1978). "Jeffreys–Bayes Posterior Odds Ratio and the Akaike Information Criterion for Discriminating Between Models," *Economics Letters*, **1**, 337–342.

Zellner, A. and Chung-ki Min (1993). "Bayesian Analysis, Model Selection and Prediction," in *Physics and Probability: Essays in Honor of Edwin T. Jaynes*, Cambridge, Cambridge University Press, 195–206.

COMPLEMENT TO CHAPTER 15: PROOF OF THEOREM 15.1

Recall that the quantity in Equation (15.39) is $R \equiv p(\boldsymbol{X}, \boldsymbol{\Lambda}, \boldsymbol{F}, \boldsymbol{\Psi} \mid M = m)$. If we integrate Equation (15.39) with respect to $\boldsymbol{\Lambda}, \boldsymbol{F}, \boldsymbol{\Psi}$ and multiply by $P\{M = m\}$, by Bayes' theorem, the result will be proportional to the posterior density for the number of factors. Accordingly, we carry out these three integrations.

Integration with Respect to $\boldsymbol{\Psi}$

We integrate R first with respect to $\boldsymbol{\Psi}$. In terms of $\boldsymbol{\Psi}$ the integration involves an inverted Wishart density integrand. Thus,

$$R_1 \equiv \int R d\boldsymbol{\Psi} = K_1 \int \frac{\exp\{-0.5 \operatorname{tr} \boldsymbol{\Psi}^{-1} \boldsymbol{G}\}}{|\,\boldsymbol{\Psi}\,|^{0.5(N+n+v)}} d\boldsymbol{\Psi}$$

where

$$K_1 \equiv C_0(v, p) \,|\, \boldsymbol{B} \,|^{(v-p-1)} \,|\, \boldsymbol{H} \,|^{0.5p}\, p(\boldsymbol{F} \mid M)(2\pi)^{-0.5(Np+mp)},$$

$$\boldsymbol{G} \equiv \boldsymbol{B} + (\boldsymbol{\Lambda} - \boldsymbol{\Lambda}_0)\boldsymbol{H}(\boldsymbol{\Lambda} - \boldsymbol{\Lambda}_0)' + (\boldsymbol{X} - \boldsymbol{F}\boldsymbol{\Lambda}')'(\boldsymbol{X} - \boldsymbol{F}\boldsymbol{\Lambda}).$$

Then, evaluating gives

$$R_1 = p(\boldsymbol{X}, \boldsymbol{\Lambda}, \boldsymbol{F} \mid M = m) = \frac{K_2 p(\boldsymbol{F} \mid M)}{|\,\boldsymbol{G}\,|^{0.5(N+m+v-p-1)}},$$

where

$$K_2 \equiv \frac{2^{0.5(v-p-1)} C_0(v, p) \,|\, \boldsymbol{B} \,|^{0.5(v-p-1)} \,|\, \boldsymbol{H} \,|^{0.5p} \prod_{j=1}^{p} \Gamma[0.5(N + m + v - p - j)]}{\pi^{0.5p(N+m-0.5p+0.5)}}$$

Integration with Respect to $\boldsymbol{\Lambda}$

Now define

$$R_2 = p(\boldsymbol{X}, \boldsymbol{F} \mid M = m) = \int R_1 d\boldsymbol{\Lambda} = K_2 p(\boldsymbol{F} \mid M) \int \frac{d\boldsymbol{\Lambda}}{|\,\boldsymbol{G}\,|^{0.5\gamma}},$$

where $\gamma \equiv N + m + v - p - 1$. Combining terms that depend upon $\boldsymbol{\Lambda}$ and $\boldsymbol{G}$ gives

$$R_2 = K_2 p(\boldsymbol{F} \mid M = m) \int \frac{d\boldsymbol{\Lambda}}{|\, \boldsymbol{R}_F + (\boldsymbol{\Lambda} - \boldsymbol{\Lambda}_F \boldsymbol{Q}_F(\boldsymbol{\Lambda} - \boldsymbol{\Lambda}_F)' \,|^{0.5\gamma}},$$

where

$$\boldsymbol{Q}_F \equiv \boldsymbol{H} + \boldsymbol{F}'\boldsymbol{F},$$

$$\boldsymbol{\Lambda}_F \equiv (\boldsymbol{X}'\boldsymbol{F} + \boldsymbol{\Lambda}_0 \boldsymbol{H})(\boldsymbol{H} + \boldsymbol{F}'\boldsymbol{F})^{-1},$$

$$\boldsymbol{R}_F \equiv \boldsymbol{X}'\boldsymbol{X} + \boldsymbol{B} + \boldsymbol{\Lambda}_0 \boldsymbol{H} \boldsymbol{\Lambda}_0' - (\boldsymbol{X}'\boldsymbol{F} + \boldsymbol{\Lambda}_0 \boldsymbol{H}) \boldsymbol{Q}_F^{-1} (\boldsymbol{X}'\boldsymbol{F} + \boldsymbol{\Lambda}_0 \boldsymbol{H})'.$$

Integrating with respect to $\boldsymbol{\Lambda}$ involves integrating a matrix $\boldsymbol{T}$-density. Therefore,

$$R_2 = \frac{K_3 p(\boldsymbol{F} \mid M)}{|\,\boldsymbol{Q}_F\,|^{0.5p} \,|\, \boldsymbol{R}_F \,|^{0.5(\gamma-m)}},$$

where

$$K_3 \equiv \frac{K_2 \pi^{0.5(mp)} \Gamma_m[0.5(\gamma - p)]}{\Gamma_m[0.5\gamma]},$$

and $\Gamma_m(\alpha)$ denotes the m-dimensional gamma function of argument α.

Integration with Respect to F

The unconditional likelihood function is given by

$$p(X \mid M = m) = \int R_2 dF = K_3 \int \frac{p(F \mid M = m)dF}{\mid Q_F \mid^{0.5p} \mid R_F \mid^{0.5(\gamma - m)}}.$$

Define

$$\underset{(p\times p)}{W} \equiv B + X'X + \Lambda_0 H \Lambda_0',$$

$$\underset{(m\times m)}{A} \equiv H - H'\Lambda_0 W^{-1}\Lambda_0 H - (H'\Lambda_0' W^{-1} X')^{-1}(I_N - X W^{-1} X) X W^{-1} \Lambda H,$$

$$\underset{(N\times m)}{\hat{F}} \equiv (I_N - X W^{-1} X) X W^{-1} \Lambda_0 H.$$

After some algebra, we can find that the last integral is expressible as

$$p(X \mid M = m) = \frac{K_3}{\mid W \mid^{0.5(\gamma - m)}} \int \frac{\mid H + F'F \mid^{0.5(\gamma - m - p)} p(F \mid M) dF}{\mid A + (F - \hat{F})'(I_N - XW^{-1}X')(F - \hat{F}) \mid^{0.5(\gamma - m)}}.$$

Next, we make the traditional factor analysis assumption that the f_js are mutually independent, and $f_j \sim N(\mathbf{0}, I_m)$, $j = 1, \ldots, N$. Then

$$p(F \mid M = m) = p(f_1, \ldots, f_N) = \frac{\exp\{-0.5 \operatorname{tr} F'F\}}{(2\pi)^{0.5(Nm)}}.$$

Now, a simple approximation will yield a closed-form solution. Accordingly, note, as in Section 3.5.8, from the weak law of large numbers that if the f_js are i.i.d. $N(\mathbf{0}, I)$, the average value of $(f_j f_j')$ tends, in probability, to the identity matrix; that is, approximately,

$$F'F \cong N I_m.$$

Conceptually, we are replacing "squared factors" by their average values. Note that the quality of this approximation was evaluated by Gibbs sampling and it was found that the approximation is very good even in moderate samples. Alternatively, the more precise integration of Equation (15.39) by Gibbs sampling could be used when even very small samples are available.

Using the approximation in order to obtain a closed-form solution we obtain

$$p(\boldsymbol{X} \mid M = m) \cong \frac{K_3 \mid \boldsymbol{H} + N\boldsymbol{I}_m \mid^{0.5(\gamma-m-p)}}{\mid \boldsymbol{W} \mid^{0.5(\gamma-m)}} \cdot \int \frac{p(\boldsymbol{F} \mid M = m)d\boldsymbol{F}}{\mid A + (\boldsymbol{F} - \hat{\boldsymbol{F}})'(\boldsymbol{I}_N - \boldsymbol{X}\boldsymbol{W}^{-1}\boldsymbol{X}')(\boldsymbol{F} - \hat{\boldsymbol{F}}) \mid^{0.5(\gamma-m)}}.$$

Assuming N is large, we find

$$p(\boldsymbol{F} \mid M = m) \cong \frac{\exp\{-0.5NM\}}{(2\pi)^{0.5(Nm)}} = (2\pi e)^{-0.5Nm}.$$

Substituting,

$$p(\boldsymbol{X} \mid M = m) \cong \frac{K_3 \mid \boldsymbol{H} + N\boldsymbol{I}_m \mid^{0.5(\gamma-m-p)}}{\mid \boldsymbol{W} \mid^{0.5(\gamma-m)} (2\pi e)^{0.5(Nm)}} \cdot J_m,$$

where

$$J_m \equiv \int \frac{d\boldsymbol{F}}{\mid \boldsymbol{A} + (\boldsymbol{F} - \hat{\boldsymbol{F}})'(\boldsymbol{I}_N - \boldsymbol{X}\boldsymbol{W}^{-1}\boldsymbol{X}')(\boldsymbol{F} - \hat{\boldsymbol{F}}) \mid^{0.5(\gamma-m)}}.$$

The integral remaining is that of a matrix $\boldsymbol{T}$-density. So we find

$$J_m = \frac{\pi^{0.5mN}\Gamma_N[0.5(N + v - p - m - 1)] \mid \boldsymbol{I}_N - \boldsymbol{X}\boldsymbol{W}^{-1}\boldsymbol{X}' \mid^{-0.5m} \mid \boldsymbol{A} \mid^{-0.5(v-p-1)}}{\Gamma_N[0.5(N + v - p - 1)]}.$$

Combining terms and substituting in for the constants gives

$$I(m) \equiv p(\boldsymbol{X} \mid M = m) \cong \left[\frac{\Gamma_m[0.5(N + m + v - 2p - 1)]\Gamma_N[0.5(N + v - p - m - 1)] \prod_{j=1}^{p} \Gamma[0.5(N + m + v - p - j)]}{(2e)^{0.5Nm}\pi^{0.5Np}\Gamma_m[0.5(N + m + v - p - 1)]\Gamma_N[0.5(n + v - p - 1)] \prod_{j=1}^{p} \Gamma[0.5(v - p - j)]} \right] \cdot \left[\frac{\mid \boldsymbol{B} \mid^{0.5(v-p-1)} \mid \boldsymbol{H} \mid^{0.5p} \mid \boldsymbol{H} + N\boldsymbol{I}_m \mid^{0.5(N+v-2p-1)}}{\mid \boldsymbol{W} \mid^{0.5(N+v-p-1)} \mid \boldsymbol{A} \mid^{0.5(v-p-1)} \mid \boldsymbol{I}_N - \boldsymbol{X}\boldsymbol{W}^{-1}\boldsymbol{X}' \mid^{0.5m}} \right].$$

We now obtain $P\{M = m \mid \boldsymbol{X}\}$ from Equation (15.28).

CHAPTER 16

Bayesian Inference in Classification and Discrimination

16.1 INTRODUCTION

The problem of classifying an observation vector into one of several populations from which it must have come is well known. We focus here on the predictive Bayesian approach that began with the work of de Finetti (1937), and then Geisser (1964, 1966, 1967), and Neal (1996). For a general discussion of classification that includes both frequentist and Bayesian methods, as well as a comparison of Bayesian with frequentist methods, see, for example, Press (1982). A summary is given below.

There are K populations $\pi_1, \ldots, \pi_K$, where

$$\pi_i \equiv N(\boldsymbol{\theta}_i, \boldsymbol{\Sigma}_i), \qquad i = 1, \ldots, K.$$

$(\boldsymbol{\theta}_i \boldsymbol{\Sigma}_i)$ are assumed to be unknown. $z : (p \times 1)$ is an observation known to come from one of the K populations, but we do not know which one. So the observation is to be correctly classified. (In some applications, such as in remote sensing with image processing, there are many observation vectors to be classified.)

We have available "training samples"; that is, we have p-vectors:

$$\begin{array}{ccc} x_1^{(1)}, \ldots, x_{N_1}^{(1)} & \text{are i.i.d from} & \pi_1 \\ \vdots \quad \vdots & \vdots & \vdots \, . \\ x_1^{(K)}, \ldots, x_{N_K}^{(K)} & \text{are i.i.d from} & \pi_K \end{array}$$

Approach

1. Introduce a prior on $(\boldsymbol{\theta}_i, \Sigma_i)$, and develop the posterior.
2. Find the predictive distribution for $(z \mid \pi_i)$.
3. Find the posterior probability $P\{z \in \pi_i \mid z\}$.

REMARK: When the parameters of the π_is are known (which is unusual in practice), Bayesian and frequentist procedures coincide. In this case, the indicated procedure is: Classify z into that population for which a linear or quadratic function of z is maximum (linear, if the Σ_i's are the same; quadratic, otherwise). In the sequel we assume the parameters are unknown.

Preliminaries. The MLEs are

$$\hat{\boldsymbol{\theta}}_i = \bar{\boldsymbol{x}}_i = \frac{1}{N_i}\sum_{j=1}^{N_i} \boldsymbol{x}_j^{(i)}$$

and

$$\hat{\boldsymbol{\Sigma}}_i = \frac{V_i}{N_i} = \frac{1}{N_i}\sum_{j=1}^{N_i} (\boldsymbol{x}_j^{(i)} - \bar{\boldsymbol{x}}_i)(\boldsymbol{x}_j^{(i)} - \bar{\boldsymbol{x}}_i)'.$$

The unbiased estimator of $\boldsymbol{\Sigma}_i$ is

$$\hat{\boldsymbol{\Sigma}}_i^* = \frac{V_i}{N_i - 1}.$$

Denote the "precision matrix" by

$$\boldsymbol{\Lambda}_i \equiv \boldsymbol{\Sigma}_i^{-1}, \qquad \boldsymbol{\Sigma}_i > 0,$$

and parameterize the problem in terms of $\boldsymbol{\Lambda}_i$. Then, if $p(\boldsymbol{\theta}_i, \boldsymbol{\Lambda}_i)$ denotes the prior density, the posterior density is given by

$$p(\boldsymbol{\theta}_i, \boldsymbol{\Lambda}_i \mid \bar{\boldsymbol{x}}_i, V_i, \pi_i) \propto p(\bar{\boldsymbol{x}}_i, V_i \mid \boldsymbol{\theta}_i, \boldsymbol{\Lambda}_i, \pi_i) p(\boldsymbol{\theta}_i, \boldsymbol{\Lambda}_i).$$

16.2 LIKELIHOOD FUNCTION

Since

$$(\bar{\boldsymbol{x}}_i \mid \boldsymbol{\theta}_i) \sim N\left(\boldsymbol{\theta}_i, \frac{\Sigma_i}{N_i}\right)$$

and

$$(V_i \mid \boldsymbol{\Lambda}_i, \pi_i) \sim W(\boldsymbol{\Lambda}_i^{-1}, p, n), \qquad \text{i.e., a Wishart distribution,}$$

where

$$n_i \equiv N_i - 1, \text{ and } p \leq n_i,$$

it follows that

$$p(\bar{\boldsymbol{x}}_i \mid \boldsymbol{\theta}_i, \boldsymbol{\Lambda}_i, \pi_i) \propto \mid \boldsymbol{\Lambda}_i \mid^{0.5} e^{-0.5N_i}(\bar{\boldsymbol{x}}_i - \boldsymbol{\theta})' \boldsymbol{\Lambda}_i (\bar{\boldsymbol{x}}_i - \boldsymbol{\theta}_i)$$

and

$$p(\boldsymbol{V}_i \mid \boldsymbol{\Lambda}_i, \pi_i) \propto \mid \boldsymbol{V}_i \mid^{0.5(n_i - p - 1)} \mid \boldsymbol{\Lambda}_i \mid^{0.5n_i} \exp\{-0.5\text{tr}\boldsymbol{\Lambda}_i \boldsymbol{V}_i\}$$

The likelihood function can therefore be expressed as

$$\begin{aligned} p(\bar{\boldsymbol{x}}_i, \boldsymbol{V}_i \mid \boldsymbol{\theta}_i, \boldsymbol{\Lambda}_i, \pi_i) &= p(\bar{\boldsymbol{x}}_i \mid \boldsymbol{\theta}_i, \boldsymbol{\Lambda}_i, \pi_i) p(\boldsymbol{V}_i \mid \boldsymbol{\Lambda}_i, \pi_i) \\ &\propto \mid \boldsymbol{V}_i \mid^{0.5(n_i - p - 1)} \mid \boldsymbol{\Lambda}_i \mid^{0.5N_i} \exp\left\{-0.5\text{tr}\boldsymbol{\Lambda}_i \left[\boldsymbol{V}_i + N_i(\bar{\boldsymbol{x}}_i - \boldsymbol{\theta}_i)(\bar{\boldsymbol{x}}_i - \boldsymbol{\theta}_i)'\right]\right\} \end{aligned}$$

16.3 PRIOR DENSITY

We can assess a complete prior distribution for (θ_i, Λ_i) at this point (we could use natural conjugate priors as well). Alternatively, we use a vague prior (see Chapter 5).

$$p(\boldsymbol{\theta}_i, \boldsymbol{\Lambda}_i) \propto \frac{1}{\mid \boldsymbol{\Lambda}_i \mid^{0.5(p+1)}}.$$

16.4 POSTERIOR DENSITY

The posterior density becomes

$$p(\boldsymbol{\theta}_i, \boldsymbol{\Lambda}_i \mid \bar{x}_i, \boldsymbol{V}_i, \pi_i) \propto \mid \boldsymbol{\Lambda}_i \mid^{0.5(N_i - p - 1)} \exp\left\{-0.5\text{tr}\, \boldsymbol{\Lambda}_i \left[\boldsymbol{V}_i + N_i(\bar{\boldsymbol{x}}_i - \theta_i)(\bar{x}_i - \theta_i)'\right]\right\}.$$

16.5 PREDICTIVE DENSITY

Lemma. The predictive density of $(z \mid \bar{\boldsymbol{x}}_i, \boldsymbol{V}_i, \pi_i))$ is multivariate $\boldsymbol{t}$ and is given by (see Chapter 10):

$$P(\boldsymbol{z} \mid \bar{\boldsymbol{x}}_i, \boldsymbol{V}_i, \pi_i) \propto \frac{1}{\left[1 + \frac{N_i}{N_i^2 - 1}(\boldsymbol{z} - \bar{\boldsymbol{x}}_i)' \boldsymbol{S}_i^{-1}(\boldsymbol{z} - \bar{\boldsymbol{x}})\right]^{0.5\, N_i}},$$

where

$$\boldsymbol{S}_i \equiv \frac{\boldsymbol{V}_i}{N_i - 1}.$$

Proof (*Outline*).

$$p(z \mid \bar{x}_i, V_i, \pi_i) \equiv \int\int p(z \mid \theta_i, \Lambda_i, \pi_i) p(\theta_i, \Lambda_i, \mid \bar{x}_i, V_i, \pi_i) d\theta_i d\Lambda_i.$$

But $(z \mid \theta_i, \pi_i) \sim N(\theta_i, \Lambda_i^{-1})$. Substituting the normal and posterior densities into the integral gives

$$p(z \mid \bar{x}_i V_i, \pi_i) \propto \int\int \mid \Lambda_i \mid^{0.5(N_i - p)} \exp\{-0.5 \text{tr} \Lambda_i A\} d\theta_i d\Lambda_i,$$

where

$$A \equiv V_i + N_i(\bar{x}_i - \theta_i)(\bar{x}_i - \theta_i)' + (z - \theta_i)(z - \theta_i)';$$

note that A does not depend upon Λ_i.

Integrating with respect to Λ_i by using the Wishart density properties (see for example, Press, 1982) gives

$$p(z \mid \bar{x}_i, V_i, \pi_i) \propto \int \frac{d\theta_i}{\mid A \mid^{0.5(N_i+1)}}.$$

Note that A contains two terms that are quadratic in θ_i. Completing the square in θ_i, and simplifying the matrix algebra, gives

$$p(z \mid \bar{x}_i, V_i, \pi_i) \propto \frac{1}{\mid F_i \mid^{0.5(N_i+1)}} \int \frac{d\theta_i}{[(N_j + 1)^{-2} + (\theta_i - \alpha_i)' F_i^{-1} (\theta_i - \alpha_i)]^{0.5(N_i+1)}},$$

where

$$\alpha_i = \left(\frac{N_i}{N_i + 1}\right)\bar{x}_i + \left(\frac{1}{N_i + 1}\right) z$$

and

$$F_i = (V_i + N_i \bar{x}_i \bar{x}_i' + zz') - \frac{(N_i \bar{x}_i + z)(N_i \bar{x}_i + z)'}{N_i + 1}.$$

The last integration is readily carried out by noting that the integrand is the kernel of a multivariate Student's t-density. We find

$$p(z \mid x_i, V_i, \pi_i) \propto \frac{1}{\mid F_i \mid^{0.5 N_i}}.$$

Simplifying gives the required result.

16.6 POSTERIOR CLASSIFICATION PROBABILITY

Let q_i = prior probability that z comes from π_i; that is,

$$q_i = P\{z \in \pi_i\}.$$

We preassign this value. By Bayes' theorem,

$$p\{z \in \pi_i \mid z\} = \frac{P\{z \in \pi_i\}p(z \mid \bar{x}_i, V_i, \pi_i)}{\sum_{j=1}^{K} p(z \mid \bar{x}_j, V_j, \pi_j)P\{z \in \pi_j\}}.$$

The posterior odds is given by

$$\frac{P\{z \in \pi_i \mid z\}}{P\{z \in \pi_i \mid z\}} = \left(\frac{q_i}{q_j}\right)\frac{p(z \mid \bar{x}_i, V_i, \pi_i)}{p(z \mid \bar{x}_j, V_j, \pi_j)}$$

$$= L_{ij}\frac{\left\{1 + \left(\frac{N_j}{N_j^2 - 1}\right)(z - \bar{x}_j)'S_j^{-1}(z - \bar{x}_j)\right\}^{0.5N_j}}{\left\{1 + \left(\frac{N_i}{N_i^2 - 1}\right)(z - \bar{x}_i)'S_i^{-1}(z - \bar{x}_i)\right\}^{0.5N_i}},$$

where

$$L_{ij} = \left(\frac{q_i}{q_j}\right)\frac{|(N_j - 1)S_j|^{0.5}}{|(N_i - 1)S_i|^{0.5}}\,\frac{\Gamma[0.5N_i]\Gamma[0.5(N_j - p)]}{\Gamma[0.5N_j]\Gamma[0.5(N_i - p)]}\left[\frac{N_i(N_j + 1)}{N_j(N_i + 1)}\right]^{0.5p},$$

for all $i, j = 1, \ldots, K$.

REMARKS:

1. For Bayesian classification with unknown parameters, we need only compute the ratios of pairs of Student's t-densities to see which of the two populations is more likely (see Section 16.6).
2. We need not make any assumption about equality of covariance matrices.
3. Sizes of the training samples need not be large for the procedure to be valid.
4. If natural conjugate families of priors are used, the same structural form results, except the parameters are different.
5. If $N_i = N_j$, L_{ij} reduces to

$$L_{ij} = \left(\frac{q_i}{q_j}\right)\sqrt{\frac{|S_j|}{|S_i|}}.$$

16.7 EXAMPLE: TWO POPULATIONS

Suppose there are two normal populations ($K = 2$) that are each two-dimensional ($p = 2$), with unknown parameters. We want to classify an observation z. Suppose the training samples give

$$\bar{\boldsymbol{x}}_1 = \begin{pmatrix} 1 \\ 1 \end{pmatrix}, \qquad \bar{\boldsymbol{x}}_2 = \begin{pmatrix} 0 \\ 0 \end{pmatrix},$$
$$\boldsymbol{S}_1 = \begin{pmatrix} 1 & 0.5 \\ 0.5 & 1 \end{pmatrix}, \qquad \boldsymbol{S}_2 = \begin{pmatrix} 1 & 0.25 \\ 0.25 & 0.5 \end{pmatrix},$$

based upon $N_1 = N_2 = N = 10$ observations. Since $|\boldsymbol{S}_1| = 0.75$ and $|\boldsymbol{S}_2| = 7/16$, if the prior probabilities are equal ($q_1 = q_2 = 0.5$) we obtain

$$L_{12} = 0.76.$$

So

$$\frac{P(z \in \pi_1 \mid z)}{P(z \in \pi_2 \mid z)} = (0.76)\frac{\left\{1 + \frac{10}{99}(z - \bar{\boldsymbol{x}}_2)'\boldsymbol{S}_2^{-1}(z - \bar{\boldsymbol{x}})\right\}^5}{\left\{1 + \frac{10}{99}(z - \bar{\boldsymbol{x}}_1)'\boldsymbol{S}_1^{-1}(z - \bar{\boldsymbol{x}}_1)\right\}^5}.$$

Suppose $z = \binom{0.25}{0.25}$. We then find

$$\frac{P(z \in \pi_1 \mid z)}{P(z \in \pi_2 \mid z)} = 0.57.$$

Although the prior odds ratio on π_1 compared to π_2 was $1:1$, the posterior odds are down to 0.57 to 1; that is, it is almost two to one in favor of π_2. (See Fig. 16.1 for a

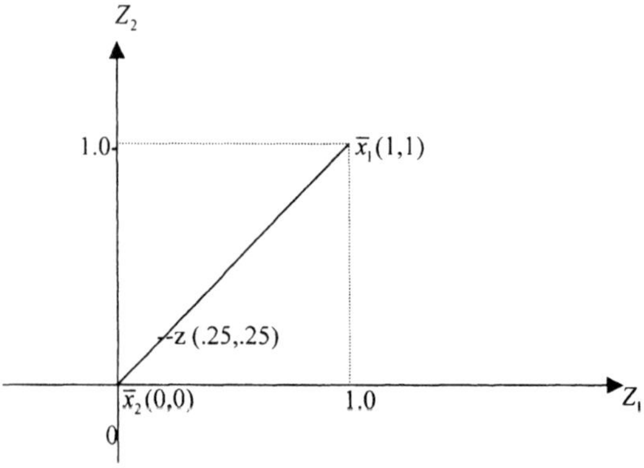

Figure 16.1 Location of z relative to the centers of two populations

pictorial presentation of the location of z relative to the centers of the two populations.)

16.8 SECOND GUESSING UNDECIDED RESPONDENTS: AN APPLICATION[1]

In this section we apply Bayesian classification procedures to a problem in sample surveys, namely, that of estimation category probabilities when we have some respondents who are difficult to classify because they claim to be "undecided."

16.8.1 Problem

In sample surveys and opinion polls involving sensitive questions, some subjects respond in the category "Don't know," "No opinion," or "Undecided," to avoid stating their true beliefs (not because they truly do not know because they are uninformed). How can these subjects be correctly classified by the survey analyst?

Solution

We assume there is a main question of interest, and we also assume that there are some subsidiary questions. In order to classify the "undecided" respondents, we propose (1) drawing upon the information obtained from subjects who respond unambiguously on both the main question and subsidiary questions and (2) taking advantage of the correlations among the responses to these two types of questions.

Suppose there are n_i subjects who respond unambiguously in category i of the main question, $i = 1, \ldots, M$. Suppose further that there are m subjects who are "undecided" on the main question but who answer unambiguously on the subsidiary questions. The joint responses of the subjects who respond unambiguously on the main question follow a multinomial distribution with category probabilities $q_1, \ldots, q_M$. We are interested in estimating the category probabilities q_i, taking into account the "undecided" group. We can give Bayesian estimators of the q_is, given all of the observed data.

The Bayesian estimators have been shown (see Press and Yang, 1974) to be (assuming vague priors) given by

$$\hat{q}_i = \frac{n_i + 1}{m + \sum_{j=1}^{M}(n_j + 1)} \sum_{j=1}^{m} \frac{h(z^{(j)} \mid \pi_i)}{\sum_{t=1}^{M}(n_t + 1)h(z^{(j)} \mid \pi_t)},$$

where $\hat{q}_i$ denotes the mean of the posterior distribution of q_i, given all of the observed data, $h(z^{(j)} \mid \pi_i)$ denotes the marginal predictive density of the response to the subsidiary questions, $z^{(j)}$, for the jth "undecided" respondent, if he/she had

[1] This section is adapted from Press and Yang, 1975.

responded in category i on the main question (i.e., he/she was in population π_i), n_i is the number of subjects who respond unambiguously in category i of the main question, and m is the number of "undecideds."

The subsidiary questions may have categorical, continuous, or mixed categorical and continuous responses. In the special case where the subsidiary questions have categorical responses, we have

$$h(z^{(j)} \mid \pi_i) = h(z^{(j)} = \boldsymbol{u}_k \mid \pi_i) = \frac{x(k \mid i) + 1}{n_i + S},$$

where u_k denotes an $S \times 1$ vector with a "one" in the kth place and zeros elsewhere, $x(k \mid i)$ denotes the number of respondents who answered unambiguously in category i on the main question and cell k of the subsidiary responses to the subsidiary question set.

The posterior variance of q_i is given by

$$\sigma_{qi}^2 = E(q_i^2 \mid \boldsymbol{Z}, \boldsymbol{N}) - [E(q_i \mid \boldsymbol{Z}, \boldsymbol{N})]^2,$$

where

$$E(q_i \mid \boldsymbol{Z}, \boldsymbol{N}) \equiv \hat{q}_i,$$

and

$$AE(q_i^2 \mid \boldsymbol{Z}, \boldsymbol{N}) = (n_i + 1)(n_i + 2) + 2(n_i + 2)H + W,$$

where

$$H \equiv \sum_{j=1}^{m} P\{\pi_i \mid z^{(j)}\} = \sum_{j=1}^{m} \left[\frac{(n_i + 1)h(z^{(j)} \mid \pi_i)}{\sum_{k=1}^{M} (n_k + 1)h(z^{(j)} \mid \pi_k)} \right],$$

$$\boldsymbol{N} \equiv (n_1, \ldots, n_{M-1}),$$

$$\boldsymbol{Z} \equiv (z^{(1)}, \ldots, z^{(m)}),$$

$$W \equiv 2 \sum_{j_1=1}^{m-1} \sum_{j_2=j_1+1}^{m} P\{\pi_i, \pi_i \mid z^{(j1)}, z^{(j2)}\},$$

$$A \equiv \left[n + \sum_{i=1}^{M} (n_i + 1) \right] \left[(m + 1) + \sum_{i=1}^{M} (n_i + 1) \right],$$

$$P\{\pi_i, \pi_i \mid z^{(j1)}, z^{(j2)}\} = C(n_i + 1)(n_i + 2)h(z^{(j1)} \mid \pi_i)h(z^{(j2)} \mid \pi_i),$$

$$C^{-1} = \sum_{i=1}^{M} (n_i + 2)(n_i + 1)h(z^{(j1)} \mid \pi_i)h(z^{(j2)} \mid \pi_i) + \sum_{i=1_{i \neq j}}^{M} \sum_{i=1}^{M} ((n_i + 1)(n_j + 1)h(z^{(j1)} \mid \pi_i)h(z^{(j2)} \mid \pi_i)).$$

Proofs of these summary results can be found in Press and Yang (1974).

16.8.2 Example

Suppose there are 100 respondents to a sensitive question, such that 20 people respond to each of three possible, unambiguous categories, and 40 are "undecided." Suppose further that there is just one subsidiary question (with four possible response categories). The "decided" group on the main question responded as shown in Table 16.1. Thus, 17 subjects responded in category 1 of the subsidiary question, given they responded in Category 1 on the main question.

For the "undecided" group, suppose 25 respond in subsidiary question Category 1, and also suppose that five respond in each of the other three categories.

Since there were an equal number of subjects who responded in each of the three categories of the main question (20 subjects in each), we have

$$h\left(z^{(j)} = \boldsymbol{u}_k \mid \pi_i\right) = \frac{x(k \mid i) + 1}{n_i + S} = \frac{x(k \mid i) + 1}{20 + 4} \qquad \text{for } i = 1, \ldots, 3.$$

The values of $h(\cdot \mid \cdot)$ are arrayed in Table 16.2. Bayesian classification yields the results shown in Table 16.3. Thus, if the undecided respondents were to be ignored, we would be making substantial errors, $(\hat{q}_i - \tilde{q}_i)/\tilde{q}_i$: $(0.062/0.333, -0.004/0.333, -0.057/0.333) = (18.6\%, -1.2/\%, -17.1/\%)$, for point estimates of the category probabilities; in addition, we have entire distributions for the category probabilities, so we can readily evaluate dispersion of our estimates and tail probabilities.

16.9 EXTENSIONS OF THE BASIC CLASSIFICATION PROBLEM

16.9.1 Classification by Bayesian Clustering

A fundamental assumption made throughout this chapter has been that the populations in which an observation is to be classified are well defined (this context for classification is sometimes called *supervised learning* because the populations have already been defined by a supervisor that teaches the system (this is pattern recognition terminology)). For example, we might assume that the observation(s) to be classified belong to one of K multivariate normal populations. But in many situations the possible populations are not well defined. We must first determine the structure of the possible populations, and then carry out the classification (this

Table 16.1 Responses of the "Decided" Group

Main Question Category	Subsidiary Question Category				Total (%)
	1	2	3	4	
1	17	1	1	1	20(33)
2	5	5	5	5	20(33)
3	1	1	1	17	20(33)

Table 16.2 Values of $h(\cdot \mid \cdot)$

Main Question Category	Subsidiary Question Category 1	2	3	4
1	$\frac{17+1}{24}$	$\frac{1+1}{24}$	$\frac{1+1}{24}$	$\frac{1+1}{24}$
2	$\frac{5+1}{24}$	$\frac{5+1}{24}$	$\frac{5+1}{24}$	$\frac{5+1}{24}$
3	$\frac{1+1}{24}$	$\frac{1+1}{24}$	$\frac{1+1}{24}$	$\frac{17+1}{24}$

context for classification is sometimes called *unsupervised learning*). How to handle such situations involves the theory of *clustering.* For a Bayesian approach to the theory of clustering see Cheeseman et al., 1995. The approach is computer intensive.

16.9.2 Classification Using Bayesian Neural Networks And Tree-Based Methods

Neal, (1996) approaches the task of Bayesian classification by using multilayer perceptron networks to define a probabilistic model. Such networks were originally structured for modeling the brain, but they are applied to multilayer hierarchical statistical modeling (see Section 10.8, and Chapter 14). The multilayer perceptron network has various inputs and several layers of hidden or latent units (functions) that are unknown and that must be determined to understand the outputs to a Bayesian learning model. The outputs are used to define the conditional classification probability (conditional on the observed input data) for each possible population into which a vector is to be classified. Bridle (1989) adopted multivariate logistic distributions for the discrete probabilities of classification. The weights used in such models are estimated from training data. The Bayesian predictive distribution (see Chapter 10) is used to find the target values of the model conditional on the observed input data. The Bayesian neural network approach to classification generalizes the predictive classification approach originally suggested by Geisser (1964, 1966, 1967).

For some types of classification we require a hierarchical type of classification model in which the variables are split into clusters at each level of the hierarchy. (For fundamental development of this topic, see Breiman et al., 1984.) There is some-

Table 16.3 Results of Bayesian Classification

Main Question Category (i)	$\tilde{q}_i$ Ignoring "Undecideds"	$\hat{q}_i$ Second Guessing the "Undecideds"	σ_{q_i}	90% Credibility Interval
1	0.333	0.395	0.062	(0.30, 0.50)
2	0.333	0.329	0.057	(0.24, 0.43)
3	0.333	0.276	0.050	(0.20, 0.36)

times a meaningful interpretability of the clusters in terms of the context. For example, we might wish to classify genes of an organism using microarray analysis in a hierarchy involving clusters that are meaningful in terms of developmental biology. In a medical diagnosis context, a physician might wish to diagnose the illness of a patient in terms of the values of symptom variables, and there is a natural hierarchy involving some symptom variables that have normal ranges and some that exceed normal bounds. Such hierarchies or *trees* (the terminology derives from "decision trees") are grown in stages by successive (binary) splitting of the dependent variable data at each stage. (See also Denison et al., 2002, Chapter 6.) Classification procedures resulting from the building of such trees can be developed from simple algorithms such as CART (Classification And Regression Trees; see Breiman et al., 1984), or a stepwise linear regression extension of it called MARS (Multivariate Adaptive Regression Splines; see Friedman, 1991). Typically, large tree models are built and then pruned back. These procedures, and some competitors, are also discussed in Hastie et al., 2001. Generally, these computer-intensive procedures are linear classifiers in that the classifications that result are separable by linear boundaries. But the classification methods can be extended to nonlinear ones using *support vector machines*.

To what extent are there Bayesian methods that correspond to these procedures? Early Bayesian tree modeling was developed by Buntine, 1992. Chipman et al. (1998, 2000, 2001b, 2003), and Denison et al. (1998) proposed models for tree-building that extend the CART/MARS approaches from a Bayesian point of view. Instead of partitioning the *dependent variable space* (sometimes also called the *feature space*) into regions in which the dependent variable is constant, Chipman et al. propose that in each partition region of the dependent variable there be a linear regression structure that better explains the variation within that region.

There are also other Bayesian unsupervised learning classification procedures that involve representing the data pdf as a mixture of normal, or other, distributional kernels, and then using natural conjugate-type prior distributions to develop posterior distributions for the classifier.

16.9.3 Contextual Bayesian Classification

In some situations there is a need to classify a collection of observation vectors that are related to one another in some way. They may be related temporally, or spatially, or both. For example, Earth satellites and high-flying aircraft often contain sensors that collect data about ground features in, say, p frequencies, simultaneously. Thus, there result p-vectors of data that jointly define ground features, and we often want to classify these data vectors to determine the nature of the ground features. Is the satellite or high-flying aircraft viewing a lake, a mountain, an urban area, or an underground missile silo, or what? We often have training data consisting of previously classified data vectors that were classified on the ground with certainty. Such so-called *ground truth* is used to estimate population parameters that define specific populations of interest. Because the satellite can cover the same areas in repeated cycles around the Earth, collections of data vectors are sometimes temporally correlated. Data vectors from large ground features also generate collections of pixels (picture elements) or voxels (volume elements) that jointly define the ground feature. Such data vectors are correlated spatially (in two, and for topographical features, three, dimensions). These spatial correlations should be taken into account in developing the appropriate Bayesian classification procedures. Such procedures are called contextual classification methods (for further details about these types of Bayesian classification procedures, see, for example, Besag, 1986; Fatti, et al., 1997; Ferryman and Press, 1996; Geman and Geman, 1984; Klein and Press, 1989a, 1989b, 1992, 1993, 1996; McLachlan, 1992; Press, 1998; Ripley, 1981).

16.9.4 Classification in Data Mining

With the advent of the availability of large storage capacity computers in the last couple of decades has come the need for understanding large databases containing, in some instances, hundreds of thousands or millions or more records (cases) of data for perhaps hundreds or more variables (see also Section 5.5). Such large databases are found to be of interest across the disciplines. How does one try to understand the relationships among the variables in these large databases (software that will handle such large databases includes Microsoft Access or Corel Paradox). Briefly, this is the objective of *data mining*. Questions that are asked of these databases are called *queries*. Oftentimes, when the database is designed, the spectrum of queries that will eventually be desired is unknown and unanticipated. So analysts often try to restructure the database by classifying the potential types of queries into groups (text classification is a special case). Moreover, using some of the records to establish classification rules and then using those rules to classify other records can lead to understanding of relationships among records, and among variables in the database. By employing Bayesian classification methods described earlier in this chapter we can bring prior information about the nature of the underlying relationships among the variables to bear on the entire process. For additional discussion of these issues, see, for example, Breiman et al., 1984; Chatfield, 1995; DuMouchel, 1999; Ripley, 1996.

SUMMARY

We have shown that by using Bayesian predictive distribution procedures for classification:

1. The methods are exact (not approximate or large sample);
2. It is unnecessary to assume equal covariance matrices (an assumption that is necessary in frequentist approaches);
3. Results can be used to improve accuracy and understanding in opinion polling.

We have also pointed out various directions in which Bayesian classification methods have been extended: to clustering procedures; to contextual classification methods; to the analysis of Bayesian neural nets; and to data mining.

EXERCISES

16.1* Suppose there are two normal populations that are each two-dimensional with unknown parameters. We want to classify an observation z. Suppose the training samples give

$$\bar{x}_1 = (2, 3)' \qquad \bar{x}_2 = (3, 1)'$$

$$S_1 = \begin{pmatrix} 2 & 1 \\ 1 & 4 \end{pmatrix}, \qquad S_2 = \begin{pmatrix} 3 & 2 \\ 2 & 4 \end{pmatrix},$$

based upon $N_1 = N_2 = 10$ observations. Let the observed vector be given by

$$z = (2, 1.5)'.$$

If the prior odds are equal, classify z into one of the two populations.

16.2* Suppose there are three bivariate normal populations with training sample statistics as in Exercise 16.1, but, in addition, we know $N_1 = N_2 = N_3 = N = 10$, and

$$\bar{x}_3 = (0, 0), \qquad S_3 = \begin{pmatrix} 3 & 1 \\ 1 & 1 \end{pmatrix}.$$

Assuming the prior odds for all three populations are the same, classify

$$z = (2, 2)'$$

into one of the three populations.

16.3* For the classification problem in Section 16.1 adopt a natural conjugate prior family for (θ_i, Λ_i), instead of the vague prior family used there, and find:

(a) The joint posterior density of θ_i, Λ_i, given $(\bar{x}_i, V_i, \pi_i)$;

(b) The predictive density of $(z \mid \bar{x}_i, V_i, \pi_i)$;

(c) The posterior classification probability for classifying z into π_i, given z, that is $P\{z \in \pi_i \mid z\}$.

16.4* Explain the classification concept behind the notion of "second guessing undecided respondents," as discussed in Section 16.8.

16.5* In what types of situations would we be likely to encounter large fractions of "undecided" respondents, in the sense presented in Section 16.8?

16.6* While it is a general principle that it is better to have more information than less, why might it be better to have very few subsidiary questions in the undecided respondents problem? (*Hint*: see Press and Yang, 1974.)

16.7 What are the arguments for and against merging observations already classified into say, π_1, with its training sample, and using them to better estimate the parameters of π_1 in order to classify new observations?

16.8 Suppose you would like to classify an unknown observation vector $z : (p \times 1)$ into one of two populations π_1 or π_2 for which both π_1 and

*Solutions for asterisked exercises may be found in Appendix 7.

π_2 are multinomial populations with probability mass functions $f_1(\boldsymbol{x})$ and $f_2(\boldsymbol{x})$. How would you develop a Bayesian classification procedure for classifying z into π_1 or π_2?

16.9 You would like to use Bayesian classification methods to classify an unknown observation vector $z : (p \times 1)$ into π_1 or π_2, where $\pi_i \sim N(\boldsymbol{\theta}_i, \boldsymbol{\Sigma}_i), i = 1, 2$. You have training data with N_1 observation vectors believed to come from π_1, and N_2 believed to come from π_2. Suppose there are errors in the training data so that each training data vector has only a 90 percent chance of having been correctly classified. How would this fact affect your rules for classifying z?

16.10 Suppose you were attempting to diagnose a patient's disease on the basis of a vector of symptoms that included some variables that were continuous and some that were discrete. You would like to use a Bayesian classification procedure to classify the vector of symptoms into one of K possible diseases. How would you set up the procedure? (*Hint*: see Section 16.8.)

FURTHER READING

Besag, J. E. (1986). "On the Statistical Analysis of Dirty Pictures," *J. Royal Statist. Soc., B*, **48**, 259–302.

Breiman, L., Friedman, J. H., Olshen, R. A., and Stone, C. J. (1984). *Classification and Regression Trees*, California, Wadsworth International Group.

Bridle, J. S. (1989). "Probabilistic Interpretation of Feedforward Classification Network Outputs with Relationships to Statistical Pattern Recognition," in F. Fouglemann-Soulie and J. Heault, Eds., *Neurocomputing: Algorithms, Architectures, and Applications*. New York, Springer-Verlag.

Buntine, W. L. (1992). " A Theory of Learning Classification Rules," Ph.D. Thesis, School of Computing Science, University of Technology, Sidney.

Chatfield, C. (1995). "Model Uncertainty, Data Mining, and Statistical Inference," with discussion, *J. Royal Statist. Soc. A*, **158**, 419–466.

Cheeseman, P. and Stutz, J. (1995). "Bayesian Classification (AutoClass): Theory and Methods," in U. M. Fayyad, G. Piatetsky-Shapiro, P. Smyth and R. Uthurusamy, (Eds). *Advances in Knowledge Discovery and Data Mining*, Menlo Park, The AAAI Press.

Chipman, H. A., George, E. I., and McCulloch, R. E. (1998). "Bayesian CART Model Search" (with discussion), *Jour. American Statist. Assn.* **93**, 935–960.

De Finetti, B. (1937). "Le Prevision: Ses Lois Logiques, Ses Sources Subjectives", *Ann. Inst. Poincare*, tome VII, fasc. 1, 1–68. Reprinted in *Studies in Subjective Probability*, Melbourne, Florida, Krieger, 1980 (English translation).

Denison, D., Mallick, B., and Smith, A. F. M. (1998). "A Bayesian CART Algorithm," Biometrika, 85, 363–377.

DuMouchel, W. (1999). "Bayesian Data Mining in Large Frequency Tables, with an Application to the FDA Spontaneous Reporting System," with discussion, *Am. Statist.*, **53**, 177–202.

Fatti, P., Klein, R. and Press, S. J. (1997). "Hierarchical Bayes Models in Contextual Spatial Classification," in S. Ghosh, W. Schueany and W. B. Smith, Eds., *Statistics for Quality: Dedicated to Donald B. Owen*, New York, Marcel Dekker, Inc. Chapter 14, 275–294.

Ferryman, T. and Press, S. J. (1996). "A Comparison of Nine Pixel Classification Algorithms," Technical Report No. 232, Department of Statistics, University of California, Riverside, June 1996.

Friedman J. (1991). "Multivariate Adaptive Regression Splines," with dicussion, *Ann. of Statistics*, **19**(1), 1–141.

Friedman, H. P. and Goldberg, J. D. (2000). "Knowledge Discovery from Databases and Data Mining: New Paradigms for Statistics and Data Analysis?," Biopharmaceutical Report, *Am. Statist. Assoc.*, **8**(2), 1–12.

Geisser, S. (1964). "Posterior Odds for Multivariate Normal Classifications," *J. Royal Statist. Soc. B*, **26**, 69–76.

Geisser, S. (1966). "Predictive Discrimination," in P. R. Krishnaiah, Ed., *Multivariate Analysis*, New York, Academic Press, 149–163.

Geisser, S. (1967). "Estimation Associated With Linear Discriminants," *Ann. Math. Statist.*, **38**, 807–817.

Geman, S. and Geman D. (1984). "Stochastic Relaxation, Gibbs Distributions and the Bayesian Restoration of Images," *IEEE Trans. Pattern Ann., Machine Intell.*, **6**, 721–741.

Hastie T., Tibshirani, and Friedman, J. (2001). *The Elements of Statistical Learning: Data Mining, Inference, and Prediction*, New York, Springer-Verlag.

Klein, R. and Press S. J. (1989a). "Contextual Bayesian Classification of Remotely Sensed Data," *Commun. in Statist. Theory and Methods*, **18**(9); 3177–3202.

Klein, R. and Press, S. J. (1989b). "Bayesian Contextual Classification With Neighbors Correlated With Training Data," in S. Geisser, J. Hodges, S. James Press and A. Zellner, Eds., *Bayesian and Likelihood Methods in Statistics and Econometrics: Essays in Honor of George A. Barnard*, New York, North Holland Publishing Co., 337–355.

Klein, R. and Press, S. J. (1992). "Adaptive Bayesian Classification of Spatial Data," *J. Am. Statist. Assoc.*, **87**(419), 844–851.

Klein, R. and Press, S. J. (1993). "Adaptive Bayesian Classification with a Locally Proportional Prior," *Comm. in Statist, Theory and Methods*, **22**(10), 2925–2940.

Klein, R. and Press, S. J. (1996). "The Directional Neighborhoods Approach to Contextual Reconstruction of Images from Noisy Data," *J. Am. Statist. Assoc.*, **91**(435), 1091–1100.

McLachlan, G. J. (1992). *Discriminant Analysis and Statistical Pattern Recognition*, New York, John Wiley and Sons, Inc.

Neal, R. M. (1996). *Bayesian Learning for Neural Networks*, New York, Springer-Verlag.

Press, S. J. (1982). *Applied Multivariate Analysis: Using Bayesian and Frequentist Methods of Inference*, Melbourne, Florida, Krieger Publishing. Co.

Press, S. J. (1998). "Comparisons of Pixel Classification Algorithms," Technical Report. No. 257, Department of Statistics, University of California, Riverside.

Press, S. J. and Yang, C. (1974). "A Bayesian Approach to Second Guessing Undecided Respondents," *J. Am. Statist. Assoc.*, **69**, 58–67.

Ripley, B. D. (1981). *Spatial Statistics*, New York; John Wiley and Sons, Inc.

Ripley, B. D. (1996). *Pattern Recognition and Neural Networks*, Cambridge, Cambridge University Press.

Appendices

Appendix 1 is an article by Hilary Seal entitled "Bayes, Thomas."[1] The article discusses and references articles that followed Bayes' essay. Those articles critique Bayes' original essay.

Appendix 2 is a biographical note on Bayes' original essay, by G. A. Barnard, Imperial College, London.[2]

Appendix 3 is a letter from Mr. Richard Price, a friend of Bayes, to Mr. John Canton, representing the Royal Society of London, submitting Thomas Bayes' original essay for publication after Bayes' death.[3]

Appendix 4 is the original essay of Thomas Bayes, "An Essay Towards Solving a Problem in the Doctrine of Chances," by Reverend Thomas Bayes.[4]

Note: What we normally think of as Bayes' theorem is given (in somewhat obscure form by today's standards) in Proposition 9, Appendix 4.

Appendix 5 is a listing of references of articles that apply methods of Bayesian statistical science in a wide variety of disciplines.[5]

Appendix 6 is an explanation of how the Bayesian Hall of Fame was developed; it also provides a listing of those who received honorable mention. Portraits of the members of the Bayesian Hall of Fame may be found following the Table of Contents. The members of the Bayesian Hall of Fame were deemed by leading Bayesian statisticians to have contributed most significantly to the development of the field.

Appendix 7 is a collection of solutions to selected exercises.

[1]Reprinted with the permission of the publishers from the *International Encyclopedia of Statistics*, edited by W. H. Kruskal and J. M. Tanur, Volume 1, The Free Press, A Division of MacMillan Publishing Company, Incorporated, New York, 1978.
[2]Reprinted with the permission of the publishers from *Biometrika*, **45**, Parts 3 and 4, December 1958, 293–295.
[3]Reprinted with the permission of the publishers from *Biometrika*, **45**, Parts 3 and 4, December 1958, 296–298.
[4]Reprinted with the permission of the publishers from *Biometrika*, **45**, Parts 3 and 4, December 1958, 298–315. The article originally appeared in *The Philosophical Transactions*, **53**, 1763, 370–418, and was reprinted in *Biometrika*.
[5]Reprinted from the Appendix in Press and Tanur, 2001. "The Subjectivity of Scientists and the Bayesian Approach," New York: John Wiley and Sons, Inc.

APPENDIX 1

Bayes, Thomas

Hilary L. Seal

Thomas Bayes (1702–1761) was the eldest son of the Reverend Joshua Bayes, one of the first nonconformist ministers to be publicly ordained in England. The younger Bayes spent the last thirty years of his comfortable, celibate life as Presbyterian minister of the meeting house, Mount Sion, in the fashionable town of town of Tunbridge Wells, Kent. Little is known about his personal history, and there is no record that he communicated with the well-known scientists of his day. Circumstantial evidence suggests that he was educated in literature, languages, and science at Coward's dissenting academy in London (Holland 1962). He was elected a fellow of the Royal Society in 1742, presumably on the basis of two metaphysical tracts he published (one of them anonymously) in 1731 and 1736 (Barnard 1958). The only mathematical work from his pen consists of two articles published posthumously in 1764 by his friend Richard Price, one of the pioneers of social security (Ogborn 1962). The first is a short note, written in the form of an undated letter, on the divergence of the Stirling (de Moivre) series ln ($z!$). It has been suggested that Bayes' remark that the use of "a proper number of the first terms of the ... series" will produce an accurate result constitutes the first recognition of the asymptotic behavior of a series expansion (see Deming's remarks in Bayes [1764] 1963). The second article is the famous "An Essay Towards Solving a Problem in the Doctrine of Chances," with Price's preface, footnotes, and appendix (followed, a year later, by a continuation and further development of some of Bayes' results).

The "Problem" posed in the Essay is: "*Given* the number of times in which an unknown event has happened and failed: *Required* the chance that the probability of its happening in a single trial lies somewhere between any two degrees of probability that can be named." A few sentences later Bayes writes: "By *chance* I mean the same as probability" ([1764]. 1963, p. 376).

If the number of successful happenings of the event is p and the failures q, and if the two named "degrees" of probability are b and f, respectively, Proposition 9 of the Essay provides the following answer expressed in terms of areas under the curve $x^p(1-x)^q$:

$$\int_b^f x^p(1-x)^q\,dx \bigg/ \int_0^1 x^p(1-x)^q\,dx. \tag{1}$$

This is based on the assumption (Bayes' "Postulate 1") that all values of the unknown probability are equally likely before the observations are made. Bayes indicated the applicability of this postulate in his famous "Scholium": "that the ... rule is the proper one to be used in the case of an event concerning the probability of which we absolutely know nothing antecedently to any trials made concerning it, seems to appear from the following consideration; viz. that concerning such an event I have no reason to think that, in a certain number of trials, it should rather happen any one possible number of times than another" (*ibid.*, pp. 392–393).

The remainder of Bayes' Essay and the supplement (half of which was written by Price) consists of attempts to evaluate (1) numerically, (a) by expansion of the integrand and (b) by integration by parts. The results are satisfactory for p and q small but the approximations for large p, q are only of historical interest (Wishart 1927).

Opinions about the intellectual and mathematical ability evidenced by the letter and the essay are extraordinarily diverse. Netto (1908), after outlining Bayes' geometrical proof, agreed with Laplace ([1812] 1820) that it is *ein wenig verwickelt* ("somewhat involved"). Todhunter (1865) thought that the résumé of probability theory that precedes Proposition 9 was "excessively obscure." Molina (in Bayes [1764] 1963, p. xi) said that "Bayes and Price ... can hardly be classed with the great mathematicians that immediately preceded or followed them," and Hogben (1957, p. 133) stated that "the ideas commonly identified with the name of Bayes are largely [Laplace's]."

On the other hand, von Wright (1951, p. 292) found Bayes' Essay "a masterpiece of mathematical elegance and free from ... obscure philosophical pretentions." Barnard (1958, p. 295) wrote that Bayes' "mathematical work ... is of the very highest quality." Fisher ([1956] 1959, p. 8) concurred with these views when he said Bayes' "mathematical contributions ... show him to have been in the first rank of independent thinkers ...".

The subsequent history of mathematicians' and philosophers' extensions and criticisms of Proposition 9—the only statement that can properly be called Bayes' theorem (or rule)—is entertaining and instructive. In his first published article on probability theory, Laplace (1774), without mentioning Bayes, introduced the principle that if p_j is the probability of an observable event resulting from "cause"

$j(j = 1, 2, 3, \ldots, n)$ then the probability that "cause" j is operative to produce the observed event is

$$p_j \Big/ \sum_{j=1}^{n} p_j^* \tag{2}$$

This is Principle III of the first (1812) edition of Laplace's probability text, and it implies that the prior (antecedent, initial) probabilities of each of the "causes" are the same. However, in the second (1814) edition Laplace added a few lines saying that if the "causes" are not equally probable a priori, (2) would become

$$\omega_j p_i \Big/ \sum_{j=1}^{n} \omega_j p_j, \tag{3}$$

where ω_j is the prior probability of cause j and p_j is now the probability of the event, given that "cause" j is operative. He gave no illustrations of this more general formula.

Laplace (1774) applied his new principle (2) to find the probability of drawing m white and n black tickets in a specified order from an urn containing an infinite number of white and black tickets in an unknown ratio and from which p white and q black tickets have already been drawn. His solution, namely,

$$\int_0^1 x^{p+m}(1-x)^{q+n}\,dx \Big/ \int_0^1 x^p(1-x)^q\,dx = \frac{(p+m)!(q+n)!(p+q+1)!}{p!q!(p+q+m+1)!}, \tag{4}$$

was later (1778–1781; 1812, chapter 6) generalized by the bare statement that if all values of x are not equally probable a factor $z(x)$ representing the a priori probability density (*facilité*) of x must appear in both integrands. However, Laplace's own views on the applicability of expressions like (4) were stated in 1778 (1778–1781, p. 264) and agree with those of Bayes' Scholium: "Lorsqu' on n'a aucune donnée *a priori* sur la possibilité d'un événement, il faut supposer toutes les possibilités, depuis zéro jusqu'à l'unité, également probables. ..." ("When nothing is given a priori as to the probability of an event, one must suppose all probabilities, from zero to one, to be equally likely. ..."). Much later Karl Pearson (1924, p. 191) pointed out that Bayes was "considering excess of one variate ... over a second ... as the determining factor of occurrence" and this led naturally to a generalization of the measure in the integrals of (1). Fisher (1956) has even suggested that Bayes himself had this possibility in mind.

Laplace's views about prior probability distributions found qualified acceptance on the Continent (von Kries 1886) but were subjected to strong criticism in England (Boole 1854; Venn 1866; Chrystal 1891; Fisher 1922), where a relative frequency definition of probability was proposed and found incompatible with the uniform prior distribution (for example, E. S. Pearson 1925). However, developments in the theory of inference (Keynes 1921; Ramsey 1923–1928; Jeffreys 1931; de Finetti 1937; Savage 1954; Good 1965) suggest that there are advantages to be gained from a "subjective" or a "logical" definition of probability and this approach gives Bayes' theorem, in its more general form, a central place in inductive procedures (Jeffreys 1939; Raiffa & Schlaifer 1961; Lindley 1965).

BIBLIOGRAPHY

BARNARD, G. A. 1958 Thomas Bayes: A Biographical Note. *Biometrika* 45: 293–295.

BAYES, THOMAS (1764) 1963 *Facsimiles of Two Papers by Bayes*. New York: Hafner ⇒ Contains "An Essay Towards Solving a Problem in the Doctrine of Chances, With Richard Price's Foreward and Discussion", with a commentary by Edward C. Molina; and "A Letter on Asymptotic Series From Bayes to John Canton," with a commentary by W. Edwards Deming. Both essays first appeared in Volume 53 of the *Philosophical Transactions*, Royal Society of London, and retain the original pagination.

BOOLE, GEORGE (1854) 1951 *An Investigation of the Laws of Thought, on Which Are Founded the Mathematical Theories of Logic and Probabilities*. New York: Dover.

CHRYSTAL, GEORGE 1891 On Some Fundamental Principles in the Theory of Probability. Actuarial Society of Edinburgh, *Transactions* 2: 419–439.

[1] DE FINETTI, BRUNO 1937 La prévision: Ses lois logiques, ses sources subjectives. Paris, Université de, Institut Henri Poincaré, *Annales* 7: 1–68.

FISHER, R. A. (1922) 1950 On the Mathematical Foundations of Theoretical Statistics. Pages 10.307a–10.368 in R. A. Fisher, *Contributions to Mathematical Statistics*. New York: Wiley. ⇒ First published in Volume 222 of the *Philosophical Transactions*, Series A, Royal Society of London.

FISHER, R. A. (1956) 1959 *Statistical Methods and Scientific Inference*. 2d ed., rev. New York: Hafner; London: Oliver & Boyd.

GOOD, IRVING J. 1965 *The Estimation of Probabilities: An Essay on Modern Bayesian Methods*. Cambridge, Mass.: M.I.T. Press.

HOGBEN, LANCELOT T. 1957 *Statistical Theory; the Relationship of Probability, Credibility and Error: An Examination of the Contemporary Crisis in Statistical Theory From a Behaviourist Viewpoint*. London: Allen & Unwin.

HOLLAND, J. D. 1962 The Reverend Thomas Bayes, F.R.S. (1702-1761). *Journal of the Royal Statistical Society* Series A 125: 451–461.

JEFFREYS, HAROLD (1931) 1957 *Scientific Inference*. 2d ed. Cambridge Univ. Press.

JEFFREYS, HAROLD (1939) 1961 *Theory of Probability*. 3d ed. Oxford: Clarendon.

KEYNES, J. M. (1921) 1952 *A Treatise on Probability*. London: Macmillan. ⇒ A paperback edition was published in 1962 by Harper.

KRIES, JOHANNES VON (1886) 1927 *Die Principien der Wahrscheinlichkeitsrechnung: Eine logische Untersuchung*. 2d ed. Tübingen (Germany): Mohr.

LAPLACE, PIERRE S. (1774) 1891 Mémoire sur la probabilité des causes par les événements. Volume 8, pages 27-65 in Pierre S. Laplace, *Oeuvres complètes de Laplace*. Paris: Gauthier-Villars.

LAPLACE, PIERRE S. (1778-1781) 1893 Mémoire sur les probabilités. Volume 9, pages 383-485 in Pierre S. Laplace, *Oeuvres complètes de Laplace*. Paris: Gauthier-Villars.

LAPLACE, PIERRE S. (1812) 1820 *Théorie analytique des probabilités*. 3d ed., rev. Paris: Courcier.

LINDLEY, DENNIS V. 1965 *Introduction to Probability and Statistics From a Bayesian Viewpoint*. 2 vols. Cambridge Univ. Press.

NETTO, E. 1908 Kombinatorik, Wahrscheinlichkeitsrechnung, Reihen-Imaginäres. Volume 4, pages 199-318, in Moritz Cantor (editor), *Vorlesungen über Geschichte der Mathematik*. Leipzig: Teubner.

OGBORN, MAURICE E. 1962 *Equitable Assurances: The Story of Life Assurance in the Experience of The Equitable Life Assurance Society, 1762-1962*. London: Allen & Unwin.

PEARSON, EGON S. 1925 Bayes' Theorem, Examined in the Light of Experimental Sampling. *Biometrika* 17: 388–442.

PEARSON, KARL 1924 Note on Bayes' Theorem. *Biometrika* 16: 190–193.

PRICE, RICHARD 1765 A Demonstration of the Second Rule in the Essay Towards a Solution of a Problem in the Doctrine of Chances. Royal Society of London, *Philosophical Transactions* 54: 296–325. ⇒ Reprinted by Johnson in 1965.

RAIFFA, HOWARD; and SCHLAIFER, ROBERT 1961 *Applied Statistical Decision Theory*. Harvard University Graduate School of Business Administration, Studies in Managerial Economics. Boston: The School.

RAMSEY, FRANK P. (1923-1928) 1950 *The Foundations of Mathematics and Other Logical Essays*. New York: Humanities.

SAVAGE, LEONARD J. (1954) 1972 *The Foundations of Statistics*. Rev. ed. New York: Dover. ⇒ Includes a new preface.

TODHUNTER, ISAAC (1865) 1949 *A History of the Mathematical Theory of Probability From the Time of Pascal to That of Laplace*. New York: Chelsea.

VENN, JOHN (1866) 1888 *The Logic of Chance: An Essay on the Foundations and Province of the Theory of Probability, With Special Reference to Its Logical Bearings and Its Application to Moral and Social Science*. 3d ed. London: Macmillan.

WISHART, JOHN 1927 On the Approximate Quadrature of Certain Skew Curves, With an Account of the Researches of Thomas Bayes. *Biometrika* 19: 1–38.

WRIGHT, GEORG H. VON 1951 *A Treatise on Induction and Probability*. London: Routledge.

POSTSCRIPT

It is difficult to stay current with the extensive literature dealing with methods flowing from Thomas Bayes' original suggestion. It is also difficult to maintain a clear mind in the profusion of discussions about whether to use so-called Bayesian techniques. One may cite the proceedings of the 1970 Symposium on the Foundations of Statistical Inference (1971) and the philosophical treatise by Stegmüller (1973). Two pairs of eminent authors, with long experience in both theory and application of statistics, have adopted very different approaches toward the Bayesian viewpoint: Box and Tiao (1973) and Kempthorne and Folks (1971).

ADDITIONAL BIBLIOGRAPHY

BOX, GEORGE E. P.; and TIAO, GEORGE C. 1973 *Bayesian Inference in Statistical Analysis.* Reading, Mass.: Addison-Wesley.

[1] DE FINETTI, BRUNO (1937) 1964 Foresight: Its Logical Laws, Its Subjective Sources. Pages 93-158 in Henry E. Kyberg, Jr. and Howard E. Smokler, *Studies in Subjective Probability.* New York: Wiley. ⇒ First published in French.

KEMPTHORNE, OSCAR; and FOLKS, LEROY 1971 *Probability, Statistics and Data Analysis,* Ames: Iowa State Univ. Press.

STEGMÜLLER, WOLFGANG 1973 *Personelle und statistische Wahrscheinlichkeit.* Volume 2: *Statistisches Schliessen, statistische Begründung, statistische Analyse.* Berlin: Springer.

SYMPOSIUM ON THE FOUNDATIONS OF STATISTICAL INFERENCE, UNIVERSITY OF WATERLOO, *1970* 1971 *Foundations of Statistical Inference: Proceedings.* Edited by V. P. Godambe and D. A. Sprott, Toronto: Holt.

APPENDIX 2

Thomas Bayes—A Biographical Note

George A. Barnard

Bayes's paper, reproduced in the following pages, must rank as one of the most famous memoirs in the history of science and the problem it discusses is still the subject of keen controversy. The intellectual stature of Bayes himself is measured by the fact that it is still of scientific as well as historical interest to know what Bayes had to say on the questions he raised. And yet such are the vagaries of historical records, that almost nothing is known about the personal history of the man. *The Dictionary of National Biography*, compiled at the end of the last century, when the whole theory of probability was in temporary eclipse in England, has an entry devoted to Bayes's father, Joshua Bayes, F.R.S., one of the first six Nonconformist ministers to be publicly ordained as such in England, but it has nothing on his much more distinguished son. Indeed, the note on Thomas Bayes which is to appear in the forthcoming new edition of the *Encyclopedia Britannica* will apparently be the first biographical note on Bayes to appear in a work of general reference since the *Imperial Dictionary of Universal Biography* was published in Glasgow in 1865. And in treatises on the history of mathematics, such as that of Loria (1933) and Cantor (1908), notice is taken of his contributions to probability theory and to mathematical analysis, but biographical details are lacking.

The Reverend Thomas Bayes, F.R.S., author of the first expression in precise, quantitative form of one of the modes of inductive inference, was born in 1702, the eldest son of Ann Bayes and Joshua Bayes, F.R.S. He was educated privately, as was usual with Nonconformists at that time, and from the fact that when Thomas was 12 Bernoulli wrote to Leibniz that 'poor de Moivre' was having to earn a living in

London by teaching mathematics, we are tempted to speculate that Bayes may have learned mathematics from one of the founders of the theory of probability. Eventually Thomas was ordained, and began his ministry by helping his father, who was at the time stated, minister of the Presbyterian meeting house in Leather Lane, off Holborn. Later the son went to minister in Tunbridge Wells at the Presbyterian Chapel on Little Mount Sion which had been opened on 1 August 1720. It is not known when Bayes went to Tunbridge Wells, but he was not the first to minister on Little Mount Sion, and he was certainly there in 1731, when he produced a tract entitled 'Divine Benevolence, or an attempt to prove that the Principal End of the Divine Providence and Government is the happiness of His Creatures'. The tract was published by John Noon and copies are in Dr. Williams's library and the British Museum. The following is a quotation:

> [p. 22]: I don't find (I am sorry to say it) any necessary connection between mere intelligence, though ever so great, and the love or approbation of kind and beneficient actions.

Bayes argued that the principal end of the Deity was the happiness of His creatures, in opposition to Balguy and Grove who had, respectively, maintained that the first spring of action of the Deity was Rectitude, and Wisdom.

In 1736 John Noon published a tract entitled 'An Introduction to the Doctrine of Fluxions, and a Defense of the Mathematicians against the objections of the Author of the Analyst'. De Morgan (1860) says: 'This very acute tract is anonymous, but it was always attributed to Bayes by the contemporaries who write in the names of the authors as I have seen in various copies, and it bears his name in other places.' The ascription to Bayes is accepted also in the British Museum catalogue.

From the copy in Dr. Williams's library we quote:

> [p. 9]: It is not the business of the Mathematician to dispute whether quantities do in fact ever vary in the manner that is supposed, but only whether the notion of their doing so be intelligible; which being allowed, he has a right to take it for granted, and then see what deductions he can make from that supposition. It is not the business of a Mathematician to show that a strait line or circle can be drawn, but he tells you what he means by these; and if you understand him, you may proceed further with him; and it would not be to the purpose to object that there is no such thing in nature as a true strait line or perfect circle, for this is none of his concern: he is not inquiring how things are in matter of fact, but supposing things to be in a certain way, what are the consequences to be deduced from them; and all that is to be demanded of him is, that his suppositions be intelligible, and his inferences just from the suppositions he makes.

[p. 48]: He [i.e. the Analyst = Bishop Berkeley] represents the disputes and controversies among mathematicians as disparaging the evidence of their methods: and, Query 51, he represents Logics and Metaphysics as proper to open their eyes, and extricate them from their difficulties. Now were ever two things thus put together? If the disputes of the professors of any science disparage the science itself, Logics and Metaphysics are much more to be disparaged than Mathematics; why, therefore, if I am half blind, must I take for my guide one that can't see at all?

[p. 50]: So far as Mathematics do not tend to make men more sober and rational thinkers, wiser and better men, they are only to be considered as an amusement, which ought not to take us off from serious business.

This tract may have had something to do with Bayes's election, in 1742, to Fellowship of the Royal Society, for which his sponsors were Earl Stanhope, Martin Folkes, James Burrow, Cromwell Mortimer, and John Eames.

William Whiston, Newton's successor in the Lucasian Chair at Cambridge, who was expelled from the University for Arianism, notes in his Memoirs (p. 390) that 'on August the 24th this year 1746, being Lord's Day, and St. Bartholomew's Day, I breakfasted at Mr Bay's, a dissenting Minister at Tunbridge Wells, and a Successor, though not immediate, to Mr Humphrey Ditton, and like him a very good mathematician also'. Whiston goes on to relate what he said to Bayes, but he gives no indication that Bayes made reply.

According to Strange (1949) Bayes wished to retire from his ministry as early as 1749, when he allowed a group of Independents to bring ministers from London to take services in his chapel week by week, except for Easter, 1750, when he refused his pulpit to one of these preachers; and in 1752 he was succeeded in his ministry by the Rev. William Johnston, A.M., who inherited Bayes's valuable library. Bayes continued to live in Tunbridge Wells until his death on 17 April 1761. His body was taken to be buried, with that of his father, mother, brothers and sisters, in the Bayes and Cotton family vault in Bunhill Fields, the Nonconformist burial ground by Moorgate. This cemetery also contains the grave of Bayes's friend, the Unitarian Rev. Richard Price, author of the *Northampton Life Table* and object of Burke's oratory and invective in *Reflections on the French Revolution*, and the graves of John Bunyan, Samuel Watts, Daniel Defoe, and many other famous men.

Bayes's will, executed on 12 December 1760, shows him to have been a man of substance. The bulk of his estate was divided among his brothers, sisters, nephews and cousins, but he left £200 equally between 'John Boyl late preacher at Newington and now at Norwich, and Richard Price now I suppose preacher at Newington Green'. He also left 'To Sarah Jeffrey daughter of John Jeffrey, living with her father at the corner of Fountains Lane near Tunbridge Wells, £500, and my watch made by Elliott and all my linen and wearing apparell and household stuff.'

Apart from the tracts already noted, and the celebrated Essay reproduced here, Bayes wrote a letter on Asymptotic Series to John Canton, published in the *Philosophical Transactions of the Royal Society* (1763, pp. 269–271). His mathematical work, though small in quantity, is of the very highest quality; both his tract on fluxions and his paper on asymptotic series contain thoughts which did not receive as clear expression again until almost a century had elapsed.

Since copies of the volume in which Bayes's essay first appeared are not rare, and copies of a photographic reprint issued by the Department of Agriculture, Washington, D.C., U.S.A., are fairly widely dispersed, the view has been taken that in preparing Bayes's paper for publication here some editing is permissible. In particular, the notation has been modernized, some of the archaisms have been removed and what seem to be obvious printer's errors have been corrected. Sometimes, when a word has been omitted in the original, a suggestion has been supplied, enclosed in square brackets. Otherwise, however, nothing has been changed, and we hope that while the present text should in no sense be regarded as definitive, it will be easier to read on that account. All the work of preparing the text for the printer was most painstakingly and expertly carried out by Mr M. Gilbert, B.Sc., A.R.C.S. Thanks are also due to the Royal Society for permission to reproduce the Essay in its present form.

In writing the biographical notes the present author has had the friendly help of many persons, including especially Dr A. Fletcher and Mr R. L. Plackett, of the University of Liverpool, Mr J. F. C. Willder, of the Department of Pathology, Guy's Hospital Medical School, and Mr M. E. Ogborn, F.I.A., of the Equitable Life Assurance Society. He would also like to thank Sir Ronald Fisher, for some initial prodding which set him moving, and Prof. E. S. Pearson, for patient encouragement to see the matter through to completion.

REFERENCES

ANDERSON J. G. (1941). *Mathematical Gazette*, **25**, 160–2.

CANTOR, M. (1908). *Geschichte der Mathematik*, vol. IV. (Article by Netto.)

DE MORGAN, A. (1860). *Notes and Queries*, 7 Jan. 1860.

LORIA, G. (1933), *Storia delle Mathematiche*, vol. III. Turin.

MACKENZIE, M. (Ed.) (1865). *Imperial Dictionary of Universal Biography*, 3 vols., Glasgow.

STRANGE, C. H. (1949). *Nonconformity in Tunbridge Wells*. Tunbridge Wells. *The Gentleman's Magazine* (1761), **31**, 188.

Notes and Queries (1941), 19 April.

APPENDIX 3

Communication of Bayes' Essay to the Philosophical Transactions of the Royal Society of London

Richard Price

Dear Sir,

I now send you an essay which I have found among the papers of our deceased friend Mr Bayes, and which, in my opinion, has great merit, and well deserves to be preserved. Experimental philosophy, you will find, is nearly interested in the subject of it; and on this account there seems to be particular reason for thinking that a communication of it to the Royal Society cannot be improper.

He had, you know, the honour of being a member of that illustrious Society, and was much esteemed by many in it as a very able mathematician. In an introduction which he has writ to this Essay, he says, that his design at first in thinking on the subject of it was, to find out a method by which we might judge concerning the probability that an event has to happen, in given circumstances, upon supposition that we know nothing concerning it but that, under the same circumstances, it has happened a certain number of times, and failed a certain other number of times. He adds, that he soon perceived that it would not be very difficult to do this, provided some rule could be found according to which we ought to estimate the chance that the probability for the happening of an event perfectly unknown, should lie between any two named degrees of probability, antecedently to any experiments made about it; and that it appeared to him that the rule must be to suppose the chance the same that it should lie between any two equidifferent degrees; which, if it were allowed, all the rest might be easily calculated in the common method of proceeding in the doctrine of chances. Accordingly, I find among his papers a very ingenious solution of this problem in this way. But he afterwards considered, that the *postulate* on which he had argued might not perhaps be looked upon by all as reasonable; and therefore he chose to lay down in another form the proposition in which he thought the solution of the problem is contained, and in a *scholium* to subjoin the reasons why he thought so, rather than to take into his mathematical reasoning any thing that might admit dispute. This, you will observe, is the method which he has pursued in this essay.

Every judicious person will be sensible that the problem now mentioned is by no means merely a curious speculation in the doctrine of chances, but necessary to be solved in order to [provide] a sure foundation for all our reasonings concerning past facts, and what is likely to be hereafter. Common sense is indeed sufficient to shew us that, from the observation of what has in former instances been the consequence of a certain cause or action, one may make a judgment what is likely to be the consequence of it another time, and that the larger [the] number of experiments we have to support a conclusion, so much the more reason we have to take it for granted. But it is certain that we cannot determine, at least not to any nicety, in what degree repeated experiments confirm a conclusion, without the particular discussion of the beforementioned problem; which, therefore, is necessary to be considered by any one who would give a clear account of the strength of *analogical* or *inductive reasoning*; concerning, which at present, we seem to know little more than that it does sometimes in fact convince us, and at other times not; and that, as it is the means of [a]cquainting us with many truths, of which otherwise we must have been ignorant; so it is, in all probability, the source of many errors, which perhaps might in some measure be avoided, if the force that this sort of reasoning ought to have with us were more distinctly and clearly understood.

These observations prove that the problem enquired after in this essay is no less important than it is curious. It may be safely added, I fancy, that it is also a problem that has never before been solved. Mr De Moivre, indeed, the great improver of this part of mathematics, has in his *Laws of Chance**, after Bernoulli, and to a greater degree of exactness, given rules to find the probability there is, that if a very great number of trials be made concerning any event, the proportion of the number of times it will happen, to the number of times it will fail in those trails, should differ less than by small assigned limits from the proportion of the probability of its happening to the probability of its failing in one single trial. But I know of no person who has shewn how to deduce the solution of the converse problem to this; namely, 'the number of times an unknown event has happened and failed being given, to find the chance that the probability of its happening should lie somewhere between any two named degrees of probability.' What Mr De Moivre has done therefore cannot be thought sufficient to make the consideration of this point unnecessary: especially, as the rules he has given are not pretended to be

* See Mr De Moivre's *Doctrine of Chances*, p. 243, etc. He has omitted the demonstrations of his rules, but these have been since supplied by Mr Simpson at the conclusion of his treatise on *The Nature and Laws of Chance*.

rigorously exact, except on supposition that the number of trials made are infinite; from whence it is not obvious how large the number of trials must be in order to make them exact enough to be depended on in practice.

Mr De Moivre calls the problem he has thus solved, the hardest that can be proposed on the subject of chance. His solution he has applied to a very important purpose, and thereby shewn that those are much mistaken who have insinuated that the Doctrine of Chances in mathematics is of trivial consequence, and cannot have a place in any serious enquiry.† The purpose I mean is, to shew what reason we have for believing that there are in the constitution of things fixt laws according to which events happen, and that, therefore, the frame of the world must be the effect of the wisdom and power of an intelligent cause; and thus to confirm the argument taken from final causes for the existence of the Deity. It will be easy to see that the converse problem solved in this essay is more directly applicable to this purpose; for it shews us, with distinctness and precision, in every case of any particular order or recurrency of events, what reason there is to think that such recurrency or order is derived from stable causes or regulations in nature, and not from any of the irregularities of chance.

The two last rules in this essay are given without the deductions of them. I have chosen to do this because these deductions, taking up a good deal of room, would swell the essay too much; and also because these rules, though of considerable use, do not answer the purpose for which they are given as perfectly as could be wished. They are, however, ready to be produced, if a communication of them should be thought proper. I have in some places writ short notes, and to the whole I have added an application of the rules in the essay to some particular cases, in order to convey a clearer idea of the nature of the problem, and to shew how far the solution of it has been carried.

I am sensible that your time is so much taken up that I cannot reasonably expect that you should minutely examine every part of what I now send you. Some of the calculations, particularly in the Appendix, no one can make without a good deal of labour. I have taken so much care about them, that I believe there can be no material error in any of them; but should there be any such errors, I am the only person who ought to be considered as answerable for them.

Mr Bayes has thought fit to begin his work with a brief demonstration of the general laws of chance. His reason for doing this, as he says in his introduction, was not merely that his reader might not have the trouble of searching elsewhere for the principles on which he has argued, but because he did not know whither to refer him for a clear demonstration of them. He has also made an apology for the peculiar definition he has given of the word *chance* or *probability*. His design herein was to cut off all dispute about the meaning of the word, which in common language is used in different senses by persons of different opinions, and according as it is applied to *past* or *future* facts. But whatever different senses it may have, all (he observes) will allow that an expectation depending on the truth of any *past* fact, or the happening of

† See his *Doctrine of Chances*, p. 252, etc.

any *future* event, ought to be estimated so much the more valuable as the fact is more likely to be true, or the event more likely to happen. Instead therefore, of the proper sense of the word *probability*, he has given that which all will allow to be its proper measure in every case where the word is used. But it is time to conclude this letter. Experimental philosophy is indebted to you for several discoveries and improvements; and, therefore, I cannot help thinking that there is a peculiar propriety in directing to you the following essay and appendix. That your enquiries may be rewarded with many further successes, and that you may enjoy every valuable blessing, is the sincere wish of, Sir,

your very humble servant,

RICHARD PRICE

Newington-Green,
10 *November* 1763

APPENDIX 4

An Essay Towards Solving a Problem in the Doctrine of Chances

Reverend Thomas Bayes

PROBLEM

Given the number of times in which an unknown event has happened and failed: *Required* the chance that the probability of its happening in a single trial lies somewhere between any two degrees of probability that can be named.

SECTION I

Definition

1. Several events are *inconsistent*, when if one of them happens, none of the rest can.
2. Two events are *contrary* when one, or other of them must; and both together cannot happen.
3. An event is said to *fail*, when it cannot happen; or, which comes to the same thing, when its contrary has happened.
4. An event is said to be determined when it has either happened or failed.
5. The *probability of any event* is the ratio between the value at which an expectation depending on the happening of the event ought to bc computed, and the value of the thing expected upon it's* happening.

* Author's Note: the spelling "it's" was correct and appropriate form in Bayes' time even though today we would use "its."

6. By *chance* I mean the same as probability.
7. Events are independent when the happening of any one of them does neither increase nor abate the probability of the rest.

Proposition 1

When several events are inconsistent the probability of the happening of one or other of them is the sum of the probabilities of each of them.

Suppose there be three such events, and whichever of them happens I am to receive N, and that the probability of the 1st, 2nd, and 3rd are respectively $a/N, b/N, c/N$. Then (by the definition of probability) the value of my expectation from the 1st will be a, from the 2nd b, and from the 3rd c. Wherefore the value of my expectations from all three will be $a + b + c$. But the sum of my expectations from all three is in this case an expectation of receiving N upon the happening of one or other of them. Wherefore (by definition 5) the probability of one or other of them is $(a + b + c)/N$ or $a/N + b/N + c/N$, the sum of the probabilities of each of them.

Corollary

If it be certain that one or other of the three events must happen, then $a + b + c = N$. For in this case all the expectations together amounting to a certain expectation of receiving N, their values together must be equal to N. And from hence it is plain that the probability of an event added to the probability of its failure (or of its contrary) is the ratio of equality. For these are two inconsistent events, one of which necessarily happens. Wherefore if the probability of an event is P/N that of it's failure will be $(N - P)/N$.

Proposition 2

If a person has an expectation depending on the happening of an event, the probability of the event is to the probability of its failure as his loss if it fails to his gain if it happens.

Suppose a person has an expectation of receiving N, depending on an event the probability of which is P/N. Then (by definition 5) the value of his expectation is P, and therefore if the event fail, he loses that which in value is P; and if it happens he receives N, but his expectation ceases. His gain therefore is $N - P$. Likewise since the probability of the event is P/N, that of its failure (by corollary prop. 1) is $(N - P)/N$. But P/N is to $(N - P)/N$ as P is to $N - P$, i.e. the probability of the event is to the probability of it's failure, as his loss if it fails to his gain if it happens.

Proposition 3

The probability that two subsequent events will both happen is a ratio compounded of the probability of the 1st, and the probability of the 2nd on supposition the 1st happens.

Suppose that, if both events happen, I am to receive N, that the probability both will happen is P/N, that the 1st will is a/N (and consequently that the 1st will not is $(N - a)/N$) and that the 2nd will happen upon supposition the 1st does is b/N. Then (by definition 5) P will be the value of my expectation, which will become b if the 1st happens. Consequently if the 1st happens, my gain by it is $b - P$, and if it fails my loss is P. Wherefore, by the foregoing proposition, a/N is to $(N - a)/N$, i.e. a is to $N - a$ as P is to $b - P$. Wherefore (*componendo inverse*) a is to N as P is to b. But the ratio of P to N is compounded of the ratio of P to b, and that of b to N. Wherefore the same ratio of P to N is compounded of the ratio of a to N and that of b to N, i.e. the probability that the two subsequent events will both happen is compounded of the probability of the 1st and the probability of the 2nd on supposition the 1st happens.

Corollary

Hence if of two subsequent events the probability of the 1st be a/N, and the probability of both together be P/N, then the probability of the 2nd on supposition the 1st happens is P/a.

Proposition 4

If there be two subsequent events to be determined every day, and each day the probability of the 2nd is b/N and the probability of both P/N, and I am to receive N if both the events happen the first day on which the 2nd does; I say, according to these conditions, the probability of my obtaining N is P/b. For if not, let the probability of my obtaining N be x/N and let y be to x as $N - b$ to N. Then since x/N is the probability of my obtaining N (by definition 1) x is the value of my expectation. And again, because according to the foregoing conditions the first day I have an expectation of obtaining N depending on the happening of both the events together, the probability of which is P/N, the value of this expectation is P. Likewise, if this coincident should not happen I have an expectation of being reinstated in my former circumstances, i.e. of receiving that which in value is x depending on the failure of the 2nd event the probability of which (by corollary to proposition 1) is $(N - b)/N$ or y/x, because y is to x as $N - b$ to N. Wherefore since x is the thing expected and y/x the probability of obtaining it, the value of this expectation is y. But these two last expectations together are evidently the same with my original expectation, the value of which is x, and therefore $P + y = x$. But y is to x as $N - b$ is to N. Wherefore x is to P as N is to b, and x/N (the probability of my obtaining N) is P/b.

Corollary

Suppose after the expectation given me in the foregoing proposition, and before it is at all known whether the 1st event has happened or not, I should find that the 2nd event has happened; from hence I can only infer that the event is determined on which my expectation depended, and have no reason to esteem the value of my expectation either greater or less than it was before. For if I have reason to think it less, it would be reasonable for me to give something to be reinstated in my former circumstances, and this over and over again as often as I should be informed that the 2nd event had happened, which is evidently absurd. And the like absurdity plainly follows if you say I ought to set a greater value on my expectation than before, for then it would be reasonable for me to refuse something if offered me upon condition I would relinquish it, and be reinstated in my former circumstances; and this likewise over and over again as often as (nothing being known concerning the 1st event) it should appear that the 2nd had happened. Notwithstanding therefore this discovery that the 2nd event has happened, my expectation ought to be esteemed the same in value as before, i.e. x, and consequently the probability of my obtaining N is (by definition 5) still x/N or P/b.* But after this discovery the probability of my obtaining N is the probability that the 1st of two subsequent events has happened upon the supposition that the 2nd has, whose probabilities were as before specified. But the probability that an event has happened is the same as the probability I have to guess right if I guess it has happened. Wherefore the following proposition is evident.

Proposition 5

If there be two subsequent events, the probability of the 2nd b/N and the probability of both together P/N, and it being first discovered that the 2nd event has happened, from hence I guess that the 1st event has also happened, the probability I am in the right is P/b.†

Proposition 6

The probability that several independent events shall all happen is a ratio compounded of the probabilities of each.

* What is here said may perhaps be a little illustrated by considering that all that can be lost by the happening of the 2nd event is the chance I should have had of being reinstated in my former circumstances, if the event on which my expectation depended had been determined in the manner expressed in the proposition. But this chance is always as much *against* me as it is *for* me. If the 1st event happens, it is *against* me, and equal to the chance for the 2nd event's failing. If the 1st event does not happen, it is *for* me, and equal also to the chance for the 2nd event's failing. The loss of it, therefore, can be no disadvantage.

† What is proved by Mr Bayes in this and the preceding proposition is the same with the answer to the following question. What is the probability that a certain event, when it happens. will be accompanied with another to be determined at the same time? In this case, as one of the events is given, nothing can be due for the expectation of it; and, consequently, the value of an expectation depending on the happening of both events must be the same with the value of an expectation depending on the happening of one of them. In other words; the probability that, when one of two events happens, the other will, is the same with the probability of this other. Call x then the probability of this other, and if b/N be the probability of the given event, and p/N the probability of both, because $p/N = (b/N) \times x$, $x = p/b =$ the probability mentioned in these propositions.

For from the nature of independent events, the probability that any one happens is not altered by the happening or failing of any of the rest, and consequently the probability that the 2nd event happens on supposition the 1st does is the same with its original probability; but the probability that any two events happen is a ratio compounded of the probability of the 1st event, and the probability of the 2nd on supposition the 1st happens by proposition 3. Wherefore the probability that any two independent events both happen is a ratio compounded of the probability of the 1st and the probability of the 2nd. And in like manner considering the 1st and 2nd events together as one event; the probability that three independent events all happen is a ratio compounded of the probability that the two 1st both happen and the probability of the 3rd. And thus you may proceed if there be ever so many such events; from whence the proposition is manifest.

Corollary 1

If there be several independent events, the probability that the 1st happens the 2nd fails, the 3rd fails and the 4th happens, etc. is a ratio compounded of the probability of the 1st, and the probability of the failure of the 2nd, and the probability of the failure of the 3rd, and the probability of the 4th, etc. For the failure of an event may always be considered as the happening of its contrary.

Corollary 2

If there be several independent events, and the probability of each one be a, and that of its failing be b, the probability that the 1st happens and the 2nd fails, and the 3rd fails and the 4th happens, etc. will be $abba$, etc. For, according to the algebraic way of notation, if a denote any ratio and b another, $abba$ denotes the ratio compounded of the ratios a, b, b, a. This corollary therefore is only a particular case of the foregoing.

Definition

If in consequence of certain data there arises a probability that a certain event should happen, its happening or failing, in consequence of these data, I call it's happening or failing in the 1st trial. And if the same data be again repeated, the happening or failing of the event in consequence of them I call its happening or failing in the 2nd trail; and so on as often as the same data are repeated. And hence it is manifest that the happening or failing of the same event in so many diffe[rent] trials, is in reality the happening or failing of so many distinct independent events exactly similar to each other.

Proposition 7

If the probability of an event be a, and that of its failure be b in each single trial, the probability of its happening p times, and failing q times in $p + q$ trails is $Ea^p b^q$ if E

be the coefficient of the term in which occurs $a^p b^q$ when the binomial $(a+b)^{p+q}$ is expanded.

For the happening or failing of an event in different trials are so many independent events. Wherefore (by cor. 2 prop. 6) the probability that the event happens the 1st trial, fails the 2nd and 3rd, and happens the 4th, fails the 5th, etc. (thus happening and failing till the number of times it happens be p and the number if fails be q) is *abbab* etc. till the number of *a*'s be p and the number of *b*'s be q, that is; 'tis $a^p b^q$. In like manner if you consider the event as happening p times and failing q times in any other particular order, the probability for it is $a^p b^q$; but the number of different orders according to which an event may happen or fail, so as in all to happen p times and fail q, in $p+q$ trials is equal to the number of permutations that *aaaa bbb* admit of when the number of *a*'s is p, and the number of *b*'s is q. And this number is equal to E, the coefficient of the term in which occurs $a^p b^q$ when $(a+b)^{p+q}$ is expanded. The event therefore may happen p times and fail q in $p+q$ trials E different ways and no more, and its happening and failing these several different ways are so many inconsistent events, the probability for each of which is $a^p b^q$, and therefore by prop. 1 the probability that some way or other it happens p times and fails q times in $p+q$ trials is $Ea^p b^q$.

SECTION II

Postulate

1. I suppose the square table or plane *ABCD* to be so made and levelled, that if either of the balls *o* or *W* be thrown upon it, there shall be the same probability that it rests upon any one equal part of the plane as another, and that it must necessarily rest somewhere upon it.

2. I suppose that the ball *W* shall be first thrown, and through the point where it rests a line *os* shall be drawn parallel to *AD*, and meeting *CD* and *AB* in *s* and *o*; and that afterwards the ball *O* shall be thrown $p+q$ or n times, and that its resting between *AD* and *os* after a single throw be called the happening of the event *M* in a single trial. These things supposed:

Lemma 1

The probability that the point *o* will fall between any two points in the line *AB* is the ratio of the distance between the two points to the whole line *AB*.

Let any two points be named, as *f* and *b* in the line *AB*, and through them parallel to *AD* draw *fF*, *bL* meeting *CD* in *F* and *L*. Then if the rectangles *Cf*, *Fb*, *LA* are commensurable to each other, they may each be divided into the same equal parts, which being done, and the ball *W* thrown, the probability it will rest somewhere upon any number of these equal parts will be the sum of the probabilities it has to rest upon each one of them, because its resting upon any different parts of the plane *AC* are so many inconsistent events; and this sum, because the probability it should rest upon any one equal part as another is the same, is the probability it should rest upon any one equal part multiplied by the number of parts. Consequently, the probability there is that the ball *W* should rest somewhere upon *Fb* is the probability it has to rest upon one equal part multiplied by the number of equal parts in *Fb*; and the probability it rests somewhere upon *Cf* or *LA*, i.e. that it does not rest upon *FB* (because it must rest somewhere upon *AC*) is the probability it rests upon one equal part multiplied by the number of equal parts in *Cf*, *LA* taken together. Wherefore, the probability it rests upon *Fb* is to the probability it does not as the number of equal parts in *Fb* is to the number of equal parts in *Cf*, *LA* together, or as *Fb* to *Cf*, *LA* together, or as *fb* to *Bf*, *Ab* together. Wherefore the probability it rests upon *Fb* is to the probability it does not as *fb* to *Bf*, *Ab* together. And (*componendo inverse*) the probability it rests upon *Fb* is to the probability it rests upon *Fb* added to the probability it does not, as *fb* to *AB*, or as the ratio of *fb* to *AB* to the ratio of *AB* to *AB*. But the probability of any event added to the probability of its failure is the ratio of equality; wherefore, the probability it rests upon *Fb* is to the ratio of equality as the ratio of *fb* to *AB* to the ratio of *AB* to *AB*, or the ratio of equality; and therefore the probability it rests upon *Fb* is the ratio of *fb* to *AB*. But *ex hypothesi* according as the ball *W* falls upon *Fb* or not the point *o* will lie between *f* and *b* or not, and therefore the probability the point *o* will lie between *f* and *b* is the ratio of *fb* to *AB*.

Again; if the rectangles *Cf*, *Fb*, *LA* are not commensurable, yet the last mentioned probability can be neither greater nor less than the ratio of *fb* to *AB*; for, if it be less, let it be the ratio of *fc* to *AB*, and upon the line *fb* take the points *p* and *t*, so that *pt* shall be greater than *fc*, and the three lines *Bp*, *pt*, *tA* commensurable (which it is evident may be always done by dividing *AB* into equal parts less than half *cb*, and taking *p* and *t* the nearest points of division to *f* and *c* that lie upon *fb*). Then because *Bp*, *pt*, *tA* are commensurable, so are the rectangles *Cp*, *Dt*, and that

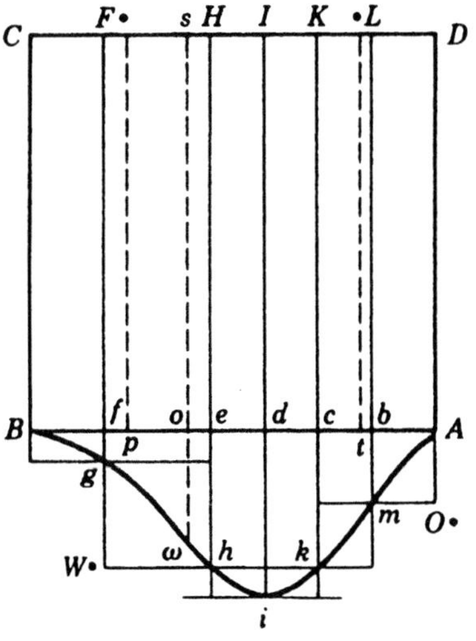

upon *pt* compleating the square *AB*. Wherefore, by what has been said, the probability that the point *o* will lie between *p* and *t* is the ratio of *pt* to *AB*. But if it lies between *p* and *t* it must lie between *f* and *b*. Wherefore, the probability it should lie between *f* and *b* cannot beless than the ratio of *pt* to *AB*, and therefore must be greater than the ratio of *fc* to *AB* (since *pt* is greater than *fc*). And after the same manner you may prove that the forementioned probability cannot be greater than the ratio of *fb* to *AB*, it must therefore be the same.

Lemma 2

The ball *W* having been thrown, and the line as drawn, the probability of the event *M* in a single trial is the ratio of *Ao* to *AB*.

For, in the same manner as in the foregoing lemma, the probability that the ball *o* being thrown shall rest somewhere upon *Do* or between *AD* and *so* is the ratio of *Ao* to *AB*. But the resting of the ball *o* between *AD* and *so* after a single throw is the happening of the event *M* in a single trials. Wherefore the lemma is manifest.

Proposition 8

If upon *BA* you erect the figure *BghikmA* whose property is this, that (the base *BA* being divided into any two parts, as *Ab*, and *Bb* and at the point of division *b* a perpendicular being erected and terminated by the figure in *m*; and y, x, r representing respectively the ratio of *bm*, *Ab*, and *Bb* to *AB*, and *E* being the coefficient of the term in which occurs $a^p b^q$ when the binomial $(a + b)^{p+q}$ is expanded) $y = Ex^p r^q$. I say that before the ball *W* is thrown, the probability the point *o* should fall between *f* and *b*, any two points named in the line *AB*, and with all that the event *M* should happen *p* times and fail *q* in $p + q$ trials, is the ratio of *fghikmb*, the part of the figure *BghikmA* intercepted between the perpendiculars *fg*, *bm* raised upon the line *AB*, to *CA* the square upon *AB*.

DEMONSTRATION

For if not; first let it be the ratio of *D* a figure greater than *fghikmb* to *CA*, and through the points e, d, c draw perpendiculars of *fb* meeting the curve *AmigB* in h, i, k; the point *d* being so placed that *di* shall be the longest of the perpendiculars terminated by the line *fb*, and the curve *AmigB*; and the points e, d, c being so many and so placed that the rectangles, *bk*, *ci*, *ei*, *fh* taken together shall differ less from *fghikmb* than *D* does; all which may be easily done by the help of the equation of the curve, and the difference between *D* and the figure *fghikmb* given. Then since *di* is the longest of the perpendicular ordinates that insist upon *fb*, the rest will gradually decrease as they are farther and farther from it it on each side, as appears from the construction of the figure, and consequently *eh* is greater than *gf* or any other ordinate that insists upon *ef*.

Now if *Ao* were equal to *Ae*, then by lem. 2 the probability of the event *M* in a single trial would be the ratio of *Ae* to *AB*, and consequently by cor. Prop. 1 the probability of it's failure would be the ratio of *Be* to *AB*. Wherefore, if x and r be the two forementioned ratios respectively, by proposition 7 the probability of the event M happening p times and failing q in $p+q$ trials would be $Ex^p r^q$. But x and r being respectively the ratios of *Ae* to *AB* and *Be* to *AB*, if y is the ratio of *eh* to *AB*, then, by construction of the figure *AiB*, $y = Ex^p r^q$. Wherefore, if *Ao* were equal to *Ae* the probability of the event M happening p times and failing q in $p+q$ trials would be y, or the ratio of *eh* to *AB*. And if *Ao* were equal to *Af*, or were any mean between *Ae* and *Af*, the last mentioned probability for the same reasons would be the ratio of *fg* or some other of the ordinates insisting upon *ef*, to *AB*. But *eh* is the greatest of all the ordinates that insist upon *ef*. Wherefore, upon supposition the point should lie anywhere between *f* and *e*, the probability that the event M happens p times and fails q in $p+q$ trials cannot be greater than the ratio of *eh* to *AB*. There then being these two subsequent events, the 1st that the point *o* will lie between *e* and *f*, the 2nd that the event M will happen p times and fail q in $p+q$ trails, and the probability of the first (by lemma 1) is the ratio of *ef* to *AB*, and upon supposition the 1st happens, by what has been now proved, the probability of the 2nd cannot be greater than the ratio of *eh* to *AB*, it evidently follows (from proposition 3) that the probability both together will happen cannot be greater than the ratio compounded of that of *ef* to *AB* and that of *eh* to *AB*, which compound ratio is the ratio of *fh* to *CA*. Wherefore, the probability that the point *o* will lie between *f* and *e*, and the event M happen p times and fail q, is not greater than the ratio of *fh* to *CA*. And in like manner the probability the point *o* will lie between *e* and *d*, and the event M happen and fails as before, cannot be greater than the ratio of *ei* to *CA*. And again, the probability the point *o* will lie between *d* and *c*, and the event M happen and fail as before, cannot be greater than the ratio of *ci* to *CA*. And lastly, the probability that the point *o* will lie between *c* and *b*, and the event M happen and fail as before, cannot be greater than the ratio of *bk* to *CA*. Add now all these several probabilities together, and their sum (by proposition 1) will be the probability that the point will lie somewhere between *f* and *b*, and the event M happen p times and fail q in $p+q$ trials. Add likewise the correspondent ratios together, and their sum will be the ratio of the sum of the antecedents to their common consequent, i.e. the ratio of *fh*, *ei*, *ci*, *bk* together to *CA*: which ratio is less than that of *D* to *CA* because *D* is greater than *fh*, *ei*, *ci*, *bk* together. And therefore, the probability that the point *o* will lie between *f* and *b*, and withal [in addition (author's note)] that the event M will happen p times and fail q in $p+q$ trials, is less than the ratio of *D* to *CA*; but it was supposed the same which is absurd. And in like manner, by inscribing rectangles within the figure, as *eg*, *dh*, *dh*, *dk*, *cm*, you may prove that the last mentioned probability is *greater* than the ratio of any figure less than *fghikmb* to *CA*.

Wherefore, that probability must be the ratio of *fghikmb* to *CA*.

Corollary

Before the ball W is thrown the probability that the point *o* will lie somewhere between *A* and *B*, or somewhere upon the line *AB*, and withal that the event M will happen p times, and fail q in $p+q$ trials is the ratio of the whole figure *AiB* to *CA*.

But it is certain that the point o will lie somewhere upon AB. Wherefore, before the ball W is thrown the probability the event M will happen p times and fail q in $p+q$ trials is the ratio of AiB to CA.

Proposition 9

If before anything is discovered concerning the place of the point o, it should appear that the event M had happened p times and failed q in $p+q$ trials, and from hence I guess that the point o lies between any two points in the line AB, as f and b, and consequently that the probability of the event M in a single trial was somewhere between the ratio of Ab to AB and that of Af to AB; the probability I am in the right is the ratio of that part of the figure AiB described as before which is intercepted between perpendiculars erected upon AB at the points f and b, to the whole figure AiB.

For, there being these two subsequent events, the first that the point o will lie between f and b; the second that the event M should happen p. times and fail q in $p+q$ trials; and (by corollary to proposition 8) the original probability of the second is the ratio of AiB to CA, and (by proposition 8) the probability of both is the ratio of $fghimb$ to CA; wherefore (by proposition 5) it being first discovered that the second has happened, and from hence I guess that the first has happened also, the probability I am in the right is the ratio of $fghimb$ to AiB, the point which was to be proved.

Corollary

The same things supposed, if I guess that the probability of the event M lies somewhere between 0 and the ratio of Ab to AB, my chance to be in the right is the ratio of Abm to AiB.

Scholium

From the preceding proposition it is plain, that in the case of such an event as I there call M, from the number of times it happens and fails in a certain number of trials, without knowing anything more concerning it, one may give a guess whereabouts it's probability is, and, by the usual methods computing the magnitudes of the areas there mentioned, see the chance that the guess is right. And that the same rule is the proper one to be used in the case of an event concerning the probability of which we absolutely know nothing antecedently to any trials made concerning it, seems to appear from the following consideration; viz. that concerning such an event I have no reason to think that, in a certain number of trials, it should rather happen any one possible number of times than another. For, on this account, I may justly reason concerning it as if its probability had been at first unfixed, and then determined in such a manner as to give me no reason to think that, in a certain number of trials, it should rather happen any one possible number of times than another. But this is exactly the case of the event M. For before the ball W is thrown, which determines it's probability in a single trial (by corollary to proposition 8), the probability it has to happen p times and fail q in $p + q$ or n trials is the ratio of AiB to CA, which ratio is the same when $p + q$ or n given, whatever number p is; as will appear by computing the magnitude of AiB by the method of fluxions.* And consequently before the place of the point o is discovered or the number of times the event M has happened in n trials, I can have no reason to think it should rather happen one possible number of times than another.

In what follows therefore I shall take for granted that the rule given concerning the event M in proposition 9 is also the rule to be used in relation to any event concerning the probability of which nothing at all is known antecedently to any trials made or observed concerning it. And such an event I shall call an unknown event. [Author's Note: The "method of fluxions" referred to in this Scholium is, of course, the Newtonian Calculus. The implication, in modern terms, of Proposition 9 and the Scholium, is given in Eq. (1), Appendix 1.]

Corollary

Hence, by supposing the ordinates in the figure AiB to be contracted in the ratio of E to one, which makes no alteration in the proportion of the parts of the figure intercepted between them, and applying what is said of the event M to an unknown event, we have the following proposition, which gives the rules for finding the probability of an event from the number of times it actually happens and fails.

Proposition 10

If a figure be described upon any base AH (Vid. Fig.) having for its equation $y = x^p r^q$; where y, x, r are respectively the ratios of an ordinate of the figure insisting on the base at right angles, of the segment of the base intercepted between the ordinate and A the beginning of the base, and of the other segment of the base lying between the ordinate and the point H, to the base as their common consequent. I say then that if an unknown event has happened p times and failed q in $p + q$ trials, and in the base AH taking any two points as f and t you erect the ordinates fC, tF at right angles with it, the chance that the probability of the event lies somewhere between the ratio of Af to AH and that of At to AH, is the ratio of $tFCf$, that part of the before-described figure which is intercepted between the two ordinates, to $ACFH$ the whole figure insisting on the base AH.

This is evident from proposition 9 and the remarks made in the foregoing scholium and corollary.

1. Now, in order to reduce the foregoing rule to practice, we must find the value of the area of the figure described and the several parts of it separated, by ordinates

* It will be proved presently in art. 4 by computing in the method here mentioned that AiB contracted in the ratio of E to 1 is to CA as 1 to $(n + 1)E$: from whence it plainly follows that, antecedently to this contraction, AiB must be to CA in the ratio of 1 to $n + 1$, which is a constant ratio when n is given, whatever p is.

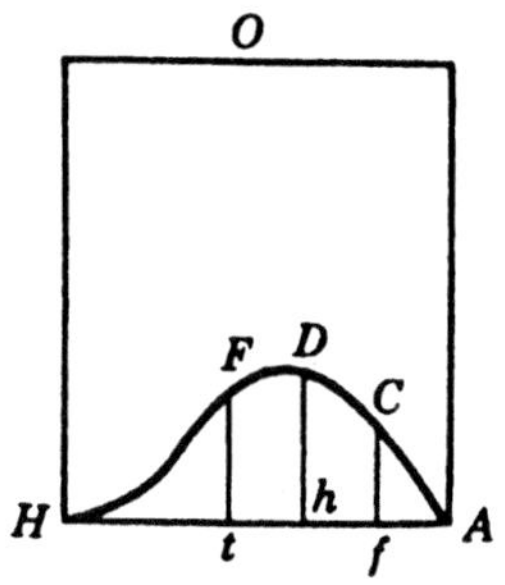

perpendicular to its base. For which purpose, suppose $AH = 1$ and HO the square upon AH likewise $= 1$, and Cf will be $= y$, and $Af = x$, and $Hf = r$, because y, x and r denote the ratios of Cf, Af, and Hf respectively to AH. And by the equation of the curve $y = x^p r^q$ and (because $Af + fH = AH$) $r + x = 1$. Wherefore

$$y = x^p(1-x)^q$$
$$= x^p - qx^{p+1} + \frac{q(q-1)x^{p+2}}{2} - \frac{q(q-1)(q-2)x^{p+3}}{2.3} + \text{etc.}$$

Now the abscisse being x and the ordinate x^p the correspondent area is $x^{p+1}/(p+1)$ (by proposition 10, cas. 1, Quadrat. Newt.)[†] and the ordinate being qx^{p+1} the area is $qx^{p+2}/(p+2)$; and in like manner of the rest. Wherefore, the abscisse being x and the ordinate y or $x^p - qx^{p+1} +$ etc. the correspondent area is

$$\frac{x^{p+1}}{p+1} - \frac{qx^{p+2}}{p+2} + \frac{q(q-1)x^{p+3}}{2(p+3)} - \frac{q(q-1)(q-2)x^{p+4}}{2.3(p+4)} + \text{etc.}$$

[Author's Note: 2.3 means 2 × 3.]
Wherefore, if $x = Af = Af/(AH)$, and $y = Cf = Cf/(AH)$, then

$$ACf = \frac{ACf}{HO} = \frac{x^{p+1}}{p+1} - \frac{qx^{p+2}}{p+2} + \frac{q(q-1)x^{p+3}}{2(p+3)} - \text{etc.}$$

[†] Tis very evident here, without having recourse to Sir Isaac Newton, that the fluxion of the area ACf being

$$y\dot{x} = x^p\dot{x} - qx^{p+1}\dot{x} + \frac{q(q-1)}{2}x^{p+2}\dot{x} - \text{etc.}$$

the fluent or area itself is

$$\frac{x^{p+1}}{p+1} - \frac{qx^{p+2}}{p+2} + \frac{q(q-1)x^{p+3}}{2(p+3)} - \text{etc.}$$

From which equation, if q be a small number, it is easy to find the value of the ratio of ACf to HO and in like manner as that was found out, it will appear that the ratio of HCf to HO is

$$\frac{r^{q+1}}{q+1}-\frac{pr^{q+2}}{q+2}+\frac{p(p-1)r^{q+3}}{2(q+3)}-\frac{p(p-1)(p-2)r^{q+4}}{2.3(q+4)}+\text{etc.}$$

which series will consist of few terms and therefore is to be used when p is small.

2. The same things supposed as before, the ratio of ACf to HO is

$$\frac{x^{p+1}r^{q}}{p+1}+\frac{qx^{p+2}r^{q-1}}{(p+1)(p+2)}+\frac{q(q-1)x^{p+3}r^{q-2}}{(p+1)(p+2)(p+3)}$$
$$+\frac{q(q-1)(q-2)x^{p+4}r^{q-3}}{(p+1)(p+2)(p+3)(p+4)}$$
$$+\text{etc.}+\frac{x^{n+1}q(q-1)\cdots 1}{(n+1)(p+1)(p+2)\cdots n},$$

where $n=p+q$. For this series is the same with $x^{p+1}/(p+1)-qx^{p+2}/(p+2)+$ etc. set down in Art. 1st as the value of the ratio of ACf to HO; as will easily be seen by putting in the former instead of r its value $1-x$, and expanding the terms and ordering them according to the powers of x. Or, more readily, by comparing the fluxions of the two series, and in the former instead of $\dot{r}$ substituting $-\dot{x}$.*

* The fluxion of the first series is

$$x^{p}r^{q}\dot{x}+\frac{qx^{p+1}r^{q-1}\dot{r}}{p+1}+\frac{qx^{p+1}r^{q-1}\dot{x}}{p+1}+\frac{q(q-1)x^{p+2}r^{q-2}\dot{r}}{(p+1)(p+2)}+\frac{q(q-1)x^{p+2}r^{q-2}\dot{x}}{(p+1)(p+2)}$$
$$+\frac{q(q-1)(q-2)x^{p+3}r^{q-3}\dot{r}}{(p+1)(p+2)(p+3)}+\text{etc.}$$

or, substituting $-\dot{x}$ for $\dot{r}$,

$$x^{p}r^{q}\dot{x}-\frac{qx^{p+1}r^{q-1}\dot{x}}{p+1}+\frac{qx^{p+1}r^{q-1}\dot{x}}{p+1}-\frac{q(q-1)x^{p+2}r^{q-2}\dot{x}}{(p+1)(p+2)}$$
$$+\frac{q(q-1)x^{p+2}r^{q-2}\dot{x}}{(p+1)(p+2)}-\text{etc.}$$

which, as all the terms after the first destroy one another, is equal to

$$x^{p}r^{q}\dot{x}=x^{p}(1-x)^{q}\dot{x}=x\left[1-qx+q\frac{(q-1)}{2}x^{2}-\text{etc.}\right]$$
$$=x^{p}x-qx^{p+1}\dot{x}+\frac{q(q-1)x^{p+2}}{2}\dot{x}-\text{etc.}$$
$$=\text{the fluxion of the latter series, or of }\frac{x^{p+1}}{p+1}-\frac{qx^{p+2}}{p+2}+\text{etc.}$$

The two series therefore are the same.

3. In like manner, the ratio of HCf to HO is

$$\frac{r^{q+1}x^p}{q+1} + \frac{pr^{q+2}x^{p-1}}{(q+1)(q+2)} + \frac{p(p-1)r^{q+3}x^{p-2}}{(q+1)(q+2)(q+3)} + \text{etc.}$$

4. If E be the coefficient of that term of the binomial $(a+b)^{p+q}$ expanded in which occurs $a^p b^q$, the ratio of the whole figure $ACFH$ to HO is $\{(n+1)E\}^{-1}$, n being $= p+q$. For, when $Af = AH$, $x = 1$, $r = 0$. Wherefore, all the terms of the series set down in Art. 2 as expressing the ratio of ACf to HO will vanish except the last, and that becomes

$$\frac{q(q-1)\cdots 1}{(n+1)(p+1)(p+2)\cdots n}.$$

But E being the coefficient of that term in the binomial $(a+b)^n$ expanded in which occurs $a^p b^q$ is equal to

$$\frac{(p+1)(p+2)\cdots n}{q(q-1)\cdots 1}.$$

And, because Af is supposed to become $= AH, ACf = ACH$. From whence this article is plain.

5. The ratio of ACf to the whole figure $ACFH$ is (by Art. 1 and 4)

$$(n+1)E\left[\frac{x^{p+1}}{p+1} - \frac{qx^{p+2}}{p+2} + \frac{q(q-1)x^{p+3}}{2(p+3)} - \text{etc.}\right]$$

and if, as x expresses the ratio of Af to AH, X should express the ratio of At to AH; the ratio of AFt to $ACFH$ would be

$$(n+1)E\left[\frac{X^{p+1}}{p+1} - \frac{qX^{p+2}}{p+2} + \frac{q(q-1)X^{p+3}}{2(p+3)} - \text{etc.}\right]$$

and consequently the ratio of $tFCf$ to $ACFH$ is $(n+1)E$ multiplied into the difference between the two series. Compare this with prop. 10 and we shall have the following practical rule.

Rule 1

If nothing is known concerning an event but that it has happened p times and failed q in $p+q$ or n trials, and from hence I guess that the probability of its happening in a single trial lies somewhere between any two degrees of probability as X and x, the chance I am in the right in my guess is $(n+1)E$ multiplied into the difference between the series

$$\frac{X^{p+1}}{p+1}\ \frac{qX^{p+2}}{p+2} + \frac{q(q-1)X^{p+3}}{2(p+3)} - \text{etc.}$$

and the series

$$\frac{X^{p+1}}{p+1} - \frac{qX^{p+2}}{p+2} + \frac{q(q-1)X^{p+3}}{2(p+3)} - \text{etc.}$$

E being the coefficient of $a^p b^q$ when $(a+b)^n$ is expanded.

This is the proper rule to be used when q is a small number; but if q is large and p small, change everywhere in the series here set down p into q and q into p and x into r or $(1-x)$, and X into $R = (1-X)$; which will not make any alteration in the difference between the two series.

Thus far Mr Bayes's essay.

With respect to the rule here given, it is further to be observed, that when both p and q are very large numbers, it will not be possible to apply it to practice on account of the multitude of terms which the series in it will contain. Mr Bayes, therefore, by an investigation which it would be too tedious to give here, has deduced from this rule another, which is as follows.

Rule 2

If nothing is known concerning an event but that it has happened p times and failed q in $p+q$ or n trials, and from hence I guess that the probability of its happening in a single trial lies between $(p/n)+z$ and $(p/n)-z$; if $m^2 = n^3/(pq)$, $a = p/n$, $b = q/n$, E the coefficient of the term in which occurs $a^p b^q$ when $(a+b)^n$ is expanded, and

$$\Sigma = \frac{(n+1)\sqrt{(2pq)}}{n\sqrt{n}} Ea^p b^q$$

multiplied by the series

$$mz - \frac{m^3 z^3}{3} + \frac{(n-2)m^5 z^5}{2n.5} - \frac{(n-2)(n-4)m^7 z^7}{2n.3n.7} + \frac{(n-2)(n-4)(n-6)m^9 z^9}{2n.3n.4n.9} - \text{etc.}$$

my chance to be in the right is greater than

$$\frac{2\Sigma}{1 + 2Ea^p b^q + 2Ea^p b^{q/n}}*$$

* In Mr Bayes's manuscript this chance is made to be greater than $2\Sigma/(1+2Ea^p b^q)$ and less than $2\Sigma/(1-2Ea^p b^q)$. The third term in the two divisors, as I have given them, being omitted. But this being evidently owing to a small oversight in the deduction of this rule, which I have reason to think Mr Bayes had himself discovered, I have ventured to correct his copy, and to give the rule as I am satisfied it ought to be given.

and less than

$$\frac{2\Sigma}{1 - 2Ea^p b^q - 2Ea^p b^q/n},$$

and if $p = q$ my chance is $2\sum$ exactly. [Author's Note: $2n.2n$ means $(2n) \times (3n)$.]

In order to render this rule fit for use in all cases it is only necessary to know how to find within sufficient nearness the value of $Ea^p b^q$ and also of the series $mz - 1/3m^3z^3 +$ etc. With respect to the former Mr Bayes has proved that, supposing K to signify the ratio of the quadrantal arc to its radius, $Ea^p b^q$ will be equal to $\frac{1}{2}\sqrt{n}/\sqrt{(Kpq)}$ multiplied by the *ratio*, [h], whose *hyberbolic* logarithm is

$$\frac{1}{12}\left[\frac{1}{n} - \frac{1}{p} - \frac{1}{q}\right] - \frac{1}{360}\left[\frac{1}{n^3} - \frac{1}{p^3} - \frac{1}{q^3}\right] + \frac{1}{1260}\left[\frac{1}{n^5} - \frac{1}{p^5} - \frac{1}{q^5}\right] - \frac{1}{1680}\left[\frac{1}{n^7} - \frac{1}{p^7} - \frac{1}{q^7}\right] + \frac{1}{1188}\left[\frac{1}{n^9} - \frac{1}{p^9} - \frac{1}{q^9}\right] - \text{etc.}^{\dagger}$$

where the numeral coefficients may be found in the following manner. Call them A, B, C, D, E etc. Then

$$A = \frac{1}{2.2.3} = \frac{1}{3.4}, B = \frac{1}{2.4.5} - \frac{A}{3}, C = \frac{1}{2.6.7} - \frac{10B + A}{5},$$

$$D = \frac{1}{2.8.9} - \frac{35C + 21B + A}{7}, E = \frac{1}{2.10.11} - \frac{126C + 84D + 36B + A}{9},$$

$$F = \frac{1}{2.12.13} - \frac{462D + 330C + 165E + 55B + A}{11} \text{ etc.}$$

where the coefficients of $B, C, D, E, F,$ etc. in the values of $D, E, F,$ etc. are the 2,3,4, etc. highest coefficients in $(a+b)^7, (a+b)^9, (a+b)^{11}$, etc. expanded; affixing in every particular value the least of these coefficients to B, the next in magnitude

† A very few terms of this series will generally give the hyperbolic logarithm to a sufficient degree of exactness. A similar series has been given by Mr De Moivre, Mr Simpson and other eminent mathematicians in an expression for the sum of the logarithms of the numbers 1, 2, 3, 4, 5, to x, which sum they have asserted to be equal to

$$\frac{1}{2}\log c + \left(x + \frac{1}{2}\right)\log x - x + \frac{1}{12x} - \frac{1}{360x^3} + \frac{1}{1260x^5} - \text{etc.}$$

c denoting the circumference of a circle whose radius is unity. But Mr Bayes, in a preceding paper in this volume, has demonstrated that, though this expression will very nearly approach to the value of this sum when only a proper number of the first terms is taken, the whole series cannot express any quantity at all, because, let x be what it will, there will always be a part of the series where it will begin to diverge. This observation, though it does not much affect the use of this series, seems well worth the notice of mathematicians.

to the furthest letter from B, the next to C, the next to the furthest but one, the next to D, the next to the furthest but two, and so on.*

With respect to the value of the series

$$mz - \frac{1}{3}m^3z^3 + \frac{(n-2)m^5z^5}{2n.5}\text{etc.}$$

he has observed that it may be calculated directly when mz is less than 1, or even not greater than $\surd 3$: but when mz is much larger it becomes impracticable to do this; in which case he shews a way of easily finding two values of it very nearly equal between which its true value must lie.

The theorem he gives for this purpose is as follows.

Let K, as before, stand for the ratio of the quadrantal arc to its radius, and H for the ratio whose hyperbolic logarithm is

$$\frac{2^2-1}{2n} - \frac{2^4-1}{360n^3} + \frac{2^6-1}{1260n^5} - \frac{2^8-1}{1680n^7} + \text{etc.}$$

Then the series $mz - 1/3m^3z^3 +$ etc. will be greater or less than the series

$$\frac{Hn\surd K}{(n+1)\surd 2} - \frac{n\left(1-\dfrac{2m^2z^2}{n}\right)^{\frac{1}{2}n+1}}{(n+2)2mz} + \frac{n^2\left(1-\dfrac{2m^2z^2}{n}\right)^{\frac{1}{2}n+2}}{(n+2)(n+4)4m^3z^3}$$
$$- \frac{3n^3\left(1-\dfrac{2m^2z^2}{n}\right)^{\frac{1}{2}n+3}}{(n+2)(n+4)(n+6)8m^5z^5}$$
$$+ \frac{3.5n^4\left(1-\dfrac{2m^2z^2}{n}\right)^{\frac{1}{2}n+4}}{(n+2)(n+4)(n+6)(n+8)16m^7z^7} - \text{etc.}$$

continued to any number of terms, according as the last term has a positive or a negative sign before it.

From substituting these values of Ea^pb^q and

$$mz - \frac{m^3z^3}{3} + \frac{(n-2)}{2n}\frac{m^5z^5}{5}\text{etc.}$$

in the second rule arises a third rule, which is the rule to be used when mz is of some considerable magnitude.

* This method of finding these coefficients I have deduced from the demonstration of the third lemma at the end of Mr Simpson's *Treatise on the Nature and Laws of Chance.*

Rule 3

If nothing is known of an event but that it has happened p times and failed q in $p+q$ or n trials, and from hence I judge that the probability of its happening in a single trial lies between $p/n+z$ and $p/n-z$ my chance to be right is *greater* than

$$\frac{\frac{1}{2}\surd(Kpq)h}{\surd(Kpq)-hn^{\frac{1}{2}}+hn^{-\frac{1}{2}}}\left\{2H-\frac{\surd 2(n+1)(1-2m^2z^2/n)^{\frac{1}{2}n+1}}{\surd K(n+2)mz}\right\}$$

and *less* than

$$\frac{\frac{1}{2}\surd(Kpq)h}{\surd(Kpq)-hn^{\frac{1}{2}}-hn^{-\frac{1}{2}}}\left\{2H-\frac{\surd 2(n+1)(1-2m^2z^2/n)^{\frac{1}{2}n+1}}{\surd K(n+2)mz}+\frac{\surd 2n(n+1)(1-2m^2z^2/n)^{\frac{1}{2}n+z}}{\surd K(n+2)(n+4)2m^3z^3}\right\}$$

where m^2, K, h and H stand for the quantities already explained.

AN APPENDIX

Containing an application of the foregoing Rules to some particular Cases

The first rule gives a direct and perfect solution in all cases; and the two following rules are only particular methods of approximating to the solution given in the first rule, when the labour of applying it becomes too great.

The first rule may be used in all cases where either p or q are nothing or not large. The second rule may be used in all cases where mz is less than $\surd 3$; and the third in all cases where m^2z^2 is greater than 1 and less than $\frac{1}{2}n$, if n is an even number and very large. If n is not large this last rule cannot be much wanted, because, m decreasing continually as n is diminished, the value of z may in this case be taken large (and therefore a considerable interval had between $p/n-z$ and $p/n+z$), and yet the operation be carried on by the second rule; or mz not exceed $\surd 3$.

But in order to shew distinctly and fully the nature of the present problem, and how far Mr Bayes has carried the solution of it; I shall give the result of this solution in a few cases, beginning with the lowest and most simple.

Let us then first suppose, of such an event as that called M in the essay, or an event about the probability of which, antecedently to trials, we know nothing, that it has happened *once*, and that is is enquired what conclusion we may draw from hence with respect to the probability of it's happening on a *second* trial.

The answer is that there would be an odds of three to one for somewhat more than an even chance that it would happen on a second trial.

For in this case, and in all others where q is nothing, the expression

$$(n+1)\left\{\frac{x^{p+1}}{p+1}-\frac{x^{p+1}}{p+1}\right\} \quad \text{or} \quad X^{p+1}-x^{p-1}$$

gives the solution, as will appear from considering the first rule. Put therefore in this expression $p+1=2, X=1$ and $x=\frac{1}{2}$ and it will be $1-(\frac{1}{2})^2$ or $\frac{3}{4}$; which shews the chance there is that the probability of an event that has happened once lies somewhere between 1 and $\frac{1}{2}$; or (which is the same) the odds that it is somewhat more than an even chance that it will happen on a second trial.*

In the same manner it will appear that if the event has happened twice, the odds now mentioned will be seven to one; if thrice, fifteen to one; and in general, if the event has happened p times, there will be an odds of $2^{p+1}-1$ to one, for *more* than an equal chance that it will happen on further trials.

Again, suppose all I know of an event to be that it has happened ten times without failing, and the enquiry to be what reason we shall have to think we are right if we guess that the probability of it's happening in a single trial lies somewhere between $\frac{16}{17}$ and $\frac{2}{3}$, or that the ratio of the causes of it's happening to those of it's failure is some ratio between that of sixteen to one and two to one.

Here $p+1=11$, $X=\frac{16}{17}$ and $x=\frac{2}{3}$ and $X^{p+1}-x^{p+1}=(\frac{16}{17})^{11}-(\frac{2}{3})^{11}=0.5013$ etc. The answer therefore is, that we shall have very nearly an equal chance for being right,

In this manner we may determine in any case what conclusion we ought to draw from a given number of experiments which are unopposed by contrary experiments. Everyone sees in general that there is reason to expect an event with more or less confidence according to the greater or less number of times in which, under given circumstances, it has happened without failing; but we here see exactly what this reason is, on what principles it is founded, and how we ought to regulate our expectations.

But it will be proper to dwell longer on this head.

Suppose a solid or die of whose number of sides and constitution we know nothing; and that we are to judge of these from experiments made in throwing it.

In this case, it should be observed, that it would be in the highest degree improbable that the solid should, in the first trial, turn any one side which could be assigned beforehand; because it would be known that some side it must turn, and that there was an infinity of other sides, or sides otherwise marked, which it was equally likely that it should turn. The first throw only shews that *it has* the side then thrown, without giving any reason to think that it has it any one number of times rather than any other. It will appear, therefore, that *after* the first throw and not before, we should be in the circumstances required by the conditions of the present problem, and that the whole effect of this throw would be to bring us into these circumstances. That is: the turning the side first thrown in any subsequent single trial would be an

* There can, I suppose, be no reason for observing that on this subject unity is always made to stand for certainty, and $\frac{1}{2}$ for an even chance.

event about the probability or improbability of which we could form no judgment, and of which we should know no more than that it lay somewhere between nothing and certainty. With the second trial then our calculations must begin; and if in that trial the supposed solid turns again the same side, there will arise the probability of three to one that it has more of that sort of sides than of *all* others; or (which comes to the same) that there is somewhat in its constitution disposing it to turn that side oftenest: And this probability will increase, in the manner already explained, with the number of times in which that side has been thrown without failing. It should not, however, be imagined that any number of such experiments can give sufficient reason for thinking that it would *never* turn any other side. For, suppose it has turned the same side in every trial a million of times. In these circumstances there would be an improbability that it has *less* than 1,400,000 more of these sides than all others; but there would also be an improbability that it had *above* 1,600,000 times more. The chance for the latter is expressed by 1,600,000/1,600,001 raised to the millioneth power subtracted from unity, which is equal to 0.4647 etc and the chance for the former is equal to 1,400,000/1,400,001 raised to the same power, or to 0.4895; which, being both less than an equal chance, proves what I have said. But though it would be thus improbable that it had *above* 1,600,000 times more or *less* than 1,400,000 times *more* of these sides than of all others, it by no means follows that we have any reason for judging that the true proportion in this case lies somewhere between that of 1,600,000 to one and 1,400,000 to one. For he that will take the pains to make the calculation will find that there is nearly the probability expressed by 0.527, or but little more than an equal chance, that it lies somewhere between that of 600,000 to one and three millions to one. It may deserve to be added, that it is more probable that this proportion lies somewhere between that of 900,000 to 1 and 1,900,000 to 1 than between any other two proportions whose antecedents are to one another as 900,000 to 1,900,000, and consequents unity.

I have made these observations chiefly because they are all strictly applicable to the events and appearances of nature. Antecedently to all experience, it would be improbable as infinite to one, that any particular event, beforehand imagined, should follow the application of any one natural object to another; because there would be an equal chance for any one of an infinity of other events. But if we had once seen any particular effects, as the burning of wood on putting it into fire, or the falling of a stone on detaching it from all contiguous objects, then the conclusions to be drawn from any number of subsequent events of the same kind would be to be determined in the same manner with the conclusions just mentioned relating to the constitution of the solid I have supposed. In other words. The first experiment supposed to be ever made on any natural object would only inform us of one event that may follow a particular change in the circumstances of those objects; but it would not suggest to us any ideas of uniformity in nature, or give us the least reason to apprehend that it was, in that instance or in any other, regular rather than irregular in its operations. But if the same event has followed without interruption in any one or more subsequent experiments, then some degree of uniformity will be observed; reason will be given to expect the same success in further experiments, and the calculations directed by the solution of this problem may be made.

One example here it will not be amiss to give.

Let us imagine to ourselves the case of a person just brought forth into this world, and left to collect from his observation of the order and course of events what powers and causes take place in it. The Sun would, probably, be the first object that would engage his attention; but after losing it the first night he would be entirely ignorant whether he should ever see it again. He would therefore be in the condition of a person making a first experiment about an event entirely unknown to him. But let him see a second appearance or one *return* of the Sun, and an expectation would be raised in him of a second return, and he might know that there was an odds of 3 to 1 for *some* probability of this. This odds would increase, as before represented, with the number of returns to which he was witness. But no finite number of returns would be sufficient to produce absolute or physical certainty. For let it be supposed that he has seen it return at regular and stated intervals a million of times. The conclusions this would warrant would be such as follow. There would be the odds of the millioneth power of 2, to one, that it was likely that it would return again at the end of the usual interval. There would be the probability expressed by 0.5352, that the odds for this was not *greater* than 1,600,000 to 1; and the probability expressed by 0.5105, that it was not less than 1,400,000 to 1.

It should be carefully remembered that these deductions suppose a previous total ignorance of nature. After having observed for some time the course of events it would be found that the operations of nature are in general regular, and that the powers and laws which prevail in it are stable and permanent. The consideration of this will cause one or a few experiments often to produce a much stronger expectation of success in further experiments than would otherwise have been reasonable; just as the frequent observation that things of a sort are disposed together in any place would lead us to conclude, upon discovering there any object of a particular sort, that there are laid up with it many others of the same sort. It is obvious that this, so far from contradicting the foregoing deductions, is only one particular case to which they are to be applied.

What has been said seems sufficient to shew us what conclusions to draw from *uniform* experience. It demonstrates, particularly, that instead of proving that events will *always* happen agreeably to it, there will be always reason against this conclusion. In other words, where the course of nature has been the most constant, we can have only reason to reckon upon a recurrency of events proportioned to the degree of this constancy; but we can have no reason for thinking that there are no causes in nature which will *ever* interfere with the operations of the causes from which this constancy is derived, or no circumstances of the world in which it will fail. And if this is true, supposing our only *data* derived from experience, we shall find additional reason for thinking thus if we apply other principles, or have recourse to such considerations as reason, independently of experience, can suggest.

But I have gone further than I intended here; and it is time to turn our thoughts to another branch of this subject; I mean, to cases where an experiment has sometimes succeeded and sometimes failed.

Here, again, in order to be as plain and explicit as possible, it will be proper to put the following case, which is the easiest and simplest I can think of.

Let us then imagine a person present at the drawing of a lottery, who knows nothing of its scheme or of the proportion of *Blanks* to *Prizes* in it. Let it further be supposed, that he is obliged to infer this from the number of *blanks* he hears drawn compared with the number of *prizes*; and that it is enquired what conclusions in these circumstances he may reasonably make.

Let him first hear *ten* blanks drawn and *one* prize, and let it be enquired what chance he will have for being right if he guesses that the proportion of *blanks* to *prize* in the lottery lies somewhere between the proportions of 9 to 1 and 11 to 1.

Here taking $X = \frac{11}{12}, x = \frac{9}{10}, p = 10, q = 1, n = 11, E = 11$, the required chance, according to first rule, is $(n+1)E$ multiplied by the difference between

$$\left\{\frac{X^{p+1}}{p+1} - \frac{qX^{p+2}}{p+1}\right\} \quad \text{and} \quad \left\{\frac{x^{p+1}}{p+1} - \frac{qx^{p+2}}{p+2}\right\} =$$

$$12.11 \cdot \left\{\left[\frac{\left(\frac{11}{12}\right)^{11}}{11} - \frac{\left(\frac{11}{12}\right)^{12}}{12}\right] - \left[\frac{\left(\frac{9}{10}\right)^{11}}{11} - \frac{\left(\frac{9}{10}\right)^{12}}{12}\right]\right\} = 0.07699 \text{ etc.}$$

There would therefore be an odds of about 923 to 76, or nearly 12 to 1 *against* his being right. Had he guessed only in general that there were less than 9 blanks to a prize, there would have been a probability of his being right equal to 0.6589, or the odds of 65 to 34.

Again, suppose that he has heard 20 *blanks* drawn and 2 *prizes*; what chance will he have for being right if he makes the same guess?

Here X and x being the same, we have $n = 22, p = 20, q = 2, E = 231$, and the required chance equal to

$$(n+1)E\left\{\left[\frac{X^{p+1}}{p+1} - \frac{qX^{p+2}}{p+2} + \frac{q(q-1)X^{p+3}}{2(p+3)}\right] - \left[\frac{x^{p+1}}{p+1} - \frac{qx^{p+2}}{p+2} + \frac{q(q-1)x^{p+3}}{2(p+3)}\right]\right\} = 0.10843 \text{ etc.}$$

He will, therefore, have a better chance for being right than in the former instance, the odds against him now being 892 to 108 or about 9 to 1. But should he only guess in general, as before, that there were less than 9 blanks to a prize, his chance for being right will be worse; for instead of 0.6589 or an odds of near two to one, it will be 0.584, or an odds of 584 to 415.

Suppose, further, that he has heard 40 *blanks* drawn and 4 *prizes*; what will the before-mentioned chances be?

The answer here is 0.1525, for the former of these chances; and 0.527, for the latter. There will, therefore, now be an odds of only $5\frac{1}{2}$ to 1 against the proportion of blanks to prizes lying between 9 to 1 and 11 to 1; and but little more than an equal chance that it is less than 9 to 1.

Once more. Suppose he has heard 100 *blanks* drawn and 10 *prizes*.

The answer here may still be found by the first rule; and the chance for a proportion of blanks to prizes *less* than 9 to 1 will be 0.44109, and for a proportion *greater* than 11 to 1, 0.3082. It would therefore be likely that there were not *fewer* than 9 or *more* than 11 blanks to a prize. But at the same time it will remain unlikely* that the true proportion should lie between 9 to 1 and 11 to 1, the chance for this being 0.2506 etc. There will therefore be still an odds of near 3 to 1 against this.

From these calculations it appears that, in the circumstances I have supposed, the chance for being right in guessing the proportion of *blanks* to *prizes* to be nearly the same with that of the number of *blanks* drawn in a given time to the number of prizes drawn, is continually increasing as these numbers increase; and that therefore, when they are considerably large, this conclusion may be looked upon as morally certain. By parity of reason, it follows universally, with respect to every event about which a great number of experiments has been made, that the causes of its happening bear the same proportion to the causes of its failing, with the number of happenings to the number of failures; and that, if an event whose causes are supposed to be known, happens oftener or seldomer than is agreeable to this conclusion, there will be reason to believe that there are some unknown causes which disturb the operations of the known ones. With respect, therefore, particularly to the course of events in nature, it appears, that there is demonstrative evidence to prove that they are derived from permanent causes, or laws originally established in the constitution of nature in order to produce that order of events which we observe, and not from any of the powers of chance.† This is just as evident as it would be, in the case I have insisted on, that the reason of drawing 10 times more *blanks* than *prizes* in millions of trials, was, that there were in the wheel about so many more *blanks* than *prizes*.

But to proceed a little further in the demonstration of this point.

We have seen that supposing a person, ignorant of the whole scheme of a lottery, should be led to conjecture, from hearing 100 *blanks* and 10 prizes drawn, that the proportion of *blanks* to *prizes* in the lottery was somewhere between 9 to 1 and 11 to 1, the chance for his being right would be 0.2506 etc. Let [us] now enquire what this chance would be in some higher cases.

Let it be supposed that *blanks* have been drawn 1000 times, and prizes 100 times in 1100 trials.

In this case the powers of X and x rise so high, and the number of terms in the two series

$$\frac{X^{p+1}}{p+1} - \frac{qX^{p+2}}{p+2}\text{etc.} \quad \text{and} \quad \frac{x^{p+1}}{p+1} - \frac{qx^{p+2}}{p+2}\text{etc.}$$

* I suppose no attentive person will find any difficulty in this. It is only saying that, supposing the interval between nothing and certainty divided into a hundred equal chances, there will be 44 of them for a less proportion of blanks to prizes than 9 to 1, 31 for a greater than 11 to 1, and 25 for some proportion between 9 to 1 and 11 to 1; in which it is obvious that, though one of these suppositions must be true, yet, having each of them more chances against them than for them, they are all separately unlikely.

† See Mr De Moivre's *Doctrine of Chances*, page 250.

become so numerous that it would require immense labour to obtain the answer by the first rule. 'Tis necessary, therefore, to have recourse to the second rule. But in order to make use of it, the interval between X and x must be a little altered. $\frac{10}{11} - \frac{9}{10}$ is $\frac{1}{110}$, and therefore the interval between $\frac{10}{11} - \frac{1}{110}$ *and* $\frac{10}{11} + \frac{1}{110}$ will be nearly the same with the interval between $\frac{9}{10}$ and $\frac{11}{12}$, only somewhat larger. If then we make the question to be; what chance there would be (supposing no more known than that blanks have been drawn 1000 times and prizes 100 times in 1100 trials) that the probability of drawing a blank in a single trial would lie somewhere between $\frac{10}{11} - \frac{1}{110}$ and $\frac{10}{11} + \frac{1}{110}$ we shall have a question of the same kind with the preceding questions, and deviate but little from the limits assigned in them.

The answer, according to the second rule, is that this chance is greater than

$$\frac{2\Sigma}{1 + 2Ea^p b^q + \dfrac{2Ea^p b^q}{n}}$$

and less than

$$\frac{2\Sigma}{1 + 2Ea^p b^q - 2E\dfrac{a^p b^q}{n}}$$

$\sum$ being

$$\frac{(n+1)\surd(2pq)}{n\surd n} Ea^p b^q \left\{ mz - \frac{m^3 z^3}{3} + \frac{(n-2)m^5 z^5}{2n.5} - \text{etc.} \right\}.$$

By making here $1000 = p$, $100 = q$, $1100 = n$, $\frac{1}{110} = z$,

$$mz = z\sqrt{\left(\frac{n^3}{pq}\right)} = 1.048808, \quad Ea^p b^q = \frac{1}{2} h \frac{\surd n}{\surd(Kpq)^r}$$

h being the ratio whose hyperbolic logarithm is

$$\frac{1}{12}\left[\frac{1}{n} - \frac{1}{p} - \frac{1}{q}\right] - \frac{1}{360}\left[\frac{1}{n^3} - \frac{1}{p^3} - \frac{1}{q^3}\right] + \frac{1}{1260}\left[\frac{1}{n^5} - \frac{1}{p^5} - \frac{1}{q^5}\right] - \text{etc.}$$

and K the ratio of the quadrantal arc to radius; the former of these expressions will be found to be 0.7953, and the latter 0.9405 etc. The chance enquired after, therefore, is greater than 0.7953, and less than 0.9405. That is; there will be an odds for being right in guessing that the proportion of blanks to prizes lies *nearly* between 9 to 1 and 11 to 1, (*or exactly* between 9 to 1 and 1111 to 99), which is greater than 4 to 1, and less than 16 to 1.

Suppose, again, that no more is known than that *blanks* have been drawn 10,000 times and *prizes* 1000 times in 11,000 trials; what will the chance now mentioned be!

Here the second as well as the first rule becomes useless, the value of mz being so great as to render it scarcely possible to calculate directly the series

$$\left\{mz - \frac{m^3z^3}{3} + \frac{(n-2)m^5z^5}{2n.5} - \text{etc.}\right\}$$

The third rule, therefore, must be used; and the information it gives us is, that the required chance is greater than 0.97421, or more than an odds of 40 to 1.

By calculations similar to these may be determined universally, what expectations are warranted by any experiments, according to the different number of times in which they have succeeded and failed; or what should be thought of the probability that any particular cause in nature, with which we have any acquaintance, will or will not, in any single trial, produce an effect that has been conjoined with it.

Most persons, probably, might expect that the chances in the specimen I have given would have been greater than I have found them. But this only shews how liable we are to error when we judge on this subject independently of calculation. One thing, however, should be remembered here; and that is, the narrowness of the interval between $\frac{9}{10}$ and $\frac{11}{12}$, or between $\frac{10}{11} + \frac{1}{110}$ and $\frac{10}{11} - \frac{1}{110}$. Had this interval been taken a little larger, there would have been a considerable difference in the results of the calculations. Thus had it been taken double, or $z = \frac{1}{55}$, it would have been found in the fourth instance that instead of odds against there were odds for being right in judging that the probability of drawing a blank in a single trial lies between $\frac{10}{11} + \frac{1}{55}$ and $\frac{10}{11} - \frac{1}{55}$.

The foregoing calculations further shew us the uses and defects of the rules laid down in the essay. 'Tis evident that the two last rules do not give us the required chances within such narrow limits as could be wished. But here again it should be considered, that these limits become narrower and narrower as q is taken larger in respect of p; and when p and q are equal, the exact solution is given in all cases by the second rule. These two rules therefore afford a direction to our judgment that may be of considerable use till some person shall discover a better approximation to the value of the two series in the first rule.*

But what most of all recommends the solution in this *Essay* is, that it is compleat in those cases where information is most wanted, and where Mr De Moivre's solution of the inverse problem can give little or no direction; I mean, in all cases where either p or q are of no considerable magnitude. In other cases, or when both p and q are very considerable, it is not difficult to perceive the truth of what has been here demonstrated, or that there is reason to believe in general that the chances for the

* Since this was written I have found out a method of considerably improving the approximation in the second and third rules by demonstrating that the expression $2\Sigma/\{1 + 2Ea^pb^q + 2Ea^pb^q/n\}$ comes almost as near to the true value wanted as there is reason to desire, only always somewhat less. It seems necessary to hint this here; though the proof of it cannot be given.

happening of an event are to the chances for its failure in the same *ratio* with that of p to q. But we shall be greatly deceived if we judge in this manner when either p or q are small. And tho' in such cases the *Data* are not sufficient to discover the exact probability of an event, yet it is very agreeable to be able to find the limits between which it is reasonable to think it must lie, and also to be able to determine the precise degree of assent which is due to any conclusions or assertions relating to them.

APPENDIX 5

Applications of Bayesian Statistical Science

A considerable amount of Bayesian statistical inference procedures that formally admit meaningful prior information in the scientific process of data analysis have had to await the advent of modern computer methods of analysis. This did not really happen in any major way until about the last couple of decades of the twentieth century. Moreover, at approximately the same time, numerical methods for implementing Bayesian solutions to complex applied problems were greatly augmented by the introduction of Markov chain Monte Carlo (MCMC) methods (Geman and Geman, 1984; Tanner & Wong, 1987; Gelfand and Smith, 1990). Since the arrival of these developments, Bayesian methods have been very usefully applied to problems in many fields, and the general approach has generated much new research. We list, in this appendix, references to various applications of the Bayesian paradigm to a wide variety of disciplines.

A5.1 BAYESIAN METHODOLOGY

Some books and articles summarizing Bayesian methodology in various fields are given below.

Anthropology and Archaeology, Buck et al. (1996), John Wiley & Sons.

Econometrics, Zellner (1971), John Wiley & Sons.

Economics, Dorfman (1997), Springer-Verlag.

Evaluation Research, Pollard (1986), Sage Publications.

Mutual Fund Management, Hulbert (1999), New York Times.

Physics and Engineering, There is a series of books on maximum entropy and Bayesian methods published by Kluwer; a few are given in the references below and listed under "Kluwer". See also Stone et al. (1999, for target tracking); Ruanaidh et al. (1996, for signal processing); Jaynes et al. (1983, for physics), Bretthorst (1988, for physics and spectral analysis); and Iba (2000, for Physics & Monte Carlo methods).

Reliability, Sander and Badoux (1991), Kluwer Academic Publishers.

A5.2 APPLICATIONS

For the convenience of readers in various fields, we have organized and presented immediately below citations of some technical papers that use the Bayesian approach to find solutions to real problems, by discipline. Sometimes a paper really belongs to several of our discipline categories, but is only listed in one. These citations have been drawn from the following four volumes entitled *Case Studies in Bayesian Statistics*, published by Springer-Verlag (New York). Volume I, published in 1993, and Volume II, published in 1995, were edited by Gatsonis, Hodges, Kass, and Singpurwalla. Volume III, published in 1997, was edited by Gatsonis, Hodges, Kass, McCulloch, Rossi, and Singpurwalla. Volume IV, published in 1999, was edited by Gatsonis, Carlin, Gelman, West, Kass, Carriquiry, and Verdinelli.

Biochemistry

Etzioni, R., B. P. Carlin, *Bayesian Analysis of the Ames Salmonella/Microsome Assay,* Vol. I, p. 311.

Business Economics

Nandram, B., and Sedransk, J. *Bayesian Inference for the Mean of a Stratified Population When There are Order Restrictions*, Vol. II, p. 309.

Ecology

Lad, F., and Brabyn, M. W. *Synchronicity of Whale Strandings with Phases of the Moon*, Vol. I, p. 362.

Raftery, A. E., and Zeh, J. E. *Estimation of Bowhead Whale, Balaena mysticetus, Population Size*, Vol. I, p. 163.

Wolfson, L. J., Kadane, J. B., and Small, M. J. *A Subjective Bayesian Approach to Environmental Sampling*, Vol. III, p. 457.

Wolpert, R. L., Steinberg, L. J., and Reckhow, K. H. *Bayesian Decision Support Using Environmental Transport-And-Fate Models*, Vol. I, p. 241.

Engineering

Andrews, R. W., Berger, J. O., and Smith, M. H. *Bayesian Estimation of Fuel Economy Potential Due to Technology Improvements*, Vol. I, p. 1

Jewell, W. S., and Shrane-Koung Chou, *Predicting Coproduct Yields in Microchip Fabrication*, Vol. I, p. 351.

O'Hagan, A. and Wells, F. S. *Use of Prior Information to Estimate Costs in a Sewerage Operation*, Vol. I, p. 118.

Short, T. H., *Restoration and Segmentation of Rail Surface Images*, Vol. I, p. 403.

Finance

Stevens, R. L., *Variable Selection Tests of Asset Pricing Models*, Vol. III, p. 271.

Forestry

Ickstadt, K., Wolpert, R. L., *Multiresolution Assessment of Forest Inhomogeneity*, Vol. III, p. 371.

Genetics

Churchill, G. A., *Accurate Restoration of DNA Sequences*, Vol. II, p. 90.

Lee, Jae Kyun, *Assessment of Deleterious Gene Models Using Predictive p-values*, Vol. III, p. 387.

Geology/Geophysics

Craig, P. S., Goldstein, M., Seheult, A. H., and Smith, J. A. *Pressure Matching for Hydrocarbon Reservoirs: A Case Study in the Use of Bayes Linear Strategies for Large Computer Experiments*, Vol. III, p. 37.

Royle, J. A., Berliner, L. M., Wikle, C. K., and Milliff, R. *A Hierarchical Spatial Model for Constructing Wind Fields from Scatterometer Data in the Labrador Sea*, Vol. IV, p. 367.

Sanso, B., and Muller, P., *Redesigning a Network of Rainfall Stations*, Vol. IV, p. 383.

Humanities

Greene, S., and Wasserman, L., *The Composition of a Composition: Just the Facts*, Vol. I, p. 337.

Marketing

Clark, L. A., Cleveland, W. S., Denby, L., and Liu, C., *Modeling Customer Survey Data*, Vol. IV, p. 3.

Hodges, J. S., Kalyanam, K. and Putler, D. S., *Estimating the Cells of a Contingency Table with Limited Information, for Use in Geodemographic Marketing*, Vol. III, p. 347.

Montgomery, A. L., *Hierarchical Bayes Models for Micro-Marketing Strategies*, Vol. III, p. 95.

Nandram, B., *Bayesian Inference for the Best Ordinal Multinomial Population in a Taste Test*, Vol. III, p. 399.

Nobile, A., Bhat, C. R., and Pas, E. I., *A Random Effects Multinomial Probit Model of Car Ownership Choice*, Vol. III, p. 419.

Rossi, P. E., McCulloch, R. E., and Allenby, G. M., *Hierarchical Modeling of Consumer Heterogeneity: An Application to Target Marketing*, Vol. II, p. 323.

Medicine

Carlin, B. P., Xia, H., Devine, O., Tolbert, P., and Mulholland, J., *Spatio-Temporal Hierarchical Models for Analyzing Atlanta Pediatric Asthma ER Visit Rates*, Vol. IV, p. 303.

Clyde, M., Muller, P. and Parmigiani, G., *Optimal Design for Heart Defibrillators*, Vol. II, p. 278.

Craig, B. A., Newton, M. A., *Modeling the History of Diabetic Retinopathy*, Vol. III, p. 305.

Crawford, S. L., Johnson, W. G., and Laird, N. M., *Bayes Analysis of Model-Based Methods for Non-ignorable Nonresponse in the Harvard Medical Practice Survey*, Vol. I, p. 78.

Crawford, S. L., Tennstedt, S. L., and McKinlay, J. B., *Longitudinal Care Patterns for Disabled Elders: A Bayesian Analysis of Missing Data*, Vol. II, p. 293.

Flournoy, N., *A Clinical Experiment in Bone Marrow Transplantation: Estimating a Percentage Point of a Quantal Response Curve*, Vol. I, p. 324.

Normand, S-L. T., Glickman, M. E., and Ryan, T. J., *Modeling Mortality Rates for Elderly Heart Attack Patients: Profiling Hospitals in the Cooperative Cardiovascular Project*, Vol. III, p. 155

Parmigiani, G., and Kamlet, M. S., *A Cost-Utility Analysis of Alternative Strategies in Screening for Breast Cancer*, Vol. I, p. 390.

Meteorology

Smith, R. L., and Robinson, P. J., *A Bayesian Approach to the Modeling of Spatial-Temporal Precipitation Data*, Vol. III, p. 237.

Neurophysiology

Genovese, C. R., and Sweeney, J. A., *Functional Connectivity in the Cortical Circuits Subserving Eye Movements*, Vol. IV, p. 59.

West, M., and Guoliang Cao, *Assessing Mechanisms of Neural Synaptic Activity*, Vol. I, p. 416.

Oncology/Genetics

Adak, S., and Sarkar, A., *Longitudinal Modeling of the Side Effects of Radiation Therapy*, Vol. IV, p. 269.

Iversen, E. S. Jr., Parmagiani, G., and Berry, D. A., *Validating Bayesian Prediction Models: A Case Study in Genetic Susceptibility to Breast Cancer*, Vol. IV, p. 321.

Palmer, J. L., and Muller, P., *Population Models for Hematologic Models*, Vol. IV, p. 355.

Parmigiani, G., Berry, D. A., Iversen, Jr., E. S., Muller, P., Schildkraut, J. M., and Winer, E. P., *Modeling Risk of Breast Cancer and Decisions about Genetic Testing*, Vol. IV, p. 133.

Slate, E. H., and Clark, L. C. *Using PSA to Detect Prostate Cancer Onset: An Application of Bayesian Retrospective and Prospective Changepoint Identification*, Vol. IV, p. 395.

Slate, E. H., and Crohin, K. A., *Changepoint Modeling of Longitudinal PSA as a Biomarker for Prostate Cancer*, Vol. III, p. 435.

Pharmacology/Pharmacokinetics

Paddock, S., West, M., Young, S. S., and Clyde, M., *Mixture Models in the Exploration of Structure–Activity Relationships in Drug Design*, Vol. IV, p. 339.

Wakefield, J., Aarons, L., and Racine-Poon, A., *The Bayesian Approach to Population Pharmacokinetic/Pharmacodynamic Modeling*, Vol. IV, p. 205.

Physiology

Brown, E. N., and Sapirstein, A., *A Bayesian Model for Organ Blood Flow Measurement with Colored Microspheres*, Vol. II, p. 1.

Political Science

Bernardo, J. M., *Probing Public Opinion: The State of Valencia*, Vol. III, p. 3.

Psychiatry

Erkanli, A., Soyer, R., and Stangl, D., *Hierarchical Bayesian Analysis for Prevalence Estimation*, Vol. III, p. 325.

Public Health

Aguilar, O., and West, M., *Analysis of Hospital Quality Monitors Using Hierarchical Time Series Models*, Vol. IV, p. 287.

Belin, T. R., Elashoff, R. M., Leung, K-M., Nisenbaum, R., Bastani, R., Nasseri, K., and Maxwell, A., *Combining Information from Multiple Sources in the Analysis of a Non-Equivalent Control Group Design*, Vol. II, p. 241.

Blattenberger, G., and Fowles, R., *Road Closure: Combining Data and Expert Opinion*, Vol. II, p. 261.

Carlin, B. P., Chaloner, K. M., Louis, T. A., and Rhame, F. S., *Elicitation, Monitoring, and Analysis for an AIDS Clinical Trial*, Vol. II, p. 48.

Malec, D., Sedransk, J., and Tompkins, L., *Bayesian Predictive Inference for Small Areas for Binary Variables in the National Health Interview Survey*, Vol. I, p. 377.

Radiology

Johnson, V., Bowsher, J., Jaszczak, R., and Turkington, T., *Analysis and Reconstruction of Medical Images Using Prior Information*, Vol. II, p. 149.

REFERENCES

Bretthorst, G, Larry (1988), *Bayesian Spectrum Analysis & Parameter Estimation*, Lecture Notes in Statistics, Vol. 48, New York: Springer-Verlag.

Buck, Caitlin E.; William G. Cavanagh, and Clifford D. Litton (1996). *Bayesian Approach to Interpreting Archaeological Data*, New York: John Wiley & Sons.

Dorfman, Jeffrey H. (1997). *Bayesian Economics Through Numerical Methods: A Guide to Econometrics and Decision-Making with Prior Information*, New York: Springer Verlag.

Gelfand, A.E. and A. F. M. Smith (1990). "Sampling Based Approaches to Calculating Marginal Densities," *Journal of the American Statistical Association*, **85**: 398–409.

Geman, S., and D. Geman (1984). "Stochastic Relaxation, Gibbs Distributions and the Bayesian Resoration of Images," *IEEE Transactions and Pattern Analysis of Machine Intelligence*," **6**: 721–741.

Hulbert, Mark (1999). "Are Fund Managers Irrelevant? An 18th Century Theory Suggests Not," *The New York Times*, Oct. 10 1999, p.26, Business.

Iba, Yukito (2000). *Bayesian Statistics, Statistical Physics, & Monte Carlo Methods*, Ph.D. Thesis, Japan.

Jaynes, E. T. (1983). *Papers on Probability, Statistics, and Statistical Physics of E.T. Jaynes*, R. D. Rosenkrantz (Ed.) Dordrecht, Holland: D. Reidel Publishing Co.

Kluwer Academic Publishers (1982). Smith, C. Ray, and Gary J. Erickson (Eds.). *Maximum-Entropy and Bayesian Spectral Analysis and Estimation Problems.*

Kluwer Academic Publishers (1991). *Maximum Entropy and Bayesian Methods, Proceedings of the Eleventh International Workshop on Maximum Entropy and Bayesian Methods, Seattle, Wash.*

Kluwer Academic Publishers (1992). *Proceedings of the Twelfth International Workshop on Maximum Entropy and Bayesian Methods, Paris, France*, A. Mohammad-Djafari, and G. Demoment (Eds.).

Kluwer Academic Publishers (1994). *Proceedings of the Fourteenth International Workshop on Maximum Entropy and Bayesian Methods, Cambridge, England*, John Skilling and Sibusiso Sibisi (Eds.).

Kluwer Academic Publishers (1995). *Proceedings of the Fifteenth International Workshop on Maximum Entropy and Bayesian Methods, Santa Fe, New Mexico*, Kenneth M. Hanson and Richard N. Silver (Eds.)

Kluwer Academic Publishers (1997). *Proceedings of the 17th International Workshop on Maximum Entropy and Bayesian methods. Boise, Idaho*, Gary J. Erickson, Joshua T. Rychert, and C. Ray Smith (Eds.)

Kluwer Academic Publishers (1999). *Maximum Entropy and Bayesian Methods, Proceedings of the 18th International Workshop on Maximum Entropy and Bayesian Methods.*

Lad, Frank (1996). *Operational Subjective Statistical Methods: A Mathematical, Philosophical, and Historical Introduction*, New York: Wiley Interscience.

Pollard, William E. (1986). *Bayesian Statistics for Evaluation Research: An Introduction*, Beverly Hills, CA: Sage Publications.

Ruanaidh, Joseph J. K. O., and William J. Fitzgerald (1996). *Numerical Bayesian Methods Applied to Signal Processing*. New York: Springer Verlag.

Sander, P., and R. Badoux (Eds.) (1991). *Bayesian Methods in Reliability*, Dordrecht, Netherlands and Boston, MA: Kluwer Academic Publishers.

Smith, C. Ray (Ed.) (1985). *Maximum-Entropy and Bayesian Methods in Inverse Problems (Fundamental Theories of Physics)*, 14, Dordrecht, Netherlands: D. Reidel Pub Co.

Stone, Lawrence D., Carl A. Barlow, and Thomas L. Corwin (1999). *Bayesian Multiple Target Tracking*, Artech House Radar Library.

Zellner, Arnold (1971). *An Introduction to Bayesian Inference in Econometrics*, New York: John Wiley and Sons, Inc.

APPENDIX 6

Selecting the Bayesian Hall of Fame

Who are the people who have been most responsible for developing the Bayesian approach to statistical science? A question like this, while fascinating, is difficult to answer, especially for any given person. So I decided to ask a panel of experts this question, and to form a composite of their responses. I recognize that there will always be differences of opinion, even among experts, depending not only upon how experts differ in how they evaluate the contributions of particular scientists, but also upon the assumptions they make in answering such a question. Moreover, different panels of experts might well differ in their opinions. Regardless, I polled some *experts*.

For the purposes of the poll, I defined *experts* as very senior Bayesian researchers who have worked in the field of Bayesian statistical science for many years, including people from different disciplines, and from different countries. Issues of concern to me included:

1. Should opinions of contemporary Bayesian researchers be weighted equally with those of Bayesian researchers who made their major contributions in earlier years?
2. Should the opinions of all experts be weighted equally regardless of age and degree of seniority?
3. Should researchers who have contributed mostly to the foundational theory of the field be weighted equally with researchers who have contributed mostly to the applications of the field?
4. Should researchers of the last few decades be excluded as possible choices since respondents might not have an appropriate historical perspective on the significance of their contributions?
5. Should personal biases of respondents be taken into account, and if so, how?
6. Should relationships among respondents be considered and compensated for in some way?

Many other questions could be raised as well. I decided not to let any such considerations be viewed as constraints in the poll. I therefore permitted respondents to make whatever assumptions they wished, except that they were instructed not to list themselves as a possible choice.

I formed a small panel of some senior members of the international Bayesian community. I did this by creating a list of some 60 Bayesian scientists whom I considered experts, and then sampled 30 randomly from this list. The list included both males and females; people from the fields of business, computer science, econometrics, physics, psychology, and statistics; and from seven countries: Brazil, Japan, The Netherlands, South Africa, Spain, the United Kingdom, and the USA. In total, 24 scientists responded and constituted the expert panel. I asked the members of this panel independently, by e-mail the question:

> I am trying to formulate a list of 12 people who constitute the most important contributors to the development of Bayesian statistics, from the earliest of times, to today, not including yourself.

Note that the respondents on the panel were not asked to rank their choices in any order of importance. Moreover, this was not a sample survey using probability sampling; it was a "sense vote" of the opinions of experts. There were 24 panelists, all of whom turned out to be university professors, but the original list of potential respondents included some nonuniversity scientists. I asked this question in two waves of respondents: the first wave started in October, 2001; the second wave started in January, 2002. Responses obtained after February 22, 2002, were not included. The reason for carrying out the poll in two waves was to see whether the cumulative results would change very much after the first wave; they did not. The respondents included the following people (Table A6.1).

Table A6.1 People who Selected the Bayesian Hall of Fame

Bayarri, Susie (Spain)	Berger, James (USA)	Bernardo, Jose (Spain)
Berry, Donald (USA)	Chaloner, Kathryn (USA)	Dawid, Philip (UK)
De Waal, Daan (South Africa)	Fienberg, Stephen (USA)	George, Edward (USA)
Geweke, John (USA)	Ghosh, Malay (USA)	Johnson, Wesley (USA)
Kadane, Joseph (Jay) (USA)	Klein, Ruben (Brazil)	Lindley, Dennis V. (UK)
O'Hagen, Anthony (UK)	Poirier, Dale (USA)	Press, S. James (USA)
Shigemasu, Kazuo (Japan)	Singpurwalla, Noser (USA)	Van Der Merwe, Abrie (South Africa)
Van Dijk, Herman (The Netherlands)	West, Mike (USA)	Zellner, Arnold (USA)

Most respondents provided the 12 choices requested, but a couple gave fewer than 12 choices and a few gave more than 12 (sometimes giving a grouping of two or three people). All names mentioned in their choices were counted equally. The 24 scientists cast a total of 287 votes. Other very worthy expert panelists could have been added to increase the panel size, and we regret that because of publication time constraints, some other, very worthy, experts were not included on the panel.

I had really not intended necessarily to end up with 12 scientists in the Hall of Fame, but had merely suggested in the question that respondents provide 12 names each so that the panelists would be encouraged to examine a broad collection of possible candidates. People who were selected by more than 75 percent of the expert panelists (at least 19 votes from the 24 experts) constitute the Hall of Fame. It turned out that just six people met this criterion. (Table No. 2). Their portraits are presented following the Table of Contents.

Table A6.2 Bayesian Hall of Fame

Bayes, Thomas	De Finetti, Bruno	DeGroot, Morris H.
Jeffreys, Harold	Lindley, Dennis V.	Savage, L. J.

There were five other scientists who were deemed worthy to receive *honorable mention*; they were selected by more than 50 percent of the respondents, but not by more than 75 percent (Table A6.3).

Table A6.3 Honorable Mention

Berger, James O.	Box, George E. P.	Laplace, Pierre Simone de
Smith, Adrian F. M.	Zellner, Arnold	

There were many other people selected by 1–10 of the panelists (there were 49 such people, 30 of whom received just one vote). Such choices were not included in our analysis.

We have no way of knowing, of course, whether other panels of experts (possibly larger and differently constituted panels, and perhaps using other criteria for selection) might arrive at a somewhat different collection of most worthy contributors to the field. Our two-wave study showed, however, that the results of this poll changed only very slightly from the first wave of respondents to the second. We therefore believe that most of the members of this Hall of Fame would be included in other such polls.

APPENDIX 7

Solutions to Selected Exercises*

CHAPTER 1

1.2 Stigler's speculation is that Nicholas Saunderson was the person who wrote Bayes' theorem. The conclusion of the article is not definitive. However, Stigler approaches the issue from an interesting perspective. He claims that as with most eponymous theorems, inequalities, and famous formulae that are named after a person, or a group of people, they are often mistakenly named (due to historical inaccuracies in crediting the right author); so must it be with Bayes' theorem, and it is just a matter of time until the truth will surface, and the identity of the person who actually wrote Bayes' theorem will be discovered.

The key proposition in Bayes' paper is Proposition 9. Here the problem solved can be expressed (in modern terms) as follows.

Given n Bernoulli trials with probability π of success on a single trial, and of s trials resulting in a success, find $P(a < \pi < b)$. Proposition 9 states:

$$P\{a < \pi < b \mid n, s, a, b\} = \frac{\int_a^b u^s(1-u)^{n-s}du}{\int_0^1 u^s(1-u)^{n-s}du}.$$

Here, Bayes is assuming a binomial problem with the parameter π unknown. Also, he uses what would be described as a vague prior for π. Bayes justified this using equivalent reasoning to what is today called Laplace's Principle of Insufficient Reason.

Since Bayes' time this has been generalized and extended greatly.

* Many solutions presented in this appendix were developed jointly with my former students: Dr. Thomas Ferryman, Dr. Mahmood Ghamsary, and Dawn Kummer.

1.4 In the book *Ars Conjectandi* the development of the binomial theorem is presented, as well as the rules for permutations and combinations. It is of particular interest because it poses the problem of inverse probability which Bayes attempted to solve 50 years later.

1.8 The statement refers to the countless judgments and decisions that scientists must make without rigid guidelines controlling those decisions. This includes decisions like selecting the sample size, a level (type I error), likelihood model (i.e., normal, Cauchy, ... and others). Selecting a prior is one more of these subjective judgments and does not represent a departure from good scientific investigation.

CHAPTER 2

2.2 Distinguish between finite and countable additivity and explain the importance of the difference.

Kolmogorov's axiom system included "countable additivity" as one of the key elements:

$$P\left\{\bigcup_{i=1}^{\infty} A_i\right\} = \sum_{i=1}^{\infty} P\{A_i\}.$$

This allows for an infinite number of disjoint events A_i.

Ramsey, Savage, de Finetti, and others argued that it was inappropriate to consider an infinite number of events, and that only a finite number should be considered because only a finite number of events are encountered in practice. As a result finite additivity axioms systems have the key element:

$$P\left\{\bigcup_{i=1}^{K} A_i\right\} = \sum_{i=1}^{K} P\{A_i\}.$$

Finite additivity reflects the logical concept that any event that is possible will have a nonzero probability. Countable additivity requires special allowance for nonmeasurable events, which cannot have a probability assigned to them. Also, some possible events may have probability zero (as in a specific value from a continuous distribution). A significant advantage of countable additivity is its mathematical tractability of being able to use asymptotic theory, continuity, and differentiability.

2.3 Explain the importance of the Rényi axiom system of probability for Bayesian inference.

The Rényi axiom system uses countable additivity and its structure is based on conditional probability from the beginning. Also, the Rényi axiom system assumes probabilities are assigned subjectively. This axiom system allows the use of continuity and asymptotic theory. It includes the Kolmogoroff axiom system as a special case. The use of conditional probability in the axioms readily allows the application of Bayes' theorem and the use of priors, including improper priors.

The Rényi axiom system of probability relies on the following axioms:

1. For any events A and B:

$$P\{A \mid B\} \geq 0; \; P\{B \mid B\} = 1.$$

2. For disjoint events $A_1, A_2, \ldots$ and some event B,

$$P\left\{ \bigcup_{i=1}^{\infty} A_i \mid B \right\} = \sum_{i=1}^{\infty} P\{A_i \mid B\}$$

3. For every event collection (A, B, C), such that $B \subseteq C, P\{B \mid C\} > 0$, we have

$$P\{A \mid B\} = \frac{P\{(A \cap B) \mid C\}}{P\{B \mid C\}}.$$

Note that since Rényi adopts a countable additivity approach, all the properties of such axiom systems are present in his set of probability axioms.

The importance of the Rényi axiom system is twofold:

1. It makes the Kolmogoroff axiom system a special case in the following sense. The Rényi system allows for conditioning from the start, hence it allows the inclusion of prior information. Conditioning on Ω (the entire space) results in the Kolmogoroff axiom system.

It also allows for a subjective interpretation of the probability without ruling out the possibility of an objective interpretation.

2. The Rényi axiom system can accommodate probabilities on the entire real line (by faking the limits of certain ratios). The importance of this property stems from the Bayesian approach that allows probability distributions to spread their densities uniformly over the entire real line. In other axiom systems such a property is not allowed since it gives rise to improper integrals that upon integration do not integrate to 1.

2.4 Subjective probability is a measure of one's degree of belief. If a person's subjective probabilities are consistent with Rényi's axiom system, his probabilities have coherence. If his probabilities are inconsistent, then he is incoherent.

Consider, for example, a six-sided die that is possibly unfairly weighted.

1. If I thought the probability of a 6 was $1/2$ and the other sides were equally likely with probability $1/10$, I, I am coherent.

2. If I thought the probability of an even number was $2/3$ and $P\{1\} = 0.1, P\{3\} = 0.2$ and $P\{5\} = 0.3$, then I am incoherent (since the probabilities of the mutually exclusive events add to more than 1, which is a violation of the axiom system).

3. If I thought the probability of an odd number was 1/2 and the probability of a 1 or 3 was 2/3, then I am incoherent (since the event {odd number} is a superset of event {1 or 3}, but the probability is less).

One way to take advantage of incoherence is to create a "dutch book"; that is to create a set of bets that appear fair based on the subjective probabilities, but will result in the person losing regardless of the outcome. Consider the following example:

Tim feels $P\{E\} = 3/4$, and $P\{\bar{E}\} = 1/2$. Create the following lotteries:

Lottery I	bet A:	bet \$3 on E to win \$1
	bet B:	bet \$1 on $\bar{E}$ to win \$3
Lottery II	bet C:	bet \$2 on $\bar{E}$ to win \$2
	bet D:	bet \$2 on E to win \$2

Tim thinks bet A is fair since he believes $P\{E\} = 3/4$, and $E(X) = (1) \times 3/4 + (-3) \times 1/4 = 0$.

Tim also thinks bet C is fair since he believes $P\{\bar{E}\} = 1/2$, and $E(X) = (2) \times 1/2 + (-2) \times 1/2 = 0$.

If Tim makes both bets, he will lose money. If E occurs he makes \$1 on bet A and loses \$2 on bet C, net −\$1. If $\bar{E}$ occurs he loses \$3 on bet A and makes \$2 on bet C, net −\$1. Therefore, Tim is incoherent.

2.5 How would you operationalize your assertion that the probability of rain tomorrow in Riverside, California, is 0.7?

The statement should reflect indifference in choosing between the following options:

1. Receiving \$1 if it rains tomorrow in Riverside, California, and \$0 otherwise.
2. Receiving \$1 if a black ball is randomly drawn from an urn containing 70 black balls and 30 whites ones, and receiving \$0 otherwise.

If after viewing these options, the speaker wanted to change the ratio of the black marbles to total marbles in the urn before he was indifferent between the options, it would result in changing the 70 percent probability in the original statement to a probability equal to the new ratio.

2.6 The probability that event E occurs could be based on the idea that in a long series of trials the portion of trials in which E occurs is $P\{E\}$. This is a fundamental concept, but it is difficult to implement in reality.

For example, consider a roulette wheel (which may be unbalanced). Estimate by the long-run frequency approach the probability of the event that the ball rests in the "9" slot. Spin the wheel and record if a 9 comes up or not. After a large number of spins the cumulative portion may be close to the true value, but it has some likelihood that it will not. Consider a truly fair roulette wheel with probability of a 9 equal to $1/38 = 0.0263$ (there is a 0, and a 00, as well as all of the numbers 1–36 on a roulette wheel). It is possible that after 100 spins on 9s will occur. This will occur 7 percent of the time. But this means that if 100 was considered a large number then the long-run probability of 9 must be set at 0.0. If 10,000 spins is considered a large number then the possibility of still setting $P\{9\} = 0.0$ is quite real, almost 1 percent. To get the estimated $P\{9\}$ to within a small value, one must spin the wheel a very large number of times. Then one only has the probability associated with this wheel, at this time, under these conditions, with this ball.

Applying the probability to other roulette wheels, conditions, and so on, is not supported by the long-run frequency approach. The long-run frequency approach, by itself, is not practical.

2.8 A frequency interpretation of probability must have the experiment repeated a number of times to allow a calculation of the empirical results to occur. Clearly, in all of time, there will be only one calendar date corresponding to tomorrow's date, and only one trial can be run to see whether it rains. The experiment is not repeatable.

2.10 Savage's axiom system includes the three axioms associated with a finite additivity axiom system and the seven axioms that may be found in the Complement to Chapter 2. The axioms associated with a finite additivity axiom system are:

1. $0 \leq P\{A\}$;
2. $P\{\text{all possible events}\} = 1$.
3. If $A_1, \ldots, A_n, \ldots$ are mutually exclusive events, then:

$$P\left\{\bigcup_{i=1}^{K} A_i\right\} = \sum_{i=1}^{K} P\{A_i\}.$$

Savage's axiom system is used as a basis for study and development of rational decision-making theory. This is a normative system, that is, it focuses on how people should behave and should make decisions, not on how they actually behave in making decisions (empirical theory). Savage's axiom system implies the existence of a utility function, such that:

1. If f and g are bounded, then f is not preferred to g if and only if the utility of f is not greater than the utility of g.
2. If f and g are bounded, and $P\{B\} > 0$, then f is not preferred to g, given B, *if* and *only if*:

$$E\{[U(g) - U(f)] \mid B\} \geq 0.$$

2.11 For any decision situation, there is a set of actions that can be taken for every event. Those actions result in consequences. A utility function is a real-valued function that assigns a real number to every consequence, such that the higher the number the more desirable the consequence. Utility theory is the study of decision making using these numbers. Often decision criteria are based on the utility numbers, such as, maximizing the expected utility for a given decision.

2.12 When Rényi's axiom system is used conditional on the entire space then the system reduces to Kolmogoroff's system. As such, Kolmogoroff's axiom system can be thought of as a special case of Rényi's system. The Kolmogoroff axiom system is the defining system of mathematical probabilities. Moreover, if the probabilities given in this system are interpreted as mathematical frequencies instead of personal degrees of belief, the system becomes a special case of the Rényi system (see also the solution to Exercise 2.3).

2.13 Normative theory relates to how people should make decisions, that is, what are reasonable and logical methods. Empirical theory deals with how people make decisions in actual real world situations. Empirical theory is often studied by psychologists, sociologists, and so on, to determine the impact on decision making of emotions, perceptions, etc. (see also the solution to Exercise 2.10).

CHAPTER 3

3.1 The likelihood principle says: Suppose $x = (x_1, \ldots, x_n), y = (y_1, \ldots, y_n)$, and x and y are distinct sets of observables, and $L(x \mid \theta) \propto L(y \mid \theta)$. Then, the MLEs are the same, and for the same prior, the posteriors are the same. The implication of this is that all information about the parameters from an experiment is contained in the likelihood function. Furthermore, consideration of data that was not collected is a violation of the likelihood principle. (For instance, one should not calculate admissibility over the full range of possible unobserved data. Rather, one should calculate it based only on the observed data.)

The moral of the likelihood principle could be summarized as: "base decisions on results that *did* occur, not on those that might have occurred."

3.8 Unbiasedness of an estimator $\hat{\theta}(x)$ is, by definition,

$$\theta = E(\hat{\theta}) = \int_{-\infty}^{+\infty} \hat{\theta} f(x \mid \theta) dx,$$

where $\hat{\theta}$ is a statistic based on x. This means integrating over values of x that have not occurred. This is a violation of the likelihood principle, which says information about the parameters must come only from data actually collected, not from data that might have been collected.

CHAPTER 4

4.1 A "prior distribution" is the subjective probability distribution assigned to an unknown quantity. It represents someone's set of degrees-of-belief about that quantity prior to observing any data in an experiment that may be performed. The assignment occurs before an experiment and the data collection. (This does not exclude data from previous experiments from being considered in assessing the subjective probability distribution.)

An experiment is conducted. Data is collected. A likelihood model is selected based upon the observed data.

A "posterior distribution" represents someone's set of degrees-of-belief about an unknown quantity subsequent to observing data that bear on its value. It is an updated belief distribution, which reflects the degree of increase in someone's knowledge regarding the unknown quantity. The posterior distribution is calculated from Bayes' theorem.

CHAPTER 5

5.9 What is meant by a "public policy prior," which is also called an "objective prior"?

After an experiment is conducted and the results are reviewed by a group of people, it is sometimes important that all members come to the same conclusion. When this is the case, either all members of the group must have the same prior or agree to use a common prior. The adoption of a common prior for the sake of ensuring harmonious results is a "public policy prior." If the group cannot agree on a common informative prior, a vague prior is often used, following Laplace's principle of insufficient reason.

In lieu of the group adopting a common prior, the statistician could provide a customized posterior for each person's prior, then some type of weighted average of the individual priors could be adopted (linear, log-linear, etc.).

5.16 The forming of a prior is necessary and appropriate for Bayesian statistical analysis. The prior reflects the summary of all available information regarding the quantity of interest prior to conducting the experiment. If one has little or no knowledge as to the value of a quantity of interest, the principle of insufficient reason tells us to adopt a prior that implies that all values are equally likely.

This is especially useful when working with issues of public policy. When a large number of people, each with his/her own prior, view an issue, the statistician can:

1. Generate a posterior for each person's prior.
2. Generate a mapping of various priors to posteriors by generating a posterior for each of a number of priors, distributed over the range of

the priors of the people viewing an issue. Thus, allowing each viewer to select the closest prior to his/hers and finding the posterior so indicated, . . .

3. Use Laplace's "principle of insufficient reason" and express a prior with little or no information; then generate its posterior.

5.27 It would be unlikely to change things very much. The fitted densities tend to be insensitive to the form of the kernel.

5.28 An expert is a person with a high degree of skill or specialized knowledge of the topic at hand. His/her opinions are carefully thought out and expressed with their commitment that the opinions come after careful research and analysis.

Experts could be defined by self-designation, that is, if someone says he is an expert, believe him/her. This is a "substantive matter expert," but that does not mean the individual understands how to express their expert opinions quantitatively. For example, a geologist expert might know locations where oil is likely to be found, but the geologist might not be able to quantify his belief about how much oil might be found in a given location.

5.29 Suppose we want to assess a prior, say $\theta_{(p\times 1)} \sim N(\phi, \Sigma)$. We could ask each expert to provide their best guesses for each of p questions and use the average of the best guesses for each of the p questions to estimate ϕ. To estimate Σ we could ask people for fractiles that would provide measures of variation on p questions.

5.30 As with any statistical analysis, the mean is greatly affected by the magnitude of the data, while the median is not. Since the mean is much greater than the median in all of Figures 5.2–5.4, this implies that the distributions are skewed to the right with very long tails. It specifically implies that there are a few subjects with relatively very large assessments of nuclear detonation as compared with the others.

5.31 Many people do not perceive the advent of a nuclear war in the near future as a likely event since most people do not have any previous experience with occurrences of a nuclear event. Very few people alive today (outside of Japan) have ever experienced nuclear war or have known an individual who has. In these situations, most people tend to underassess this probability. This is known as an availability heuristic bias.

5.32 I assume that:

1. There are about 30,000,000 people in California;
2. On average each person eats 15 chickens per year (this is about 1/3 of a chicken per week).
3. A chicken takes 1 year to mature.
4. California imports and exports the same number of chickens.

Therefore, my prior is centered about 450,000,000. I am very uncertain about this and will assume a standard deviation of 100,000,000, which reflects my opinion about the quality of my assumptions. I assume the likelihood model is normal.

If I wanted to improve the prior, I could do a little research about the chicken industry and California and come up with a much better prior by estimating the number of chickens based on numbers alive in past years, or even by just updating the numbers in the method above, or any of hundreds of other methods.

5.33 A natural conjugate distribution is one that has the property that when its pdf or pmf is multiplied by the associated likelihood function, it will result in a new distribution that is a member of the same family of distributions as the prior.

5.34 If X given λ follows a Poisson distribution, a natural conjugate prior for λ:

$$\lambda \sim \gamma(\alpha, \beta), \qquad \alpha > 0, \qquad \beta > 0,$$

or, equivalently, for $I(\cdot)$ denoting the indicator function,

$$g(\lambda) = \frac{\lambda^{\alpha-1} e^{-\lambda/\beta}}{\beta^{\alpha}\Gamma(\alpha)} I(\lambda > 0).$$

This result can be deduced by interchanging the roles of the data and parameters, normalizing the new density, and enriching it with general hyperparameters. A natural conjugate prior will result in the posterior having the same form as the prior.

5.35 The Kadane et al. method uses predictive distributions to have subjects assess observable quantities. Then the procedure suggests that we work backwards to impute the prior that the observation implied.

5.36 The density of $\boldsymbol{X}$ is:

$$f(\boldsymbol{x} \mid \theta) \propto \exp\left\{ -0.5 \sum_{i=1}^{p} (\boldsymbol{x}_i - \theta_i)'(\boldsymbol{x}_i - \theta_i)\right\}.$$

Interchanging the roles of $\boldsymbol{X}$ and θ shows that an appropriate natural conjugate family is

$$\theta \sim N(a, A),$$

where $\boldsymbol{a}$ denotes a p-vector of means, and $\boldsymbol{A}$ denotes a $p \times p$ covariance matrix.

5.37 $\mathcal{L}(X \mid \Lambda) = N(\boldsymbol{0}, \boldsymbol{\Sigma})$, with density

$$f(\boldsymbol{x} \mid \Sigma) \propto \mid \Sigma^{-0.5} \mid \exp\{-0.5\boldsymbol{x}'\Sigma^{-1}\boldsymbol{x}\}.$$

The likelihood function for n observations is given by:

$$L(\Sigma) = | \Sigma^{-0.5n} | \exp\left\{ - 0.5 \sum_{i=1}^{n} \boldsymbol{x}_i' \Sigma^{-1} \boldsymbol{x}_i \right\}.$$

Letting $V \equiv \sum_{i=1}^{n} \boldsymbol{x}_i \boldsymbol{x}_i'$, the likelihood function becomes:

$$L(\Sigma) = | \Sigma^{-1} |^{0.5n} \exp\{-0.5 \operatorname{tr} \Sigma^{-1} \boldsymbol{V}\}.$$

If we interchange the roles of Σ and the data, $\boldsymbol{V}$, we observe that Σ has an inverted Wishart density, so Λ has a Wishart density. Now enrich the Wishart density by generalizing the parameters $\boldsymbol{V}$ and n to parameters $\boldsymbol{G}$ and m, which do not depend upon the data. Then, the natural conjugate prior density family for Λ is given by:

$$g(\Lambda) \propto | \Lambda |^{0.5(m-p-1)} \exp\{-0.5 \operatorname{tr} \Lambda \boldsymbol{G}\}$$

That is, $\Lambda \sim W(\boldsymbol{G}, p, m), p \leq m$, where $\boldsymbol{G}$ denotes a positive definite matrix of dimension p with m degrees of freedom.

5.38

$$p(\boldsymbol{X} \mid \theta, \Sigma) \propto | \Sigma |^{-0.5n} \exp\{-0.5(\boldsymbol{X} - \theta)' \Sigma^{-1} (\boldsymbol{X} - \theta)\}.$$

$$p(\theta, \Lambda) = p_1(\theta \mid \Lambda) p_2(\Lambda),$$

$$\theta \mid \Lambda \sim N(\mu, c\Lambda^{-1}) \qquad \text{and} \qquad \Lambda \sim W(G, p, m).$$

But

$$p(\theta \mid \Lambda) \propto \frac{1}{| c\Lambda^{-1} |^{0.5}} \exp\left\{-0.5(\underline{\theta} - \underline{\mu})' \frac{A}{c} (\underline{\theta} - \underline{\mu})\right\},$$

and

$$p(\Lambda) \propto | \Lambda |^{0.5(m-p-1)} e^{-0.5 \operatorname{tr} G^{-1} \Lambda},$$

so wc havc

$$p(\theta, \Lambda) \propto | \Lambda |^{0.5(m-p-1)} \exp\{-0.5 \operatorname{tr} \Lambda [G^{-1} + (\underline{\theta} - \underline{\mu})' \Lambda (\underline{\theta} - \underline{\mu})]\}.$$

CHAPTER 6

6.5 For solution, see Section 6.3.1.

CHAPTER 7

7.1 Natural conjugate prior is a gamma with parameters a and b and density

$$g(\beta) = \frac{1}{\Gamma(a)b^a}\,\beta^{a-1}e^{-b\beta}, \qquad \beta > 0.$$

Likelihood: The likelihood is given by

$$L(x_1, \ldots, x_n \mid \beta) = \beta^n e^{-\beta \sum x_i},$$

and the log-likelihood is given by

$$l(\beta) = n \log \beta - \beta \textstyle\sum x_i.$$

(a) The posterior density will be

$$h(\beta \mid x) = \frac{f(x \mid \beta)g(\beta)}{\int f(x \mid \beta)g(\beta)d\beta} \propto \beta^{n+a-1}e^{-(b+\sum x_i)\beta},$$

which is a gamma distribution with parameters $n + a$ and $1/\beta + \sum x_i$, i.e.,

$$\beta \mid x \sim \gamma\left(n + a, \frac{1}{b + \sum X_i}\right).$$

(b) By differentiation of $l(\beta) = n\ell n\beta - \beta \sum x_i$, we get $(\partial/\partial\beta)l(\beta) = n/\beta - \sum x_i$. We then set it equal to 0 to get $\hat{\beta} = 1/\hat{x}$.
Now evaluate the second derivative at β and we get

$$\frac{\partial^2}{\partial\beta^2}l(\beta) = -\frac{n}{\beta^2}\bigg|_{\hat{\beta}=1/x} \Rightarrow \Lambda = \frac{n}{1/\tilde{x}^2} = n\tilde{x}^2.$$

Take $n\tilde{x}^2$ to equal the reciprocal variance in the asymptotic normal distribution. By Theorem 7.1, $(\beta - \hat{\beta})/\sqrt{\hat{\Lambda}} \mid \text{data} \rightarrow N(0, 1)$ or equivalently $\beta \mid \text{data} \rightarrow N(1/\tilde{x}, 1/n\tilde{x}^2)$, which means the asymptotic posterior distribution of β is normal, and centered at $1/\tilde{x}$, with variance $1/(n\tilde{x}^2)$. It is interesting to note that this distribution *does not depend on the prior*, that is, in large samples, the data totally dominate the prior beliefs.

7.2 Let $\mu(\beta) = \beta$. Then $\partial\mu/\partial\beta = 1$, $\partial^2\mu/\partial\beta^2 = 0$. From Exercise 7.1 we have $\hat{\beta} = 1/\tilde{x}$ and $\hat{\sigma}^2 = 1/n\tilde{x}^2$, and $g(\beta) \propto \beta^a e^{-b\beta}$ which means $\rho(\beta) = \log c + (a - 1)\ln\beta - b\beta$.

Also

$$\frac{\partial \rho}{\partial \beta} = \frac{a-1}{\beta} - b \bigg|_{\hat{\beta}=1/\tilde{x}} = (a-1)\tilde{x} - b \quad \text{and} \quad \frac{\partial^3 \ell}{\partial \beta^3} = \frac{2n}{\beta^3}\bigg|_{\hat{\beta}=1/\tilde{x}} = 2n\tilde{x}^3.$$

Since the dimension of the parameter space is only 1, we use Equation 7.4. After some simplification we get

$$\mathcal{I}(x_1, \ldots, x_n) \cong \mu(\hat{\beta}) + \left\{ \frac{\partial \rho}{\partial \beta}\bigg|_{\hat{\beta}=1/\tilde{x}} \right\} \hat{\sigma}^2 + 0.5 \left\{ \frac{\partial^3 \ell}{\partial \beta^3}\bigg|_{\hat{\beta}=\frac{1}{\tilde{x}}} \right\} \hat{\sigma}^4.$$

Substituting the above gives

$$E(\hat{\beta} \mid x_1, \ldots, x_n) \cong \frac{1}{\tilde{x}} + \{(a-1)\tilde{x} - b\} \frac{1}{n\tilde{x}^2} + 0.5(2n\tilde{x}^3)\left(\frac{1}{n\tilde{x}^2}\right)^2,$$

or equivalently

$$E(\hat{\beta} \mid x_1, \ldots, x_n) = \frac{1}{\tilde{x}} + \{(a-1)\tilde{x} - b\} \frac{1}{n\tilde{x}^2} + \frac{1}{n\tilde{x}}.$$

7.3 The log of the kernel of the likelihood and prior are given by

$$\ell(\beta) = n \log \beta - \beta \sum x_i,$$

and

$$\rho(\beta) = (a-1) \log \beta - b\beta,$$

respectively.

Also we have

$$n\mathcal{T}(\beta) = \ell(\beta) + \rho(\beta) = (n+a-1) \log \beta - \left(b + \sum x_i\right),$$

and from

$$\frac{\partial(n\mathcal{T}(\beta))}{\partial \beta} = \frac{n+a-1}{\beta} - \left(b + \sum x_i\right) = 0 \Rightarrow \hat{\beta} = \frac{n+a-1}{b + \sum x_i},$$

or

$$\hat{\beta} = \frac{\tilde{a}-1}{\tilde{b}},$$

where$\tilde{a} = n + a$ and $\tilde{b} = b + \sum x_i$. $\hat{\beta} = (\tilde{a} - 1)/\tilde{b}$ maximizes $n\mathcal{T}(\beta)$ since $\partial^2(n\mathcal{T}(n))/\partial\beta^2 = (n + a - 1)/\beta^2 < 0$.

Similarly

$$n\mathcal{T}^*(\beta) = \log\beta + \ell(\beta) + \rho(\beta) = (n + a)\log\beta - (b + \sum x_i),$$

and

$$\frac{\partial(n\mathcal{T}^*(\beta))}{\partial\beta} = \frac{n + a}{b} - (b + \sum x_i) = 0 \Rightarrow \hat{\beta}^* = \frac{n + a}{b + \sum x_i},$$

and

$$\hat{\beta}^* = \frac{n + a}{b + \sum x_i} = \frac{\tilde{a}}{\tilde{b}}$$

maximizes $n\mathcal{T}^*(\beta)$, since

$$\frac{\partial^2(n\mathcal{T}(\beta))}{\partial\beta^2} = -\frac{n + a}{\beta^2} < 0.$$

The maximum of $n\mathcal{T}(\beta)$ and $n\mathcal{T}^*(\beta)$ at $\hat{\beta}$ are given by

$$n\mathcal{T}(\hat{\beta}) = (\tilde{a} - 1)\log\frac{\tilde{a} - 1}{\tilde{b}} - (\tilde{a} - 1),$$

and

$$n\mathcal{T}^*(\hat{\beta}^*) = \tilde{a}\log\frac{\tilde{a}}{\tilde{b}} - \tilde{a},$$

respectively. The estimates of the variances are given by:

$$\sigma^{-2} = \left.\frac{\partial^2(n\mathcal{T}(\beta))}{\partial\beta^2}\right|_{\beta=\hat{\beta}} = \frac{\tilde{b}^2}{\tilde{a} - 1} \Rightarrow = \sigma^2\frac{\tilde{a} - 1}{\tilde{b}^2},$$

$$\sigma^{*-2} = \left.\frac{\partial^2(n\mathcal{T}^*(\beta))}{\partial\beta^2}\right|_{\beta=\hat{\beta}} = \frac{\tilde{b}^2}{\tilde{a}} \Rightarrow \sigma^2 = \frac{\tilde{a}}{\tilde{b}^2}.$$

So we have $E(\beta \mid x_1, \ldots, x_n) \cong I(\underline{x}) = (\sigma^*/\sigma)\exp\{n\mathcal{T}^*(\beta) - n\mathcal{T}(\beta)\}$. After substitution we get

$$E(\beta \mid x_1, \ldots, x_n) = \sqrt{\frac{\tilde{a}}{\tilde{a} - 1}}\exp\left\{\log\left(\frac{\tilde{a}}{\tilde{b}}\right)^{\tilde{a}} - \log\left(\frac{\tilde{a} - 1}{\tilde{b}}\right)^{\tilde{a}-1} - 1\right\}.$$

Similarly we can approximate $E(\beta^2 \mid x)$.

The only difference in the computation is $u(\beta) = \beta^2$.

$$nT^*(\beta) = \log \mu(\beta) + \ell(\beta) + \rho(\beta),$$
$$nT^*(\beta) = 2\log\beta + n\log\beta - \beta\sum x_i + (a-1)\log(\beta) - b\beta,$$
$$nT^*(\beta) = (n+a+1)\log\beta - (b + \sum x_i)\beta.$$
$$\frac{\partial}{\partial\beta}[nT^*(\beta)] = \frac{n+a+1}{\beta} - (b+\sum x_i) = 0 \Rightarrow \hat{\beta} = \frac{n+a+1}{\sum x_i + \tilde{b}} = \frac{\tilde{a}+1}{\tilde{b}},$$

where $\tilde{a} = n + a$ and $\tilde{b} = b + \sum x_i$.

The estimate of variance is given by

$$\sigma^{*-2} = -\frac{\partial^2}{\partial\beta^2}[nT^*(\beta)] = \frac{n+a+1}{\beta^2} \quad \text{at } \hat{\beta}^* = \frac{\tilde{a}+1}{\tilde{b}}.$$

$$\sigma^{*-2} = \frac{\tilde{b}^2}{\tilde{a}-1} \quad \text{or} \quad \sigma^* = \frac{\sqrt{\tilde{a}+1}}{\tilde{b}}.$$

But we want the second moment, which is given by

$$\tilde{E}(\beta^2 \mid \underline{x}) \cong \frac{\sigma^*}{\sigma} \exp\{nT^*(\beta) - nT(\beta)\}.$$
$$\tilde{E}(\beta^2 \mid \underline{x}) = \sqrt{\frac{\tilde{a}+1}{\tilde{a}-1}} \exp\left\{(\tilde{a}+1)\log\left(\frac{\tilde{a}+1}{\tilde{b}}\right) - (\tilde{a}+1) - (\tilde{a}-1)\log\left(\frac{\tilde{a}-1}{\tilde{b}}\right) - (\tilde{a}-1)\right\},$$
$$\tilde{E}(\beta^2 \mid \underline{x}) = \sqrt{\frac{\tilde{a}+1}{\tilde{a}-1}} \exp\left\{\log\left(\frac{\tilde{a}+1}{\tilde{b}}\right)^{\tilde{a}+1} - \log\left(\frac{\tilde{a}-1}{\tilde{b}}\right)^{\tilde{a}-1} - 2\tilde{a}\right\}.$$

The variance is given by $V(\hat{\beta} \mid x) = \hat{E}(\hat{\beta}^2 \mid x) - (\hat{E}(\hat{\beta} \mid x))^2$. Now

$$V(\hat{\beta} \mid x) = \sqrt{\frac{\tilde{a}+1}{\tilde{a}-1}} \exp\left\{\log\left(\frac{\tilde{a}+1}{\tilde{b}}\right)^{\tilde{a}+1} - \log\left(\frac{\tilde{a}-1}{\tilde{b}}\right)^{\tilde{a}-1} - 2\tilde{a}\right\} - \frac{\tilde{a}}{\tilde{a}-1} \exp\left\{\log\left(\frac{\tilde{a}}{\tilde{b}}\right)^{2\tilde{a}} - \log\left(\frac{\tilde{a}-1}{\tilde{b}}\right)^{2\tilde{a}-2} - 2\right\}.$$

It is interesting to compare this result and the one found in Exercise 7.2 numerically.

For $a = b = 5$, and $\bar{x} = 2.5$ we have:

$n = 10$, then from Exercise 7.2 we gain $E(\beta \mid x) = 0.52$ and from the above we get $E(\beta \mid x) = 0.500$.

$n = 40$, then from Exercise 7.2 we gain $E(\beta \mid x) = 0.43$ and from the above we get $E(\beta \mid x) = 0.4286$.

$n = 100$, then from Exercise 7.2 we gain $E(\beta \mid x) = 0.412$ and from the above we get $E(\beta \mid x) = 0.4118$.

Note that as n increases, the values get close to each other.

7.4

$$\text{For } n = 3 \text{ we have} \left\{ \begin{array}{c} x_1 = 0, \quad x_2 = 1.2247449, \quad x_3 = -1.2247449 \\ a_1 = 1.1816359, \quad a_2 = a_3 = 0.29540100 \end{array} \right\}.$$

Therefore $\int_{-\infty}^{\infty} e^{-x^2} f(x)dx = \sum_{k=1}^{n} a_k f(x_k)$, where $f(x) = 1/(1+x^2)^{10}$. Thus $I = a_1 f(x_1) + a_2 f(x_2) + a_3 f(x_3) = 1.181642$.

7.5 When the order of integration (no. of parameters) exceeds 5 or 6, greater precision can be achieved with Monte Carlo integration and importance sampling. Suppose we want to evaluate

$$I(\underline{x}) = \int_{R^k} u(\beta) h(\beta \mid \underline{x}) d\beta \tag{1}$$

where $\beta_{k \times 1}$ denotes the parameter vector of a sampling distribution. Let $g^*(\beta)$ denote a "generating density," which is called the importance function. This k-dimensional density will be used to generate M points whose ordinates will be averaged to approximate $I(\underline{x})$.

Let $\beta_1, \ldots, \beta_M$ be M points generated independently from $g^*(\beta)$. This is known as the "importance sample."

$g^*(\beta)$ is generally chosen to approximate the posterior density, but it is also chosen so that the βs can be generated easily.

Equation (1) can be rewritten as:

$$I(x) = \frac{\int u(\beta) f(\underline{x} \mid \beta) \rho(\beta) d\beta}{f(\underline{x} \mid \beta) \rho(\beta) d\beta} = \int u(\beta) \omega(\beta) d\beta,$$

where

$$\omega(\beta) = \frac{f(\underline{x} \mid \beta) \rho(\beta) d\beta}{\int f(\underline{x} \mid \beta) \rho(\beta) d\beta}.$$

As an approximation to $I(\underline{x})$ we take the weighted average of the $u(\beta)$s,

$$\hat{I}(\underline{x}) = \sum_{m=1}^{M} \hat{W}(\beta_m) u(\beta_m),$$

where

$$\hat{W}(\underset{\sim}{x}) = 1 \frac{f(\underset{\sim}{x} \mid \beta_m)\rho(\beta_m)/g^*(\beta_m)}{\sum_{m=1}^{M}(f(\underset{\sim}{x} \mid \beta_m)\rho(\beta_m)/g^*(\beta_m))}, \qquad \text{which sums to 1.}$$

$$\hat{I}(\underset{\sim}{x}) \overset{a.s.}{\rightarrow} I(\underset{\sim}{x}) \qquad \text{as } M \rightarrow \infty.$$

The precision of the approximation will be heavily dependent on the choice of $g^*(\beta)$.

7.6 Since the ε_is are normally distributed, the likelihood can be written as follows:

Likelihood:

$$L(x_1, \ldots, x_n \mid \beta) = \frac{1}{(2\pi\sigma^2)^{0.5a}} \exp\left\{\frac{1}{2\sigma^2}\sum_{i=1}^{n}(y_i - a - bx_i)^2\right\}.$$

Posterior:

$$h(a, b, \sigma^2 \mid \text{data}) \propto L(x_1, \ldots, x_n \mid \beta)g(\beta)$$
$$\propto \frac{1}{(\sigma^2)^{0.5n}} \exp\left\{\frac{1}{2\sigma^2}\sum_{i=1}^{n}(y_i - a - bx_i)^2\right\}\frac{1}{2\sigma^2}.$$

Or, equivalently

$$h(a, b, \sigma^2 \mid \text{data}) \propto \frac{1}{(\sigma^2)^{0.5n+1}} e^{-(A/2\sigma^2)},$$

where

$$A = \sum_{i=1}^{n}(y_i - a - bx_i)^2.$$

Rewrite A as:

$$A = \sum_{i=1}^{n}[(y_i - \hat{a} - \hat{b}x_i) - (a - \hat{a}) - (b - \hat{b})x_i]^2.$$

By orthogonality properties of estimators we have,

$$A = \left\{(n-2)s^2 + n(a - \hat{a})^2 + (b - \hat{b})^2\sum_{i=1}^{n}x_i^2 + 2(a - \hat{a})(b - \hat{b})\sum_{i=1}^{n}x_i\right\},$$

where

$$s^2 = \frac{1}{n-2}\sum_{i=1}^{n}(y_i - \hat{a} - \hat{b}x_i)^2.$$

The posterior marginal density of σ^2 is given by

$$h(\sigma^2 \mid \text{data}) \propto \int\int h(a, b, \sigma^2 \mid \text{data})\, da\, db$$

$$h(\sigma^2 \mid \text{data}) \propto \frac{1}{(\sigma^2)^{0.5n+1}} \int\int \exp\left\{-\frac{A}{2\sigma^2}\right\} da\, db$$

$$h(\sigma^2 \mid \text{data}) \propto \frac{e^{-[(n-2)s^2/2\sigma^2]}}{(\sigma^2)^{0.5n+1}}$$

$$\overbrace{\int\int \exp\left\{-\frac{1}{2\sigma^2}\left[n(a-\hat{a})^2 + (b-\hat{b})^2 \sum_{i=1}^{n} x_i^2 + 2(a-\hat{a})(b-\hat{b}) \sum_{i=1}^{n} x_i\right]\right\} da\, db}^{\propto \text{ constant}}$$

$$h(\sigma^2 \mid \text{data}) \propto \frac{1}{(\sigma^2)^{0.5n+1}} e^{-[(n-2)^2/2\sigma^2]}, \qquad (*)$$

which is the kernel of an inverted gamma distribution.

Now to estimate σ^3 we do as follows:

Generate a sample $\Lambda_1, \ldots, \Lambda_k$ from this posterior (*). For each $\Lambda_i = \sigma_i^2$ calculate $\sigma_i^3 = (\sigma_i^2)^{3/2}$, $i = 1, 2, \ldots, k$.

Using this sample we may plot a histogram. Then from the histogram the distribution of σ^3 can be approximated.

CHAPTER 8

8.3 Often one is interested in a reasonable interval that might contain the parameter, as opposed to fully specified density for a parameter or a point estimate of the parameter's value. Highest posterior probability credibility intervals provide an interval, $[a,\ b]$, for which $1 - \alpha$ is the desired probability that θ is in the interval. That is,

$$P\{a \le \theta \le b \mid x\} = 1 - \alpha.$$

But also, it is the highest such density. Thus, in addition to the probability requirement stated above, we say

$$P(\theta_1 \mid X_1, \ldots, X_n) \ge P(\theta_2 \mid X_1, \ldots, X_n) \qquad \forall \theta_1 \in [a,\ b] \text{ and } \theta_2 \notin [a,\ b].$$

This result is the "best" interval that has probability $1 - \alpha$ of having θ in the interval, based on the data $X_1, \ldots, X_n$.

It is possible that the HPD may be multiple disjoint intervals that collectively have probability $1 - \alpha$ of containing θ.

In all cases, the value of the posterior density for every point inside the interval is greater or equal to every point outside the interval.

8.5 If the hyperparameters of the prior density are based on the current data set, the "posterior" is an empirical Bayes' posterior and all estimation based on the posterior is empirical Bayes' estimation. One should note that the dependence of the prior and likelihood that results from basing the prior on the data means that empirical Bayes' estimation does not obey the laws of probability. Nevertheless, it has proved useful in many applications, in which actual Bayesian inference is difficult.

8.6 Credibility intervals are intervals that have a clear probability-based interpretation about the parameter θ. The credibility interval $[a, b]$ can be interpreted as:

$$P(a \leq \theta \leq b \mid \underset{\sim}{x}\} = 1 - \alpha.$$

Confidence intervals can only be interpreted as a statement about a procedure, that if it were conducted many times, it would result in a $(1 - \alpha)$ portion of the intervals spanning the true θ, and a α portion not including θ. This can only be interpreted in a long-run probability sense.

In a Bayesian framework a and b are considered to be fixed as a function of x (and could be denoted $a(x)$ and $b(x)$) and θ is random. The probability is the subjective probability that θ falls into the interval $[a, b]$. The classical approach considers the unknown θ to be fixed, the interval $[a, b]$ to be random, based on the random sample drawn.

8.12 (a) The posterior is:

$$\begin{aligned} h(\theta \mid \underset{\sim}{x}) &\propto f\underset{\sim}{X}(\underset{\sim}{x})g(\theta) \\ &\propto f\bar{X}(\bar{x})g(\theta) \\ &\propto \exp\left\{-\frac{3(\theta - \bar{x})^2}{2}\right\}. \end{aligned}$$

In other words,

$$\theta \mid \bar{x} \sim N(3, \tfrac{1}{3}).$$

(b) The two-tailed 95 percent credibility interval for θ is:

$$3 \pm 1.96\frac{1}{\sqrt{3}} = (1.8684,\ 4.1316).$$

CHAPTER 9

9.6 The classical and Bayesian methods of hypothesis testing can result in different conclusions for the same problem, as shown by Lindley (1957). An example follows.

Consider testing $H\colon \mu = 0$ vs. $A\colon \mu \neq 0$.

Assume

$$y = (x_1 + \cdots + x_n)/n$$

$$y \mid \mu \sim N(\mu, 1/n)$$

$$P(\mu = 0) = P(\mu \neq 0)$$

If $\mu \neq 0$, then the prior of μ is uniform $(-M/2,\ M/2)$.

The classical statistician would reject H at the $\alpha = 0.05$ level if $N_y > (1.96)^2$. So if $N = 40{,}000$ and $y = 0.01$, he would reject H. The Bayesian would calculate his posterior odds ratio and accept H:

$$\text{P.O.R.} = \frac{P(H) \cdot P(X \mid H)}{P(A) \cdot P(X \mid A)} = \frac{e^{-(0.5n)y^2}}{\int_{-M/2}^{M/2} \frac{1}{M} \cdot e^{-(0.5n) \cdot (\mu - \omega)^2}\, d\omega}.$$

So if $M = 1$, $n = 40{,}000$ and $y = 0.01$, P.O.R. $= 11$, favoring H fairly strongly.

This shows that different conclusions may be reached by different statisticians based on the same data, and a lump of probability mass is placed on a point while the remaining mass is spread out smoothly over the remaining points. To many, this is discomforting and, hence, a paradox.

9.12 Posterior odds ratio:

Define $\rho = \dfrac{P(H \mid T)}{P(A \mid T)} = \dfrac{P(T \mid H)}{P(T \mid A)} \cdot \dfrac{P(H)}{P(A)}$.

For this problem,

$$P(T \mid H) = \binom{10}{3}(0.2)^3(0.8)^7,$$

$$P(T \mid A) = \binom{10}{3}(0.8)^3(0.2)^7,$$

$$P(H) = P(A) = \frac{1}{2}.$$

So

$$\rho = \frac{(0.8)^4}{(0.2)^4} \cdot \frac{1/2}{1/2} = 256.$$

Since ρ is much greater than 1, $H\colon p = 0.2$ is much more likely than the alternative. Therefore, $H\colon \rho = 0.2$ is accepted and the alternative is rejected.

CHAPTER 10

10.2

$$f^*(y \mid x) \propto \int f(y \mid \theta)h(\theta \mid x)d\theta,$$

where

$$\begin{aligned} h(\theta \mid x) &= kg(\theta)f(x \mid \theta) \\ &= k\left\{\frac{1}{\sqrt{2\pi}}e^{-(0.5\theta^2)}\right\}\left\{\frac{1}{\sqrt{2\pi}}e^{-[0.5(x-\theta)^2]}\right\} \propto e^{-(\theta-x/2)^2}, \quad \text{i.e.} \, N\left(\frac{x}{2}, \frac{1}{2}\right), \end{aligned}$$

which means

$$\begin{aligned} f^*(y \mid x) &= \int \left(\frac{1}{\sqrt{2\pi}}e^{-[0.5(y-\theta)^2]}\right)\left(\frac{1}{\sqrt{\pi}}e^{-[(\theta-0.5s)^2]}\right)d\theta \\ &= \frac{1}{\sqrt{2\pi(3/2)}}e^{-(1/[2(3/2)])[y-0.5x)]^2}, \end{aligned}$$

which means $y \sim N(x/2, 3/2)$.

10.5 **Definition 10.1.** A sequence of random variables $\theta_1, \theta_2, \ldots, \theta_q$ is said to be exchangeable if the joint distribution of any finite subset of the θ_is are invariant under permutations of the subscripts.

If we consider the θ_is to be i.i.d., then they are certainly exchangeable.

If $\theta_1, \theta_2, \ldots, \theta_q$ are exchangeable, then θ_1, θ_2 are exchangeable and thus all pairwise θ_i, θ_js are exchangeable. If a finitely exchangeable sequence $\{\theta_i\}_{i=1}^q$ drawn from an infinitely exchangeable sequence can be imbedded into an exchangeable sequence $\{\theta_1, \theta_2, \ldots, \theta_k\}$ of length $k > q$, then the original sequence $\{\theta_1, \theta_2, \ldots, \theta_q\}$ is almost a mixture of i.i.d. variables with an error going to zero as $k \to \infty$ (Diaconis, 1977).

10.9 The mathematical definitions of partial exchangeability are:

1. Every exchangeable probability assignment on sequences of length N is a unique mixture of draws without replacement from the $N + 1$ extremal urns. By a mixture of urns we simply mean a probability assignment over the $N + 1$ possible urns. Any exchangeable assignment on sequences of length N can be realized by first choosing an urn and then drawing without replacement until the urn is empty. The extremal urns, representing certain knowledge of the total number of ones, seem like unnatural probability assignments in most cases. While such situations arise (for example, when drawing a sample from a finite population), the more usual situation is that of the coin. While we are only considering ten spins, in principle it seems possible to extend the ten to an arbitrarily large number. In this case there

is a stronger form that restricts the probability assignment within the class of exchangeable assignments (Diaconis and Freedman, 1978).

2. Let $\boldsymbol{Z} = (Z_0, Z_1, \ldots)$ be a sequence of random variables taking values in a countable state space I. We use a generalization of exchangeability called partial exchangeability. $\boldsymbol{Z}$ is partially exchangeable if for two sequences $\sigma, \tau \in I^{n+1}$ which have the same starting state and the same transition counts, $P(Z_0 = \sigma_0,\ Z_1 = \sigma_1, \ldots, Z_n = \sigma_n) = P(Z_0 = \tau_0,\ Z_1 = \tau_1, \ldots, Z_n = \tau_n)$. The main result is that for recurrent processes, $\boldsymbol{Z}$ is a mixture of Markov chains if and only if $\mathbf{Z}$ is partially exchangeable (Diaconis and Freedman, 1980).

Example 10.1. Consider that two people toss a particular coin and record the outcome. If we consider that the ways in which the two people tossed the coin were completely different, then we have two separate sequences of exchangeable events that have no relationship. However, if we believe that there is a clear influence of how the two people are tossing the coin (an effort that intentionally produces, say, a long series of tails), then we still obtain two separate sequences of exchangeable events, but with dependence across the classes. The events are then called partially exchangeable events.

For other examples see Diaconis and Freedman (1978).

10.12 **Theorem 6.1.** (de Finetti) Let $\{X_i\}_{i=1}^{\infty}$ be an infinite sequence of random variables with $\{X_i\}_{i=1}^{n}$ exchangeable for each n; then there is a unique probability measure μ on [0, 1] such that for each fixed sequence of zeros and ones $\{e_i\}_{i=1}^{n}$, we have

$$p(X_1 = e_1, \ldots, X_n = e_n) = \int_0^1 p^k(1-p)^{n-k}\, d\mu(p) \quad \text{where } k = \textstyle\sum_{i=1}^{n} e_i.$$

A condition of de Finetti's theorem is that it requires an infinite sequence of exchangeable events and is shown to be easily falsified if the sequence is finite.

Example 10.2. (Diaconis, 1977) Let $P(X_1 = 1, X_2 = 0) = P(X_1 = 0, X_2 = 1) = \frac{1}{2}$ and $P(X_1 = 0, X_2 = 0) = P(X_1 = 1, X_2 = 1) = 0$.

The pair X_1, X_2 is exchangeable, but if there were a prior probability measure μ such that $0 = P(X_1 = 1, X_2 = 1) = \int_0^1 p^2\, d\mu(p)$, then necessarily μ puts mass 1 at the point 0, so we cannot have $0 = P(X_1 = 0, X_2 = 0) = \int_0^1 (1-p)^2 d\mu(p)$.

The approximation to the infinite case is given by the following theorems.

Theorem 10.2. (Diaconis, 1977) Let S be a finite set of cardinality c. Let P be an exchangeable probability on S^n. Then there exists a probability μ on the Borel subsets of S^* such that $\| P_k - P_{\mu,k} \| \leq 2ck/n$ for all $k \leq n$.

Theorem 10.3. (Diaconis, 1977) Suppose an urn U contains n balls, each marked by one or another element of the set S, whose cardinality c is finite. Let H_{Uk} be the distribution of k draws made at random without replacement from U, and M_{Uk} be the distribution of k draws made at random with replacement. (H stands for hypergeometric; M for multinomial.) Thus, H_{Uk} and M_{Uk} are two probabilities on S^k. Then $\| H_{Uk} - M_{Uk} \| \leq 2ck/n$.

Theorem 10.2 can be modified to handle S of infinite cardinality. If (S, Υ) is a measurable space and P is an exchangeable probability on (S^n, Υ^n), then Theorem 10.2 shows there is a probability μ on S^* such that $\| P_k - P_{\mu k} \| \leq k(k-1)/n$.

10.16 Model.

$$y_{ij} = \mu_{(p\times 1)} + \alpha_i + \mu_{ij}, \qquad i = 1, 2, \ldots, q, \qquad j = 1, 2, \ldots, n_i,$$

$$u_{ij} \sim N(0, \Sigma),$$

$\alpha_i \sim N(\xi^*, \Phi^*)$ exchangeable, independent.

Let $Y' = [(y_{11}, \ldots, y_{1n_1}), \ldots, (y_{q1}, \ldots, y_{qn_q})]_{(p\times N)}$,

$$U' = [(u_{11}, \ldots, u_{1n_1}), \ldots, (u_{q1}, \ldots, u_{qn_q})]_{(p\times N)},$$

$$\theta_i = \mu + \alpha_i,$$

$$B = [\theta_1, \ldots, \theta_q]_{(p\times q)},$$

$$X = \begin{bmatrix} e_{n_1} & 0 & \cdots & 0 \\ 0 & & \ddots & 0 \\ 0 & & \cdots & e_{n_q} \end{bmatrix}_{(N\times q)},$$

$$e_n' = [1, 1, \ldots, 1],$$

$$Y = XB + U.$$

The likelihood becomes:

$$P(Y \mid X, B, \Sigma) \propto |\, \Sigma \,|^{0.5N} e^{-0.5\,\mathrm{tr}\{V + (B-\hat{B})'S(B-\hat{B})\}\Sigma^{-1}},$$

where $\hat{B} = (X'X)^{-1}X'Y, \quad S_{(q\times q)} = (X'X), \quad V_{(p\times p)} = (Y - X\hat{B})'(Y - X\hat{B})$.

(a) If you adopt a prior for μ as being a constant, then the prior for θ_i is given by

$$(\theta_i) \sim N(\mu_0 + \xi^*, \Phi^*),$$

$$p(\theta_i) \propto |\Phi^*|^{-0.5q} \exp\left\{ -\frac{1}{2\Phi^*} \sum_{i=1}^{q} (\theta_i - \mu_0 - \xi^*)^2 \right\}.$$

(b) Then the prior for B is

$$p(B) \propto |\Phi^*|^{-0.5q} \exp\left\{ -0.5 \sum_{i=1}^{q} (\theta_i - \mu_0 - \xi^*)' \Phi^{*-1} (\theta_i - \mu_0 - \xi^*) \right\}.$$

Let $\theta_{(pq\times 1)} = (\theta_1, \ldots, \theta_q)'$, $\theta^*_{(pq\times 1)} = (\mu_0 + \xi^*, \ldots, \mu_0 + \xi^*)'$, and

$$I_q \otimes \Phi^{*-1} = \begin{bmatrix} \Phi^{*-1} & 0 & \cdots & 0 \\ 0 & & \ddots & 0 \\ 0 & & \cdots & \Phi^{*-1} \end{bmatrix}_{(q\times q)}.$$

$$p(B) \propto |\Phi^*|^{-0.5q} \exp\{-0.5(\theta - \theta^*)'(I_q \otimes \Phi^{*-1})(\theta - \theta^*)\}.$$

The prior for Σ is $p(\Sigma) \propto |\Sigma|^{-0.5\nu} \exp\{-0.5 \operatorname{tr} \Sigma^{-1} H\}$.

(c) $$p(B, \Sigma) \propto \frac{1}{|\Sigma|^{0.5N} |\Sigma|^{0.5\mu} |\Phi^*|^{0.5q}} \exp\Big[-0.5 \operatorname{tr} \Sigma^{-1}[(V + H) + (B - \hat{B})' S (B - \hat{B})] + (\theta - \theta^*)' I_q \otimes \Phi^{*-1} (\theta - \theta^*)\Big].$$

Another solution

If you adopt a prior for μ, as $\mu \sim N(\mu_0, \Sigma_0)$, and if μ and α_i are independent, then $\theta_i \sim N(\mu_0 + \xi^*, \Sigma_0 + \Phi^*)$.

Thus the prior for B is

$$p(B) = \prod_{i=1}^{q} p(\theta_i) \propto \prod_{i=1}^{q} |\Sigma_0 + \Phi^*|^{-0.5} \times \exp[-0.5(\theta_i - \mu_0 - \xi^*)'(\Sigma_0 + \Phi^*)^{-1}(\theta_i - \mu_0 - \xi^*)],$$

$$p(B) \propto |\Sigma_0 + \Phi^*|^{0.5q} \exp\left[-0.5 \sum_{i=1}^{q} (\theta_i - \mu_0 - \xi^*)'(\Sigma_0 + \Phi^*)^{-1}(\theta_i - \mu_0 - \xi^*) \right].$$

Using the same prior for Σ as above,

$$p(B, \Sigma) \propto \frac{1}{|\Sigma|^{0.5N} |\Sigma_0 + \Phi^*|^{0.5q} |\Sigma|^{0.5v}}.$$

CHAPTER 11

11.4 The posterior density function for a parameter summarizes all the information that a researcher has about that parameter. Sometimes it is of interest to the researcher to decide on just one value, that is, a point estimate of the parameter, and the question that arises is, which is the best value for the point estimate of that parameter?

The Bayesian framework approaches this problem in the following manner:

Assume there is a loss function $L(\hat{\theta}, \theta) > 0$ that describes the consequences of basing a decision on $\hat{\theta}$ when the real value is θ.
Choose $\hat{\theta}$ that minimizes the expected loss when the expectation is taken with respect to the posterior density for θ.

Suppose then that we have the quadratic loss function $L = k(\hat{\theta} - \theta)^2$, where k is any positive constant. We minimize the average loss over all possible values of θ, with the various values of θ being weighted by the posterior density for θ, that is,

$$\text{Min}_{\hat{\theta}}\, E[L(\hat{\theta}, \theta)] = \int k(\hat{\theta} - \theta)^2 h(\theta \mid x) d\theta.$$

The above integral (or sum in the discrete case) is called Bayes risk and has an important link to the prior density, $g(\theta)$; that is, as long as $g(\theta)$ is strictly proper (i.e., the integral converges), then Bayesian estimators are unique and admissible.

By admissibility we mean that an estimator of θ, say $\hat{\theta}$, is admissible, if there is not another estimator, say $\hat{\theta}_B$, for which:

1. $E\{L(\hat{\theta}, \theta)\} \leq E\{L(\hat{\theta}_B, \theta)\}$, for all θ.
2. $E\{L(\hat{\theta}, \theta_0)\} < E\{L(\hat{\theta}_B, \theta_0)\}$, for some θ_0.

The advantage of a Bayes' estimator is that given a proper prior density, it always yields an admissible estimator of θ.

In summary, proper prior distributions will always result in Bayesian estimators being admissible. Admissible estimators will always have the lowest expected loss, and consequently the highest expected utility.

11.7 The principle of admissibility is dependent on observations not yet taken, but this violates the likelihood principle. Usually, a (Bayesian) statistician does not care about what happens on the average over many possible samples; he/she cares principally about the current sample.

11.11 (a) As we have derived, the distributrion of β_2 is t with $df = n - 2$. Thus the density of β_2 will be given by

$$f(\beta_2) \propto \frac{1}{\left[(n-2) + \frac{(\beta_2 - \hat{\beta}_2)^2}{\sigma_{22}}\right]^{0.5(n-1)}},$$

which is centered at $\hat{\beta}_2$. This is the mode with the estimated value of $\hat{\beta}_2 = 3.00$.

(b) As we know, $E(X - a)^2$ is minimum at $a = E(X)$, the mean of X. So the estimate of β_2 is given by $\hat{\beta}_2 = 3.00$.

(c) As we know, $E \mid X - a \mid$ is minimum at $a =$ median of X. So the estimate of β_2 is again given by $\hat{\beta}_2 = 3.00$.

CHAPTER 12

For solutions in this chapter it will be helpful to first give some definitions and lemmas.

Definition 12.1. If X has a gamma distribution with parameters α and β, then its density is given by $f(x{:}\ \alpha, \beta) = [\beta^a/\Gamma(\alpha)]x^{\alpha-1}e^{-\beta x}$; $x > 0$. Also, by assuming $\Gamma(\alpha)$ is a constant,

$$\int_0^\infty x^{\alpha-1}e^{-\beta a}dx = \frac{\Gamma(\alpha)}{\beta^a} \propto \frac{1}{\beta^a}.$$

Definition 12.2. A positive random variable Y has a distribution that is called inverted gamma with parameters a and b, and is denoted by $IG(a, b)$, if its density is given by

$$f(y) \propto \frac{1}{(y)^{a+1}} e^{-b/y}.$$

Note that $1/Y$ is distributed as gamma with parameters (a, b).

Also

$$E(Y) = \frac{b}{a-1}, \qquad V(Y) = \frac{b^2}{(a-1)^2(a-2)},$$

$$\int \frac{1}{y^{a+1}} e^{-b/y}\, dy \propto \frac{1}{b^a}. \qquad \text{(See Definition 5.1.)}$$

Lemma 12.1. If $p(\sigma^2 \mid \text{data}) \propto 1/(\sigma^2)^{0.5n} e^{-(A/2\sigma^2)}$, then σ^2 has an inverted gamma distribution. Furthermore $\int_0^\infty [1/(\sigma^2)^{0.5n}] e^{-(A/2\sigma^2)} \propto 1/A^{(0.5n)-1}$, (i.e., $1/\sigma^2$ has a gamma distribution).

Proof

Let $t = 1/2\sigma^2$. Then, $dt = -(1/2\sigma^4)\, d\sigma^2 \Rightarrow d\sigma^2 = -2\sigma^4\, dt$.

Also $\sigma^2 = (1/2t)$, so we get the following:

$$\int_0^\infty \frac{1}{(\sigma^2)^{0.5n}} e^{-(A/2\sigma^2)}\, d\sigma^2 = \int_\infty^0 (2t)^{0.5n} e^{-At}\left(-\frac{1}{4t^2}\right) dt \propto \int_0^\infty t^{(0.5n)-2} e^{-At}\, dt.$$

By definition 12.1 we have

$$\int_0^\infty t^{(0.5n)-2} e^{-At} dt = \int_0^\infty t^{[(0.5n)-1]-1} e^{-At} dt \propto \frac{1}{A^{(0.5n)-1}}.$$

Another alternative is to use the integral in definition 12.2 with $a = (0.5n) - 1$ and $b = 0.5A$.

Definition 12.3. Let V denote a $p \times p$, symmetric, positive definite matrix of random variables. V follows a Wishart distribution if the distinct elements are jointly continuous with multivariate probability density function:

$$p(V \mid \sigma, n) \propto \frac{\mid V \mid^{0.5(n-p-1)}}{\mid \Sigma \mid^{0.5n}} \exp\{-0.5 \operatorname{tr} \Sigma^{-1} V\}.$$

We write $V \sim W(\Sigma, p, n)$, where Σ is a scalar matrix, $n =$ degrees of freedom, and $p =$ the dimension of Σ.

Definition 12.4.

(a) The random matrix Y has a distribution called inverted Wishart if Y^{-1} follows a Wishart distribution.

(b) Assume that Σ follows an "inverted Wishart distribution." Then Σ^{-1} follows a Wishart distribution. The density of Σ is given by

$$p(\Sigma) \propto \frac{1}{|\Sigma|^{0.5v}} \exp\{-0.5 \operatorname{tr} \Sigma^{-1} H\},$$

and is denoted by $W^{-1}(H, P, v)$ for some hyperparameters (v, H).

Lemma 12.2. By the notations of definition 12.4 we have:

$$\int_{\Sigma>0} \frac{1}{|\Sigma|^{0.5(n+p+1)}} e^{-0.5(\operatorname{tr} \Sigma^{-1} A)} d\Sigma \propto \frac{1}{|A|^{0.5n}}.$$

Note that Σ in the integrand is a matrix and direct computation of these integrals is not so simple.

Lemma 12.3. Let $t_{p\times 1} = (t_1', t_2')$, t_1: $(p_1 \times 1)$, so that $p_1 + p_2 = p$, be a multivariate t distribution with density

$$g(t) \propto \frac{|\Sigma|^{-0.5}}{\{n + (t-\theta)'\Sigma^{-1}(t-\theta)\}^{0.5(n+p)}}.$$

Then t_1 is also a multivariate t distribution with density

$$g(t_1) \propto \frac{|\Sigma_{11}|^{-0.5}}{\{n + (t_1 - \theta_1)'\Sigma_{11}^{-1}(t_1 - \theta_1)\}^{0.5(n+p_1)}}$$

where

$$\Sigma = \begin{pmatrix} \Sigma_{11} & \Sigma_{12} \\ \Sigma_{21} & \Sigma_{22} \end{pmatrix}.$$

Lemma 12.4. (Orthogonality property of least-square estimators) Let $\hat{\beta}$ denote the LSE of B, then:

$$(Y - XB)'(Y - XB) = (Y - X\hat{B})'(Y - X\hat{B}) + (B - \hat{B})'(X'X)(B - \hat{B}),$$

or

$$(Y - XB)'(Y - XB) = V + (B - \hat{B})'(X'X)(B - \hat{B}),$$

where

$$V = (Y - X\hat{B})'(Y - X\hat{B}).$$

Proof
See the text.

12.1
Prior. Use a vague prior so $p(\underline{\beta}, \sigma^2) = p(\underline{\beta} \mid \sigma^2)p(\sigma^2) \propto 1/\sigma^2$. Since $p(\beta_i \mid \sigma^2) \propto$ constant $\forall_i = 1, \ldots, p$ and $p(\sigma^2) \propto 1/\sigma^2$.

Likelihood.

$$L(\underline{y} \mid X, \underline{\beta}, \sigma^2) \propto \frac{1}{(\sigma^2)^{0.5n}} \exp\left\{ -\frac{1}{2\sigma^2} (\underline{y} - X\underline{\beta})'(\underline{y} - X\underline{\beta}) \right\}.$$

(a) **Posterior.**

$$h(\underline{\beta}, \sigma^2 \mid \underline{y}, X) \propto L(\underline{y} \mid X, \underline{\beta}, \sigma^2)p(\underline{\beta}, \sigma^2).$$

So we get

$$h(\underline{\beta}, \sigma^2 \mid \underline{y}, X) \propto \frac{1}{(\sigma^2)^{0.5n}} \exp\left\{ -\frac{1}{2\sigma^2} (\underline{y} - X\underline{\beta})'(\underline{y} - X\underline{\beta})) \right\}.$$

(b) The marginal posterior density of $\underline{\beta}$ is given by:

$$h_1(\underline{\beta} \mid \text{data}) = \int_0^\infty h(\underline{\beta}, \sigma^2) d\sigma^2 \propto \int_0^\infty \frac{1}{(\sigma^2)^{0.5(n+2)}} e^{-(1/2\sigma^2)A} \, d\sigma^2.$$

Or $h_1(\underline{\beta} \mid \text{data}) \propto 1/(A)^{0.5n}$, by using Lemma (12.1), where $A = (\underline{y} - X\overline{\underline{\beta}})'(\underline{y} - X\underline{\beta})$. But by Lemma (12.4) we can write

$$A = V + (\underline{\beta} - \hat{\underline{\beta}})'(\underline{\beta} - \hat{\underline{\beta}}),$$

where $V = ns^2$.
Thus we get

$$h_1(\underline{\beta} \mid \text{data}) \propto \frac{1}{\{ns^2 + (\underline{\beta} - \hat{\underline{\beta}})'X'X(\underline{\beta} - \hat{\underline{\beta}})\}^{0.5n}}$$

Now let us write this density into standard form as follows:

$$h_1(\underline{\beta} \mid \text{data}) \propto \frac{1}{\left\{\dfrac{ns^2}{n-p-1}\right\}^{0.5n}\left\{(n-p-1)+\dfrac{n-p-1}{ns^2}(\underline{\beta}-\hat{\underline{\beta}})'X'X(\underline{\beta}-\hat{\underline{\beta}})\right\}^{0.5n}},$$

$$h_1(\underline{\beta} \mid \text{data}) \propto \frac{1}{\left\{(n-p-1)+\dfrac{n-p-1}{ns^2}(\underline{\beta}-\hat{\underline{\beta}})'X'X(\underline{\beta}-\hat{\underline{\beta}})\right\}^{0.5n}}.$$

Let $\Sigma^{-1} = [(n-p-1/ns^3)X'X$. Then we have:

$$h_1(\underline{\beta} \mid \text{data}) \propto \frac{1}{\{(n-p-1)+(\underline{\beta}-\hat{\underline{\beta}})'\Sigma^{-1}(\underline{\beta}-\hat{\underline{\beta}})\}^{0.5n}},$$

which is a multivariate t-distribution with $df = n-p-1$ or equivalently $(\underline{\beta} \mid \text{data}) \sim t_{(n-p-1)}$.

(c) To obtain the density of σ^2 we have,

$$h_2(\sigma^2 \mid \text{data}) = \int_{R^{p+1}} h(\underline{\beta}, \sigma^2 \mid \text{data})\, d\underline{\beta}$$

$$h_2(\sigma^2 \mid \text{data}) \propto \int_{R^{p+1}} \frac{1}{(\sigma^2)^{0.5(n+2)}} \exp\left\{-\frac{1}{2\sigma^2}(\underline{y}-X\underline{\beta})'(\underline{y}-X\underline{\beta})\right\} d\beta,$$

$$h_2(\sigma^2 \mid \text{data}) \propto \int_{R^{p+1}} \frac{1}{(\sigma^2)^{0.5(n+2)}} e^{-(1/2\sigma^2)A}\, d\underline{\beta},$$

where $A = \{(\underline{y}-X\underline{\beta})'(\underline{y}-X\underline{\beta})\}$. By applying Lemma (12.4) we have,

$$A = V + (\underline{\beta}-\hat{\underline{\beta}})'X'X(\underline{\beta}-\hat{\underline{\beta}}).$$

Let $\Sigma^{-1} = X'X/\sigma^2$. Then

$$h_2(\sigma^2 \mid \text{data}) \propto \frac{1}{(\sigma^2)^{0.5(n+2)}} e^{-(1/2\sigma^2)V} \int_{p^{n+1}} \exp\left\{-0.5(\underline{\beta}-\hat{\underline{\beta}})'\Sigma^{-1}(\underline{\beta}-\hat{\underline{\beta}})\right\} d\underline{\beta},$$

$$h_2(\sigma^2 \mid \text{data}) \propto \frac{1}{(\sigma^2)^{0.5n+2}} e^{-(1/2\sigma^2)V}$$

$$\times \mid \Sigma \mid^{0.5} \overbrace{\int_{p^{n+1}} \frac{1}{\mid \Sigma \mid^{0.5}} \exp\{-0.5(\underline{\beta}-\hat{\underline{\beta}})'\Sigma^{-1}(\underline{\beta}-\hat{\underline{\beta}})\}\, d\underline{\beta}}^{=1},$$

and

$$\mid \Sigma \mid^{0.5} \propto (\sigma^2)^{0.5(p+1)}$$

Thus

$$h_2(\sigma^2 \mid \text{data}) \propto \frac{1}{(\sigma^2)^{0.5(n+p+1)}} e^{-(1/\sigma^2)[1/(2/V)]},$$

which is the kernel of the density of an inverted gamma distribution. That is, $\sigma^2 \sim \gamma^{-1}[0.5(n-p+1),\ 2/V]$, or equivalently $1/\sigma^2 \sim \gamma\,[0.5(n-p),\ 2/V)$.

(d) To get the posterior density of β_1 we use the results found in part (b) as follows:

$$p(\underline{\beta} \mid \underline{y}, X) \propto \frac{1}{[(n-p-1)+(\beta-\hat{\beta})'\Sigma^{-1}(\beta-\hat{\beta})]^{0.5n}}.$$

By applying Lemma 12.3 we have:

$$h(\beta_i \mid \underline{y}, X) \propto \frac{1}{[(n-p-1)+(\beta_i-\hat{\beta}_i)'\Sigma_{ii}^{-1}(\beta_i-\hat{\beta}_i)]^{0.5(n-p)}},$$

where $\Sigma_{ii} = \sigma_{ii}$, or

$$h(\beta_i \mid \underline{y}, X) \propto \frac{1}{\left[(n-p-1)+\dfrac{(\beta_i-\hat{\beta}_i)^2}{\sigma_{ii}}\right]^{0.5(n-p)}},$$

which is a univariate t-distribution, (i.e., $\beta_i \sim t_{n-p-1}$), where

$$\Sigma = \begin{pmatrix} \sigma_{00} & \sigma_{01} & \cdots & \sigma_{0p} \\ \sigma_{10} & \sigma_{11} & \cdots & \sigma_{1p} \\ \cdots & & & \\ \cdots & \cdots & \cdots & \sigma_{pp} \end{pmatrix}$$

and $\sigma_{ii} = \text{variance } (\beta_i),\ i = 0, 1, \ldots, p.$

(e) The predictive density of y^* given x^* is given by

$$p(y^* \mid x^*) = \int_{R^{p+1}} \int_0^\infty p(y^* \mid x^*, \underline{\beta}, \sigma^2) \cdot p(\underline{\beta} \cdot \sigma^2 \mid \text{data}) \underline{\beta}\, d\sigma^2$$

or equivalently

$$p(y^* \mid x^*) \propto \int_{R^{p+1}} \int_0^\infty \frac{1}{\sigma} \exp\left\{-\frac{1}{2\sigma^2}(y^* - \underline{\beta}x^*)^2\right\}$$

$$\times \frac{1}{\sigma^{n+2}} \exp\left\{-\frac{1}{2\sigma^2}(y - X\underline{\beta})'(y - X\underline{\beta})]\right\} d\underline{\beta}\, d\sigma^2.$$

By combining the terms we get:

$$p(y^* \mid x^*) \propto \int_{R^{p+1}} \int_0^{\infty} \frac{1}{(\sigma^2)^{0.5(n+3)}} e^{-(A/2\sigma^2)} \, d\beta \, d\sigma^2.$$

where $A = \{v + (\beta - \hat{\beta})' X'X(\beta - \hat{\beta}) + (y^* - \underline{\beta} x^{*\prime})'(y^* - \underline{\beta}\underline{x}^*)\}$.

Let $X^* = (1, x^{*\prime})$. By using Lemma (12.1), we get:

$$p(y^* \mid x^*) \propto \int \frac{1}{A^{0.5(n+1)}} d\underline{\beta}.$$

Now we write A into a quadratic equation in terms of b:

$$A = \beta'(X'X + X^{*\prime}X^*)\beta - 2(\hat{\beta}'X'X + y^{*\prime}X^{*\prime})\beta + (v + \hat{\beta}'X'X\beta + y^{*2}),$$

or equivalently

$$A = \beta' A^* \beta - 2\gamma'\beta + \delta = (\beta - \beta^*)' A^* (\beta - \beta^*) + \delta^*,$$

where $\delta^* = \delta - \beta^* A^* B^*$, $B^* = (A^*)^{-1}\gamma$, $A^* = X'X + X^{*\prime}X^*$, $\gamma' = \hat{\beta}'X'X + \hat{y}^{*\prime}X^{*\prime}$, and $\delta = v + \hat{\beta}'X'X\beta + y^{*2}$. So after substitution we get:

$$p(y^* \mid x^*) \propto \int \frac{1}{[\delta^* + (\beta - \beta^*) A^* (\beta - \beta^*)]^{0.5(n+1)}} d\underline{\beta}.$$

To compute this integral, we rewrite it in standard format as follows;

$$p(y^* \mid x^*) \propto \left(\frac{n}{\delta^*}\right)^{0.5(n+1)} \int \frac{1}{(n/\delta^*)^{0.5(n+1)}} \frac{1}{[\delta^* + (\beta - \beta^*)' A^* (\beta - \beta^*)]^{0.5(n+1)}} d\beta,$$

$$p(y^* \mid x^*) \propto \left(\frac{n}{\delta^*}\right)^{0.5(n+1)} \mid \Omega \mid^{0.5} \overbrace{\int \frac{1}{\mid \Omega \mid^{0.5}} \frac{1}{[n + (\beta - \beta^*)' \Omega^{-1} (\beta - \beta^*)]^{0.5(n+1)}} d\beta}^{=1},$$

where $\Omega^{-1} = (n/\delta^*) A^*$.

However, $\mid \Omega \mid^{0.5} \propto 1/(\delta^*)^{0.5(n+1)}$ or $\mid \Omega \mid^{0.5} \propto (\delta^*)^{0.5(p+1)}$. since dim $A^* = p + 1$. Thus we have

$$p(y^* \mid x^*) \propto \left(\frac{n}{\delta^*}\right)^{0.5(n+1)} (\delta^*)^{0.5(n+1)} \propto \frac{1}{(\delta^*)^{0.5(n-p)}},$$

where $\delta^* = v + \hat{\beta}'X'X\beta + y^{*2} - \beta^{*\prime} A^* \beta$.

Lemma 12.5. Show that

$$p(y^* \mid x^*) \propto \frac{1}{\left\{(n-p-1) + \dfrac{[y^* - (b/a)]^2}{d}\right\}^{0.5(n-p)}},$$

for appropriate choices of a, b, and d.

Proof
By completing the square in $\delta^* = v + \hat{\beta}'X'X\beta + y^{*2} - \beta^{*\prime}A^*B^*$ we obtain:

$$\begin{aligned}\delta^* &= y^{*\prime}[1 - X^{*\prime}(A^*)^{-1}X^*]y^* - 2y^*[X^{*\prime}(A^*)^{-1}X'X\hat{\beta}] \\ &\quad + [v + \hat{\beta}X'X\hat{\beta} - \hat{\beta}'X'X(A^*)^{-1}X'X\hat{\beta}],\end{aligned}$$

where, as before $\beta^* = (A^*)^{-1}\gamma$, $\gamma' = \hat{\beta}'X'X + y^{*\prime}X^{*\prime}$. So we can write:

$$\begin{aligned}\delta^* &= a(y^*)^2 - 2by^* + c = a\left\{\left(y^* - \frac{b}{a}\right)^2 - \frac{b^2}{a^2} + \frac{c}{a}\right\} \\ &= a\left\{\left(\frac{c}{a} - \frac{b^2}{a^2}\right) + \left(y^* - \frac{b}{a}\right)^2\right\},\end{aligned}$$

where $a = [1 - X^{*\prime}A^{-1}X^*]$, $b = [X^{*\prime}A^{-1}X'X\hat{\beta}]$, $c = [v + \hat{\beta}X'X\hat{\beta} - \hat{\beta}'XA^{-1}X'X\hat{\beta}]$, and $c_0^2 = [(c/a) - (b^2/a^2)]$, $\delta^* = a\{c_0^2 + [(y^* - (b/a)]^2\}$. So

$$p(y^* \mid x^*) \propto \frac{1}{\{c_0^2 + [y^* - (b/a)]^2\}^{0.5(n-p)}}.$$

or equivalently,

$$p(y^* \mid x^*) \propto \frac{[(n-p-1)/c_0^2]^{0.5(n-p)}}{[(n-p-1)/c_0^2]^{0.5(n-p)}} \frac{1}{\{c_0^2 + [y^* - (b/a)]^2\}^{0.5(n-p)}},$$

$$p(y^* \mid x^*) \propto \frac{1}{\{(n-p-1) + [(n-p-1)/c_0^2][y^* - (b/a)]^2\}^{0.5(n-p)}},$$

$$p(y^* \mid x^*) \propto \frac{1}{\left\{(n-p-1) + \dfrac{[y^* - (b/a)]^2}{d}\right\}^{0.5(n-p)}},$$

where $d = c_0^2/(n-p-1)$
Thus $y^* \mid x^* \sim t_{n-p-1}$, a univariate t-distribution with $df = n - p - 1$.

12.2 The model is given by:

$$y_i \mid x_i = \beta_1 + \beta_2 x_i + u_i \qquad i = 1, \ldots, n \qquad u_i \sim N(0, \sigma^2).$$

Let

$$\underline{y} = (y_1, y_2, \ldots, y_n)', \qquad X_{n\times 2} = \begin{pmatrix} 1 & x_1 \\ 1 & x_2 \\ \vdots & \vdots \\ 1 & x_n \end{pmatrix},$$

$$\underline{\beta} = (\beta_1, \beta_2)', \qquad \text{and } \underline{u} = (u_1, u_2).$$

Then

$$\underline{Y} \mid X = X\underline{\beta} + u,$$

where

$$\underline{u} \sim N(\underline{0}, \sigma^2 I_{2\times 2}).$$

Likelihood. Since the error term is $N(\underline{0}, \sigma^2 I)$, then $\underline{y} \sim N(X\underline{\beta}, \sigma^2 I)$ with the density

$$f(\underline{y} \mid X, \underline{\beta}, \sigma^2) \propto \frac{1}{(\sigma^2)^{0.5n}} \exp\left\{\frac{-1}{2\sigma^2}(\underline{Y} - X\underline{\beta})'(\underline{Y} - X\underline{\beta})\right\}.$$

By Lemma 12.4, we can write

$$(\underline{Y} - X\underline{\beta})'(\underline{Y} - X\underline{\beta}) = V + (\underline{\beta} - \hat{\underline{\beta}})'(X'X)(\underline{\beta} - \hat{\underline{\beta}}),$$

where

$$V = (\underline{Y} - X\hat{\underline{\beta}})'(\underline{Y} - X\hat{\underline{\beta}}).$$

So the likelihood can be written as

$$f(\underline{y} \mid X, \underline{\beta}, \sigma^2) \propto \frac{1}{(\sigma^2)^{0.5n}} \exp\left\{\frac{-1}{2\sigma^2}[V + (\underline{\beta} - \hat{\underline{\beta}})'(X'X)(\underline{\beta} - \hat{\underline{\beta}})]\right\}.$$

Prior. $p(\beta, \sigma^2) = p(\beta \mid \sigma^2)p(\sigma^2)$.

Let as choose $\beta \mid \sigma^2 \sim N(\beta^*, k\sigma^2 I)$ so we get:

$$p(\beta \mid \sigma^2) \propto \frac{1}{(\sigma^2)^{0.5a}} \exp\left\{\frac{-1}{2k\sigma^2}[(\underline{\beta} - \underline{\beta}^*)'(\underline{\beta} - \underline{\beta}^*)]\right\}.$$

Also let $\sigma^2 \sim \gamma^{-1}(a, b)$ (that is, inverted gamma) so the density is given by:

$$p(\sigma^2) \propto \frac{1}{(\sigma^2)^{0.5n}} e^{-(b/2\sigma^2)}.$$

Thus

$$p(\beta, \sigma^2) \propto \frac{1}{(\sigma^2)^{0.5(a+2)}} \exp\left\{\frac{-1}{2\sigma^2}\left[(\underline{\beta} - \underline{\beta}^*)' + \frac{1}{k} I(\underline{\beta} - \underline{\beta}^*) + b\right]\right\}.$$

(a) **Posterior.** $h(\underline{\beta}, \sigma^2) \propto f(y \mid X, \underline{\beta}, \sigma^2)^* \cdot p(\underline{\beta}, \sigma^2)$,

$$h(\underline{\beta}, \sigma^2) \propto \frac{1}{(\sigma^2)^{0.5(a+n+2)}} \exp\left\{\frac{-1}{2\sigma^2}\left\{[V + (\underline{\beta} - \hat{\underline{\beta}})'(X'X)(\hat{\underline{\beta}})] + \frac{1}{k}[(\underline{\beta} - \underline{\beta}^*)'(\underline{\beta} - \underline{\beta}^*)] + b\right\}\right\}.$$

So

$$h(\underline{\beta}, \sigma^2) \propto \frac{1}{(\sigma^2)^{0.5(a+n+2)}} \exp \frac{-A}{2\sigma^2},$$

where

$$A = \left\{(V + b) + (\underline{\beta} - \hat{\underline{\beta}})'(X'X)(\underline{\beta} - \hat{\underline{\beta}}) + \frac{(\underline{\beta} - \underline{\beta}^*)'(\underline{\beta} - \underline{\beta}^*)}{k}\right\}.$$

(b) To get the marginal density of $\beta = (\beta_1, \beta_2)$, we integrate $h(\beta, \sigma^2)$ w.r.t. σ^2. By using Lemma 12.1 we get

$$p(\beta_1, \beta_2) \propto \int_0^\infty \frac{1}{(\sigma^2)^{0.5(a+n+2)}} \exp\left\{\frac{-A}{2\sigma^2}\right\} d\sigma^2 \propto \frac{1}{(A)^{0.5(n+a)}}.$$

Now let us write

$$A = (V + b) + (\underline{\beta} - \hat{\underline{\beta}})'(X'X)(\underline{\beta} - \hat{\underline{\beta}}) + \frac{(\underline{\beta} - \underline{\beta}^*)'(\underline{\beta} - \underline{\beta}^*)}{k}$$

as

$$A = \underline{\beta}' A_1 \underline{\beta} - 2A_2 \underline{\beta} + A_3,$$

where

$$A_1 = X'X + \frac{1}{k} I, \qquad A_2 = \hat{\underline{\beta}}' X'X + \frac{1}{k} \underline{\beta}^{*\prime},$$

and

$$A_3 = V + b + \hat{\underline{\beta}}' X'X \hat{\underline{\beta}} + \frac{1}{k} \underline{\beta}^{*\prime} \beta^{*}.$$

Thus

$$A = (\underline{\beta} - \tilde{\underline{\beta}})' D (\underline{\beta} - \tilde{\underline{\beta}}) + C,$$

where $D = X'X + (1/k)I$, $\tilde{\underline{\beta}} = D^{-1} A_2$, and $C = A_3 - \tilde{\underline{\beta}}' D \tilde{\underline{\beta}}$.

Therefore

$$f(\beta_1, \beta_2 \mid \text{data}) \propto \frac{1}{[C + (\underline{\beta} - \tilde{\underline{\beta}})' D (\underline{\beta} - \tilde{\underline{\beta}})]^{0.5(n+a)}},$$

which is the density of a multivariate t-distribution.

(c) To get marginal density of σ^2, we integrate the joint posterior w.r.t. β_1 and β_2 as follows:

$$h(\sigma^2 \mid X, \underline{y}) = \int\int f(\beta_1, \beta_2, \sigma^2 \mid \text{data})\, d\beta_1\, d\beta_2.$$

Let $\Omega^{-1} = D/\sigma^2$, then we have

$$h(\sigma^2 \mid X, y) \propto \frac{1}{(\sigma^2)^{0.5(n+a+2)}} e^{-(C/2\sigma^2)} \times \mid \Omega \mid^{0.5} \underbrace{\int_{-\infty}^{\infty} \int_{-\infty}^{\infty} \frac{\exp\{-0.5[(\underline{\beta} - \tilde{\underline{\beta}})' \Omega^{-1} (\underline{\beta} - \tilde{\underline{\beta}})]\}}{\mid \Omega \mid^{0.5}} d\beta_1\, d\beta_2}_{=1}.$$

Also we know that since D is a constant, $\mid \Omega \mid^{0.5} \propto (\sigma^2)^{2/2}$ since $\dim(X'X) = 2$, so

$$h(\sigma^2 \mid X, \underline{y}) \propto \frac{1}{(\sigma^2)^{0.5(n+a)}} - e^{-(C/2\sigma^2)},$$

which is the kernel of an inverted gamma distribution.

(d) From part (b) we have

$$h(\underline{\beta} \mid X, \underline{y}) \propto \frac{1}{C} \frac{1}{\left\{1 + (\underline{\beta} - \underline{\tilde{\beta}})(\Omega - 1/C)(\underline{\beta} - \underline{\tilde{\beta}}\right\}^{0.5(n+a)}},$$

where $\underline{\beta} = (\beta_1, \beta_2)$.

$$h(\beta_1, \beta_2 \mid X, \underline{y}) \propto \frac{1}{\{(n+a-2) + (\beta - \underline{\tilde{\beta}})^{(n+a-2)\Omega^{-1}/C}(\beta - \tilde{\beta})\}^{0.5(n+a)}},$$

$$h(\beta_1, \beta_2 \mid X, \underline{y}) \propto \frac{1}{\{(n+a-2) + (\underline{\beta} - \underline{\tilde{\beta}})\Sigma^{-1}(\underline{\beta} - \underline{\tilde{\beta}})\}^{0.5(n+a)}},$$

where $\Sigma^{-1} = (n+a-2)\beta^{-1}/C$. This is a multivariate t-distribution with $df = n + a - 2$. Now by applying Lemma 12.2, we get:

$$h(\beta_2 \mid X, y) \propto \frac{\mid \Sigma_{22} \mid^{-0.5}}{\{(n+a-2) + (\beta_2 - \tilde{\beta}_2)'\Sigma_{22}^{-1}(\beta_2 - \tilde{\beta}_2)\}^{0.5(n+a-1)}},$$

which is a univariate t-distribution. Since Σ is a 2×2 matrix and

$$\Sigma = \begin{pmatrix} \Sigma_{11} & \Sigma_{12} \\ \Sigma_{21} & \Sigma_{22} \end{pmatrix} = \begin{pmatrix} \sigma_{11} & \sigma_{12} \\ \sigma_{21} & \sigma_{22} \end{pmatrix},$$

$$h(\beta_2 \mid X, \underline{y}) \propto \frac{1}{\{(n+a-2) + [(\beta_2 - \tilde{\beta}_2)^2/\sigma_{22}]\}^{0.5(n+a-1)}}.$$

(e) The predictive density is defined as:

$$p(y^* \mid x^*, X, \underline{y}) = \int p(y^* \mid x^* \underline{\beta}, \sigma^2) h(\beta, \sigma^2 \mid X, \underline{y})\, d\underline{\beta}\, d\sigma^2.$$

But $(y^* \mid x^*, X, \underline{y}) \sim N(\beta_1 + \beta_2 x^*, \sigma^2)$ with the density as follows:

$$p(y^* \mid x^*, \underline{\beta}, \sigma^2) \propto \frac{1}{(\sigma^2)^{0.5}} \exp\left\{-\frac{1}{2\sigma^2}(y^* - X^*\underline{\beta})^2\right\},$$

where $X^*_{1\times 2} \triangleq (1 \quad x^*)$, and $\underline{\beta}_{1\times 2} = \binom{\beta_1}{\beta_2}$.

Now

$$p(y^* \mid x^*, X, y) \propto \int \frac{1}{(\sigma^2)} \exp\left\{-\frac{1}{2\sigma^2}(y^* - X^*\underline{\beta})^2\right\}$$
$$\times \frac{1}{(\sigma^2)^{0.5(n+a)}} \exp -\frac{1}{2\sigma^2}A_\beta \, d\underline{\beta} d\sigma^2,$$
$$p(y^* \mid x^*, X, y) \propto \int \frac{1}{(\sigma^2)^{0.5(n+a+1)}} \exp\left\{-\frac{1}{2\sigma^2}\left[(y^* - X^*\underline{\beta})^2 + A\right]\right\} d\underline{\beta} d\sigma^2.$$

where, as before, $A = (\underline{\beta} - \tilde{\underline{\beta}})'D(\underline{\beta} - \tilde{\underline{\beta}}) + C$.

Now complete the square in the exponent:

$$(y^* - X^*\beta)^2 + (\beta - \tilde{\beta})'D(\beta - \tilde{\beta}) + C = \beta' H\beta - 2\gamma'\beta + \delta + C,$$

where $H = D + X^{*\prime}X^*$, $\gamma' = y^*X^* + \tilde{\beta}D$, and $\delta = y^{*2} + \tilde{\beta}'D\tilde{\beta}$. But $\beta' H\beta - 2\gamma'\beta + \delta = (\beta - H^{-1}\gamma)'H(\beta - H^{-1}\gamma) + \delta - \gamma' H^{-1}\gamma)$.

So, after combining terms, we obtain:

$$p(y^* \mid x^*, X, \underline{y}) \propto \int \frac{1}{(\sigma^2)^{0.5(n+a+1)}} \exp\left\{-\frac{C + (\delta - \gamma' H^{-1}\gamma)}{2\sigma^2}\right\}$$
$$\times \iint \exp\left\{-\frac{1}{2\sigma^2}[(\beta - H^{-1}\gamma)'H(\beta - H^{-1}\gamma)]\right\} d\beta_1 \, d\beta_2 \, d\sigma^2,$$
$$p(y^* \mid x^*, X, \underline{y}) \propto \int \frac{1}{(\sigma^2)^{0.5(n+a+1)}} \exp\left\{-\frac{C + (\delta - \gamma' H^{-1}\gamma)}{2\sigma^2}\right\} K \, d\sigma^2,$$

where

$$K = \mid H^* \mid^{0.5} \iint \frac{1}{\mid H^* \mid^{0.5}} \exp\left\{-\frac{1}{2\sigma^2}[(\beta - H^{-1}\gamma)'H(\beta - H^{-1}\gamma)]\right\} d\beta_1 \, d\beta_2,$$

and

$$(H^*)^{-1} = \frac{H}{\sigma^2}.$$

But now we have

$$\iint \frac{1}{\mid H^* \mid^{0.5}} \exp\{-0.5[(\beta - H^{-1}\gamma)'H(\beta - H^{-1}\gamma)]\} \, d\beta_1 \, d\beta_2 = 1.$$

Also we have

$$\mid H^* \mid^{0.5} \propto (\sigma^2)^{2/2} = \sigma^2.$$

So after combining the above, we get

$$p(y^* \mid x^*, X, \underline{y}) \propto \int \frac{1}{(\sigma^2)^{0.5(n+a)}} \exp\left\{-\frac{C + (\delta - \gamma' H^{-1}\gamma)}{2\sigma^2}\right\} d\sigma^2.$$

By applying Lemma 12.1, we get

$$p(y^* \mid x^*, X, \underline{y}) \propto \int \frac{1}{(\sigma^2)^{0.5(n+a+1)}} e^{-(q^*/2\sigma^2)} d\sigma^2 \propto \frac{1}{(q^*)^{0.5(n+a-1)}}.$$

where $q^* = C + (\delta - \gamma' H^{-1}\gamma)$.

Now let us write q^* into a quadratic equation in terms of y^*:

$$q^* = C + (\delta - \gamma' H^{-1}\gamma) = \phi y^{*2} - 2\theta y^* + \eta,$$

where

$$\phi = 1 - X^* H^{-1} X^{*\prime}, \qquad \theta = \tilde{\beta}' D H^{-1} X^{*\prime},$$
$$\text{and } \eta = C + \tilde{\beta}' D \tilde{\beta} + \tilde{\beta}' D H^{-1} D \tilde{\beta}.$$

Now

$$\begin{aligned} \phi y^{*2} - 2\theta y^* + \eta &= \phi\left[y^{*2} - 2\left(\frac{\theta}{\phi}\right) y^* + \frac{\eta}{\phi}\right] \\ &= \phi\left[\left(y^* - \frac{\theta}{\phi}\right)^2 + \left(\frac{\eta}{\phi} - \frac{\theta^2}{\phi^2}\right)\right]. \end{aligned}$$

Thus

$$\begin{aligned} p(y^* \mid x^*, X, \underline{y}) &\propto \frac{1}{\left\{\phi\left[\left(y^* - \frac{\theta}{\phi}\right)^2 + \left(\frac{\eta}{\phi} - \frac{\theta^2}{\phi^2}\right)\right]\right\}^{0.5(n+a-1)}} \\ &\propto \frac{1}{\left[\omega^2 + \left(y^* - \frac{\theta}{\phi}\right)^2\right]^{0.5(n+a-1)}}, \end{aligned}$$

where $\omega^2 = [(\eta/\phi) - (\theta^2/\phi^2)]$.

$$\begin{aligned} p(y^* \mid x^*, X, \underline{y}) &\propto \frac{1}{\left[\omega^2 + \left(y^* - \frac{\theta}{\phi}\right)^2\right]^{0.5(n+a-1)}} \\ &\propto \frac{1}{\left[(n+a-2) + \frac{(y^* - y_0^*)^2}{\lambda^2}\right]^{0.5(n+a-1)}}, \end{aligned}$$

where

$$\lambda^2 = \frac{\omega^2}{n+a-2} \qquad \text{and} \qquad y_0^* = \frac{\theta}{\phi}.$$

Thus $(y^* \mid x^*, X, \underline{y}) \sim t_{n+a-2}$ (i.e., univariate t-distribution with $df = n + a - 2$), with density as follows:

$$p(y^* \mid x^*, X, \underline{y}) \propto \frac{1}{\left[(n+a-2) + \frac{(y^* - y_0^*)^2}{\lambda^2}\right]^{0.5(n+a-1)}}.$$

12.3
$$y_i \mid x_i = \beta_1 + \beta_2 x_i + u_i, \qquad i = 1, \ldots, n$$
$$u_i \sim N(0, \sigma_i^2) = N(0, \sigma^2 x_i).$$

Rewrite the new model as follows:

$$\frac{y_i}{\sqrt{y_i}} \bigg| x_i = \frac{\beta_1}{\sqrt{x_i}} + \frac{\beta_2}{\sqrt{x_i}} x_i + \frac{u_i}{\sqrt{x_i}}, \qquad i = 1, \ldots, n,$$

where $u_i \sim N(0, \sigma_i^2) = N(0, \sigma^2 x_i)$.

So the new model can be written as

$$y^* \mid X^* = X^* \tilde{\beta} + v,$$

where $v = (v_1, \ldots, v_n)$ and $v_i = u_i / \sqrt{x_i}$, and where, in this new model:

$$X^*_{n\times 2} = \begin{pmatrix} \frac{1}{\sqrt{x_1}} & \sqrt{x_1} \\ \vdots & \vdots \\ \frac{1}{\sqrt{x_n}} & \sqrt{x_n} \end{pmatrix}, \qquad \tilde{\beta} = \begin{pmatrix} \tilde{\beta}_1 \\ \tilde{\beta}_2 \end{pmatrix}, \qquad y^* = \frac{y_i}{\sqrt{x_i}} \; v \sim N(0, \sigma^2 I_n).$$

So all the results from Exercise 12.2 are identical, we must now change y to y^* and X to X^*.

12.4 First Approach. The model is given by

$$y \mid X = X\beta + u, \qquad \beta = \begin{pmatrix} \beta_1 \\ \beta_2 \end{pmatrix},$$

$$X_{n\times 2} = \begin{pmatrix} 1 & x_1 \\ \vdots & \vdots \\ 1 & x_n \end{pmatrix}, \qquad \text{and} \qquad \text{var}(u) = \Sigma_{n\times n} = \sigma^2 R_{n\times n},$$

where

$$R_{n\times n} = \begin{pmatrix} 1 & & \rho \\ & \ddots & \\ \rho & & 1 \end{pmatrix},$$

that is, $R_{n\times n} = (r_{ij})$, where

$$r_{ij} = \begin{cases} 1, & i = j \\ \rho, & i \neq j \end{cases},$$

$\text{corr}(u_i, u_j) = \rho, 0 < \rho < 1, i \neq j$, and $\text{var}(u_i) = \sigma^2$.

(a) **Likelihood.** The error term is normally distributed, so we have:

$$p(y \mid X, \beta, \sigma^2, \rho) \propto \frac{1}{\mid \Sigma \mid^{0.5}} \exp\{-0.5(y - X\beta)'(\Sigma^{-1}(y - X\beta)\}.$$

Prior. $p(\beta, \sigma^2, \rho) = p(\beta)p(\sigma^2)p(\rho)$.

But we have adopted vague priors, so $p(\beta) \propto$ (constant) and $p(\rho) \propto$ (constant). Thus, we obtain

$$p(\beta, \sigma^2, \rho) \propto \frac{1}{\sigma^2}.$$

Posterior. $p(\beta, \sigma^2, \rho \mid X, y) \propto p(y \mid \beta, \sigma^2, \rho, X)p(\beta, \sigma^2, \rho)$. Then,

$$p(\beta, \sigma^2, \rho \mid X, y) \propto \frac{1}{\sigma^2 \mid \Sigma \mid^{0.5}} \exp\{-0.5(y - X\beta)'\Sigma^{-1}(y - X\beta)\},$$

where $\mid \Sigma \mid^{0.5} = \mid R \mid^{0.5}(\sigma^2)^{0.5n}$. Then,

$$p(\beta, \sigma^2, \rho \mid X, y) \propto \frac{1}{(\sigma^2)^{0.5(n+2)} \mid R \mid^{0.5}} \exp\{-0.5(y - X\beta)'\Sigma^{-1}(y - X\beta)\},$$

where $\mid R \mid = [1 + (n-1)\rho](1-\rho)^{n-1}$ and

$$\Sigma^{-1} = \frac{I_n}{\sigma^2(1-\rho)} - \frac{pee'}{\sigma^2(1-\rho)[1 + (n-1)\rho]},$$

where $ee' = (1)_{n\times n}$ is an $n \times n$ matrix with all entries 1. (See Equation 2.5.9, Press, *Applied Multivariate Analysis*, 1982.)

(b) To find the marginal density of ρ, we first find the joint density of (σ^2, ρ) as follows:

$$p(\sigma^2, \rho \mid X, y) \propto \int_{-\infty}^{\infty}\int_{-\infty}^{\infty} p(\beta, \sigma^2, \rho \mid X, Y)\, d\beta_1\, d\beta_2,$$

$$p(\sigma^2, \rho \mid X, Y) \propto \int_{-\infty}^{\infty}\int_{-\infty}^{\infty} \frac{1}{(\sigma^2)^{0.5(n+2)} \mid R \mid^{0.5}} \times \exp\{-0.5(y - X\beta)'\Sigma^{-1}(y - X\beta)\}\, d\beta_1\, d\beta_2,$$

$$p(\sigma^2, \rho \mid X, Y) \propto \frac{1}{(\sigma^2)^{0.5(n+2)} \mid R \mid^{0.5}} \int_{-\infty}^{\infty}\int_{-\infty}^{\infty} \times \exp\{-0.5(y - X\beta)'\Sigma^{-1}(y - X\beta)\}\, d\beta_1\, d\beta_2.$$

Let

$$J = \int_{-\infty}^{\infty}\int_{-\infty}^{\infty} e^{-0.5Q}\, d\beta_1\, d\beta_2,$$

where

$$Q = [(y - X\beta)'\Sigma^{-1}(y - X\beta)].$$

But we can write:

$$Q = \beta'(X'\Sigma^{-1}X)\beta - 2\beta'\Sigma^{-1}y + y'\Sigma^{-1}y.$$

Let $\tilde{\beta} = (X'\Sigma^{-1}X)^{-1}X'\Sigma^{-1}y$ be the GLSE of β. Then we have

$$Q = (\beta - \tilde{\beta})'(X'\Sigma^{-1}X)(\beta - \tilde{\beta}) - \tilde{\beta}'(X'\Sigma^{-1}X)\tilde{\beta} + y'\Sigma^{-1}y.$$

After substitution we obtain:

$$\begin{aligned} J &= \int_{-\infty}^{\infty}\int_{-\infty}^{\infty} \exp\{-0.5[(\beta - \tilde{\beta})'(X'\Sigma^{-1}X)(\beta - \tilde{\beta}) \\ &\quad - \tilde{\beta}'(X'\Sigma^{-1}X)\tilde{\beta} + y'\Sigma^{-1}y]\}\, d\beta_1\, d\beta_2 \\ &= \exp\left\{-0.5[y'\Sigma^{-1}y - \tilde{\beta}'(X'\Sigma^{-1}X)\tilde{\beta}]\right\} \int_{-\infty}^{\infty}\int_{-\infty}^{\infty} \\ &\quad \times \exp\{-0.5[(\beta - \tilde{\beta})'(X'\Sigma^{-1}X)(\beta - \tilde{\beta})]\}\, d\beta_1\, d\beta_2. \end{aligned}$$

Let $\Omega^{-1} = X'\Sigma^{-1}X$. Then we can write:

$$J = \exp\{-0.5[y'\Sigma^{-1}y - \tilde{\beta}'(X'\Sigma^{-1}X)\tilde{\beta}]\}$$

$$\cdot \mid \Omega \mid^{0.5} \overbrace{\int_{-\infty}^{\infty}\int_{-\infty}^{\infty} \frac{1}{\mid \Omega \mid^{0.5}} \exp\left\{-0.5\{(\beta - \tilde{\beta})'(X'\Sigma^{-1}X)(\beta - \tilde{\beta})\}\right\} d\beta_1\, d\beta_2}^{=1}.$$

Thus we obtain

$$J \propto \mid \Omega \mid^{0.5} \exp\{-0.5[y'\Sigma^{-1}y - \tilde{\beta}'(X'\Sigma^{-1}X)\tilde{\beta}]\},$$

where

$$\mid \Omega \mid^{0.5} \propto \frac{1}{\mid X'\Sigma^{-1}X \mid^{0.5}} = \frac{1}{\mid X'(R - 1/\sigma^2)X \mid^{0.5}} = \frac{(\sigma^2)^{2/2}}{\mid X'R^{-1}X \mid^{0.5}},$$

since $\dim(X'R^{-1}X) = 2$. Thus

$$J \propto \mid \Omega \mid^{0.5} \exp\{-0.5[y'\Sigma^{-1}y - \tilde{\beta}'(X'\Sigma^{-1}X)\tilde{\beta}]\}$$

$$\propto \frac{\sigma^2}{\mid X'R^{-1}X \mid^{0.5}} \exp\left\{-\frac{1}{2\sigma^2}[y'R^{-1}y - \tilde{\beta}'(X'R^{-1}X)\tilde{\beta}]\right\}.$$

Note that $[y'R^{-1}y - \tilde{\beta}'(X'R^{-1}X)\tilde{\beta}] = \operatorname{tr} GR^{-1}$.
So

$$p(\sigma^2, \rho \mid X, y) \propto \frac{e^{-[(1/2\sigma^2)\operatorname{tr} GR^{-1}]}}{(\sigma^2)^{0.5n} \mid X'R^{-1}X \mid^{0.5} \mid R \mid^{0.5}}.$$

Now to get the density of ρ, we integrate above w.r.t. σ^2.

$$p(\rho \mid X, y) \propto \int_0^{\infty} p(\sigma^2, \rho) d\sigma^2,$$

$$\propto \int_0^{\infty} \frac{e^{-[(1/2\sigma^2)\operatorname{tr} GR^{-1}]}}{(\sigma^2)^{0.5n} \mid X'R^{-1}X \mid^{0.5} \mid R \mid^{0.5}} d\sigma^2,$$

$$= \frac{1}{\mid X'R^{-1}X \mid^{0.5} \mid R \mid^{0.5}} \int_0^{\infty} \frac{e^{-[(1/2\sigma^2)\operatorname{tr} GR^{-1}]}}{(\sigma^2)^{0.5n}} d\sigma^2.$$

Now we again apply Lemma 12.1:

$$\int_0^\infty \frac{e^{-[(1/2\sigma^2)\alpha]}}{(\sigma^2)^{0.5n}}\, d\sigma^2 \propto \frac{1}{\alpha^{(0.5n)-1}},$$

where $\alpha = \operatorname{tr} GR^{-1}$. So we will get

$$p(\rho \mid X, Y) \propto \frac{1}{\mid R \mid^{0.5} \mid X'R^{-1}X \mid^{0.5} [\operatorname{tr}(GR^{-1})]^{(0.5n)-1}}.$$

The above is a function of R and therefore a function of ρ.

(c) To find the density of $\underline{\beta}$ and then the density of β_1, it is better to find the joint density of (β, ρ) first, as follows:

$$p(\beta, \rho \mid X, y) = \int_0^\infty p(\beta, \sigma^2, \rho \mid X, Y)\, d\sigma^2,$$

$$p(\beta, \rho \mid X, y) \propto \int_0^\infty \frac{1}{(\sigma^2)^{0.5(n+2)} \mid R \mid^{0.5}} \exp\{-0.5(y - X\beta)'\Sigma^{-1}(y - X\beta)\}\, d\sigma^2,$$

$$p(\beta, \rho \mid X, y) \propto \frac{1}{\mid R \mid^{0.5}} \int_0^\infty \frac{1}{(\sigma^2)^{(0.5n)+1}} \exp\left\{-\frac{1}{2\sigma^2}(y - X\beta)'R^{-1}(y - X\beta)\right\} d\sigma^2.$$

By applying Lemma 12.1 we obtain:

$$p(\beta, \rho \mid X, y) \propto \frac{1}{\mid R \mid^{0.5} [(y - X\beta)'R^{-1}(y - X\beta)]^{0.5n}}.$$

Now we integrate w.r.t. ρ to get the density of β:

$$p(\beta \mid X, y) = \int_{-1}^{1} p(\beta, \rho \mid X, y)\, d\rho,$$

$$\propto \int_{-1}^{1} \frac{1}{\mid R \mid^{0.5}[(y - X\beta)'R^{-1}(y - X\beta)]^{0.5n}}\, d\rho.$$

The above integral can be evaluated by numerical methods, and from there we can estimate the density of β_1.

(d) As we have in part (b):

$$p(\sigma^2, \rho \mid X, y) \propto \frac{e^{-[(1/2\sigma^2)\operatorname{tr} GR^{-1}]}}{(\sigma^2)^{0.5n} \mid X'R^{-1}X \mid^{0.5} \mid R \mid^{0.5}}.$$

From this we obtain:

$$p(\sigma^2 \mid X, y) \propto \frac{1}{(\sigma^2)^{0.5n}} \int_{-1}^{1} \frac{e^{-[(1/2\sigma^2)\,\mathrm{tr}\, GR^{-1}]}}{\mid X'R^{-1}X \mid^{0.5} \mid R \mid^{0.5}} \, d\rho.$$

But this integral cannot be evaluated by ordinary techniques of integration, so numerical methods must be used.

(e) The predictive density is given by

$$p(y^* \mid X^*, X, y) = \int p(y^* \mid x^*, \beta, \sigma^2) p(\beta, \sigma^2, \rho \mid X, y) \, d\beta \, d\sigma^2 \, d\rho.$$

Let $X^*_{1\times 2} = (1, x^*)$, so we have

$$p(y^* \mid x^*, \beta, \sigma^2) = \frac{1}{(\sigma^2)^{0.5}} \exp\left\{ -\frac{1}{2\sigma^2} [(y - X^*\beta)'(y - X^*\beta)] \right\},$$

and from the posterior we have

$$p(\beta, \sigma^2, \rho \mid X, y) \propto \frac{1}{(\sigma^2)^{(0.5n)+1} \mid R \mid^{0.5}} \exp\left\{ -\frac{1}{2\sigma^2} [(y - X\beta)'(y - X\beta)] \right\}.$$

After combining we obtain:

$$p(y^* \mid x^*, X, y) = \int_{R^4} \frac{1}{(\sigma^2)^{0.5(n+3)} \mid R \mid^{0.5}} e^{-[(1/2\sigma^2)Q]} \, d\beta \, d\sigma^2 \, d\rho,$$

where $Q = (y - X\beta)'R^{-1}(y - X\beta) + (y - X^*\beta)'(y - X^*\beta)$, or equivalently, we have

$$\begin{aligned} Q &= \beta'(X'R^{-1}X + X^{*\prime}X^*)\beta - 2\beta'(X'R^{-1}y + X^{*\prime}y^*) + (y'R^{-1}y + y^{*2}) \\ &= (\beta - \beta^*)'\Delta^{-1}(\beta - \beta^*) - \beta^{*\prime}\Delta^{-1}\beta^* + y'R^{-1}y + y^{*2}, \end{aligned}$$

where $\beta^* = \Delta^{-1}\gamma'$, $\gamma' = X'R^{-1}y + X^{*\prime}y^*$, and $\Delta^{-1} = X'R^{-1}X + X^{*\prime}X^*$.

Let $\Omega^{-1} = \Delta^{-1}/\sigma^2$. Then we obtain

$$p(y^* \mid x^*, X, y) \propto \int_{R^4} \frac{1}{(\sigma^2)^{0.5(n+2)} \mid R \mid^{0.5}} \cdot \exp\left\{-\frac{1}{2\sigma^2}[(\beta - \beta^*)'\Delta^{-1}(\beta - \beta^*) - \beta^{*\prime}\Delta^{-1}\beta^* + y'R^{-1}y + y^{*2}]\right\} d\beta\, d\sigma^2\, d\rho$$

$$= \int_{R^2} \frac{1}{(\sigma^2)^{0.5(n+2)} \mid R \mid^{0.5}} \exp\left\{-\frac{1}{2\sigma^2}(-\beta^{*\prime}\Delta^{-1}\beta^* + y'R^{-1}y + y^{*2})\right\}$$

$$\cdot \mid \Omega \mid^{0.5} \overbrace{\int_{-\infty}^{\infty}\int_{-\infty}^{\infty} \frac{1}{\mid \Omega \mid^{0.5}} \exp\{-0.5[(\beta - \beta^*)'\Omega^{-1}(\beta - \beta^*)]\}\, d\beta\, d\sigma^2\, d\rho}^{=1}.$$

But we have

$$\mid \Omega \mid^{0.5} \propto \frac{(\sigma^2)^{2/2}}{\mid \Delta^{-1} \mid^{0.5}},$$

since $\dim(\delta^{-1}) = 2$.

So, by substitution, we obtain:

$$p(y^* \mid x^*, X, y) \propto \int_{-1}^{1} \frac{1}{\mid R \mid^{0.5} \mid \Delta^{-1} \mid^{0.5}} \int_0^{\infty} \frac{1}{(\sigma^2)^{0.5(n+1)}} \times \exp\left\{-\frac{1}{2\sigma^2}(-\beta^{*\prime}\Delta^{-1}\beta^* + y'R^{-1}y + y^{*2})\right\} d\sigma^2\, d\rho.$$

Now if we apply Lemma 12.1 to the inner integral, we have

$$\int_0^{\infty} \frac{1}{(\sigma^2)^{0.5(n+1)}} \exp\left\{-\frac{1}{2\sigma^2}(-\beta^{*\prime}\Delta^{-1}\beta^* + y'R^{-1}y + y^{*2})\right\} d\sigma^2 \propto \frac{1}{(-\beta^{*\prime}\Delta^{-1}\beta^* + y'R^{-1}y + y^{*2})^{0.5n}}.$$

Hence

$$p(y^* \mid x^*, X, y) \propto \int_{-1}^{1} \frac{1}{\mid R \mid^{0.5} \mid \Delta^{-1} \mid^{0.5} (-\beta^{*\prime}\Delta^{-1}\beta^* + y'R^{-1}y + y^{*2})^{0.5n}}\, d\rho.$$

Once again the above integral can only be evaluated by numerical methods.

Second Approach.

$$y_{n\times 1} = X_{n\times 2}\beta_{2\times 1} + u_{n\times 1}, \qquad u \sim N(0, \sigma^2 R),$$

and

$$R^{-0.5}y = R^{-0.5}X\beta + R^{-0.5}u,$$

where

$$R = \begin{pmatrix} 1 & & \rho \\ & \ddots & \\ \rho & & 1 \end{pmatrix}_{n\times n},$$

as defined before.

So the new model can be written as

$$y_0 = X_0\beta + u_0, \qquad u_0 \sim N(0, \sigma^2 I_n),$$

where

$$y_0 = R^{-0.5}y, \qquad X_0 = R^{-0.5}X, \qquad \text{and} \qquad u_0 = R^{-0.5}u.$$

But $\tilde{\beta} = (X_0'X_0)^{-1}X_0'y_0$ is the least square estimate of this model, so by substitution we get

$$\tilde{\beta} = (X'R^{-1}X)^{-1}X'R^{-1}y.$$

Likelihood.

$$p(y \mid X, \beta, \sigma^2, R) \propto \frac{1}{\mid \sigma^2 R \mid^{0.5}} \exp\left\{-\frac{1}{2\sigma^2}[(y - X\beta)'R^{-1}(y - X\beta)]\right\}.$$

By Lemma 12.4, we have

$$(y - X\beta)'R^{-1}(y - X\beta) = (y - X\hat{\beta})'R^{-1}(y - X\hat{\beta}) + (\beta - \tilde{\beta})'X'R^{-1}(\beta - \tilde{\beta}).$$

Therefore

$$p(y \mid X, \beta, \sigma^2, \rho) \propto \frac{1}{\mid \sigma^2 R \mid^{0.5}} \exp\left\{-\frac{1}{2\sigma^2}[(y - X\hat{\beta})'R^{-1}(y - X\hat{\beta}) + (\beta - \tilde{\beta})'X'R^{-1}X(\beta - \tilde{\beta})]\right\}.$$

Prior.
$p(\beta, \sigma^2, R) = p(\beta)p(\sigma^2)p(R) \propto \text{constant} \cdot 1/\sigma^2 \cdot \text{constant} \propto 1/\sigma^2.$

(a) **Posterior.**

$$p(\beta, \sigma^2, \rho \mid y, X) \propto \frac{1}{(\sigma^2)^{(0.5n)+1} \mid R \mid^{0.5}} \exp\left\{-\frac{1}{2\sigma^2}[(y - X\hat{\beta})'R^{-1}(y - X\hat{\beta}) + (\beta - \tilde{\beta})'X'R^{-1}X(\beta - \tilde{\beta})]\right\}.$$

Let $W = (y - X\hat{\beta})'R^{-1}(y - X\hat{\beta})$ and $\Omega^{-1} = X'R^{-1}X/\sigma^2$.

Then,

$$p(\beta, \sigma^2, \rho \mid y, X) \propto \frac{e^{-[(1/2\sigma^2)W]}}{(\sigma^2)^{(0.5n)+1} \mid R \mid^{0.5}} \exp\left\{-\frac{1}{2\sigma^2}[(\beta - \tilde{\beta})'\Omega^{-1}(\beta - \tilde{\beta})]\right\},$$

$$p(\sigma^2, \rho \mid y, X) \propto \frac{e^{-[(1/2\sigma^2)W]}}{(\sigma^2)^{(0.5n)+1} \mid R \mid^{0.5}} \mid \Omega \mid^{0.5} \times \overbrace{\int_{-\infty}^{\infty}\int_{-\infty}^{\infty} \frac{\exp\left\{\frac{1}{2\sigma^2}[(\beta - \tilde{\beta})'\Omega^{-1}(\beta - \tilde{\beta})]\right\}}{\mid \Omega \mid^{0.5}}\, d\beta_1\, d\beta_2}^{=1},$$

$$p(\sigma^2, \rho \mid y, X) \propto \frac{e^{-[(1/2\sigma^2)W]}}{(\sigma^2)^{(0.5n)+1} \mid R \mid^{0.5}} \mid \Omega \mid^{0.5},$$

where

$$\mid \Omega \mid^{0.5} = \frac{1}{\mid X'R^{-1}X/\sigma^2 \mid^{0.5}}.$$

$$p(\sigma^2, \rho \mid y, X) \propto \frac{e^{-[(1/2\sigma^2)W]}}{(\sigma^2)^{(0.5n)+1} \mid R \mid^{0.5} \mid X'R^{-1}X/\sigma^2 \mid^{0.5}} = \frac{e^{-[(1/2\sigma^2)W]}}{(\sigma^2)^{0.5n} \mid X'R^{-1}X \mid^{0.5} \mid R \mid^{0.5}},$$

$$p(\rho \mid y, X) \propto \int_0^{\infty} \frac{e^{-[(1/2\sigma^2)W]}}{(\sigma^2)^{0.5n} \mid X'R^{-1}X \mid^{0.5} \mid R \mid^{0.5}}\, d\sigma^2 = \frac{1}{\mid X'R^{-1}X \mid^{0.5} \mid R \mid^{0.5}} \int_0^{\infty} \frac{e^{-[(1/2\sigma^2)W]}}{(\sigma^2)^{0.5n}}\, d\sigma^2.$$

By applying Lemma 12.1, we have

$$p(\rho \mid y, X) \propto \frac{1}{\mid X'R^{-1}X \mid^{0.5} \mid R \mid^{0.5} [(y - X\hat{\beta})'R^{-1}(y - X\hat{\beta})]^{(0.5n)-1}}.$$

Third Approach (Linear Algebra Approach).
$Y = X\beta + U$, where $U \sim N(\underline{0}, \Sigma)$, and $\Sigma = \sigma^2 R$, where R is defined as before. Let

$$\Gamma_{n \times n} = \frac{1}{\sqrt{n}} \begin{pmatrix} \underline{e}' \\ \vdots \\ \underline{e}' \end{pmatrix}$$

be a matrix such that $\Gamma'\Gamma = I_n$.

Then $Y\Gamma = \Gamma X\beta + \Gamma U = X_0\beta + U_0$, where $U_0 \sim N(\underline{0}, D)$ and $D = (d_{ij})$ where

$$d_{ij} = \begin{cases} \alpha, & \text{if } i = j = 1 \\ \gamma, & \text{if } i = j \neq 1 \\ 0, & \text{otherwise} \end{cases}$$

So D is a diagonal matrix with $\alpha = \sigma^2[1 + (n-1)\rho]$ and $\gamma = \sigma^2(1 - p)$ being the eigenvalues of the matrix R.

Define $Y_0 = \begin{pmatrix} z_1 \\ z_2 \end{pmatrix}$ where z_1 is 1×1 and z_2 is $(n-1) \times 1$.

Also, let $X_0\beta = \begin{pmatrix} X_1 & 0 \\ 0 & X_2 \end{pmatrix} \begin{pmatrix} \delta_1 \\ \delta_2 \end{pmatrix} = \begin{pmatrix} X_1\delta_1 \\ X_2\delta_2 \end{pmatrix}.$

So, we have two new models:

$$\begin{cases} z_1 = X_1\delta_1 + u_1, & u_1 \sim N(0, \alpha), \text{ i.e., univariate} \\ z_2 = X_2\delta_2 + u_2, & u_2 \sim N(0, \gamma I_{n-1}), \text{ i.e., multivariate} \end{cases}$$

Now we use the results of Exercise 12.2 for the two models.

12.11 **Prior.** $p(\beta, \sigma^2) = p(\beta \mid \sigma^2)p(\sigma^2)$ where $p(\sigma^2) \propto 1/\sigma^2$ and the distribution of $\beta \mid \sigma^2$ is given by

$$p(\beta \mid \sigma^2) \propto \frac{1}{(\sigma^2)^{0.5(p+1)}} \exp\left\{ -\frac{g}{2\sigma^2} (\underline{\beta} - \bar{\beta})'(X'X)(\underline{\beta} - \bar{\beta}) \right\}.$$

Likelihood.

$$L(\underline{y} \mid X, \underline{\beta}, \sigma^2) \propto \frac{1}{(\sigma^2)^{0.5n}} \exp\left\{ -\frac{1}{2\sigma^2} (\underline{y} - \underline{\beta})'(\underline{y} - \underline{\beta}) \right\}.$$

(a) **Posterior.**

$$p(\beta, \sigma^2 \mid \text{data}) \propto \frac{1}{(\sigma^2)^{0.5(n+k)+1}} \exp\left\{-\frac{1}{2\sigma^2} A\right\},$$

where

$$A = \left[(\underline{\beta} - \bar{\beta})'(X'X)(\underline{\beta} - \bar{\beta}) + (\underline{y} - \beta)'g(\underline{y} - \beta)\right],$$

and $k \equiv p + 1$.

(b) The marginal posterior density of β is given by

$$p_1(\beta \mid \text{data}) \propto \int_0^\infty p(\beta, \sigma^2 \mid D) d\sigma^2 \propto \frac{1}{A^{0.5(n+k)+1}},$$

by using Lemma 12.1, where

$$A = [(\underline{\beta} - \bar{\beta})'(X'X)(\underline{\beta} - \bar{\beta}) + (\underline{y} - \beta)'g(\underline{y} - \beta)].$$

After completing the square, we will get

$$A = (\beta - \beta^*)'Q(\beta - \beta^*) + H,$$

where

$$\beta^8 = (X'X)^{-1}\frac{(X'y + gX'X\bar{\beta})}{1 + g}, \qquad Q = (1 + g)X'X,$$

and $H = y'y + g\bar{\beta}'X'X\bar{\beta}$.

So we have

$$p_1(\beta \mid \text{data}) \propto \frac{1}{[H + (\beta - \beta^*)'Q(\beta - \beta^*)]^{0.5(n+k)}},$$

$$p_1(\beta \mid \text{data}) \propto \frac{1}{[n + (\beta - \beta^*)'Q^*(\beta - \beta^*)]^{0.5(n+k)}},$$

where $Q^* = (Q/H)n$.

Thus the distribution of $\beta \mid$ data is multivariate t with $df = n$.

Also the marginal distribution of $\beta_i, i = 1, 2, \ldots, p$ is multivariate t with appropriate degrees of freedom.

(c) To get the marginal density of σ^2, simplify the posterior density as follows:

$$p(\beta, \sigma^2 \mid \text{data}) \propto \frac{1}{(\sigma^2)^{0.5(n+k)+1}} \exp\left\{-\frac{1}{2\sigma^2}[(\beta - \beta^*)' Q^{-1}(\beta - \beta^*) + H^*]\right\},$$

$$p(\sigma^2) \propto \int \frac{1}{(\sigma^2)^{0.5(n+k)+1}} \exp\left\{-\frac{1}{2\sigma^2}[(\beta - \beta^*)' Q^{-1}(\beta - \beta^*) + H]\right\} d\beta,$$

$$p(\sigma^2 \mid \text{data}) \propto \frac{e^{-(H/2\sigma^2)}}{(\sigma^2)^{0.5(n+k)+1}} \int \exp\left\{-\frac{1}{2\sigma^2}[(\beta - \beta^*)' Q^{-1}(\beta - \beta^*)]\right\} d\beta,$$

$$p(\sigma^2 \mid \text{data}) \propto \frac{e^{-(H/2\sigma^2)} \mid \Sigma \mid^{0.5}}{(\sigma^2)^{0.5(n+k)+1}} \int \frac{1}{\mid \Sigma \mid^{0.5}} \exp\left\{-\frac{1}{2\sigma^2}[(\beta - \beta^*)' Q^{-1}(\beta - \beta^*)]\right\} d\beta,$$

$$p(\sigma^2 \mid \text{data}) \propto \frac{e^{-(H/2\sigma^2)} \mid \Sigma \mid^{0.5}}{(\sigma^2)^{0.5(n+k)+1}} \underbrace{\int \frac{1}{\mid \Sigma \mid^{0.5}} \exp\left\{-\frac{1}{2\sigma^2}[(\beta - \beta^*)' Q^{-1}(\beta - \beta^*)]\right\} d\beta,}_{=1}$$

We also have: $\mid \Sigma \mid^{0.5} = (\sigma^2)^{0.5(p+1)}$. So, we will get,

$$p(\sigma^2 \mid \text{data}) \propto \frac{e^{-(H/2\sigma^2)}}{(\sigma^2)^{0.5(n+k)+1}} (\sigma^2)^{0.5(n+1)} = \frac{e^{-(H/2\sigma^2)}}{(\sigma^2)^{0.5(n+k-p+1)}}.$$

Or, equivalently,

$$p(\sigma^2 \mid \text{data}) \propto \frac{e^{-(H/2\sigma^2)}}{(\sigma^2)^{0.5(n+k-p+1)}},$$

which is the kernel of the density of an inverted gamma distribution.

(d) The marginal density of β_1 is univariate t, as was observed in part (b).

(e) The predictive density is given by

$$p(y^* \mid X^*) = \iint p(y^* \mid X^*, \beta, \sigma^2) p(\beta, \sigma^2 \mid \text{data}) d\beta d\sigma^2.$$

After substitution we get

$$p(y^* \mid X^*) \propto \iint \frac{1}{(\sigma^2)^{0.5(n+k)+2}} \exp\left\{-\frac{1}{2\sigma^2} A\right\} d\beta\, d\sigma^2,$$

where

$$A = (y^* - \beta X^*)'(y^* - \beta X^*) + (\beta - \beta^*)' Q^{-1}(\beta - \beta^*) + H.$$

Now let us rewrite the quantity A as follows:

$$\begin{aligned} A &= (y^* - \beta X^*)'(y^* - \beta X^*) + (\beta - \beta^*)'Q^{-1}(\beta - \beta^*) + H \\ &= (\beta - \check{\beta})'(X'X + Q^{-1})(\beta - \check{\beta}) + G, \end{aligned}$$

where

$$\check{\beta} = (X'X + Q^{-1})^{-1}(X^{*\prime}y^* + Q^{-1}\beta^*),$$

and

$$G = H + y^{*\prime}y^* + \beta^{*\prime}Q^{-1}\beta^* - \check{\beta}'(X'X + Q^{-1})\check{\beta},$$

$$\begin{aligned} p(y^* \mid X^*) \propto \iint \frac{1}{(\sigma^2)^{0.5(n+k)+2}} \exp\bigg\{ -\frac{1}{2\sigma^2} [(\beta - \check{\beta})'(X'X \\ + Q^{-1})(\beta - \check{\beta}) + G] \bigg\} \, d\beta \, d\sigma^2. \end{aligned}$$

Let $\Sigma^{-1} = (X'X + Q^{-1})/\sigma^2$. Then we have

$$\begin{aligned} p(y^* \mid X^*) \propto \iint \frac{e^{-(G/2\sigma^2)} \mid \Sigma \mid^{0.5}}{(\sigma^2)^{0.5(n+k)+2} \mid \Sigma \mid^{0.5}} \\ \times \exp\bigg\{ -\frac{1}{2\sigma^2} [(\beta - \check{\beta})'\Sigma^{-1}(\beta - \check{\beta})] \bigg\} \, d\beta \, d\sigma^2. \end{aligned}$$

Now

$$\int \exp\bigg\{ -\frac{1}{2\sigma^2} [(\beta - \check{\beta})'\Sigma^{-1}(\beta - \check{\beta})] \bigg\} \, d\beta \propto \mid \Sigma^{-1} \mid \propto (\sigma^2)^{0.5(p+1)}.$$

Thus

$$p(y^* \mid X^*) \propto \int \frac{e^{-(G/2\sigma^2)}}{(\sigma^2)^{0.5(n+k)+2}} (\sigma^2)^{0.5(p+1)} \, d\sigma^2 \propto \frac{1}{[G]^{0.5(n+k-p)}}.$$

Use $\check{\beta} = (X'X + Q^{-1})^{-1}(X^{*\prime}y^* + Q^{-1}\beta^*)$ in $G = H + y^{*\prime}y^* + \beta^{*\prime}Q^{-1}\beta^* - \check{\beta}'(X'X + Q^{-1})\check{\beta}$. After some algebra, we get

$$G = [1 - X^*(X'X + Q^{-1})X^{*\prime}]\{(y^* - \tilde{y})^2 - \tilde{y}^2\} + H + \beta^{*\prime}Q^{-1}\check{\beta}^*,$$

or

$$G = \left\{ H^* + \frac{(y^* - \tilde{y})^2}{s^2} \right\},$$

for some appropriate choice of $s, \tilde{y}$, and H^*.

Thus the density of $y^* \mid X^*$ is given by:

$$p(y^* \mid X^*) \propto \frac{1}{\{H^* + [(y^* - \tilde{y})^2/s^2]\}^{0.5(n+k-p)}},$$

which is the kernel of a univariate t-distribution.

12.15 Model.

$$y_{ijk} = \mu + \alpha_i + \beta_j + \gamma_{ij} + u_{ijk},$$

$$i = 1, 2, \ldots, I, \quad j = 1, 2, \ldots, J, \quad k = 1, 2, \ldots, n_{ij}.$$

Let $N = \sum_{i=1}^{I} \sum_{j=1}^{J} n_{ij}$,

$$Y' = \left[y_{111}, y_{112}, \ldots, y_{11n_{1J}}, \ldots, y_{211}, \ldots, y_{IJn_{IJ}}\right]_{(p\times N)}$$

$$X = \begin{bmatrix} e_{n_{11}} & 0 & \ldots & 0 \\ 0 & \ddots & & 0 \\ 0 & \ldots & & e_{n_{IJ}} \end{bmatrix}_{(N\times q)},$$

$$U' = [u_{111}, u_{112}, \ldots, u_{11n_{IJ}}, \ldots, u_{211}, \ldots, u_{IJn_{IJ}}]_{(p\times N)},$$

$$\theta_{ij} = \mu_{(p\times 1)} + \alpha_i + \beta_j + \gamma_{ij}, \qquad B' = [\theta_{11}, \theta_{12}, \ldots, \theta_{IJ}]'_{(p\times q)}, \qquad q = IJ.$$

Model.

$$Y \mid X = XB + U, \qquad u_{ijk} \overset{iid}{\sim} N(0, \Sigma).$$

Then apply the techniques for a one-way layout. If you want to make inferences about the main effects or the interaction effects, then define $\psi = c_1 B c_2$, where c_1 and c_2 are fixed matrices, and study the posterior distribution of ψ.

12.16 To generalize the one-way MANOVA to a one-way MANOCOVA we can consider the model

$$y_{ij} = \mu_{(p\times 1)} + \alpha_i + \Lambda'_{(p\times K)} R_{ij(K\times 1)} + u_{ij},$$
$$i = 1, 2, \ldots, q, \qquad j = 1, 2, \ldots, n_i,$$

where

K = number of covariates,
p = number of response variables,
q = number of populations,
R_{ij} is a vector of concomitant variables,
Λ is a matrix of unknown coefficients that are assumed to be the same for every i and j, and
$N = \sum_{i=1}^{q} n_i$.

Assume $u_{ij} \overset{iid}{\sim} N(0, \Sigma)$.

$$\text{Define } X_1 = \begin{bmatrix} e_{n_1} & 0 & \ldots & 0 \\ 0 & \ddots & & 0 \\ 0 & \ldots & & e_{n_q} \end{bmatrix}_{(N\times q)}$$

$$X_2' = [R_{11}, \ldots, R_{1n_1}, R_{21}, \ldots, R_{qn_q}]_{(K\times N)}, \qquad \theta_{i(p\times 1)} = \mu + \alpha_i,$$
$$B' = [\theta_1, \ldots, \theta_q]_{(p\times q)}, \qquad X = (X_1, X_2), \qquad G = (B', \Lambda'),$$
$$Y = [y_{11}, \ldots, y_{qn_q}]'_{(N\times p)}, \qquad U = [u_{11}, \ldots, u_{qn_q}]'_{(N\times p)}.$$

Then, the model becomes: $Y = X_1 B + X_2 \Lambda + U = XG + U$, with likelihood:

$$P(Y \mid X, G, \Sigma) \propto |\Sigma|^{-0.5N} \exp[-0.5 \operatorname{tr}(W + (G - \hat{G})'(X'X)(G - \hat{G}))\Sigma^{-1}],$$

where $W = (Y - X\hat{G})'(Y - X\hat{G})$.

12.17 In the example of section 12.4.8 there was a sample size of 100. When we generally have a large sample size, the prior information that was used is overpowered by the information given in the data.

Recall that

$$\tilde{\theta} = K\theta^* + (I - K)\hat{\theta},$$

where $\tilde{\theta}$ is the Bayesian estimate, θ^* is the prior, and $\hat{\theta}$ is the MLE.

In this example, the Bayesian estimators were influenced very little by the prior information since the values of K were relatively small. The calculation of K in this example is as follows:

$$\hat{\Sigma}^{-1} = \begin{bmatrix} 0.0001750 & -0.001 \\ -0.001 & 0.002 \end{bmatrix},$$

$$\Phi\hat{\Sigma}^{-1} = \begin{bmatrix} 2500 & 1250 \\ 1250 & 2500 \end{bmatrix} \begin{bmatrix} 0.0001750 & -0.001 \\ -0.001 & 0.002 \end{bmatrix} = \begin{bmatrix} 0.3125 & -0.03125 \\ 0.0000 & 0.37500 \end{bmatrix},$$

$$I_p + n\Phi\hat{\Sigma}^{-1} = \begin{bmatrix} 32.250 & -3.125 \\ 0.000 & 38.500 \end{bmatrix},$$

$$[I_p + n\Phi\hat{\Sigma}^{-1}]^{-1} = \begin{bmatrix} 0.031007752 & 0.002516863 \\ 0.000000000 & 0.025974026 \end{bmatrix} = \psi,$$

$$K = I_q \otimes [I_p + n\Phi\hat{\Sigma}^{-1}]^{-1} = \begin{bmatrix} \psi & 0 & 0 \\ 0 & \psi & 0 \\ 0 & 0 & \psi \end{bmatrix}_{(6\times 6)},$$

$$K\theta^* = \begin{bmatrix} 14.2480 & 0 & 0 \\ 11.0390 & 0 & 0 \\ 0 & 14.2480 & 0 \\ 0 & 11.0390 & 0 \\ 0 & 0 & 14.2480 \\ 0 & 0 & 11.0390 \end{bmatrix},$$

$$(I - K)\hat{\theta} = \begin{bmatrix} 444.604 & 0 & 0 \\ 438.312 & 0 & 0 \\ 0 & 405.895 & 0 \\ 0 & 418.831 & 0 \\ 0 & 0 & 376.749 \\ 0 & 0 & 448.052 \end{bmatrix}.$$

Note that the MLEs are a big determinator of the Bayesian estimates as compared to the priors that were used.

CHAPTER 14

14.1 Define the vector of data $\mathbf{y} = (y_1, y_2, \ldots, y_n)'$. Marginally, $y_i \mid \mu \sim N(\mu, \sigma^2 + \tau^2)$. The sufficient statistic is $\bar{y} \mid \mu \sim N(\mu, (\sigma^2 + \tau^2)/n)$, so with a flat prior, $\mu \mid \mathbf{y} \sim N(\bar{y}, (\sigma^2 + \tau^2)/n)$. Now since $y_i \mid \theta_i, \mu \sim N(\theta_i, \sigma^2)$ and $\theta_i \mid \mu \sim N(\mu, \tau^2)$, then the posterior conditional distribution is the usual precision-weighted distribution

$$\theta_i \mid \mathbf{y}, \mu \sim N[(\sigma^{-2} y_i + \tau^{-2}\mu)/(\sigma^{-2} + \tau^{-2}), 1/(\sigma^{-2} + \tau^{-2})].$$

We finally integrate over the posterior distribution of $\mu \mid \mathbf{y}$, using

$$\mathbf{E}\theta_i \mid \mathbf{y} = \mathbf{E}_\mu \mathbf{E}\theta_i \mid \mathbf{y}, \mu$$

and

$$\text{var } \theta_i \mid \mathbf{y} = \text{var}_\mu(\mathbf{E}\theta_i \mid \mathbf{y}, \mu) + \mathbf{E}_\mu(\text{var } \theta_i \mid \mathbf{y}, \mu).$$

The expectation is obtained by substitution. The first term of the variance is $[\tau^{-2}/(\sigma^{-2} + \tau^{-2})]^2$ var $\mu \mid \mathbf{y}$ and the second is the expectation of the constant $1/(\sigma^{-2} + \tau^{-2})$. Substitute and simplify and use the fact that a normal mixture of normals is also normal.

The sequence of steps can be abstracted as follows: (1) marginal distribution of data given hyperparameters, (2) posterior distribution of hyperparameters given data, (3) posterior distribution of parameters given data, hyperparameters, (4) integrate out hyperparameters from (3) using (2). These steps are widely applicable for inference in hierarchical models of this general form.

14.3 Order the levels of (i, j) with the first index varying fastest, that is, as (1, 1), (2, 1), (1, 2), (2, 2).

(a) Units sharing a common value of i also have the same value of α_i with variance τ_α^2. Similarly, units sharing a common value of j also have the same value of γ_j with variance τ_γ^2. Hence

$$\mathbf{J} = \begin{pmatrix} 1 & 0 & 1 & 0 \\ 0 & 1 & 0 & 1 \\ 1 & 0 & 1 & 0 \\ 1 & 0 & 1 & 0 \end{pmatrix} \quad \text{and} \quad \mathbf{K} = \begin{pmatrix} 1 & 1 & 0 & 0 \\ 1 & 1 & 0 & 0 \\ 0 & 0 & 1 & 1 \\ 0 & 0 & 1 & 1 \end{pmatrix}.$$

Finally, δ_{ij} is unique for each (i, j) pair so L is the identity matrix.

(b) Extend the ordering of units as in (a), that is $(1, 1, 1)$, $(2, 1, 1), \ldots, (2, 2, 2)$. Combine the matrices in (a), and then replicate it in blocks for $k = 1, 2$ and add a term on the diagonal for the independent error ε_{ijk}, to obtain a matrix whose first two columns are

$$\begin{pmatrix} \tau_\alpha^2+\tau_\gamma^2+\tau_\delta^2+\sigma^2 & \tau_\gamma^2 \\ \tau_\gamma^2 & \tau_\alpha^2+\tau_\gamma^2+\tau_\delta^2+\sigma^2 \\ \tau_\alpha^2 & 0 \\ 0 & \tau_\alpha^2 \\ \tau_\alpha^2+\tau_\gamma^2+\tau_\delta^2 & \tau_\gamma^2 \\ \tau_\gamma^2 & \tau_\alpha^2+\tau_\gamma^2+\tau_\delta^2 \\ \tau_\alpha^2 & 0 \\ 0 & \tau_\gamma^2 \end{pmatrix}$$

14.5 (a) Define (a, b) the hyperparameters of the Beta distribution, θ_i the probability of a true heads with coin i, δ_{ij} the result of the jth toss of coin i (heads = 1, tails = 0), and y_{ij} the recorded value for that toss.

(c) $\theta_i \mid (a, b) \sim \text{Beta}(a, b)$, $\delta_{ij} \mid \theta_i, (a, b) \sim \text{Bernoulli}(\theta_i)$, and

$$y_{ij} \mid \delta_{ij} = 1, \theta_i, (a, b) \sim \text{Bernoulli}(p),$$

$$y_{ij} \mid \delta_{ij} = 0, \theta_i, (a, b) \sim \text{Bernoulli}(1 - p).$$

(d) Applying Bayes' theorem,

$$P(\delta_{ij} = 1 \mid y_{ij} = 1, \theta_i)$$

$$= \frac{P(y_{ij} = 1 \mid \delta_{ij} = 1, \theta_i)P(\delta_{ij} = 1 \mid \theta_i)}{P(y_{ij} = 1 \mid \delta_{ij} = 1, \theta_i)P(\delta_{ij} = 1 \mid \theta_i) + P(y_{ij} = 1 \mid \delta_{ij} = 0, \theta_i)P(\delta_{ij} = 0 \mid \theta_i)}$$

$$= \frac{p\theta_i}{p\theta_i + (1 - p)(1 - \theta_i)}.$$

(e) $f(\theta) = p\theta + (1 - p)(1 - \theta)$.

(f) $p(\theta) = f(\theta)^2(1 - f(\theta))^2/C$, where C is the integral of the numerator of the expression.

14.8 (a) $L(\beta) = \exp\left(\frac{1}{2\sigma^2}(\mathbf{y}_i - \mathbf{z}_i\beta_i)'(\mathbf{y}_i - \mathbf{z}_i\beta_i)\right)$ where $\mathbf{y}_i = (y_{i1}, y_{i2}, \ldots)'$.

(b) $\beta_i \mid \mathbf{y}_i, \lambda, \sigma^2, \mathbf{T} \sim N(\mathbf{V}[(\mathbf{z}_i'\mathbf{z}_i/\sigma^2)\hat{\beta}_i + \mathbf{T}^{-1}\mathbf{x}_i\lambda], \mathbf{V})$ where

$$\mathbf{V} = \left((\mathbf{z}_i'\mathbf{z}_i/\sigma^2) + \mathbf{T}^{-1}\right)^{-1}.$$

(c) Write $\mathbf{B} = \mathbf{V}\mathbf{T}^{-1}$. Note that if prior information is weak ($\mathbf{T}$ is big) then $\tilde{\beta}$ approaches $\hat{\beta}$.

CHAPTER 15

15.3 If we standardize the data before carrying out a factor analysis, we gain the advantage that the covariance matrix becomes a correlation matrix, so the factor loadings become readily interpretable as correlations between the latent factors and the data vectors.

15.5 When there does not seem to be anything real that corresponds to the latent factors, you should strongly suspect that the factors that have been identified by the model may be artifacts of the analysis; that is, they may not be useful in helping to understand the structure of the data, nor in helping to reduce the dimensionality of the model.

15.7 Unbiasedness is a frequentist property. It is of less interest to the Bayesian because it depends upon data that has never been observed. To define unbiasedness we must take an expectation over all the possible data points, both those actually observed, and those that have not been observed.

15.8 By using MCMC to estimate the parameters of the Bayesian factor analysis model, we can estimate the hyperparameters at the same time, and consequently develop an "exact solution" (in the sense that a computer-generated solution can approximate the mathematical solution). But the solution presented in Sections 15.3 and 15.4 has the advantage of providing insights into the behavior of the factor loadings, factor scores, and disturbance covariances because that solution is based upon simple Student *t*-distributions, even though that solution requires a large sample approximation.

CHAPTER 16

16.1 The problem is a direct application of the mathematics developed in Section 16.6. In fact, since $N = N_1 = N_2 = 10$, the calculation of L_{ij} reduces to:

$$L_{ij} = \frac{q_i}{q_j}\sqrt{\frac{|S_j|}{|S_i|}},$$

where q_i and q_j are the prior probabilities of being classified in population i and j, respectively.

In this problem, we let $i = 1$ and $j = 2$ and obtain:

$$L_{ij} = \sqrt{\frac{|S_j|}{|S_i|}} = \sqrt{\frac{8}{7}},$$

$$\frac{P(z \in \pi_1 \mid \boldsymbol{Z})}{P(z \in \pi_2 \mid z)} = L_{ij}\frac{\left[1 + \left(\frac{N_j}{N_j^2 - 1}\right)(z - \bar{x}_j)'S_j^{-1}(z - \bar{x}_j)\right]^{0.5N_j}}{\left[1 + \left(\frac{N_i}{N_i^2 - 1}\right)(z - \bar{x}_i)'S_i^{-1}(z - \bar{x}_i)\right]^{0.5N_i}}$$

$$= \sqrt{\frac{8}{7}}\frac{\left[1 + \frac{10}{99}\omega_j\right]^5}{\left[1 + \frac{10}{99}\omega_i\right]^5},$$

where

$$\omega_i = (z - \bar{x}_i)'S_i^{-1}(z - \bar{x}_i), \qquad \omega_j = (z - \bar{x}_j)'S_j^{-1}(z - \bar{x}_j).$$

After substituting values, we obtain:

$$\frac{P(z \in \pi_1 \mid z)}{P(z \in \pi_2 \mid z)} = \sqrt{\frac{8}{7}}\,\frac{\left[1 + \frac{10}{99}\cdot\frac{27}{32}\right]^5}{\left[1 + \frac{10}{99}\cdot\frac{9}{14}\right]^5} = 1.17485.$$

Therefore, classify $z \in \pi_1$.

16.2 Equal priors means $q_1 = q_2 = q_3 = \frac{1}{3}$. Substitution of the above values leads to:

$$L_{12} = \sqrt{\frac{8}{7}}, \qquad L_{13} = \sqrt{\frac{2}{7}}, \qquad L_{23} = \sqrt{\frac{2}{8}}.$$

Let

$$\rho_{ij} = \frac{P(z \in \pi_i \mid z)}{P(z \in \pi_j \mid z)}.$$

Then

$$\rho_{ij} = L_{ij}\frac{\left[1 + \left(\frac{N_j}{N_j^2 - 1}\right)(z - \bar{x}_j)'S_j^{-1}(z - \bar{x}_j)\right]^{0.5N_j}}{\left[1 + \left(\frac{N_i}{N_i^2 - 1}\right)(z - \bar{x}_j)'S_j^{-1}(z - \bar{x}_i)\right]^{0.5N_i}}.$$

Substituting the values for this problem, we obtain:

$$\rho_{12} = \sqrt{\frac{8}{7}}\,\frac{\left[1 + \frac{10}{99}\cdot\frac{11}{8}\right]^5}{\left[1 + \frac{10}{99}\cdot\frac{2}{7}\right]^5} = 1.77673,$$

$$\rho_{13} = \sqrt{\frac{2}{7}}\,\frac{\left[1 + \frac{10}{99}\cdot\frac{8}{2}\right]^5}{\left[1 + \frac{10}{99}\cdot\frac{2}{7}\right]^5} = 2.52978,$$

$$\rho_{23} = \sqrt{\frac{2}{8}}\,\frac{\left[1 + \frac{10}{99}\cdot\frac{8}{22}\right]^5}{\left[1 + \frac{10}{99}\cdot\frac{11}{8}\right]^5} = 1.42384.$$

Since $\rho_{12} > 1$ and $\rho_{13} > 1$, π_1 is more probable than either π_2 or π_3. Therefore, classify $z \in \pi_1$.

16.3 (a) Find the joint posterior density of $(\boldsymbol{\theta}_i, \boldsymbol{\Lambda}_i) \mid (\bar{\boldsymbol{x}}_i, \mathbf{V}_i, \pi_i)$. Denote the training data from population i as:

$$\boldsymbol{x}_1^{(i)}, \ldots, \boldsymbol{x}_{N_i}^{(i)} \sim N(\boldsymbol{\theta}_i, \Sigma_i), \qquad i = 1, \ldots, K.$$

Note: $\boldsymbol{x}_i$ are $(p \times 1)$ vectors. $\sum_i$ are $(p \times p)$ matrices.
Express the Likelihood. Let

$$\bar{\boldsymbol{x}}_i = \frac{1}{N_i} \sum_{j=1}^{N_i} \boldsymbol{x}_j^{(i)}, \qquad \forall_i = 1, \ldots, K$$

$$\boldsymbol{V}_i = \sum_{j=1}^{N} (\boldsymbol{x}_j^{(i)} - \bar{x}i)(\boldsymbol{x}_j^{(i)} - \bar{\boldsymbol{x}}_i)',$$

$$\boldsymbol{\Lambda}_i = \Sigma_{j=1}^{-1}, \qquad \forall i = 1, \ldots, K.$$

Since $\bar{\boldsymbol{x}}_i$ and $\boldsymbol{V}_i$ are independent of one another $\forall i$ and the distribution of $\boldsymbol{V}_i$ does not depend on $\boldsymbol{\theta}_i$:

$$p(\bar{\boldsymbol{x}}_i, \boldsymbol{V}_i \mid \boldsymbol{\theta}_i, \Sigma_i, \pi_i) = p(\bar{\boldsymbol{x}}_i \mid \boldsymbol{\theta}_i, \pi_i) p(\boldsymbol{V}_i \mid \Sigma_i, \pi_i).$$

It is assumed that

$$(\bar{\boldsymbol{x}}_i \mid \boldsymbol{\theta}_i, \pi_i) \sim N\left(\boldsymbol{\theta}_i, \frac{1}{N_i} \Sigma_i\right),$$

$$(\boldsymbol{V}_i \mid \Sigma_i, \pi_i) \sim W(\Sigma_i, p, N_i - 1), \qquad \text{(i.e., Wishart)}.$$

Therefore,

$$p(\bar{\boldsymbol{x}}_i, \boldsymbol{V}_i \mid \boldsymbol{\theta}_i, \Sigma_i, \pi_i) \propto \left[\mid \boldsymbol{\Lambda}_i \mid^{0.5} \exp\left\{ -\frac{N_i}{2} (\bar{\boldsymbol{x}}_i - \boldsymbol{\theta}_i)' \boldsymbol{\Lambda}_i (\bar{\boldsymbol{x}}_i - \boldsymbol{\theta}_i) \right\} \right]$$
$$\times [\mid \boldsymbol{V}_i \mid^{0.5(N_i - p - 2)} \mid \boldsymbol{\Lambda}_i \mid^{0.5(N_i - 1)} \exp\{-0.5 \operatorname{tr} \boldsymbol{\Lambda}_i \boldsymbol{V}_i\}].$$

Simplifying, we get the likelihood to be:

$$p(\bar{\boldsymbol{x}}_i, \boldsymbol{V}_i \mid \boldsymbol{\theta}_i, \Sigma_i, \pi_i) \propto \mid \boldsymbol{\Lambda}_i \mid^{0.5(N_i - p - 2)} \mid \boldsymbol{\Lambda}_i \mid^{0.5 N_i} \exp\{-0.5 \operatorname{tr} \boldsymbol{\Lambda}_i A_L\},$$

where

$$A_L = V_i + N_i(\bar{x}_i - \theta_i)(\bar{x}_i - \theta_i)'.$$

Express the Prior. The conjugate prior for (θ_i, Λ_i) can be one of several forms. The one used in this solution is as follows:

$$p(\theta_i, \Lambda_i \mid \pi_i) = p(\theta_i \mid \Lambda_i, \pi_i)p(\Lambda_i \mid \pi_i),$$

and

$$p(\theta_i \mid \Lambda_i, \pi_i) \propto |\Lambda_i|^{0.5} \exp\left\{-\frac{k_i}{2}(\theta_i - \mu_i)'\Lambda_i(\theta_i - \mu_i)\right\};$$

that is, $(\theta_i, \Lambda_i \mid \pi_i) \sim N(\mu_i, k_i\Lambda_i^{-1})$, where μ_i is a $(p \times 1)$ hyperparameter vector, k_i is a scalar hyperparameter, and

$$p(\Lambda_i \mid \pi_i) \propto |\Lambda_i|^{0.5(n_i-p-1)} |\Psi_i|^{0.5n_i} e^{-\text{tr}\,\Lambda_i\Psi^{-1}};$$

that is, $(\Lambda_i \mid \pi_i) \sim W(\Psi_i, p, \eta_i)$, where Ψ_i is a $(p \times p)$ hyperparameter matrix and η_i is a scalar hyperparameter. Therefore, the prior is:

$$p(\theta_i, \Lambda_i \mid \pi_i) \propto |\Lambda_i|^{0.5(n_i-p)} e^{-\text{tr}\,\Lambda_i A_P},$$

where

$$A_P = \Psi^{-1} + k_i(\theta_i - \mu_i)(\theta_i - \mu_i)'.$$

Calculate the Posterior.

$$\begin{aligned} p(\theta_i, \Lambda_i \mid \bar{x}_i, V_i, \pi_i) &\propto \text{prior} \cdot \text{likelihood} \\ &\propto \{|\Lambda_i|^{0.5n_i-p} e^{-0.5\,\text{tr}\,\Lambda_i A_P}\} \\ &\quad \times \{|V_i|^{0.5(N_i-p-2)} |\Lambda_i|^{0.5n_i} e^{-0.5\,\text{tr}\,\Lambda_i A_L}\} \\ &\propto |\Lambda_i|^{0.5(N_i-N_j-p)} e^{-0.5\,\text{tr}\,\Lambda_i(A_P+A_L)}. \end{aligned}$$

(b) Find the predictive density of $(z \mid \bar{x}_i, V_i, \pi_i)$.

$$p(z \mid \bar{x}_i, V_i, \pi_i) = \iint p(z \mid \theta_i, \Lambda_i, \pi_i) \cdot p(\theta_i, \Lambda_i, \mid \bar{x}_i, V_i, \pi_i)\, d\theta_i\, d\Lambda_i,$$

where

$$p(z \mid \theta_i, \Lambda_i, \pi_i) \propto |\Lambda_i|^{0.5} \exp\{-0.5(z - \theta_i)'\Lambda_i(z - \theta_i)\},$$

since $(z \mid \theta_i, \Lambda_i, \pi_i) \sim N(\theta_i, \Lambda_i^{-1})$.

Using the posterior from above, we obtain:

$$p(z \mid \bar{x}_i, V_i, \pi_i) \propto \iint \mid \Lambda_i \mid^{0.5(N_i+n_i-p+1)} e^{-0.5 \operatorname{tr} \Lambda_i A} \, d\theta_i \, d\Lambda_i,$$

where $A = (A_L + A_P + (z - \theta_i)(z - \theta_i)')$.

Using Lemma 12.2 of Appendix 7, and integrating Λ_i out, we obtain:

$$p(z \mid \bar{x}_i, V_i, \pi_i) \propto \int \frac{d\theta_i}{\mid A \mid^{0.5(N_i+n_i-2p)}}.$$

Now, we complete the square in terms of θ_i. Recall first that:

$$A = A_L + A_P + (z - \theta_i)(z - \theta_i)',$$

and for A_P and A_L,

$$\begin{aligned}
A &= \{\Psi^{-1} + k_i(\theta_i - \mu)(\theta_i - \mu)'\} \\
&\quad + \{V_i + N_i(\bar{x}_i - \theta_i)(\bar{x}_i) - \theta_i)'\} + \{(z - \theta_i)(z - \theta_i)'\}, \\
A &= \theta_i\theta_i' + k\theta_i\theta_i' + N_i\theta_i\theta_i' - z\theta_i' - \theta_i z' - k_i\mu_i\theta_i' \\
&\quad - k_i\theta_i\mu' - N_i\bar{x}_i\theta_i' - N_i\theta_i\bar{x}_i' + zz' + \Psi_i + k_i\mu_i\mu_i' + V_i + N_i\bar{x}_i\bar{x}_i', \\
A &= c_1(\theta_i - \hat{\theta}_i)(\theta_i - \hat{\theta}_i)' + C_2,
\end{aligned}$$

where

$$c_1 = 1 + k_i + N_i, \qquad \hat{\theta}_i = \frac{1}{c_1}(z + k_i\mu_i + N_i\bar{x}_i),$$

$$\begin{aligned}
C_2 &= (zz' + \Psi_i + k_i\mu_i\mu_i' + V_i + N_i\bar{x}_i\bar{x}_i') \\
&\quad - \frac{1}{c_1}(z + k_i\mu_i + N_i\bar{x}_i)(z + k_i\mu_i + N_i\bar{x}_i)'.
\end{aligned}$$

Now integrate θ_i out:

$$p(z \mid \bar{x}_i, V_i, \pi_i) \propto \int \frac{d\hat{\theta}_i}{\mid c_1(\theta_i - \hat{\theta}_i)(\theta_i - \hat{\theta}_i)' + C_2 \mid^{0.5(v_1+p)}},$$

where

$$v_i = N_i + \eta_i - 3p.$$

Recall that Sylvester's theorem for matrices (Press, 1982, *Applied Multivariate Analysis*, p. 20) says:

$$\mid I_p + AB \mid = \mid I_q + BA \mid.$$

Working with the denominator of the integral above, we have:

$$\begin{aligned} | C_2 + c_1(\boldsymbol{\theta}_i - \hat{\boldsymbol{\theta}}_i)(\boldsymbol{\theta}_i - \hat{\boldsymbol{\theta}}_i)' | &= | C_2 | \, | I + c_1 C_2^{-1}(\boldsymbol{\theta}_i - \hat{\boldsymbol{\theta}}_i)(\boldsymbol{\theta}_i - \hat{\boldsymbol{\theta}}_i)' | \\ &= | C_2 | \, [1 + c_1(\boldsymbol{\theta}_i - \hat{\boldsymbol{\theta}}_i)' C_2^{-1}(\boldsymbol{\theta}_i - \hat{\boldsymbol{\theta}}_i)]. \end{aligned}$$

Note that $I + c_1(\boldsymbol{\theta}_i - \hat{\boldsymbol{\theta}}_i)' C_2^{-1}(\boldsymbol{\theta}_i - \hat{\boldsymbol{\theta}}_i)$ is a scalar. Therefore, the integral is:

$$p(\boldsymbol{z} \mid \bar{\boldsymbol{x}}_i, \boldsymbol{V}_i, \pi_i) \propto \frac{| c_1 v_i C_2^{-1} |^{0.5}}{| C_2 |^{0.5(v_i+p)}} \int \frac{| c_1 v_i C_2^{-1} |^{-0.5} \, d\hat{\boldsymbol{\theta}}_i}{[v_i + c_1 v_i(\boldsymbol{\theta}_i - \hat{\boldsymbol{\theta}}_i)' C_2^{-1}(\boldsymbol{\theta}_i - \hat{\boldsymbol{\theta}}_i)]^{0.5(v_i+p)}}.$$

The integrand is the kernel of a multivariate T density. Therefore, the integral equals a constant. Thus

$$p(\boldsymbol{z} \mid \bar{\boldsymbol{x}}_i, \boldsymbol{V}_i, \pi_i) \propto \frac{1}{| C_2 |^{0.5(v_i+p+1)}} = \frac{1}{| C_2 |^{0.5(N_1+\eta_i-2p+1)}}.$$

To facilitate understanding the nature of this density, let us complete the square of C_2 in $\boldsymbol{z}$:

$$\begin{aligned} C_2 = \boldsymbol{z}\boldsymbol{z}' - \frac{1}{c_1}\boldsymbol{z}\boldsymbol{z}' - \frac{1}{c_1}\boldsymbol{z}(k_i\boldsymbol{\mu}_i + N_i\bar{\boldsymbol{x}}_i)' - \frac{1}{c_1}(k_i\boldsymbol{\mu}_i + N_i\bar{\boldsymbol{x}}_i)\boldsymbol{z}' \\ + \boldsymbol{\Psi}_i + k_i\boldsymbol{\mu}_i\boldsymbol{\mu}_i' + \boldsymbol{V}_i + N_i\bar{\boldsymbol{x}}_i\bar{\boldsymbol{x}}_i' - \frac{1}{c_1}(k_i\boldsymbol{\mu}_i + N_i\bar{\boldsymbol{x}}_i)(k_i\boldsymbol{\mu}_i + N_i\bar{\boldsymbol{x}}_i)'. \end{aligned}$$

Let

$$C_2 = C_4 + c_3(\boldsymbol{z} - \hat{\boldsymbol{z}})(\boldsymbol{z} - \hat{\boldsymbol{z}})',$$

where

$$c_3 = 1 - \frac{1}{c_1}, \qquad \hat{\boldsymbol{z}} = \frac{1}{c_1 c_3}(k_i\boldsymbol{\mu}_i + N_i\bar{\boldsymbol{x}}_i),$$

$$\begin{aligned} C_4 = \boldsymbol{\Psi}_i + k_i\boldsymbol{\mu}_i\boldsymbol{\mu}_i' + \boldsymbol{V}_i + N_i\bar{\boldsymbol{x}}_i\bar{\boldsymbol{x}}_i' \\ - \frac{1}{c_1}(k_i\boldsymbol{\mu}_i + N_i\bar{\boldsymbol{x}}_i)(k_i\boldsymbol{\mu}_i + N_i\bar{\boldsymbol{x}}_i)' - c_3\hat{\boldsymbol{z}}\hat{\boldsymbol{z}}', \end{aligned}$$

$$p(\boldsymbol{z} \mid \boldsymbol{x}_i, \boldsymbol{V}_i, \pi_i) \propto \frac{1}{| C_4 |^{0.5(v_1+p+1)} | I + c_3 C_4^{-1}(\boldsymbol{z} - \hat{\boldsymbol{z}})(\boldsymbol{z} - \hat{\boldsymbol{z}})' |^{0.5(v_i+p+1)}}.$$

Using Sylvester's theorem again, we obtain:

$$p(z \mid x_i, V_i, \pi_i) \propto \frac{1}{\mid C_4 \mid^{0.5(\nu_i+p+1)} \left[1 + c_3(z - \hat{z})' C_4^{-1}(z - \hat{z})\right]^{0.5(\nu_i+p+1)}}.$$

All variables on the RHS (except of course z) are functions of the hyperparameters and the training data. Putting the equation in final form gives:

$$p(z \mid x, V_i, \pi_i) \propto \frac{1}{[\nu_i + (z - \hat{z})'(c_3 \nu_i C_4^{-1})(z - \hat{z})]^{0.5\nu_i+p+1)}},$$

the kernel of a matrix T density with ν_i degrees of freedom.

(c) Find the posterior classification probability for classifying z into π_i, given z. That is, find $P(z \in \pi_i \mid z)$.

$$P(z \in \pi_i \mid z) \propto \frac{P(z \in \pi_i)}{[\nu_i + (z - \hat{z})'(c_3 \nu_i C_4^{-1})(z - \hat{z})]^{0.5(\nu_i+p+1)}}.$$

16.4 Consider a survey with a main question and a set of subsidiary questions. Assume some respondents answer the main question and some do not. If one infers the portion of the population to have various answers based solely on the responses of those who did provide an answer to the main question, one might get a biased estimate that might also have a larger than necessary variance.

The approach discussed in Section 16.8 is one good way to get an improved population estimate. This approach rests on the assumption of a correlation or dependence between the main question and the subsidiary questions. Properly selected subsidiary questions will provide this dependence under most circumstances.

The respondents that did answer the main question and the subsidiary questions form the training group. Responses to the main question identify populations π_k; while subsidiary question responses form the data X. Respondents that did not answer the main question but did answer the subsidiary questions can then have the probability predicted that each of them belongs to population k; that is, $P\{\text{person } j \in \pi_k \mid z_j, X, \pi\}$.

Accumulating the predicted probabilities for the individuals results in a probability distribution for a new set of responses to the main question, which reflects the information provided by all the respondents. This can be condensed to a point estimate by taking the expected value of the responses, if desired.

16.5 One might expect to get a large fraction of "undecided" in situations where the response might be associated with some risk, either socially, legally, or otherwise.

It is considered inappropriate for a judge to comment on a hypothetical case since he might have to hear a case similar to the hypothetical one. (This has been a common response of the U.S. Supreme Court appointees, in response to a Senate inquiry.)

Often one finds oneself in a social setting where most or all of the other people have expressed a distinct political view. When queried about one's opinion, one might simply claim "undecided" rather than lie or have to defend one's views to a group of opinionated people. (In some totalitarian countries, this may be a survival skill. This may also occur in many business settings.)

Another situation might be wholly introspectively based. One may not have any desire to open up what one considers secret and private; such as one's illicit drug experience, sexual preferences, or criminal behavior.

Another situation could be where one has not fully concluded a decision. This may be one in which conflicting benefits and penalties exist. Selection of the "lesser of evils" *has to be delayed*, and yet a correlation to the decided subsidiary question may still exist.

16.6 First, one should follow the general guidelines that models should be parsimonious to avoid spurious correlations that create false correlations and misleading results.

There is a second reason specific to this problem when dealing with discrete responses. This is clearly expressed in the following excerpt from Press and Yang, 1974 (see text for the full reference).

> ... Another consideration, however, involves a peculiarity of the Dirichlet class of prior distributions assumed for the case of discrete-response subsidiary questions (there is no problem in the case of continuous-response subsidiary questions). The problem is that unless the prior information is very dogmatic, or the sample size is extremely large, the sample information may not be very useful.
>
> For example, for ten subsidiary questions, each with two possible responses, $S = 2^{10} = 1024$. If the sample size is 100, and a vague prior is adopted, the predictive density becomes
>
> $$h(\mu_K : x_i; H_i) = \frac{x_{k|i} + 1}{n_i + S}.$$
>
> Because there are over 1000 cells for responses, c_k will be zero for most cells, sometimes one, and will rarely be larger. So $h(\cdot)$ will vary only slightly from cell to cell.
>
> This unstable situation is caused by sparseness of the $x_{k|i}$ data and can be avoided in various ways.

The reader is referred to Press and Yang, 1974, for details.

Bibliography

Abramson, J. H. and Abramson, Z. H. (1999). *Survey Methods in Community Medicine*, 5th Edition, Churchill Livingston.

Akaike, H. (1973). "Information Theory and an Extension of the Maximum Likelihood Principle," in B. N. Petrov and F. Csaki Eds., *2nd Int. Symp. on Information Theory*, Budapest, Akademiai Kiado, 267–281.

Akaike, H. (1977). "On Entropy Maximization Principle," in P. R. Krishnaiah Ed., *Proceedings of the Symposiumi on Applications of Statistics*, Amsterdam, North Holland Publishing Co., 27–47.

Akaike, H. (1978). "A Bayesian Analysis of the Minimum AIC Procedure," *Ann. Inst. Statist. Math., Part A*, **30**, 9–14.

Akaike, H. (1987). "Factor Analysis and AIC," *Psychometrika*, **52**(37), 317–332.

Albert, J. and Chib, S. (1993). "Bayesian Analysis of Binary and Polychotomous Response Data," *J. Am. Statist. Assoc.*, **88**, 669–679.

Aldous, D. (1981). "Representations for Partially Exchangeable Arrays," *J. Multivariate Analysis*, 581–598.

Aldrich, J. (1997). "R. A. Fisher and the Making of Maximum Likelihood 1912–1922," *Statist. Sci.*, **12**, 162–176.

Allais, M. (1953). "Le Comportement de L'Homme Rationnel Devant Le Risque: Critique Des Postulats et Axiomes de L'Ecole Americaine," *Econometrica*, **21**, 503–546. Translated in Allais, M. and Hagen, D., Eds., *Expected Utility Hypotheses and the Allais Paradox*, Dordrecht, Reidel, 1979.

Anderson, J. and Rosenfeld, E. (Eds.) (1988). *Neurocomputing: Foundations of Research*, Cambridge, MA, MIT Press.

Anscombe, F. J. and Aumann, R. J. (1963). "A Definition of Subjective Probability," *Ann. Math. Stat.*, **34**, 199–205.

Arnold, B. C., Castillo, E. and Sarabia, J. M. (1999). *Conditional Specification of Statistical Models*, New York, Springer Verlag.

Athreya, K. B. (1991). "Entropy Maximization," manuscript, Iowa State University, Department of Mathematics, Math. Subject Classification 60 E.

Baldus, D. C., Woodworth, G. and Pulaski, C. A., Jr. (1989). "Charging and Sentencing of Murder and Voluntary Manslaughter Cases in Georgia 1973–1979." Data and documentation: http://www.icpsr.umich.edu/cgi/archive.prl?path = NACJD&format = tb&num = 9264.

Basu, D., see Ghosh, J. K. (1988).

Bayes, T. (1763). "An Essay Towards Solving a Problem in the Doctrine of Chances," *Philos. Trans. Royal Soc. London*, V**61.53**, 370–418. Reprinted in Appendix 4.

Berger, J. O. (1980). *Statistical Decision Theory: Foundations, Concepts and Methods*, New York, Springer-Verlag.

Berger, J. O. (1985). *Statistical Decision Theory and Bayesian Analysis*, Second Edition, New York, Springer-Verlag.

Berger, J. O. and Berliner, L. M. (1986). "Robust Bayes and Empirical Bayes Analysis with ε-Contaminated Priors," *Ann. Statist.*, **14**, 461–486.

Berger, J. O. and Bernardo, J. M. (1989). "Estimating a Product of Means: Bayesian Analysis with Reference Priors," *J. Am. Statist. Assoc.*, **84**, 200–207.

Berger, J. O. and Pericchi, L. R. (1993). "The Intrinsic Bayes Factor for Model Selection and Prediction," Technical Report. #93–43C, Department of Statistics, Purdue University.

Berger, J. O. and Pericchi, L. R. (1996). "The Intrinsic Bayes Factor for Model Selection and Prediction," *J. Am. Statist. Assoc.*, **91**, 109–122.

Berger, J. O. and Pericchi, L. R. (1996). "The Intrinsic Bayes Factor for Linear Models," in Bernardo, J. M., Berger, J. O., Dawid, A. P. and A. F. M. Smith, Eds., *Proceedings of the Fifth International Meeting On Bayesian Statistics*, Oxford, Oxford University Press, 25–44.

Berger, J. O. and Pericchi, L. R. (2001). "Objective Bayesian Methods for Model Selection: Introduction and Comparison," in *IMS Lecture Notes—Monograph Series*, **38**, 135–193.

Berger, J. O. and Selke, T. (1987). "Testing a Point Null Hypothesis: The Irreconcilability of p-Values and Evidence," *J. Am. Statist. Assoc.*, **82**(397), 112–122.

Berger, J. O., Dawid, A. P. and Smith, A. F. M. (Eds.), Oxford, Oxford University Press, 25–44.

Berger, J. O. and Wolpert, R. L. (1984). *The Likelihood Principle*, Lecture Notes Monograph Series, Vol. 6, Shanti S. Gupta (Ed.), Institute of Mathematical Statistics, Hayward, CA.

Bernardo, J. M. (1979). "Reference Posterior Distributions for Bayesian Inference," *J. Royal Statist. Soc. (B)*, **41**(2), 113–147.

Bernardo, J. M. and Smith A. F. M. (1994). *Bayesian Theory*, New York, John Wiley and Sons, Inc.

Bernoulli, J. (1713). *Ars Conjectandi*, Book 4, Baseae, Impensis Thurnisiorum.

Besag, J. E. (1986). "On the Statistical Analysis of Dirty Pictures," *J. Royal Statist. Soc.* (*Series B*), **48**, 259–302.

Besag, J., Green, E., Higdon, D. and Mengersen, K. L. (1995). "Bayesian Computation and Stochastic Systems (with discussion)," *Statist. Sci.*, **10**, 3–66.

Birnbaum, A. (1962). "On the Foundations of Statistical Inference," with discussion, *J. Am. Statist. Assoc.*, **57**, 269–306.

Blackwell, D. and Gershick, M. A. (1954). *Theory of Games and Statistical Decisions*, New York, John Wiley and Sons, Inc.

Bliss, G. A. (1946). *Lectures on the Calculus of Variations*, Chicago, University of Chicago Press.

Box, G. E. P. and Tiao, G. C. (1973). *Bayesian Inference in Statistical Analysis*, Reading, MA, Addison-Wesley Publishing Co.

Bozdogan, H. (1987). "Model Selection and Akaike's Information Criterion (AIC): The General Theory and its Analytical Extensions," *Psychometrika*, **52**(3), 345–370.

Bradlow, E. T. and Zaslavsky, A. M. (1999). "A Hierarchical Latent Variable Model for Ordinal Data from a Customer Satisfaction Survey with 'No Answer' Responses," *J. Am. Statist. Assoc.*, **94**, 43–52.

Breiman, L., Friedman, J. H., Olshen, R. A., and Stone, C. J. (1984). *Classification and Regression Trees*, California, Wadsworth International Group.

Bridle, J. S. (1989). "Probabilistic Interpretation of Feedforward Classification Network Outputs with Relationships to Statistical Pattern Recognition," in F. Fouglemann-Soulie and J. Heault, Eds., *Neurocomputing: Algorithms, Architectures, and Applications*, New York, Springer-Verlag.

Brier, G. W. (1950). "Verification of Forecasts Expressed in Terms of Probabilities," *Monthly Weather Review*, **78**, 1–3.

Broemeling, L. D. (1985). *Bayesian Analysis of Linear Models*, New York, Marcel Dekker.

Brown, L. D. (1966). "On the Admissibility of Invariant Estimators of One or More Location Parameters," *Ann. Math. Statist.*, **37**, 1087–1136.

Buntine, W. L. (1992). "A Theory of Learning Classification Rules," Ph.D. Thesis, School of Computing Science, University of Technology, Sydney.

Cacoullos, T. (1966). "Estimation of a Multivariate Density," *Ann. Inst. Statist. Math.*, **2**(2), 179–189.

Carlin, B. P. and Polson N. G. (1991). "Inference for Non-Conjugate Bayesian Models Using the Gibbs Sampler," *Canadian J. Statist.*, **19**, 399–405.

Carlin, B. P. and Chib S. (1995). "Bayesian Model Choice via Markov Chain Monte Carlo Methods," *J. Royal Statist. Soc. B*, **57**, 473–484.

Carlin, B. P. and Louis, T. A. (2000). *Bayes and Empirical Bayes Methods for Data Analysis*, 2nd Edition, Boca Raton, Florida, Chapman and Hall.

Carnap, R. and Jeffrey, R. C., (Eds.) (1971). *Studies in Inductive Logic and Probability*, Vol. 1, Berkeley, University of California Press.

Casella, G. and Berger R. L., (1987). "Reconciling Bayesian and Frequentist Evidence in the One–Sided Testing Problem," *J. Am. Statist. Assoc.*, **82**(397), 106–111.

Casella, G. and Robert, C. P. (1996). "Rao-Blackwellization of Sampling Schemes," *Biometrika*, **83**, 81–94.

Chatfield, C. (1995). "Model Uncertainty, Data Mining, and Statistical Inference," with discussion, *J. Royal Statist. Soc. (A)*, **158**, 419–466.

Chatterjee S. and S. (1987). "On Combining Expert Opinions." *Am. J. Math. Mgmt. Sci.*, 271–295.

Cheeseman, P. and Stutz, J. (1995). "Bayesian Classification (AutoClass): Theory and Methods," in U. M. Fayyad, G. Piatetsky-Shapiro, P. Smyth, and R. Uthurusamy, Eds., *Advances in Knowledge Discovery and Data Mining*, Menlo Park, The AAAI Press.

Chen, M-H. and Shao, Q-M. (1997). "On Monte Carlo Methods for Estimating Ratios of Normalizing Constants," *Ann. Statis.*, **25**, 1563–1594.

Chernoff, H. and Moses, L. E. (1959). *Elementary Decision Theory*, New York, John Wiley and Sons, Inc.

Chib, S. (1995). "Marginal Likelihood from the Gibbs Output," *J. Am. Statist. Assoc.*, **90**, 1313–1321.

Chib, S and Greenberg E. (1995). "Understanding the Metropolis–Hastings Algorithm," *Am. Statist.*, **49**, 327–335.

Chib, S. and Carlin B. P. (1999). "On MCMC Sampling in Hierarchical Longitudinal Models," *Statist. Compu.*, **9**, 17–26.

Chib, S. and Jeliazkov I., (2001). "Marginal Likelihood from the Metropolis–Hastings Output," *J. Am. Statist. Assoc.*, **96**, 270–281.

Chipman, H. (1996). "Bayesian Variable Selection with Related Predictors," *Canadian J. Statist.*, **24**, 17–36.

Chipman, H., George, E. and McCulloch, R. (2001). "The Practical Implementation of Bayesian Model Selection," in *IMS Lecture Notes—Monograph Series*, **38**, 65–134.

Chow, Y. S. and Teicher, H. (1978). *Probability Theory*, Berlin, Springer-Verlag.

Christiansen, C. L. and Morris, C. N. (1997). "Hierarchical Poisson Regression Modeling," *J. Am. Statist. Assoc.*, **92**, 618–632.

Christiansen, C. L. and Morris, C. N. (1997). "Improving the Statistical Approach to Health Care Provider Profiling," *Ann. Internal Med.*, **127**, 764–768.

Citro, C. F. and Kalton, G. (Eds.) (2000). "Small-Area Income and Poverty Estimates: Evaluation of Current Methodology," Washington, D.C., National Academy Press.

Clyde, M. (2001). "Discussion of 'The Practical Implementation of Bayesian Model Selection'," in *IMS Lecture Notes—Monograph Series*, **38**, 117–124.

Clyde. M. and George, E. (2000). "Flexible Empirical Bayes Estimation for Wavelets," *J. Royal Statist. Soc., B*, **62**, 681–698.

Cohen, A. M., Cutts, J. F., Fielder, R., Jones, D. E., Ribbans, J. and Stuart, E. (1973). *Numerical Analysis*, New York, Halsted Press (John Wiley and Sons, Inc.).

Corbeil, R. R. and Searle, S. R. (1976). "Restricted Maximum Likelihood (REML) Estimation of Variance Components in the Mixed Model," *Technometrics*, **18**, 31–38.

Courant, R. and Hilbert, D. (1953). *Methods of Mathematical Physics*, Vol. 1, in English, New York, Interscience Publishers, Inc.

Cover, T. M. and Thomas J. A. (1991). *Elements of Information Theory*, New York, John Wiley and Sons, Inc.

Dalal, S. and Hall, W. J. (1983). "Approximating Priors by Mixtures of Natural Conjugate Priors," *J. Royal Statist. Soc. B*, **45**, 278–286.

Damien, P., Wakefield J. and Walker S. (1999). "Gibbs Sampling for Bayesian Nonconjugate and Hierarchical Models Using Auxiliary Variables," *J. Royal Statist. Soc., B*, **61**, 331–344.

Davis, P. J. and Rabinowitz, P. (1967). *Numerical Integration*, Waltham, MA, Blaisdell.

Dawid, A. P. (1978). "Extendability of Spherical Matrix Distributions," *J. Multivariate Anal.*, **8**, 559–566.

Dawid, A. and Lauritzen, S. (2000). "Compatible Prior Distributions," Techical Report.

Dawid, A. P., Stone, M. and Zidik, J. V. (1973). "Marginalization Paradoxes in Bayesian and Structural Inference," *J. Royal Statist. Soc. (B)*, **35**, 189–233.

Deely, J. J. and Lindley, D. V. (1981). "Bayes Empirical Bayes," *J. Am. Statist. Assoc.*, **76**, 833–841.

De Finetti, B. (1931). "Funzioni Caxatteristica di un Fenomeno Aleatorio," *Atti della R. Academia Nazionale dei Lincii*, Ser. 6, Memorie, Classe di Scienze, Fisiche, Matematiche e Naturali, Vol. 4, 251–299.

De Finetti, B. (1937). "La Prevision: Ses Lois Logique, Ses Sources Subjectives," *Ann. de l'Institut Henri Poincare*, **7**, 1–68, Translated in H. Kyberg and H. Smokler, Eds., *Studies in Subjective Probability*, New York, John Wiley and Sons, 1964.

De Finetti, B. (1962). "Does it Make Sense to Speak of 'Good Probability Appraisers'?," in I. J. Good. Ed., *The Scientist Speculates—An Anthology of Partly-Baked Ideas*, London, Heinemann, 357–364.

De Finetti, B. (1965). "Methods for Discriminating Levels of Partial Knowledge Concerning a Test Item," *B. J. Math. Statist. Psychol.*, **18**, 87–123.

De Finetti, B. (1974). *Theory of Probability*, Vols. 1 and 2, New York, John Wiley and Sons, Inc.

De Finetti, B. (1993). Statistica, Anno LII, Supplemento al n.3, 1992, *Probabilita E Inducione: Bruno de Finetti*, Bibliotecadi Statistica, Cooperativa Libraria Universitaria Editrice Bologna, 40126, Bologna-Via Marsala 24, (in English and Italian).

De Finetti, B. (1993). Biblioteca di *Statistica*, Cooperativa Libraria Universitaria Editrice Bologna, 40126, Bologna-Via Marsala 24. (in English and Italian).

DeGroot, M. H. (1970). *Optimal Statistical Decisions*, New York, McGraw-Hill.

Dellaportas, P., Forster, J. J. and Ntzoufras I., (2000). "On Bayesian Model and Variable Selection Using MCMC," *Statistics and Computing*, in press.

Dempster, A. P. (1980). "Bayesian Inference in Applied Statistics," in J. M. Bernardo, M. H. DeGroot, D. V. Lindley and A. F. M. Smith, Eds., *Bayesian Statistics*, Valencia, Spain, University Press, 266–291.

Dempster, A. P., Laird, N. M. and Rubin, D. B. (1977). "Maximum Likelihood From Incomplete Data via the EM Algorithm," *J. Royal Statist. Soc. (B)*, **39**, 1–38.

Denison, D., Holmes, G. T., Mallick, C. C., Bani, K. and Smith, A. F. (2002). *Bayesian Methods for Nonlinear Classification and Regression*, England, John Wiley and Sons, Ltd.

Dey, D. K., Gelfand, A. E. and Peng, F. (1997). "Overdispersed Generalized Linear Models," *J. Statist. Planning Inference*, **64**, 93–107.

Diaconis, P. (1977). "Finite Forms of de Finetti's Theorem on Exchangeability," *Synthese*, **36**, 271–281.

Diaconis, P. (1987). "Application of the Method Of Moments in Probability and Statistics," *Proc. of Symposia in Applied Math.*, **37**, in H. J. Landau Ed., *Moments in Mathematics*, 125–142.

Diaconis, P. (1994). Personal Communication.

Diaconis, P. and Freedman, D. (1978). "De Finetti's Generalizations of Exchangeability," in R. Jeffrey, Ed., *Studies in Inductive Logic and Probability*.

Diaconis, P. and Freedman, D. (1981). "Partial Exchangeability and Sufficiency," in *Proc. of the Indian Statistical Institute Golden Jubilee International Conference on Statistics and New Directions*, Calcutta, Indian Statistical Institute, 205–236.

Diaconis, P. and Ylvisaker, D. (1979). "Conjugate Priors for Exponential Families," *Ann. Statist.*, **7**, 269–281.

Diaconis, P. and Ylvisaker, D. (1985). "Quantifying Prior Opinion," in J. M. Bernardo, M. H. DeGroot, D. V. Lindley, and A. F. M. Smith, Eds., *Bayesian Statistics 2*, Amsterdam, North Holland, 133–156.

DiCiccio, T. J., Kass, R. E., Raftery, A. E. and Wasserman, L. (1997). "Computing Bayes Factors by Combining Simulation and Asymptotic Approximations," *J. Am. Statist. Assoc.*, **92**, 903–915.

Dickey, J. M. (1967a). "Expansions of t-Densities and Related Complete Integrals," *Ann. Math. Statist.*, **38**(2), 503–510.

Dickey, J. M. (1967b). "Multivariate Generalizations of the Multivariate t-Distribution and the Inverted Multivariate t-Distribution," *Ann. Math. Statist.*, **38**, 511–518.

Dickey, J. M. (1968). "Three Dimensional Integral-Identities with Bayesian Applications," *Ann. Math. Statist.*, **39**(5), 1615–1627.

Dickey, J. M. (1973). "Scientific Reporting and Personal Problems: Student's Hypothesis," *J. Royal Statist. Soc. B*, **35**, 285–305.

Draper, D. (1995). "Assessment and Propagation of Model Uncertainty (Disc: P71–97)," *J. Royal Statist. Soc. B*, **57**, 45–70.

Draper, D. (1999). "Comment on 'Bayesian Model Averaging: A Tutorial'," *Statist. Sci.*, **14**, 405–409.

Dreze, J. H. (1977). "Bayesian Regression Analysis Using the Poly-t Distribution," in A. Aykac and C. Brumat, Eds., *New Developments in the Application of Bayesian Methods*, New York, North-Holland.

DuMouchel, W. (1999). "Bayesian Data Mining in Large Frequency Tables, with and Application to the FDA Spontaneous Reporting System," with discussion, *Am. Statist.*, **53**, 177–202.

Edwards, A. W. F. (1972). *Likelihood*, Cambridge, Cambridge University Press.

Edwards, A. W. F. (1974). "The History of Likelihood," *In. Statist. Rev.*, **42**, 9–15.

Edwards, W. (1982). "Conservatism in Human Information Processing," in D. Kahneman, P. Slovic and A Tversky, Eds., *Judgment Under Uncertainty: Heuristics and Biases*, Cambridge, Cambridge University Press, 359–369; excerpted from a paper that appeared in B. Kleinmuntz Ed., *Formal Representation of Human Judgment*, New York, John Wiley and Sons, Inc.

Edwards, W., Lindman, H. and Savage, L. J. (1963). "Bayesian Statistical Inference for Psychological Research," *Psychol. Rev.*, **70**, 193–242.

Efron, B. and Morris, C. (1973). "Stein's Estimation Rule and its Competitors—An Empirical Bayes Approach," *J. Am. Statist. Assoc.*, **68**, 117–130.

Efron, B. and Morris, C. (1975). "Data Analysis Using Stein's Estimator and its Generalizations," *J. Am. Statist. Assoc.*, **70**, 311–319.

Erdelyi, A. (1956). *Asymptotic Expansions*, New York, Dover Publications.

Euverman, T. J. and Vermulst, A. A. (1983). *Bayesian Factor Analysis*, Department of Social Sciences, State University of Groningen, The Netherlands.

Fahrmeir, L and Lutz G. (1994). *Multivariate Statistical Modeling Based on Generalized Linear Models*, New York, Springer-Verlag.

Fatti, P., Klein, R. and Press S. J. (1997). "Hierarchical Bayes Models in Contextual Spatial Classification," in S. Ghosh, W. Schueany and W. B. Smith, Eds., *Statistics for Quality*: Dedicated to Donald B. Owen, New York, Marcel Dekker, Inc., Chapter 14, 275–294.

Fay, R. E. and Herriot, R. A. (1979). "Estimates of Income for Small Places: An Application of James–Stein Procedures to Census Data," *J. Am. Statist. Assoc.*, **74**, 269–277.

Feller, W. (1971). *An Introduction To Probability Theory and Its Applications*, Vol II, 2nd Edition, New York, John Wiley and Sons, Inc.

Fernandez, C., Ley, E. and Steel, M. F. (2001). "Benchmark Priors for Bayesian Model Averaging," *J. Econometrics*, 381–427.

Ferryman, T. and Press, S. J. (1996). "A Comparison of Nine Pixel Classification Algorithms," Technical Report No. 232, Department of Statistics, University of California, Riverside.

Fishburn, P. C. (1981). "Subjective Expected Utility: A Review of Normative Theories," *Theory and Decision*, **13**, 139–199.

Fishburn, P. C. (1986). "The Axioms of Subjective Probability," *Statist. Sci.*, **1**(3), 335–358.

Fisher, R. A. (1921). "On the 'Probable Error' of a Coefficient of Correlation Deduced From a Small Sample," *Metron*, **1**, 3–32.

Fisher, R. A. (1922). "On the Mathematical Foundations of Theoretical Statistics," *Philosophical Trans. Royal Soc. A*, **222**, 309–368.

Fisher, R. A. (1930). "Inverse Probability," *Proc. Camb. Phil. Soc.*, **26**, 528–535.

Fisher, R. A. (1925) 1970. *Statistical Methods for Research Workers*, 14th Edition, New York, Hafner, Edinburgh, Oliver and Boyd.

Florens, J. P., Mouchart, M., and Rolin J. M. (1990). *Elements of Bayesian Statistics*, New York, Marcel Dekker, Inc.

Foster, D. P. and George, E. I. (1994). "The Risk Inflation Criterion for Multiple Regression," *Ann. Statis.*, **22**, 1947–1975.

Freedman, D. (1962). "Invariants Under Mixing Which Generalize de Finetti's Theorem," *Ann. Math. Statist.*, **33**, 916–923.

Freudenburg, W. R. (1988). "Perceived Risk, Real Risk: Social Science and the Art of Probabilistic Risk Assessment," *Science*, **242**(70), 44–49.

Friedman, H. P. and Goldberg, J. D. (2000). "Knowledge Discovery from Databases and Data Mining: New Paradigms for Statistics and Data Analysis?" Biopharmaceutical Report, *Am. Statistical Assoc.*, **8**(2), 1–12.

Furnival, G. M. and Wilson, R. W. J. (1974). "Regression by Leaps and Bounds," *Technometrics*, **16**, 499–511.

Gamerman, D. (1997). *Markov Chain Monte Carlo: Stochastic Simulation for Bayesian Inference*, London, Chapman and Hall.

Gandolfo, G. (1987). "de Finetti, Bruno," in J. Eatwell, M. Milgate, and P. Newman, Eds., *The New Palgrave: A Dictionary of Economics*, Vol. 1, 765–766.

Garthwaite, P. H. and Dickey, J. M. (1988). "Quantifying Expert Opinion in Linear Regression Problems," *J. Royal Statist. Soc. B*, **50**(3), 462–474.

Garvan, C. W. and Ghosh, M. (1999). "On the Property of Posteriors for Dispersion Models," *J. of Statist. Planning and Inference*, **78**, 229–241.

Gatsonis, C., Hodges, J. S., Kass, R. E.; and Singpurwalla, N. D. (Eds.) (1993). *Case Studies in Bayesian Statistics*, New York, Springer-Verlag.

Gatsonis, C., Hodges, J. S.; Kass, R. E., and Singpurwalla, N. D. (Eds.) (1995). *Case Studies in Bayesian Statistics*, Vol. II, New York, Springer-Verlag.

Gatsonis, C., Hodges, J. S., Kass, R. E., McCulloch, R.; Rossi, P.; and Singpurwalla, N. D. (1997). *Case Studies in Bayesian Statistics*, Vol. III, New York, Springer-Verlag.

Gatsonis, C., Kass, R. E., Carlin, B., Carriquiry, A., Gelman, A., Verdinelli, I., and West, M. (1999). *Case Studies in Bayesian Statistics*, Vol. IV, New York, Springer-Verlag.

Gatsonis, C., Kass, R. E., Carriquiry, A., Gelman, A., Higdon, D., Pauler, D. K. and Verdinelli, I. (2002). *Case Studies in Bayesian Statistics*, Vol. VI, New York, Springer Verlag.

Geiger, D. and Heckerman, D. (1999). "Parameter Priors for Directed Acyclic Graphical Models and the Characterization of Several Probability Distributions," Technical Report. MSR-TR-98-67, Microsoft Research.

Geiger, D., Heckerman, D. and Meek, C. (1996). "Asymptotic Model Selection for Directed Networks with Hidden Variables," Technical Report. MSR-TR-96-07, Microsoft Research.

Geisser, S. (1964). "Posterior Odds for Multivariate Normal Classifications," *J. Royal Statist. Soc. B*, **26**, 69–76.

Geisser, S. (1966). "Predictive Discrimination," in P. R. Krishnaiah, Ed., *Multivariate Analysis*, New York, Academic Press, 149–163.

Geisser, S. (1967). "Estimation Associated With Linear Discriminants," *Ann. Math. Statist.*, **38**, 807–817.

Geisser, S. (1993). *Predictive Inference: An Introduction*, New York, Chapman and Hall.

Geisser, S. and Cornfield, J. (1963). "Posterior Distributions for Multivariate Normal Parameters," *J. Royal Statist. Soc. (B)*, **25**, 368–376.

Geisser, S. and Eddy, W. F. (1979). "A Predictive Approach to Model Selection," *J. Am. Statist. Assoc.*, **74**(365), 153–160.

Gelfand, A. E., Hills, S. E., Racine-Poon, A. and Smith A. F. M. (1990). "Illustration of Bayesian Inference in Normal Data Models Using Gibbs Sampling," *J. Am. Statist. Asso., 85, 972–985.*

Gelfand, A., Mallick, B. and Dey, D. (1995). "Modeling Expert Opinion Arising as a Partial Probabilistic Specification," *J. Am. Statist. Assoc.*, 598–604.

Gelfand, A. E., Sahu, S. K. and Carlin, B. P. (1995), "Efficient Parameterizations for Normal Linear Mixed Models," *Biometrika*, **82**, 479–488.

Gelfand, A. E. and Smith, A. F. M. (1990). "Sampling-Based Approaches to Calculating Marginal Densities," *J. Am. Statist. Assoc.*, **85**, 398–409.

Gelman, A., Carlin, J. B., Stern, H. S. and Rubin D. B. (1995). *Bayesian Data Analysis*, London, Chapman and Hall.

Geman, S. and Geman, D. (1984). "Stochastic Relaxation, Gibbs Distributions and the Bayesian Restoration of Images," *IEEE Trans. Pattern Ann., Machine Intell.*, **6**, 721–741.

Genest, C. and Schervish, M. (1985). "Modeling Expert Judgments for Bayesian Updating," *Annals Statist.*, 1198–1212.

George, E. I. and Foster, D. P. (2000). "Calibration and Empirical Bayes Variable Selection," To appear in *Biometrika*.

George, E. I. and McCulloch, R. E. (1993). "Variable Selection Via Gibbs Sampling," *J. Am. Statist. Assoc.*, **88**, 881–889.

George, E. I. and McCulloch, R. E. (1997). "Approaches for Bayesian Variable Selection," *Statistica Sinica*, **7**, 339–374.

Ghosh, J. K. (1988). *Statistical Information and Likelihood: A Collection of Critical Essays by Dr. D. Basu*, New York, Springer-Verlag.

Ghosh, M. and Rao, J. N. K. (1994). "Small Area Estimation: An Appraisal," *Statist. Sci.*, **9**, 55–76.

Ghosh, M., Natarajan, K., Stroud, T. W. F., and Carlin B. P. (1998). "Generalized Linear Models for Small-Area Estimation," *J. Am. Statist. Assoc.*, **93**, 273–282.

Gibbons, R. D., Hedeker, D., Charles, S. C. and Frisch, P. (1994). "A Random-Effects Probit Model for Predicting Medical Malpractice Claims," *J. Am. Statist. Assoc.*, **89**, 760–767.

Gilks, W. R., Richardson, S. and Spieglehalter, D. J. (Eds.) (1996). *Markov Chain Monte Carlo in Practice*, London, Chapman and Hall.

Godsill, S. J. (1999), "On the Relationship Between Model Uncertainty Methods," Technical Report, Signal Processing Group, Cambridge University.

Gokhale, D. B. (1975). "Maximum Entropy Characterizations of Some Distributions," in G. Patil et al., Eds., *Statistical Distributions in Scientific Work*, Vol. 3, Dordrecht, Netherlands, Reidel Pub. Co., 299–304.

Gokhale, D. B. and Kullback S. (1978). *The Information in Contingency Tables*, New York, Marcel Dekker, Inc.

Goldstein, H. (1995). *Multilevel Statistical Models* (Second edition), London, Edward Arnold Publishers Ltd.

Goldstein, H. and Spiegelhalter, D. J. (1996). "League Tables and Their Limitations: Statistical Issues in Comparisons of Institutional Performance," *J. Royal Statist. Soc., A*, **159**, 385–409.

Good, I. J. (1950). *Probability and the Weighting of Evidence*, London, Charles Griffin and Co., Ltd.

Good, I. J. (1952). "Rational Decisions," *J. Royal Statist. Soc. (B)*, **14**, 107–114.

Good, I. J. (1965). *The Estimation of Probabilities: An Essay on Modern Bayesian Methods*, Research Monograph #30, Cambridge, MA, The MIT Press.

Good, I. J. (1983). *Good Thinking: The Foundations of Probability and its Applications*, Minneapolis, The University of Minnesota Press.

Green, P. E. (1995). "Reversible Jump Markov Chain Monte Carlo Computation and Bayesian Model Determination," *Biometrika*, **82**, 711–732.

Guttman, L. (1953). "Image Theory for the Structure of Quantitative Variates," *Psychometrika*, **18**(4), 277–296.

Haavelmo, T. (1947). "Methods of Measuring the Marginal Propensity to Consume," *J. Am. Statist. Assoc.*, **42**, 88.

Halmos, P. A. (1950). *Measure Theory*, Princeton, New Jersey, D. Van Nostrand Co., Inc.

Harris, B. (1982), "Entropy," in S. Kotz, N. L. Johnson and C. B. Read, Eds., *Encyclopedia of Statistical Sciences*, V61.2, 512–516.

Hastie, T., Tibshirani, R. and Friedman, J. (2001). *The Elements of Statistical Learning: Data Mining, Inference, and Prediction*, New York, Springer-Verlag.

Hartigan, J. A. (1964). "Invariant Prior Distributions," *Ann. Math. Stat.*, **35**, 836–845.

Hastings, W. K. (1970). "Monte Carlo Sampling Methods Using Markov Chains and Their Applications," *Biometrika*, **57**, 97–109.

Hausdorff, F. (1921). "Summationsmethoden und Momentfolgen," *Math. Zeit.*, **9**, 74–109, 281–299.

Heath, D. and Sudderth W. (1976). "de Finetti's Theorem on Exchangeable Variables," *Am. Stat.*, **30**, 188–189.

Heckerman, D. and Chickering, D. (1996). "A Comparison of Scientific and Engineering Criteria for Bayesian Model Selection," Technical Report. MSR-TR-96-07, Microsoft Research.

Heckerman, D., Geiger, D. and Chickering, D. (1994). Learning Bayesian Networks: The Combination of Knowledge and Statistical Data," Technical Report. MSR-TR-94-09, Microsoft Research.

Heckerman, D., Meek, C. and Cooper, G. (1997). "A Bayesian Approach to Causal Discovery," Technical Report. MSR-TR-97-05, Microsoft Research.

Henderson, C. R. (1975). "Best Linear Unbiased Estimation and Prediction Under a Selection Model," *Biometrics*, **31**, 423–447.

Hewitt, E. and Savage, I. J. (1955). "Symmetric Measures on Cartesian Products," *Trans. Am. Math. Soc.*, **80**. 470–501.

Hobert, J. P. and Casella, G. (1996). "The Effect of Improper Priors on Gibbs Sampling in Hierarchical Linear Mixed Models," *J. Am. Statist. Assoc.*, **91**, 1461–1473.

Hodges, J. S. (1987). "Uncertainty, Policy Analysis and Statistics" (C/R: *P276–291), Statist. Sci.*, **2**, 259–275.

Hoeting, H. A., Madigan, D., Raftery, A. E., and Volinsky, C. T. (1999). "Bayesian Model Averaging: A Tutorial (with Discussion)," *Statist. Sci.*, **14**, 382–417.

Hoeting, J., Raftery, A. E. and Madigan, D. (1996). "A Method for Simultaneous Variable Selection and Outlier Identification in Linear Regression," *Comput. Statist. Data Analysis*, **22**, 251–270.

Hogarth, R. (1980). *Judgment and Choice*, New York, John Wiley and Sons, Inc.

Howson, C. and Urbach P. (1989). *Scientific Reasoning*, La Salle, Illinois, Open Court Publishing Co.

Hulting, F. L., and Harville, D. A. (1991). "Some Bayesian and Non-Bayesian Procedures for the Analysis of Comparative Experiments and for Small-Area Estimation: Computational Aspects, Frequentist Properties, and Relationships," *J. Am. Statist. Assoc.* **86**, 557–568.

Hyashi, K. (1997). *The Press-Shigemasu Factor Analysis Model With Estimated Hyperparameters*, Ph.D. Thesis, Department of Psychology, University of North Carolina, Chapel Hill.

Hyde, C. C. and Johnstone, I. M. (1979). "On Asymptotic Posterior Normality for Stochastic Processes," *J. Royal Statist. Soc. B*, **41**, 184–189.

Insua, D. R. and Ruggeri, F. (Eds.) (2000). *Robust Bayesian Analysis*, New York, Springer-Verlag.

James W. and Stein, C. (1961). "Estimation With Quadratic Loss," J. Neyman and E. L. Scott, Eds., *Proceedings of the Fourth Berkeley Symposium*, **1** Berkeley, University of California Press, 361–379.

Jaynes, E. (1957). "Information Theory and Statistical Mechanics I," *Phys. Rev.*, **106**, 620–6301.

Jaynes, E. T. (1983). *Papers on Probability, Statistics and Statistical Physics of E. T. Jaynes*, R. D. Rosenkranz Ed., Dordrecht, Holland, D. Reidel Publishing Co.

Jaynes, E. (1986). "Some Applications and Extensions of the de Finetti Representation Theorem," in P. Goel and A. Zellner, Eds., *Bayesian Inference and Decision Techniques with Applications: Essays in Honor of Bruno de Finetti*, Amsterdam, North Holland Publishing Co.

Jaynes, E. T., and Bretthorst, G. Larry (2003). *Probability Theory: The Logic of Science*, Cambridge, Cambridge University Press.

Jeffrey, R. C. (Ed.) (1980). *Studies in Inductive Logic and Probability*, Vol. 2, Berkeley, The University of California Press.

Jeffreys, H. (1939, 1st edition; 1948, 2nd edition; 1961, 3rd edition). *Theory of Probability*, Oxford, Oxford University Press and Clarendon Press.

Jelenkowska, T. H. (1989), *Bayesian Multivariate Analysis*, 19 Coll. Metodl. z Agrobiom. PAN, Warszawa.

Jelenkowska, T. H. and Press S. J. (1997). "Bayesian Inference in the Multivariate Mixed Model MANOVA," in Multivariate Statistical Analysis in Honor of Minoru Siotani on His 70th Birthday, *Am. J. Math. Mgmt. Sci.*, 97–116, vol. III.

Jensen, F. V. (1996). *Introduction to Bayesian Networks*, New York, Springer-Verlag.

Jensen, F. V. and Jensen, F. B. (2001). *Bayesian Networks and Decision Graphs*, New York, Springer-Verlag.

Jensen, J. L. W. V. (1906). "Sur les Functions Convexes et les Inegalites Entre les Valeurs Moyennes," *Acta Math.*, **30**, 175–193.

Johnson, N. L., and Kotz, S. (1969). *Distributions In Statistics: Discrete Distributions*, New York, Houghton Mifflin Co.

Kadane, J. B., Dickey, J. M., Winkler, R. L., Smith, W. S. and Peters, S. C. (1980). "Interactive Elicitation of Opinion of a Normal Linear Model," *J. Am. Statist. Assoc.*, **75**, 845–854.

Kadane, J. B., Schervish, M. J. and Seidenfeld, T. (1999). *Rethinking the Foundations of Statistics*, Cambridge, Cambridge University Press.

Kagan, A. M., Linnik, Y. V. and Rao, C. R. (1973). *Characterization Problems in Mathematical Statistics*, New York, John Wiley and Sons, Inc.

Kahneman, D., Slovic, P. and Tversky, A. (1982). *Judgment Under Uncertainty: Heuristics and Biases*, Cambridge, Cambridge University Press.

Kapur, J. N. (1989). *Maximum Entropy Models in Science and Engineering*, New York, John Wiley and Sons, Inc.

Kass, R. E. and Raftery, A. E. (1995). "Bayes Factors," *J. Am. Statist. Assoc.*, **90**, 73–79.

Kass, R. and Wasserman, L. (1996). "The Selection of Prior Distributions by Formal Rules," *J. Am. Statist. Assoc.*, **91**, 1343–1370.

Kaufman, G. M. and Press, S. J. (1973a). "Bayesian Factor Analysis," Report No. 73t–25, *Bull. Inst. Math. Statist.*, **2**, (2, 7).

Kaufman, G. M. and Press, S. J. (1973b). "Bayesian Factor Analysis," Report No. 7322 Center for Mathematical Studies in Business and Economics, University of Chicago.

Kaufman, G. M. and Press, S. J. (1976), "Bayesian Factor Analysis," Working Paper No. 413, Department of Commerce and Business Administration, University of British Columbia, Vancouver, B. C. Canada.

Kendall, M. (1980), *Multivariate Analysis*, Second Edition, Charles Griffin Pub, 34.

Kendall, M. G. and Stuart, A. (1967). *The Advanced Theory of Statistics, Volume 2: Inference and Relationship*. 2nd Edition, New York, Hafner Publishing Company.

Keynes, J. M. (1921). *A Treatise on Probability*, London, Macmillan.

Kingman, J. F. C. (1978). "Uses of Exchangeability," *Ann. Prob.*, **6**, 183–197.

Klein, R. and Press, S. J. (1989a). "Contextual Bayesian Classification of Remotely Sensed Data," *Commun. Statist., Theory and Methods*, **18**(9), 3177–3202.

Klein, R. and Press S. J. (1989b). "Bayesian Contextual Classification with Neighbors Correlated With Training Data," in S. Geisser, J. Hodges, S. J. Press and A. Zellner, Eds., *Bayesian and Likelihood Methods in Statistics and Econometrics: Essays in Honor of George A. Barnard*, New York, North Holland Publishing Co., 337–355.

Klein, R. and Press, S. J. (1992). "Adaptive Bayesian Classification of Spatial Data," *J. Am. Statist. Assoc.*, **87**(419), 844–851.

Klein, R. and Press, S. J. (1993). "Adaptive Bayesian Classification with a Locally Proportional Prior," *Comm. Statist., Theory and Methods*, **22**(10), 2925–2940.

Klein, R. and Press, S. J. (1996). "The Directional Neighborhoods Approach to Contextual Reconstruction of Images from Noisy Data," *J. Am. Statist. Assoc.*, **91**(435), 1091–1100.

Kloek, T. and Van Dijk, H. K. (1978). "Bayesian Estimates of Equation System Parameters: An Application of Integration by Monte Carlo," *Econometrica*, **46**, 1–19.

Kolmogorov, A. N. (1933). *Foundations of the Theory of Probability*, translated from German, New York, Chelsey, 1950.

Kshirsagar, A. M. (1960). "Some Extensions of the Multivariate t-Distribution and the Multivariate Generalization of the Distribution of the Regression Coefficient," *Proc. Cambridge Phil. Soc.*, **57**, 80–85.

Kullback, S. (1959). *Information Theory and Statistics*, New York, John Wiley and Sons, Inc.

Laird, N. M. and Louis, T. A. (1987). "Empirical Bayes Confidence Intervals Based on Bootstrap Samples," *J. Am. Statist. Assoc.*, **82**, 739–750.

Laplace, P. S. (1774, 1814). *Essai Philosophique sur les Probabilites*. Paris. This book went through five editions (the fifth was in 1825) revised by Laplace. The sixth edition appeared in English translation by Dover Publications, New York, in 1951. While this philosophical essay appeared separately in 1814, it also appeared as a preface to his earlier work, *Theorie Analytique des Probabilites*.

Lauritzen, S. L. (1982). *Statistical Models As Extremal Families*, Aalborg, Aalborg University Press.

Lauritzen, S. L. (1996). *Graphical Models*, Oxford, Oxford University Press.

LaValle, I. H. (1978). *Fundamentals of Decision Analysis*, New York, Holt, Rinehart and Winston.

Le Cam, L. (1956). "On the Asymptotic Theory of Estimation and Testing Hypotheses," *Proc. Third Berkeley Symp. on Math. Statist. And Prob.*, **1**, Berkeley, University of California Press, 129–156.

Leamer, E. E. (1978). *Specification Searches*, New York, John Wiley and Sons, Inc.

Lee, S. Y. (1981). "A Bayesian Approach to Confirmatory Factor Analysis," *Psychometrika*, **46**(2), 153–159.

Lee, S. E. (1994). *Robustness of Bayesian Factor Analysis Estimates*, Ph.D. Thesis, Department of Statistics, University of California, Riverside.

Lee, S. E. and Press, S. J. (1998). "Robustness of Bayesian Factor Analysis Estimates," *Commun. Statist., Theory and Methods*, **27**(8), 1871–1893.

Lehmann, E. L. (1959). *Testing Statistical Hypotheses*, New York, John Wiley and Sons, Inc.

Lempers, F. B. (1971). *Posterior Probabilities of Alternative Linear Models*, Rotterdam.

Levine, R. D. and Tribus, M., Eds. (1979). *The Maximum Entropy Formalism*, Cambridge, MA, The MIT Press.

Lindley, D. V. (1957). "A Statistical Paradox," *Biometrika*, **44**, 187–192.

Lindley, D. V. (1965). *Introduction to Probability and Statistics, Part 2: Inference*, Cambridge, Cambridge University Press.

Lindley, D. V. (1976). "Bayesian Statistics," in W. L. Harper and C.A. Hooker, Eds., *Foundations of Probability Theory, Statistical Inference, and Statistical Theories of Science*, Vol. II, Boston, Reidel, 353–363.

Lindley, D. V. (1980). "Approximate Bayesian Methods," in J. M. Bernardo, M. H. DeGroot, D. V. Lindley, and A. F. M. Smith, Eds., *Bayesian Statistics*, Valencia, Spain, University Press, 223–245.

Lindley, D. V. (1982). "Scoring Rules and the Inevitability of Probability," *Int. Statist. Rev.*, **50**, 1–26.

Lindley, D. V. (1983). "Reconciliation of Probability Distributions," *Opns. Res.*, **31**, 866–880.

Lindley, D. V. (1985). *Making Decisions*, London, John Wiley and Sons, Inc.

Lindley, D. V. and Phillips, L. D. (1976). "Inference for a Bernoulli Process: A Bayesian View," *Am. Stat.*, **30**(3), 112–119.

Lindley, D. V. and Smith, A. F. M. (1972). "Bayes Estimates for the Linear Model," *J. Royal Statist. Soc., B*, **34**, 1–41.

Lisman, J. H. C. and van Zuylen, M. C. A. (1972). "Note on the Generation of Most Probable Frequency Distributions," *Statistica Neerlandica*, **26**, 19–23.

Liu, J. S. (1994). "The Collapsed Gibbs Sampler in Bayesian Computations with Applications to a Gene Regulation Problem," *J. Am. Statist. Assoc.*, **89**, 958–966.

Liu, J. S., Wong, W. H. and Kong, A. (1994), "Covariance Structure of the Gibbs Sampler with Applications to the Comparisons of Estimators and Data Augmentation Schemes," *Biometrika*, **81**, 27–40.

Longford, N. T. (1993). *Random Coefficient Models*, Oxford, Clarendon Press.

Maistrov, L. E. (1974). *Probability Theory: A Historical Sketch*, S. Kotz, Trans. and Ed., New York, Academic Press.

Mallows, C. L. (1973). "Some Comments On Cp," *Technometrics*, **15**, 661–675.

Maritz, J. S. (1970). *Empirical Bayes Methods*, London, Methuen.

Martin, J. J. (1967). *Bayesian Decision Problems and Markov Chains*, New York, John Wiley and Sons, Inc.

Martin, J. K. and McDonald, R. P. (1975). "Bayesian Estimation in Unrestricted Factor Analysis: A Treatment for Heywood Cases," *Psychometrika*, **40**(4), 505–517.

Marx, D. G., and Nel, D. G. (1985). "Matric-t Distribution," in S. Kotz, N. L. Johnson, C. B. Reed, Eds., *Encyclopedia of Statistical Sciences*, **5**, 316–320.

Mathews, J. and Walker R. L. (1965). *Mathematical Methods of Physics*, New York, W. A. Benjamin, Inc.

Matthews, R. A. J. (1998a). "Fact versus Factions: The Use and Abuse of Subjectivity in Scientific Research," ESEF Working Paper 2/98, *The European Science and Environment Forum*, Cambridge, England, September, 1998.

Mayekawa, S. (1985). "Bayesian Factor Analysis," ONR Technical Report No. 85–3, School of Education, The University of Iowa, Iowa City, Iowa.

McCullagh, P. and Nelder, J. A. (1989). *Generalized Linear Models* (Second Edition), London, Chapman and Hall Ltd.

McCulloch, W. and Pitts, W. (1943). "A Logical Calculus of the Ideas Immanent in Nervous Activity," *Bull. Math. Biophys.*, **5**, 115–133, 96–104, reprinted in Anderson and Rosenfeld, 1988.

McLachlan, G. J. (1992). *Discriminant Analysis and Statistical Pattern Recognition*, New York, John Wiley and Sons, Inc.

Metropolis, N., Rosenbluth, A. W., Rosenbluth, M. N., Teller, A. H. and Teller, E. (1953). "Equation of State Calculations by Fast Computing Machine," *J. Chem. Phys.*, **21**, 1087–1091.

Montgomery, D. C., Peck E. A. and Vining G. G. (2000). *Introduction to Linear Regression Analysis*, 3rd edition, New York, John Wiley and Sons.

Morgan, B. W. (1968). *An Introduction to Bayesian Statistical Decision Processes*, Englewood Cliffs, New Jersey, Prentice-Hall, Inc.

Morris, C. N. (1982). "Natural Exponential Families With Quadratic Variance Functions," *Ann. Statist.*, **10**, 65–80.

Morris, C. N. (1983). "Exponential Families With Quadratic Variance Functions: Statistical Theory," *Ann. Statist.*, **11**(2), 515–529.

Morris, P. A. (1977). "Combining Expert Judgments: A Bayesian Approach," *Mgmt Sci.*, **23**(7), 679–693.

Morrison, D. F. (1967). *Multivariate Statistical Methods*, New York, McGraw-Hill Book Co.

Natarajan, R. and McCulloch, C. E. (1998). "Gibbs Sampling with Diffuse Proper Priors: A Valid Approach to Data-Driven Inference?" *J. Comput. and Graphical Statist.*, **7**, 267–277.

Naylor, J. C. (1982). "Some Numerical Aspects of Bayesian Inference," Unpublished Ph.D. Thesis, University of Nottingham.

Naylor, J. C. and Smith, A. F. M. (1982). "Applications of a Method for the Efficient Computation of Posterior Distributions," *Appl. Statist.*, **31**(3), 214–225.

Neal, R. M. (1996). *Bayesian Learning for Neural Networks*, New York, Springer-Verlag.

Neyman, J. and Pearson E. S. (1933). "On the Testing of Statistical Hypotheses in Relation to Probability A Priori," *Proc. of the Cambridge Phil. Soc.*, **29**, 492–510.

Neyman, J. and Pearson E. S. (1966). *Joint Statistical Papers of J. Neyman and E. S. Pearson*, Berkeley, CA, University of California Press (10 papers).

Neyman, J. and Scott, E. L. (1948). "Consistent Estimates Based on Partially Consistent Observations," *Econometrica*, **16**, 1.

Normand, S-L. T., Glickman, M. E. and Gatsonis, C. A. (1997). "Statistical Methods for Profiling Providers of Medical Care: Issues and Applications," *J. Am. Statis. Assoc.*, **92**, 803–814.

O'Hagen, A. (1993). "Fractional Bayes Factors for Model Comparison," Statistical Research Report 93–6, University of Nottingham.

Olshen, R. (1973). "A Note on Exchangeable Sequences," *Z. Wahrscheinlichkeitstheorie und Verw. Gebeite*, **28**, 317–321.

Parmigiani, G. (2002). *Modeling in Medical Decision-Making: A Bayesian Approach (Statistics in Practice)*, New York, John Wiley and Sons, Inc.

Parzen, E. (1962). "On Estimation of a Probability Density Function and Mode," *Ann. Math. Statist.*, **33**, 1065–1076.

Pearl, J. (2001). *Causality, Reasoning, and Inference*, Cambridge, Cambridge University Press.

Pearson, K. (1892). *The Grammar of Science*, London, Adam and Charles Black.

Pearson, E. (1978). *The History of Statistics in the 17th and 18th Centuries*, New York, Macmillan.

Pinheiro, J. C. and Bates, D. M. (2000). *Mixed Effects Models in S and S-Plus*, New York, Springer.

Polasek, W. (1997). "Factor Analysis and Outliers: A Bayesian Approach," University of Basel Discussion Paper.

Popper, K. (1935), (1959), (1968). *The Logic of Scientific Discovery*, New York, Basic Books; London, Hutchinson, New York, Harper Torchbooks, Harper and Row, Publishers.

Prasad, N. G. N. and Rao, J. N. K. (1990). "The Estimation of the Mean Squared Error of Small-Area Estimators," *J. Am. Statist. Assoc.*, **85**, 163–171.

Pratt, J. W., Raiffa, H. and Schlaifer, R. (1965). *Introduction to Statistical Decision Theory*, New York, McGraw-Hill Book Co., Inc.

Press, S. J. (1972). *Applied Multivariate Analysis*, New York, Holt Rinehart, and Winston, Inc.

Press, S. J. (1978). "Qualitative Controlled Feedback for Forming Group Judgments and Making Decisions," *J. Am. Statis. Assoc.*, **73**(363), 526–535.

Press, S. J. (1979). "Bayesian Computer Programs," in A. Zellner Ed., *Studies in Bayesian Econometrics and Statistics in Honor of Harold Jeffreys*, Amsterdam, North-Holland Publishing Co.

Press, S. J. (1980). "Bayesian. Inference in MANOVA," in P. R. Kishnaiah, Ed. *Handbook of Statistics*, Vol. I., New York, North-Holland Publishing Co.

Press, S. J. (1982). *Applied Multivariate Analysis: Including Bayesian and Frequentist Methods of Inference*, Malabar, Florida, Robert E. Krieger Publishing Co.

Press, S. J. (1989). *Bayesian Statistics: Principles Models, and Applications*, New York, John Wiley and Sons.

Press, S. J. (1998). "Comparisons of Pixel Classification Algorithms," Technical Report No. 257, Department of Statistics, University of California, Riverside.

Press, S. J. (2003). "The Role of Bayesian and Frequentist Multivariate Modeling in Statistical Data Mining," in the Proceedings of the C. Warren Neel Conference on the New Frontiers of Statistical Data Mining (DM), Knowledge Discovery (KD), and e-Business, New York, John Wiley and Sons, Inc. (In press, to be published on-line).

Press, S. J. and Davis A. W. (1987). "Asymptotics for the ratio of Multiple t-densities," in A. Gelfand, Ed., *Contributions to the Theory and Applications of Statistics*, New York, Academic Press, 155–177.

Press, S. J. and Shigemasu, K. (1985). "Bayesian MANOVA and MANOCOVA Under Exchangeability," *Commun. in Statist.: Theory and Methods, A*, **14**(5), 1053–1078.

Press, S. J. and Shigemasu K. (1989). "Bayesian Inference in Factor Analysis," in L. J. Gleser, M. D. Perlman, S. J. Press and A. R. Sampson, Eds., *Contributions to Probability and Statistics: Essays in Honor of Ingram Olkin*, New York, Springer Verlag, 271–287.

Press, S. J. and Shigemasu, K. (1997). "Bayesian Inference in Factor Analysis—Revised," with an appendix by Daniel B. Rowe, Technical Report No. 243, Department of Statistics, University of California, Riverside, CA, May, 1997.

Press, S. J. and K. Shigemasu (1999). "A Note on Choosing the Number of Factors," *Comm. Statist.-Theory Meth*, **28**(7), 1653–1670.

Press, S. James and Tanur, J. M. (2001a). "The Respondent-Generated Intervals Approach to Sample Surveys: From Theory to Experiment," *Proceedings of the International Meeting of the Psychometric Society-2001*, University of Osaka, Japan.

Press, S. J. and Tanur, J. M. (2001b). *The Subjectivity of Scientists and the Bayesian Approach*, New York, John Wiley and Sons, Inc.

Press, S. J. and Yang C. (1975). "A Bayesian Approach to Second Guessing Undecided Respondents," *J. Am. Statist. Assoc.*, **69**, 58–67.

Raiffa, H. (1968). *Decision Analysis: Introductory Lectures on Choices Under Uncertainty*, Reading, MA, Addison-Wesley.

Raiffa, H. and Schlaifer, R. (1961). *Applied Statistical Decision Theory*, Boston, MA, Harvard University, Graduate School of Business Administration.

Rajagopalan, M. and Broemeling, L. (1983). "Bayesian Inference for the Variance Components of the General Mixed Models," *Commun. in Statist.*, **12**(6).

Ramsey, F. P. (1926). "Truth and Probability," in R. B. Braithwaite, Ed., *The Foundations of Mathematics and Other Logical Essays*, 1931, by permission of the Humanities Press, New York, and Routledge and Kegan Paul Ltd., London, reprinted in Kyberg and Smokler, 1980, *Studies in Subjective Probability*, Melbourne, Florida, Krieger Publishing Co.

Ramsey, F. and Schafer, D. (2002). *The Statistical Sleuth: A Course in Methods of Data Analysis*, Duxbury Press.

Rao, B. L. S. P. (1992). *Identifiability in Stochastic Models*, San Diego, CA, Academic Press, Inc.

Rao, C. R. and Kleffe J. (1980). "Estimation of Variance Components," in P. R. Krishnaiah, Ed., *Handbook of Statistics 1: Analysis of Variance*, New York, North Holland Publishing Co.

Rauden, S. W. and Bryk, A. S. (2001). *Hierarchical Linear Models: Applications and Data Analysis Methods*, Second Edition, Thousand Oaks, CA, Sage Pub., Inc.

Raudenbush, S. W., Bryk, Anthony, S., Cheong, Yuk Fai and Congdon R. T., Jr. (2000). *HLM5: Hierarchical Linear and Nonlinear Modeling*, Lincolnwood, Illinois, Scientific Software International, Inc.

Reid, N. (2000). "Likelihood," *J. Am. Statist. Assoc.*, **95**(452), 1335–1340.

Renyi, A. (1970). *Probability Theory*, New York, American Elsevier.

Rice, J. A. (1994). *Mathematical Statistics and Data Analysis*, Wadsworth Publishing Co.

Ripley, B. D. (1981). *Spatial Statistics*, New York, John Wiley and Sons, Inc.

Ripley, B. D. (1987). *Stochastic Simulation*, New York, John Wiley and Sons, Inc.

Ripley, B. D. (1996). *Pattern Recognition and Neural Networks*, Cambridge, Cambridge University Press.

Robbins, H. (1955). "An Empirical Bayes Approach to Statistics," in *Proceedings of the Third Berkeley Symposium on Mathematical Statistics and Probability*, **1**, Berkeley, University of California Press, 157–164.

Robbins, H. (1977). "Prediction and Estimation for the Compound Poisson Distribution," *Proceedings of the National Academy of Sciences*, **74**, 2670–2671.

Robert, C. (2001). *The Bayesian Choice: From Decision—Theoretic Foundations to Computational Implementation*, Second Edition, New York, Springer-Verlag.

Roberts, G. O. and Sahu, S. K. (1997). "Updating Schemes, Correlation Structure, Blocking, and Parameterization for the Gibbs Sampler," *J. Royal Statis. Soc., B*, **59**, 291–317.

Robinson, G. K. (1991). "That BLUP is a Good Thing: The Estimation of Random Effects (with discussion)," *Statist. Sci.*, **6**, 15–51.

Roby, T. B. (1965). "Belief States and Uses of Evidence," *Behav. Sci.*, **10**, 255–270.

Rohatgi, V. K. (1976). *An Introduction to Probability Theory and Mathematical Statistics*, New York, John Wiley and Sons, Inc.

Rowe, D. B. (1998). *Correlated Bayesian Factor Analysis*, Ph.D. Thesis, Department of Statistics, University of California, Riverside.

Rowe, D. B. and Press, S. J. (1998). "Gibbs Sampling and Hill Climbing in Bayesian Factor Analysis," Technical Report No. 255, Department of Statistics, University of California, Riverside, CA.

Rubin, D. B. (1980). "Using Empirical Bayes Techniques in the Law School Validity Studies," *J. Am. Statist. Assoc.*, **75**, 801–816.

Salzer, H. E., Zucker, R. and Capuano, R. (1952). "Tables of the Zeros and Weight Factors of the First Twenty Hermite Polynomials," *J. Research, National Bureau of Standards*, **48**, 111–116.

Sarndal, C. E., Swensson, B. and Wretman, J. (1992). *Model Assisted Survey Sampling*, New York, Springer-Verlag, Inc.

Savage, L. J. (1954). *The Foundations of Statistics*, New York, John Wiley and Sons, Inc.

Schafer, J. L. and Yucel, R. M. (2002). "Computational Strategies for Multivariate Linear Mixed-Effects Models With Missing Values," *J. Comput. and Graphical Statistics*, in press.

Scheffe, H. (1959). *The Analysis of Variance*, New York, John Wiley and Sons, Inc.

Schlaifer, R. (1959). *Probability and Statistics for Business Decisions: An Introduction to Managerial Economics Under Uncertainty*, New York, McGraw-Hill Book Co., Inc.

Schlaifer, R. (1961). *Introduction to Statistics for Business Decisions*, New York, McGraw-Hill Book Co., Inc.

Schlaifer, R. (1969). *Analysis of Decisions Under Uncertainty*, New York, McGraw-Hill, Inc.

Schwarz, G. (1978). "Estimating the Dimension of the Model," *Ann. Statist.* **6**, 461–464.

Seife, C. (2000). *Zero: The Biography of a Dangerous Idea*, New York, Penguin Books.

Shafer, G. (1988). "The St. Petersburg Paradox," in S. Kotz and N. Johnson, Eds., *Encyclopedia of Statistical Sciences* Vol. 8, 865–870.

Shannon, C. E. (1948). "The Mathematical Theory of Communication," *Bell System Technical Journal*, July–October 1948, reprinted in C. E. Shannon and W. Weaver, *The Mathematical Theory of Communication*, University of Illinois Press, 1949, 3–91.

Shera, D. M. and Ibrahim J. G. (1998). "Prior Elicitation and Computation for Bayesian Factor Analysis," submitted manuscript, Department of Biostatistics, Harvard School of Public Health, September 28, 1998.

Shigemasu, K. (1986). "A Bayesian Estimation Procedure with Informative Prior in Factor Analysis Model," Tokyo Institute of Technology, Jinmon-Ronso, No 11, 67–77.

Shigemasu, K. Nakamura, T. and Shinzaki, S. (1993). "A Bayesian Numerical Estimation Procedure in Factor Analysis Model," Educational Science and Technology Research Group, Report 93-6, Tokyo Institute of Technology, December 1993.

Silverman, B. W. (1986). *Density Estimation for Statistics and Data Analysis*, New York, Chapman and Hall.

Skyrms, B. (1984). *Pragmatics and Empiricism*, New Haven, CT, Yale University Press.

Smith, A. F. M. (1981). "On Random Sequences with Centered Spherical Symmetry," *J. Royal Statist. Soc., B*, **43** 208–2091.

Smith, A. F. M., Skene, A. M., Shaw, J. E. H., Naylor, J. C. and Dransfield, M. (1985). "The Implementation of the Bayesian Paradigm," *Commu. in Statist. Theory and Methods*, **14**(5), 1079–1102.

Smith, M. and Kohn, R. (1996). "Nonparametric Regression Using Bayesian Variable Selection," *J. Econometrics*, **75**, 317–343.

Snijders, T. A. B. and Bosker, R. J. (1999). *Multilevel Analysis*, London, Sage Publications.

Soofi, E. S. (1992). "Information Theory and Bayesian Statistics," manuscript, School of Business Administration, University of Wisconsin, Milwaukee, WI.

Soofi, E. S. (1993). "Capturing the Intangible Concept of Information," manuscript School of Business Administration, University of Wisconsin, Milwaukee, WI.

Spiegelhalter, D., Thomas, A. and Best, N. (2000) WinBUGS Version 1.3 User Manual. MRC Biostatistics Unit, Institute of Public Health, London.

Stael von Holstein, C. A. S. (1970). *Assessment and Evaluation of Subjective Probability Distributions*, Stockholm, The Economic Research Institute, Stockholm School of Economics.

Statistica (1993). Anno LII, Supplemento al n.3, 1992, *Probabilita E Inducione (Induction and Probability): Bruno de Finetti*, translated by Mara Khale, Editrice, Bologna.

Stein, C. (1956). "Inadmissibility of the Usual Estimator for the Mean of a Multivariate Normal Distribution," in Proceedings of the Third Berkeley Symposium on Mathematical Statistics and Probability, University of California Press, Berkeley, 197–206.

Stigler, S. M. (1982). "Thomas Bayes and Bayesian Inference," *J. Roy. Statist. Soc., (A)*, **145**(2) 250–258.

Stigler, S. M. (1983). "Who Discovered Bayes' Theorem," *Amer. Stat.*, **37**(4), 290–296.

Stigler, S. M. (1986). *The History of Statistics*, Cambridge, MA, The Belknap Press of Harvard University Press.

"Student" (William Sealy Gosset) (1908). *Biometrika*, **6**, 1–25. (Paper on the Student t-distribution.)

Swallow, W. H. and Searle, S. R. (1978). "Minimum Variance Quadratic Unbiased Estimation (MIVQUE) of Variance Components," *Technometrics*, **20**, 265–272.

Taipia, R. A. and Thompson, J. R. (1982). *Nonparametric Probability Density Estimation*, Baltimore, MD, The Johns Hopkins University Press.

Tanner, M. A. and Wong, W. H. (1987). "The Calculation of Posterior Distributions by Data Augmentation," *J. Am. Statis. Assoc.*, **82**, 528–540.

Teicher, H. (1961). "Identifiability of Mixtures," *Ann. Math. Statist.*, **32**, 244–248.

Thiesson, B., Meek, C., Chickering, D. and Heckerman, D. (1997). "Learning Mixtures of DAG Models," Technical Report. MSR-TR-97-30, Microsoft Research.

Tiao, G. C. and Zellner A. (1964). "Bayes' Theorem and the Use of Prior Knowledge in Regression Analysis," *Biometrics*, **51**, 219–230.

Tierney, L. (1994), "Markov Chains for Exploring Posterior Distributions (with discussion)," *Ann. Statist.*, **22**, 1701–1762.

Tierney, L. and Kadane, J. B. (1986). "Accurate Approximations for Posterior Moments and Marginals," *J. Am. Statist. Assoc.*, **81**, 82–86.

Tierney, L., Kass, R. E. and Kadane, J. B. (1989). "Fully Exponential Laplace Approximations of Expectations and Variances of Non-Positive Functions," *J. Am. Statist. Assoc.*, **84**, 710–716.

Toda, M. (1963). "Measurement of Subjective Probability Distributions," Report #3, Division of Mathematical Psychology, Institute for Research, State College Pennsylvania.

Tversky A. and Kahneman, D. (1974). "Judgment Under Uncertainty: Heuristics and Biases," *Science*, **185**, 1124–1131.

Varian, H. R. (1975). "A Bayesian Approach to Real Estate Assessment," in S. Fienberg and A. Zellner, Eds., *Studies in Bayesian Econometrics and Statistics in Honor of Leonard J. Savage*. Amsterdam, North Holland Publishing Co., 195–208.

Villegas, C. (1969). "On the A Priori Distribution of the Covariance Matrix," *Ann. Math. Statist.*, **40**, 1098–1099.

von Mises, R. (1928). *Probability, Statistics and Truth*, Translated from German, London, Allen and Unwin, 1957.

Von Neumann, J. and Morgenstern, O. (1st Edition, 1944; 2nd Edition, 1947; 3rd Edition, 1953; renewed copyright, 1972), *Theory of Games and Economic Behavior*, Princeton, New Jersey, Princeton University Press.

Wakefield, J. C., Smith, A. F. M., Racine Poon, A. and Gelfand, A. E. (1994). "Bayesian Analysis of Linear and Non-Linear Population Models by Using the Gibbs Sampler," *Appl. Statist.*, **43**, 201–221.

Wald, A. (1939). "Contributions to the Theory of Statistical Estimation and Testing Hypotheses," *Ann. Math. Statist.*, **10**, 299–326.

Wald, A. (1950). *Statistical Decision Functions*, New York, John Wiley and Sons, Inc.

Wambaugh, J. (1989). *The Blooding*, New York, Morrow Publishing Co.

Weisberg, S. (1985). *Applied Linear Regression (Second Edition)*, New York, John Wiley and Sons, Inc.

West, M. (1988), "Modeling Expert Opinion," in *Bayesian Statistics 3*, 493–508.

West, M. and Harrison J. (1997). *Bayesian Forecasting and Dynamic Models*, 2nd Edition, New York, Springer-Verlag.

Winkler, R. L. (1967). "The Quantification of Judgment: Some Methodological Suggestions," *J. Am. Statist. Assoc.*, **62**, 1105–1120.

Winkler, R. L. and Murphy, A. H. (1968). "'Good' Probability Assessors," *J. Appl. Meteorology*, **7**, 751–758.

Winkler, R. (1972). *An Introduction to Bayesian Inference and Decision*, New York, Holt, Rinehart and Winston, Inc.

Wolpert, L. (1992). *The Unnatural Nature of Science*, Boston, Harvard University Press.

Wong, G. (1980). "A Bayesian Approach to Factor Analysis," Program in Statistical Technical Report, No 80-9, Educational Testing Service, Princeton, N.J., August, 1980.

Yang, R. and Berger, J. O. (1996). "A Catalog of Noninformative Priors," Technical Report. #97-42, Duke University, Institute of Statistics and Decision Sciences.

Yates, J. F. (1990). *Judgment and Decision Making*, Englewood Cliffs, New Jersey, Prentice-Hall, Inc.

Zaslavsky, A. M. (1997). "Comment on Meng and van Dyk, "The EM algorithm—An Old Folk Song Sung to a Fast New Tune." *J. Royal Statist. Soc., B*, **59**, 559.

Zeger, S. L., Liang, K.-Y. and Albert, P. S. (1988). "Models for Longitudinal Data: A Generalized Estimating Equation Approach," *Biometrics*, **44**, 1049–1060.

Zellner, A. (1971). *An Introduction to Bayesian Inference in Econometrics*, New York, John Wiley and Sons, Inc.

Zellner, A. (1977). "Maximal Data Information Prior Distributions," in A. Aykac and C. Brumat, Eds., *New Developments in the Applications of Bayesian Methods*, Amsterdam, North Holland Publishing Co., 211–232.

Zellner, A (1978). "Jeffreys–Bayes Posterior Odds Ratio and the Akaike Information Criterion for Discriminating Between Models," *Economics Letters*, **1**, 337–342.

Zellner, A. (1985). "Bayesian Econometrics," *Econometrica*, **53**(2), 253–269.

Zellner, A. (1986). "On Assessing Prior Distributions and Bayesian Regression Analysis with g-Prior Distributions," in P. Goel and A. Zellner, Eds., *Bayesian Inference and Decision Techniques: Essays in Honor of Bruno de Finetti*, New York, North Holland Publishing Co., 233–243.

Zellner, A (1988). "Optimal Information Processing and Bayes' Theorem," *Am. Stat.*, **42**(4), 278–284.

Zellner, A and Chung-ki Min (1993). "Bayesian Analysis, Model Selection and Prediction," in *Physics and Probability: Essays in Honor of Edwin T. Jaynes*, Cambridge, Cambridge University Press, 195–206.

Index

WILEY SERIES IN PROBABILITY AND STATISTICS

ESTABLISHED BY WALTER A. SHEWHART AND SAMUEL S. WILKS

Editors: *David J. Balding, Peter Bloomfield, Noel A. C. Cressie, Nicholas I. Fisher, Iain M. Johnstone, J. B. Kadane, Louise M. Ryan, David W. Scott, Adrian F. M. Smith, Jozef L. Teugels*
Editors Emeriti: *Vic Barnett, J. Stuart Hunter, David G. Kendall*

The ***Wiley Series in Probability and Statistics*** is well established and authoritative. It covers many topics of current research interest in both pure and applied statistics and probability theory. Written by leading statisticians and institutions, the titles span both state-of-the-art developments in the field and classical methods.

Reflecting the wide range of current research in statistics, the series encompasses applied, methodological and theoretical statistics, ranging from applications and new techniques made possible by advances in computerized practice to rigorous treatment of theoretical approaches.

This series provides essential and invaluable reading for all statisticians, whether in academia, industry, government, or research.

ABRAHAM and LEDOLTER · Statistical Methods for Forecasting
AGRESTI · Analysis of Ordinal Categorical Data
AGRESTI · An Introduction to Categorical Data Analysis
AGRESTI · Categorical Data Analysis, *Second Edition*
ANDĚL · Mathematics of Chance
ANDERSON · An Introduction to Multivariate Statistical Analysis, *Second Edition*
*ANDERSON · The Statistical Analysis of Time Series
ANDERSON, AUQUIER, HAUCK, OAKES, VANDAELE, and WEISBERG · Statistical Methods for Comparative Studies
ANDERSON and LOYNES · The Teaching of Practical Statistics
ARMITAGE and DAVID (editors) · Advances in Biometry
ARNOLD, BALAKRISHNAN, and NAGARAJA · Records
*ARTHANARI and DODGE · Mathematical Programming in Statistics
*BAILEY · The Elements of Stochastic Processes with Applications to the Natural Sciences
BALAKRISHNAN and KOUTRAS · Runs and Scans with Applications
BARNETT · Comparative Statistical Inference, *Third Edition*
BARNETT and LEWIS · Outliers in Statistical Data, *Third Edition*
BARTOSZYNSKI and NIEWIADOMSKA-BUGAJ · Probability and Statistical Inference
BASILEVSKY · Statistical Factor Analysis and Related Methods: Theory and Applications
BASU and RIGDON · Statistical Methods for the Reliability of Repairable Systems
BATES and WATTS · Nonlinear Regression Analysis and Its Applications
BECHHOFER, SANTNER, and GOLDSMAN · Design and Analysis of Experiments for Statistical Selection, Screening, and Multiple Comparisons
BELSLEY · Conditioning Diagnostics: Collinearity and Weak Data in Regression
BELSLEY, KUH, and WELSCH · Regression Diagnostics: Identifying Influential Data and Sources of Collinearity
BENDAT and PIERSOL · Random Data: Analysis and Measurement Procedures, *Third Edition*

*Now available in a lower priced paperback edition in the Wiley Classics Library.

BERRY, CHALONER, and GEWEKE · Bayesian Analysis in Statistics and Econometrics: Essays in Honor of Arnold Zellner
BERNARDO and SMITH · Bayesian Theory
BHAT and MILLER · Elements of Applied Stochastic Processes, *Third Edition*
BHATTACHARYA and JOHNSON · Statistical Concepts and Methods
BHATTACHARYA and WAYMIRE · Stochastic Processes with Applications
BILLINGSLEY · Convergence of Probability Measures, *Second Edition*
BILLINGSLEY · Probability and Measure, *Third Edition*
BIRKES and DODGE · Alternative Methods of Regression
BLISCHKE AND MURTHY · Reliability: Modeling, Prediction, and Optimization
BLOOMFIELD · Fourier Analysis of Time Series: An Introduction, *Second Edition*
BOLLEN · Structural Equations with Latent Variables
BOROVKOV · Ergodicity and Stability of Stochastic Processes
BOULEAU · Numerical Methods for Stochastic Processes
BOX · Bayesian Inference in Statistical Analysis
BOX · R. A. Fisher, the Life of a Scientist
BOX and DRAPER · Empirical Model-Building and Response Surfaces
*BOX and DRAPER · Evolutionary Operation: A Statistical Method for Process Improvement
BOX, HUNTER, and HUNTER · Statistics for Experimenters: An Introduction to Design, Data Analysis, and Model Building
BOX and LUCEÑO · Statistical Control by Monitoring and Feedback Adjustment
BRANDIMARTE · Numerical Methods in Finance: A MATLAB-Based Introduction
BROWN and HOLLANDER · Statistics: A Biomedical Introduction
BRUNNER, DOMHOF, and LANGER · Nonparametric Analysis of Longitudinal Data in Factorial Experiments
BUCKLEW · Large Deviation Techniques in Decision, Simulation, and Estimation
CAIROLI and DALANG · Sequential Stochastic Optimization
CHAN · Time Series: Applications to Finance
CHATTERJEE and HADI · Sensitivity Analysis in Linear Regression
CHATTERJEE and PRICE · Regression Analysis by Example, *Third Edition*
CHERNICK · Bootstrap Methods: A Practitioner's Guide
CHILÈS and DELFINER · Geostatistics: Modeling Spatial Uncertainty
CHOW and LIU · Design and Analysis of Clinical Trials: Concepts and Methodologies
CLARKE and DISNEY · Probability and Random Processes: A First Course with Applications, *Second Edition*
*COCHRAN and COX · Experimental Designs, *Second Edition*
CONGDON · Bayesian Statistical Modelling
CONOVER · Practical Nonparametric Statistics, *Second Edition*
COOK · Regression Graphics
COOK and WEISBERG · Applied Regression Including Computing and Graphics
COOK and WEISBERG · An Introduction to Regression Graphics
CORNELL · Experiments with Mixtures, Designs, Models, and the Analysis of Mixture Data, *Third Edition*
COVER and THOMAS · Elements of Information Theory
COX · A Handbook of Introductory Statistical Methods
*COX · Planning of Experiments
CRESSIE · Statistics for Spatial Data, *Revised Edition*
CSÖRGŐ and HORVÁTH · Limit Theorems in Change Point Analysis
DANIEL · Applications of Statistics to Industrial Experimentation
DANIEL · Biostatistics: A Foundation for Analysis in the Health Sciences, *Sixth Edition*
*DANIEL · Fitting Equations to Data: Computer Analysis of Multifactor Data, *Second Edition*

*Now available in a lower priced paperback edition in the Wiley Classics Library.

DAVID · Order Statistics, *Second Edition*
*DEGROOT, FIENBERG, and KADANE · Statistics and the Law
DEL CASTILLO · Statistical Process Adjustment for Quality Control
DETTE and STUDDEN · The Theory of Canonical Moments with Applications in Statistics, Probability, and Analysis
DEY and MUKERJEE · Fractional Factorial Plans
DILLON and GOLDSTEIN · Multivariate Analysis: Methods and Applications
DODGE · Alternative Methods of Regression
*DODGE and ROMIG · Sampling Inspection Tables, *Second Edition*
*DOOB · Stochastic Processes
DOWDY and WEARDEN · Statistics for Research, *Second Edition*
DRAPER and SMITH · Applied Regression Analysis, *Third Edition*
DRYDEN and MARDIA · Statistical Shape Analysis
DUDEWICZ and MISHRA · Modern Mathematical Statistics
DUNN and CLARK · Applied Statistics: Analysis of Variance and Regression, *Second Edition*
DUNN and CLARK · Basic Statistics: A Primer for the Biomedical Sciences, *Third Edition*
DUPUIS and ELLIS · A Weak Convergence Approach to the Theory of Large Deviations
*ELANDT-JOHNSON and JOHNSON · Survival Models and Data Analysis
ETHIER and KURTZ · Markov Processes: Characterization and Convergence
EVANS, HASTINGS, and PEACOCK · Statistical Distributions, *Third Edition*
FELLER · An Introduction to Probability Theory and Its Applications, Volume I, *Third Edition,* Revised; Volume II, *Second Edition*
FISHER and VAN BELLE · Biostatistics: A Methodology for the Health Sciences
*FLEISS · The Design and Analysis of Clinical Experiments
FLEISS · Statistical Methods for Rates and Proportions, *Second Edition*
FLEMING and HARRINGTON · Counting Processes and Survival Analysis
FULLER · Introduction to Statistical Time Series, *Second Edition*
FULLER · Measurement Error Models
GALLANT · Nonlinear Statistical Models
GHOSH, MUKHOPADHYAY, and SEN · Sequential Estimation
GIFI · Nonlinear Multivariate Analysis
GLASSERMAN and YAO · Monotone Structure in Discrete-Event Systems
GNANADESIKAN · Methods for Statistical Data Analysis of Multivariate Observations, *Second Edition*
GOLDSTEIN and LEWIS · Assessment: Problems, Development, and Statistical Issues
GREENWOOD and NIKULIN · A Guide to Chi-Squared Testing
GROSS and HARRIS · Fundamentals of Queueing Theory, *Third Edition*
*HAHN · Statistical Models in Engineering
HAHN and MEEKER · Statistical Intervals: A Guide for Practitioners
HALD · A History of Probability and Statistics and their Applications Before 1750
HALD · A History of Mathematical Statistics from 1750 to 1930
HAMPEL · Robust Statistics: The Approach Based on Influence Functions
HANNAN and DEISTLER · The Statistical Theory of Linear Systems
HEIBERGER · Computation for the Analysis of Designed Experiments
HEDAYAT and SINHA · Design and Inference in Finite Population Sampling
HELLER · MACSYMA for Statisticians
HINKELMAN and KEMPTHORNE: · Design and Analysis of Experiments, Volume 1: Introduction to Experimental Design
HOAGLIN, MOSTELLER, and TUKEY · Exploratory Approach to Analysis of Variance
HOAGLIN, MOSTELLER, and TUKEY · Exploring Data Tables, Trends and Shapes

*Now available in a lower priced paperback edition in the Wiley Classics Library.

*HOAGLIN, MOSTELLER, and TUKEY · Understanding Robust and Exploratory Data Analysis
HOCHBERG and TAMHANE · Multiple Comparison Procedures
HOCKING · Methods and Applications of Linear Models: Regression and the Analysis of Variables
HOEL · Introduction to Mathematical Statistics, *Fifth Edition*
HOGG and KLUGMAN · Loss Distributions
HOLLANDER and WOLFE · Nonparametric Statistical Methods, *Second Edition*
HOSMER and LEMESHOW · Applied Logistic Regression, *Second Edition*
HOSMER and LEMESHOW · Applied Survival Analysis: Regression Modeling of Time to Event Data
HØYLAND and RAUSAND · System Reliability Theory: Models and Statistical Methods
HUBER · Robust Statistics
HUBERTY · Applied Discriminant Analysis
HUNT and KENNEDY · Financial Derivatives in Theory and Practice
HUSKOVA, BERAN, and DUPAC · Collected Works of Jaroslav Hajek—with Commentary
IMAN and CONOVER · A Modern Approach to Statistics
JACKSON · A User's Guide to Principle Components
JOHN · Statistical Methods in Engineering and Quality Assurance
JOHNSON · Multivariate Statistical Simulation
JOHNSON and BALAKRISHNAN · Advances in the Theory and Practice of Statistics: A Volume in Honor of Samuel Kotz
JUDGE, GRIFFITHS, HILL, LÜTKEPOHL, and LEE · The Theory and Practice of Econometrics, *Second Edition*
JOHNSON and KOTZ · Distributions in Statistics
JOHNSON and KOTZ (editors) · Leading Personalities in Statistical Sciences: From the Seventeenth Century to the Present
JOHNSON, KOTZ, and BALAKRISHNAN · Continuous Univariate Distributions, Volume 1, *Second Edition*
JOHNSON, KOTZ, and BALAKRISHNAN · Continuous Univariate Distributions, Volume 2, *Second Edition*
JOHNSON, KOTZ, and BALAKRISHNAN · Discrete Multivariate Distributions
JOHNSON, KOTZ, and KEMP · Univariate Discrete Distributions, *Second Edition*
JUREČKOVÁ and SEN · Robust Statistical Procedures: Aymptotics and Interrelations
JUREK and MASON · Operator-Limit Distributions in Probability Theory
KADANE · Bayesian Methods and Ethics in a Clinical Trial Design
KADANE AND SCHUM · A Probabilistic Analysis of the Sacco and Vanzetti Evidence
KALBFLEISCH and PRENTICE · The Statistical Analysis of Failure Time Data, *Second Edition*
KASS and VOS · Geometrical Foundations of Asymptotic Inference
KAUFMAN and ROUSSEEUW · Finding Groups in Data: An Introduction to Cluster Analysis
KEDEM and FOKIANOS · Regression Models for Time Series Analysis
KENDALL, BARDEN, CARNE, and LE · Shape and Shape Theory
KHURI · Advanced Calculus with Applications in Statistics, *Second Edition*
KHURI, MATHEW, and SINHA · Statistical Tests for Mixed Linear Models
KLUGMAN, PANJER, and WILLMOT · Loss Models: From Data to Decisions
KLUGMAN, PANJER, and WILLMOT · Solutions Manual to Accompany Loss Models: From Data to Decisions
KOTZ, BALAKRISHNAN, and JOHNSON · Continuous Multivariate Distributions, Volume 1, *Second Edition*
KOTZ and JOHNSON (editors) · Encyclopedia of Statistical Sciences: Volumes 1 to 9 with Index

*Now available in a lower priced paperback edition in the Wiley Classics Library.

KOTZ and JOHNSON (editors) · Encyclopedia of Statistical Sciences: Supplement Volume
KOTZ, READ, and BANKS (editors) · Encyclopedia of Statistical Sciences: Update Volume 1
KOTZ, READ, and BANKS (editors) · Encyclopedia of Statistical Sciences: Update Volume 2
KOVALENKO, KUZNETZOV, and PEGG · Mathematical Theory of Reliability of Time-Dependent Systems with Practical Applications
LACHIN · Biostatistical Methods: The Assessment of Relative Risks
LAD · Operational Subjective Statistical Methods: A Mathematical, Philosophical, and Historical Introduction
LAMPERTI · Probability: A Survey of the Mathematical Theory, *Second Edition*
LANGE, RYAN, BILLARD, BRILLINGER, CONQUEST, and GREENHOUSE · Case Studies in Biometry
LARSON · Introduction to Probability Theory and Statistical Inference, *Third Edition*
LAWLESS · Statistical Models and Methods for Lifetime Data
LAWSON · Statistical Methods in Spatial Epidemiology
LE · Applied Categorical Data Analysis
LE · Applied Survival Analysis
LEE and WANG · Statistical Methods for Survival Data Analysis, *Third Edition*
LePAGE and BILLARD · Exploring the Limits of Bootstrap
LEYLAND and GOLDSTEIN (editors) · Multilevel Modelling of Health Statistics
LIAO · Statistical Group Comparison
LINDVALL · Lectures on the Coupling Method
LINHART and ZUCCHINI · Model Selection
LITTLE and RUBIN · Statistical Analysis with Missing Data, *Second Edition*
LLOYD · The Statistical Analysis of Categorical Data
MAGNUS and NEUDECKER · Matrix Differential Calculus with Applications in Statistics and Econometrics, *Revised Edition*
MALLER and ZHOU · Survival Analysis with Long Term Survivors
MALLOWS · Design, Data, and Analysis by Some Friends of Cuthbert Daniel
MANN, SCHAFER, and SINGPURWALLA · Methods for Statistical Analysis of Reliability and Life Data
MANTON, WOODBURY, and TOLLEY · Statistical Applications Using Fuzzy Sets
MARDIA and JUPP · Directional Statistics
MASON, GUNST, and HESS · Statistical Design and Analysis of Experiments with Applications to Engineering and Science
McCULLOCH and SEARLE · Generalized, Linear, and Mixed Models
McFADDEN · Management of Data in Clinical Trials
McLACHLAN · Discriminant Analysis and Statistical Pattern Recognition
McLACHLAN and KRISHNAN · The EM Algorithm and Extensions
McLACHLAN and PEEL · Finite Mixture Models
McNEIL · Epidemiological Research Methods
MEEKER and ESCOBAR · Statistical Methods for Reliability Data
MEERSCHAERT and SCHEFFLER · Limit Distributions for Sums of Independent Random Vectors: Heavy Tails in Theory and Practice
*MILLER · Survival Analysis, *Second Edition*
MONTGOMERY, PECK, and VINING · Introduction to Linear Regression Analysis, *Third Edition*
MORGENTHALER and TUKEY · Configural Polysampling: A Route to Practical Robustness
MUIRHEAD · Aspects of Multivariate Statistical Theory
MURRAY · X-STAT 2.0 Statistical Experimentation, Design Data Analysis, and Nonlinear Optimization

*Now available in a lower priced paperback edition in the Wiley Classics Library.

MYERS and MONTGOMERY · Response Surface Methodology: Process and Product Optimization Using Designed Experiments, *Second Edition*
MYERS, MONTGOMERY, and VINING · Generalized Linear Models. With Applications in Engineering and the Sciences
NELSON · Accelerated Testing, Statistical Models, Test Plans, and Data Analyses
NELSON · Applied Life Data Analysis
NEWMAN · Biostatistical Methods in Epidemiology
OCHI · Applied Probability and Stochastic Processes in Engineering and Physical Sciences
OKABE, BOOTS, SUGIHARA, and CHIU · Spatial Tesselations: Concepts and Applications of Voronoi Diagrams, *Second Edition*
OLIVER and SMITH · Influence Diagrams, Belief Nets and Decision Analysis
PANKRATZ · Forecasting with Dynamic Regression Models
PANKRATZ · Forecasting with Univariate Box-Jenkins Models: Concepts and Cases
*PARZEN · Modern Probability Theory and Its Applications
PEÑA, TIAO, and TSAY · A Course in Time Series Analysis
PIANTADOSI · Clinical Trials: A Methodologic Perspective
PORT · Theoretical Probability for Applications
POURAHMADI · Foundations of Time Series Analysis and Prediction Theory
PRESS · Bayesian Statistics: Principles, Models, and Applications
PRESS · Subjective and Objective Bayesian Statistics, *Second Edition*
PRESS and TANUR · The Subjectivity of Scientists and the Bayesian Approach
PUKELSHEIM · Optimal Experimental Design
PURI, VILAPLANA, and WERTZ · New Perspectives in Theoretical and Applied Statistics
PUTERMAN · Markov Decision Processes: Discrete Stochastic Dynamic Programming
*RAO · Linear Statistical Inference and Its Applications, *Second Edition*
RENCHER · Linear Models in Statistics
RENCHER · Methods of Multivariate Analysis, *Second Edition*
RENCHER · Multivariate Statistical Inference with Applications
RIPLEY · Spatial Statistics
RIPLEY · Stochastic Simulation
ROBINSON · Practical Strategies for Experimenting
ROHATGI and SALEH · An Introduction to Probability and Statistics, *Second Edition*
ROLSKI, SCHMIDLI, SCHMIDT, and TEUGELS · Stochastic Processes for Insurance and Finance
ROSENBERGER and LACHIN · Randomization in Clinical Trials: Theory and Practice
ROSS · Introduction to Probability and Statistics for Engineers and Scientists
ROUSSEEUW and LEROY · Robust Regression and Outlier Detection
RUBIN · Multiple Imputation for Nonresponse in Surveys
RUBINSTEIN · Simulation and the Monte Carlo Method
RUBINSTEIN and MELAMED · Modern Simulation and Modeling
RYAN · Modern Regression Methods
RYAN · Statistical Methods for Quality Improvement, *Second Edition*
SALTELLI, CHAN, and SCOTT (editors) · Sensitivity Analysis
*SCHEFFE · The Analysis of Variance
SCHIMEK · Smoothing and Regression: Approaches, Computation, and Application
SCHOTT · Matrix Analysis for Statistics
SCHUSS · Theory and Applications of Stochastic Differential Equations
SCOTT · Multivariate Density Estimation: Theory, Practice, and Visualization
*SEARLE · Linear Models
SEARLE · Linear Models for Unbalanced Data
SEARLE · Matrix Algebra Useful for Statistics
SEARLE, CASELLA, and McCULLOCH · Variance Components
SEARLE and WILLETT · Matrix Algebra for Applied Economics

*Now available in a lower priced paperback edition in the Wiley Classics Library.

SEBER · Linear Regression Analysis
SEBER · Multivariate Observations
SEBER and WILD · Nonlinear Regression
SENNOTT · Stochastic Dynamic Programming and the Control of Queueing Systems
*SERFLING · Approximation Theorems of Mathematical Statistics
SHAFER and VOVK · Probability and Finance: It's Only a Game!
SMALL and McLEISH · Hilbert Space Methods in Probability and Statistical Inference
SRIVASTAVA · Methods of Multivariate Statistics
STAPLETON · Linear Statistical Models
STAUDTE and SHEATHER · Robust Estimation and Testing
STOYAN, KENDALL, and MECKE · Stochastic Geometry and Its Applications, *Second Edition*
STOYAN and STOYAN · Fractals, Random Shapes and Point Fields: Methods of Geometrical Statistics
STYAN · The Collected Papers of T. W. Anderson: 1943–1985
SUTTON, ABRAMS, JONES, SHELDON, and SONG · Methods for Meta-Analysis in Medical Research
TANAKA · Time Series Analysis: Nonstationary and Noninvertible Distribution Theory
THOMPSON · Empirical Model Building
THOMPSON · Sampling, *Second Edition*
THOMPSON · Simulation: A Modeler's Approach
THOMPSON and SEBER · Adaptive Sampling
THOMPSON, WILLIAMS, and FINDLAY · Models for Investors in Real World Markets
TIAO, BISGAARD, HILL, PEÑA, and STIGLER (editors) · Box on Quality and Discovery: with Design, Control, and Robustness
TIERNEY · LISP-STAT: An Object-Oriented Environment for Statistical Computing and Dynamic Graphics
TSAY · Analysis of Financial Time Series
UPTON and FINGLETON · Spatial Data Analysis by Example, Volume II: Categorical and Directional Data
VAN BELLE · Statistical Rules of Thumb
VIDAKOVIC · Statistical Modeling by Wavelets
WEISBERG · Applied Linear Regression, *Second Edition*
WELSH · Aspects of Statistical Inference
WESTFALL and YOUNG · Resampling-Based Multiple Testing: Examples and Methods for *p*-Value Adjustment
WHITTAKER · Graphical Models in Applied Multivariate Statistics
WINKER · Optimization Heuristics in Economics: Applications of Threshold Accepting
WONNACOTT and WONNACOTT · Econometrics, *Second Edition*
WOODING · Planning Pharmaceutical Clinical Trials: Basic Statistical Principles
WOOLSON and CLARKE · Statistical Methods for the Analysis of Biomedical Data, *Second Edition*
WU and HAMADA · Experiments: Planning, Analysis, and Parameter Design Optimization
YANG · The Construction Theory of Denumerable Markov Processes
*ZELLNER · An Introduction to Bayesian Inference in Econometrics
ZHOU, OBUCHOWSKI, and McCLISH · Statistical Methods in Diagnostic Medicine

*Now available in a lower priced paperback edition in the Wiley Classics Library.